2024

IECC®

INTERNATIONAL ENERGY CONSERVATION CODE®

2024 International Energy Conservation Code®

First Printing: May 2024

ISBN: 978-1-959851-69-1 (soft-cover edition)
ISBN: 978-1-959851-70-7 (PDF download)

T030367

PRINTED IN THE USA

NEW DESIGN FOR THE 2024 INTERNATIONAL CODES

IBC IRC IFC IPC IMC IECC IEBC IFGC IPMC IPSDC IWUIC IZC ICCPC IgCC ISPSC

The 2024 International Codes® (I-Codes®) have undergone substantial formatting changes as part of the digital transformation strategy of the International Code Council® (ICC®) to improve the user experience. The resulting product better aligns the print and PDF versions of the I-Codes with the ICC's Digital Codes® content.

The changes, promoting a cleaner, more modern look and enhancing readability and sustainability, include:

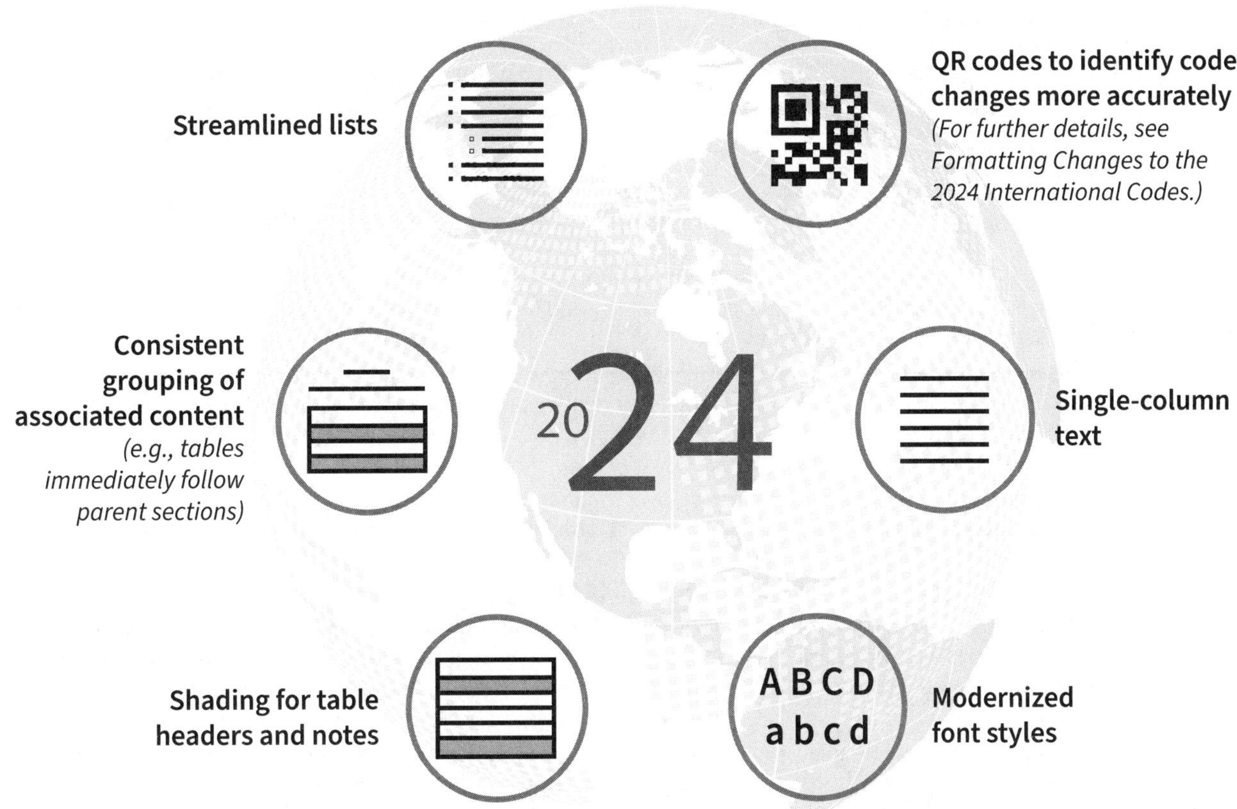

More information can be found at iccsafe.org/design-updates.

PREFACE

FORMATTING CHANGES TO THE 2024 INTERNATIONAL CODES

The 2024 International Codes® (I-Codes®) have undergone substantial formatting changes as part of the digital transformation strategy of the International Code Council® (ICC®) to improve the user experience. The resulting product better aligns the print and PDF versions of the I-Codes with the ICC's Digital Codes content. Additional information can be found at iccsafe.org/design-updates.

Replacement of Marginal Markings with QR Codes

Through 2021, print editions of the I-Codes identified technical changes from prior code cycles with marginal markings [solid vertical lines for new text, deletion arrows (➡), asterisks for relocations (*)]. The 2024 I-Code print editions replace the marginal markings with QR codes to identify code changes more precisely.

A QR code is placed at the beginning of any section that has undergone technical revision. If there is no QR code, there are no technical changes to that section.

In the following example from the 2024 *International Energy Conservation Code* ® (IECC®), a QR code indicates there are changes to Section C101 from the 2021 IECC. Note that the change may occur in the main section or in one or more subsections of the main section.

SECTION C101—SCOPE AND GENERAL REQUIREMENTS

C101.1 Title. This code shall be known as the *Energy Conservation Code* of [NAME OF JURISDICTION] and shall be cited as such. It is referred to herein as "this code."

C101.2 Scope. This code applies to the design and construction of buildings not covered by the scope of the IECC—Residential Provisions.

C101.2.1 Appendices. Provisions in the appendices shall not apply unless specifically adopted.

C101.3 Intent. The IECC—Commercial Provisions provide market-driven, enforceable requirements for the design and construction of *commercial buildings*, providing minimum efficiency requirements for buildings that result in the maximum level of energy efficiency that is safe, technologically feasible, and life cycle cost effective, considering economic feasibility, including potential costs and savings for consumers and building owners, and return on investment. Additionally, the code provides jurisdictions with supplemental requirements, including ASHRAE 90.1, and

Scan for Changes

7c8d893

To see the code changes, the user need only scan the QR code with a smart device. If scanning a QR code is not an option, changes can be accessed by entering the 7-digit code beneath the QR code at the end of the following URL: qr.iccsafe.org/ (in the above example,"qr.iccsafe.org/7c8d893"). Those viewing the code book via PDF can click on the QR code.

All methods take the user to the appropriate section on ICC's Digital Codes website, where technical changes from the prior cycle can be viewed. Digital Codes Premium subscribers who are logged in will be automatically directed to the Premium view. All other users will be directed to the Digital Codes Basic free view. Both views show new code language in blue text along with deletion arrows for deleted text and relocation markers for relocated text.

Digital Codes Premium offers additional ways to enhance code compliance research, including revision histories, commentary by code experts and an advanced search function. A full list of features can be found at codes.iccsafe.org/premium-features.

ABOUT THE I-CODES

The 2024 I-Codes, published by the ICC, are 15 fully compatible titles intended to establish provisions that adequately protect public health, safety and welfare; that do not unnecessarily increase construction costs; that do not restrict the use of new materials, products or methods of construction; and that do not give preferential treatment to particular types or classes of materials, products or methods of construction.

The I-Codes are updated on a 3-year cycle to allow for new construction methods and technologies to be incorporated into the codes. Alternative materials, designs and methods not specifically addressed in the I-Code can be approved by the building official where the proposed materials, designs or methods comply with the intent of the provisions of the code.

The I-Codes are used as the basis of laws and regulations in communities across the US and in other countries. They are also used in a variety of nonregulatory settings, including:

- Voluntary compliance programs.
- The insurance industry.
- Certification and credentialing for building design, construction and safety professionals.
- Certification of building and construction-related products.
- Facilities management.
- "Best practices" benchmarks for designers and builders.
- College, university and professional school textbooks and curricula.
- Reference works related to building design and construction.

Code Development Process

The code development process regularly provides an international forum for building professionals to discuss requirements for building design, construction methods, safety, performance, technological advances and new products. Proposed changes to the I-Codes, submitted by code enforcement officials, industry representatives, design professionals and other interested parties, are deliberated through an open code development process in which all interested and affected parties may participate.

Openness, transparency, balance, due process and consensus are the guiding principles of both the ICC Code Development Process and OMB Circular A-119, which governs the federal government's use of private-sector standards. The ICC process is open to anyone without cost. Remote participation is available through cdpAccess®, the ICC's cloud-based app.

In order to ensure that organizations with a direct and material interest in the codes have a voice in the process, the ICC has developed partnerships with key industry segments that support the ICC's important public safety mission. Some code development committee members were nominated by the following industry partners and approved by the ICC Board:

- American Gas Association (AGA)
- American Institute of Architects (AIA)
- American Society of Plumbing Engineers (ASPE)
- International Association of Fire Chiefs (IAFC)
- National Association of Home Builders (NAHB)
- National Association of State Fire Marshals (NASFM)
- National Council of Structural Engineers Association (NCSEA)
- National Multifamily Housing Council (NMHC)
- Plumbing Heating and Cooling Contractors (PHCC)
- Pool and Hot Tub Alliance (PHTA), formerly The Association of Pool and Spa Professionals (APSP)

Code development committees evaluate and make recommendations regarding proposed changes to the codes. Their recommendations are then subject to public comment and council-wide votes. The ICC's governmental members—public safety officials who have no financial or business interest in the outcome—cast the final votes on proposed changes.

The I-Codes are subject to change through future code development cycles and by any governmental entity that enacts the code into law. For more information regarding the code development process, contact the Codes and Standards Development Department of the ICC at iccsafe.org/products-and-services/i-codes/code-development/.

While the I-Code development procedure is thorough and comprehensive, the ICC, its members and those participating in the development of the codes expressly disclaim any liability resulting from the publication or use of the I-Codes, or from compliance or noncompliance with their provisions. NO WARRANTY OF ANY KIND, IMPLIED, EXPRESSED OR STATUTORY, IS GIVEN WITH RESPECT TO THE I-CODES. The ICC does not have the power or authority to police or enforce compliance with the contents of the I-Codes.

Coordination of the I-Codes

The coordination of technical provisions allows the I-Codes to be used as a complete set of complementary documents. Individual codes can also be used in subsets or as stand-alone documents. Some technical provisions that are relevant to more than one subject area are duplicated in multiple model codes.

Italicized Terms

Words and terms defined in Chapter 2, Definitions, are italicized where they appear in code text and the Chapter 2 definitions apply. Although care has been taken to ensure applicable terms are italicized, there may be instances where a defined term has not been italicized or where a term is italicized but the definition found in Chapter 2 is not applicable. For example, Chapter 2 of the *International Building Code*® (IBC®) contains a definition for "*Listed*" that is applicable to equipment, products and services. The term "listed" is also used in that code to refer to a list of items within the code or within a referenced document. For the latter, the Chapter 2 definition would not be applicable.

Adoption of International Code Council Codes and Standards

The International Code Council maintains a copyright in all of its codes and standards. Maintaining copyright allows the Code Council to fund its mission through sales of books in both print and digital format. The Code Council welcomes incorporation by reference of its codes and standards by jurisdictions that recognize and acknowledge the Code Council's copyright in the codes and standards, and further acknowledge the substantial shared value of the public/private partnership for code development between jurisdictions and the Code Council. By making its codes and standards available for incorporation by reference, the Code Council does not waive its copyright in its codes and standards.

The Code Council's codes and standards may only be adopted by incorporation by reference in an ordinance passed by the governing body of the jurisdiction. "Incorporation by reference" means that in the adopting ordinance, the governing body cites only the title, edition, relevant sections or subsections (where applicable), and publishing information of the model code or standard, and the actual text of the model code or standard is not included in the ordinance (see graphic, "Adoption of International Code Council

Codes and Standards"). The Code Council does not consent to the reproduction of the text of its codes or standards in any ordinance. If the governing body enacts any changes, only the text of those changes or amendments may be included in the ordinance.

The Code Council also recognizes the need for jurisdictions to make laws accessible to the public. Accordingly, all I-Codes and I-Standards, along with the laws of many jurisdictions, are available to view for free at codes.iccsafe.org/codes/i-codes. These documents may also be purchased, in both digital and print versions, at shop.iccsafe.org.

To facilitate adoption, some I-Code sections contain blanks for fill-in information that needs to be supplied by the adopting jurisdiction as part of the adoption legislation. For example, the IECC contains:

Section C101.1. Insert: **[NAME OF JURISDICTION]**

Section R101.1. Insert: **[NAME OF JURISDICTION]**

For further information or assistance with adoption, including a sample ordinance, jurisdictions should contact the Code Council at incorporation@iccsafe.org.

For a list of frequently asked questions (FAQs) addressing a range of foundational topics about the adoption of model codes by jurisdictions and to learn more about the Code Council's code adoption resources, scan the QR code or visit iccsafe.org/code-adoption-resources.

INTRODUCTION TO THE INTERNATIONAL ENERGY CONSERVATION CODE

The standards development process regularly provides an international forum for building professionals to discuss requirements for building design, construction methods, safety, performance, technological advances and new products. Proposed changes to the I-Codes developed through ICC Standards Consensus Procedures, submitted by code enforcement officials, industry representatives, design professionals and other interested parties, are deliberated through an open standards development process in which all interested and affected parties may participate.

Openness, transparency, balance, due process and consensus are the guiding principles of both the ICC Codes and Standards Development Processes and OMB Circular A-119, which governs the federal government's use of private-sector standards. The ICC process is open to anyone without cost. Remote participation is available through cdpAccess®, the ICC's cloud-based app.

In order to ensure that organizations with a direct and material interest in the codes have a voice in the process, the ICC has encouraged participation of key industry segments that support the ICC's important public safety mission.

Code development committees, using the Standards Consensus Procedures, evaluate proposed changes to the codes. After public comments are reviewed, the committee members vote to approve changes.

ARRANGEMENT AND FORMAT OF THE 2024 IECC

The IECC contains two separate sets of provisions—one for commercial buildings and one for residential buildings. Each set of provisions is applied separately to buildings within their scope. The IECC—Commercial Provisions apply to all buildings except for residential buildings three stories or less in height. The IECC—Residential Provisions apply to detached one- and two-family dwellings and multiple single-family dwellings as well as Group R-2, R-3 and R-4 buildings three stories or less in height. These scopes are based on the definitions of "Commercial building" and "Residential building," respectively, in Chapter 2 of each set of provisions. Note that the IECC—Commercial Provisions therefore contain provisions for residential buildings four stories or greater in height.

The following table shows how the IECC is divided. The chapter synopses detail the scope and intent of the provisions of the IECC.

CHAPTER TOPICS

Chapter	Subjects
1 and 2	Administration and definitions
3	Climate zones and general materials requirements
4	Energy efficiency requirements
5	Existing buildings
6	Referenced standards
Appendices CA/RA	Board of appeals
Appendices CB/RB	Solar-ready zone
Appendices CC/RC	Zero energy building provisions
Appendix CD	The 2030 glide path
Appendix CE	Required HVAC total system performance ratio (TSPR)
Appendix CF	Energy credits
Appendices CG/RE	Electric vehicle charging infrastructure
Appendices CH/RK	Electric-ready building provisions
Appendices CI/RJ	Demand responsive controls
Appendices CJ/RD	Electric energy storage provisions
Appendix RF	Alternative building thermal envelope insulation *R*-value options
Appendix RG	2024 IECC stretch code
Appendix RH	Operational carbon rating and energy reporting
Appendix RI	On-site renewable energy
Appendix RL	Renewable energy infrastructure

Chapter 1 Scope and Administration.

Chapters 1 [CE] and 1 [RE] establish the limits of applicability of the code and describe how the code is to be applied and enforced. The provisions of Chapter 1 establish the authority and duties of the code official appointed by the authority having jurisdiction and also establish the rights and privileges of the design professional, contractor and property owner.

Chapter 2 Definitions.

Chapters 2 [CE] and 2 [RE] are the repository of the definitions of terms used in the body of the code. The user of the code should be familiar with and consult these chapters because the definitions are essential to the correct interpretation of the code and because the user may not be aware that a term is defined.

Chapter 3 General Requirements.

Chapters 3 [CE] and 3 [RE] specify the climate zones that will serve to establish the exterior design conditions. In addition, Chapter 3 provides interior design conditions that are used as a basis for assumptions in heating and cooling load calculations, and provides basic material requirements for insulation materials and fenestration materials. Climate has a major impact on the energy use of most buildings. The code establishes many requirements such as wall and roof insulation *R*-values, window and door thermal transmittance (*U*-factors) and provisions that affect the mechanical systems based on the climate where the building is located. This chapter contains information that will be used to properly assign the building location into the correct climate zone and is used as the basis for establishing or eliminating requirements.

Chapter 4 Energy Efficiency.

Chapter 4 [CE] contains the energy-efficiency-related requirements for the design and construction of most types of commercial buildings and residential buildings greater than three stories in height above grade. This chapter defines requirements for the portions of the building and building systems that impact energy use in new commercial construction and new residential construction greater than three stories in height, and promotes the effective use of energy. In addition to energy conservation requirements for the building envelope, this chapter contains requirements that impact energy efficiency for the HVAC systems, the electrical systems and the plumbing systems. It should be noted, however, that requirements are contained in other codes that have an impact on energy conservation. For instance, requirements for water flow rates are regulated by the *International Plumbing Code*.

Chapter 4 [RE] contains the energy-efficiency-related requirements for the design and construction of residential buildings regulated under this code. It should be noted that the definition of a residential building in this code is unique for this code. In this code, residential buildings include detached one- and two-family dwellings and multiple single-family dwellings as well as R-2, R-3 or R-4 buildings three stories or less in height. All other buildings, including residential buildings greater than three stories in height, are regulated by the energy conservation requirements in the IECC—Commercial Provisions. The applicable portions of a residential building must comply with the provisions within this chapter for energy efficiency. This chapter defines requirements for the portions of the building and building systems that impact energy use in new residential construction and promotes the effective use of energy. The provisions within the chapter promote energy efficiency in the building envelope, the heating and cooling system and the service water-heating system of the building.

Chapter 5 Existing Buildings.

Chapters 5 [CE] and 5 [RE] contain the technical energy efficiency requirements for existing buildings. Chapter 5 provisions address the maintenance of buildings in compliance with the code as well as how additions, alterations, repairs and changes of occupancy need to be addressed from the standpoint of energy efficiency. Specific provisions are provided for historic buildings.

Chapter 6 Referenced Standards.

Chapters 6 [CE] and 6 [RE] list all of the product and installation standards and codes that are referenced throughout Chapters 1 through 5 and include identification of the promulgators and the section numbers in which the standards and codes are referenced. As stated in Sections C102.4 and R102.4, these standards and codes become an enforceable part of the code (to the prescribed extent of the reference) as if printed in the body of the code.

Appendices.

The appendices, while not part of the code, can become part of the code when specifically included in the adopting ordinance.

Chapter 1 requires the establishment of a board of appeals to hear appeals regarding determinations made by the code official.

Appendices CA and RA provide qualification standards for members of the board as well as operational procedures of such board.

Appendices CB, RB and RL address provisions for solar capacity in new structures.

Appendices CC and RC provide requirements intended bring about net zero annual energy consumption in their respective structures.

Appendix CD provides adopting jurisdictions a compliance path toward zero net energy construction by the 2030 adoption cycle.

Appendix CE provides a stretch code through HVAC incentives to Section C403.

Appendix CF provides advanced energy credit package requirements to improve efficiency requirements in Section C406.

Appendices CG and RE provide guidance for an authority having jurisdiction wishing to provide electric vehicle readiness provisions.

Appendices CH and RK provide guidance on how to prepare commercial and residential buildings to be electric ready.

Appendices CI and RJ provide guidance for demand responsive controls for building appliances and systems.

Appendices CJ and RD provide requirements for electric energy storage readiness provisions.

The purpose of Appendix RF is to provide expanded *R*-value options for determining compliance with the *U*-factor criteria in Section R402.

Similar to Appendix CD, Appendix RG provides requirements for residential buildings intended to lower energy consumption beyond the requirements of the 2024 IECC.

Appendix RH provides a means to evaluate a building's greenhouse gas performance in accordance with ANSI/RESNET/ICC 301.

Appendix RI describes requirements for prescriptive solar PV to be installed at the time of construction.

RELOCATION OF TEXT OR TABLES

The following tables indicate the relocation of sections and tables in the 2024 edition of the IECC from the 2021 edition.

IECC [CE] RELOCATIONS	
2024 LOCATION	**2021 LOCATION**
C101.4	C101.5
C101.4.1	C101.5.1
C102.1	C101.4
C102.1.1	C101.4.1
C102.2	C108.3
C102.3	C108.2
C102.4	C108.1
C102.4.1	C108.1.1
C102.4.2	C108.1.2
C102.5	C107.1
C104	C102
C104.1	C102.1
C104.1.1	C102.1.1
C105	C103
C105.1	C103.1
C105.2	C103.2
C105.2.1	C103.2.1
C105.3	C103.3
C105.3.1	C103.3.1
C105.3.2	C103.3.2
C105.3.3	C103.3
C105.4	C103.4
C105.5	C103.5
C105.6	C103.6
C105.6.1	C103.6.1
C105.6.2	C103.6.2
C105.6.3	C103.6.3
C106	C104
C106.1	C104.1
C106.2	C104.2
C106.4	C104.3
C106.5	C104.4
C106.6	C104.5
C107	C105
C107.1	C105.1
C107.2	C105.2
C107.2.1	C105.2.1

IECC [CE] RELOCATIONS—continued	
2024 LOCATION	**2021 LOCATION**
C107.2.2	C105.2.2
C107.2.3	C105.2.3
C107.2.4	C105.2.4
C107.2.5	C105.2.5
C107.2.6	C105.2.6
C107.3	C105.3
C107.4	C105.4
C107.5	C105.5
C107.6	C105.6
C108	C106
C108.1	C106.1
C108.2	C106.2
C109	C110
C109.1	C110.1
C109.2	C110.2
C109.3	C110.3
C110	C109
C110.1	C109.1
C110.2	C109.2
C110.3	C109.3
C110.4	C109.4
C402.1.1.2	C402.1.1.1
Table C402.1.1.2	Table C402.1.1.1
C402.1.1.3	C402.1.2
C402.1.2	C402.1.4
Table C402.1.2	Table C402.1.4
C402.1.2.1.1	C402.1.4.1.1
C402.1.2.1.2	C402.1.4.1.2
C402.1.2.1.6	C402.1.4.2
C402.1.3.3	C402.2.1.3
C402.1.4	C402.1.5
C402.1.5	C402.5.5
C402.2.1.1	C402.2.1.4
C402.2.1.2	C402.2.1.5
C402.2.4	C402.2.4.1
C402.4	C402.3
Table C402.4	Table C402.3
C402.4.1	C402.3.1
C402.5	C402.4
Table C402.5	Table C402.4
C402.5.1	C402.4.1
C402.5.1.1	C402.4.1.1
C402.5.1.2	C402.4.1.2
C402.5.2	C402.4.2
C402.5.2.1	C402.4.2.1
C402.5.2.2	C402.4.2.2

IECC [CE] RELOCATIONS—continued	
2024 LOCATION	**2021 LOCATION**
C402.5.3	C402.4.3
C402.5.3.1	C402.4.3.1
C402.5.3.2	C402.4.3.2
C402.5.3.3	C402.4.3.3
C402.5.3.4	C402.4.3.4
C402.5.4	C402.4.4
C402.5.5	C402.4.5
C402.5.5.1	C402.4.5.1
C402.5.5.2	C402.4.5.2
C402.6	C402.5
C402.6.1	C402.5.1
C402.6.1.2	C402.5.1.1
C402.6.1.2.1	C402.5.10
C402.6.2	C402.5.1.2
C402.6.2.1	C402.5.3
C402.6.2.2	C402.5.2
C402.6.2.3	C402.5.1.5
C402.6.2.3.1	C402.5.1.3
C402.6.2.3.2	C402.5.1.4
C402.6.3	C402.5.4
Table C402.6.3	Table C402.5.4
C402.6.4	C402.5.6
C402.6.5	C402.5.7
C402.6.6	C402.5.9
C402.6.7	C402.5.8
C403.3.4.2	C403.3.4
Table C403.3.4.2	Table C403.3.4
C403.4.1.4	C403.4.1.3
C403.4.1.5	C403.4.1.4
C403.4.1.6	C403.4.1.5
C403.4.7	C402.6.11
C403.11	C403.10
C403.11.1	C403.10.1
C403.11.2	C403.10.2
C403.11.3	C403.10.3
C403.11.4	C403.10.4
C403.11.5	C403.10.5
C403.11.6	C403.10.6
C403.12	C403.11
C403.12.1	C403.11.1
Table C403.12.1	Table C403.11.1
C403.12.2	C403.11.2
C403.12.2.1	C403.11.2.1
Table C403.12.2.1(1)	Table C403.11.2.1(1)
Table C403.12.2.1(2)	Table C403.11.2.1(2)
Table C403.12.2.1(3)	Table C403.11.2.1(3)

IECC [CE] RELOCATIONS—continued	
2024 LOCATION	**2021 LOCATION**
C403.12.3	C403.11.3
C403.12.3.1	C403.11.3.1
C403.12.3.2	C403.11.3.2
C403.13	C403.12
C403.13.1	C403.12.1
C403.13.2	C403.12.2
C403.13.2.1	C403.12.2.1
C403.13.2.2	C403.12.2.2
C403.13.2.3	C403.12.2.3
C403.13.3	C403.12.3
Table C403.13.3(1)	Table C403.12.3
C403.13.3.1	C403.12.3.1
C403.14	C403.13
C403.14.1	C403.13.1
C403.14.2	C403.13.2
C403.14.4	C403.13.3
C405.2.9	C405.2.8
C405.10	C405.9
C405.10.1	C405.9.1
C405.10.2	C405.9.2
C405.10.2.1	C405.9.2.1
C405.11	C405.10
C405.12	C405.11
C405.12.1	C405.11.1
C405.13	C405.12
C405.13.1	C405.12.1
C405.13.2	C405.12.2
Table C405.13.2	Table C405.12.2
C405.13.3	C405.12.3
C405.13.4	C405.12.4
C405.13.5	C405.12.5
C407.5.1.1	C407.5
C407.5.3	C407.5.2
C407.5.4	C407.5.3

IECC [RE] RELOCATIONS	
2024 LOCATION	**2021 LOCATION**
R101.4	R101.5
R101.4.1	R101.5.1
R102.1	R101.4
R102.1.1	R101.4.1
R102.2	R108.3
R102.3	R108.2
R102.4	R108.1

IECC [RE] RELOCATIONS—continued	
2024 LOCATION	**2021 LOCATION**
R102.4.1	R108.1.1
R102.4.2	R108.1.2
R102.5	R107.1
R104	R102
R104.1	R102.1
R104.1.1	R102.1.1
R105	R103
R105.1	R103.1
R105.2	R103.2
R105.2.1	R103.2.1
R105.3	R103.3
R105.3.1	R103.3.1
R105.3.2	R103.3.2
R105.3.3	R103.3.3
R105.4	R103.4
R105.5	R103.5
R106	R104
R106.1	R104.1
R106.2	R104.2
R106.4	R104.3
R106.5	R104.4
R106.6	R104.5
R107	R105
R107.1	R105.1
R107.2	R105.2
R107.2.1	R105.2.1
R107.2.2	R105.2.2
R107.2.3	R105.2.3
R107.2.4	R105.2.4
R107.2.7	R105.2.5
R107.3	R105.3
R107.4	R105.4
R107.5	R105.5
R107.6	R105.6
R108	R106
R108.1	R106.1
R108.2	R106.2
R109	R110
R109.1	R110.1
R109.2	R110.2
R109.3	R110.3
R109.4	R110.4
R110	R109
R110.1	R109.1
R110.2	R109.2
R110.3	R109.3

IECC [RE] RELOCATIONS—continued	
2024 LOCATION	**2021 LOCATION**
R110.4	R109.4
R402.1.6	R402.4.4
R402.2.4	R402.2.3
R402.2.5	R402.2.4
R402.2.5.1	R402.2.4.1
R402.2.6	R402.2.5
R402.2.7	R402.2.6
R402.2.8	R402.2.7
R402.2.9	R402.2.8
R402.2.9.1	R402.2.8.1
R402.2.10	R402.2.9
R402.2.10.1	R402.2.9.1
R402.2.11	R402.2.10
R402.2.11.1	R402.2.10.1
R402.2.12	R402.2.11
R402.2.13	R402.2.12
R402.4	R402.3
R402.4.1	R402.3.1
R402.4.2	R402.3.2
R402.4.3	R402.3.3
R402.4.4	R402.3.4
R402.4.5	R402.3.5
R402.5	R402.4
R402.5.1	R402.4.1
R402.5.1.1	R402.4.1.1
Table R402.5.1.1	Table R402.4.1.1
R402.5.1.2	R402.4.1.2
R402.5.1.3	R402.4.1.3
R402.5.2	R402.4.2
R402.5.3	R402.4.3
R402.5.4	R402.4.5
R402.5.5	R402.4.6
R402.6	R402.5
R403.3.2	R403.3.7
R403.3.3	R403.3.1
R403.3.4	R403.3.2
R403.3.5	R403.3.3
R403.3.5.1	R403.3.3.1
R403.3.6	R403.3.4
R403.3.6.1	R403.3.4.1
R403.3.7	R403.3.5
R403.3.8	R403.3.6
R403.9.2	R403.9
R404.1.5	R404.1.2
R405.4.3	R405.5.3
R405.5.4	R405.3.2

IECC [RE] RELOCATIONS—continued	
2024 LOCATION	**2021 LOCATION**
R405.5.4.1	R405.3.2.1
R405.5.4.2	R405.3.2.2
R406.3	R406.3.1
R501.4	R501.5
R501.5	R501.6
R502.2	R502.3
R502.2.1	R502.3.1
R502.2.2	R502.3.2
R502.2.3	R502.3.3
R502.2.4	R502.3.4

ABBREVIATIONS AND NOTATIONS

The following table contains a list of common abbreviations and units of measurement used in this code. Some of the abbreviations are for terms defined in Chapter 2. Others are terms used in various tables and text of the code.

ABBREVIATIONS AND NOTATIONS	
AFUE	Annual fuel utilization efficiency
bhp	Brake horsepower (fans)
Btu	British thermal unit
Btu/h × ft^2	Btu per hour per square foot
C-factor	See Chapter 2—Definitions
CDD	Cooling degree days
cfm	Cubic feet per minute
cfm/ft^2	Cubic feet per minute per square foot
ci	Continuous insulation
COP	Coefficient of performance
DCV	Demand control ventilation
°C	Degrees Celsius
°F	Degrees Fahrenheit
DWHR	Drain water heat recovery
DX	Direct expansion
E_c	Combustion efficiency
E_v	Ventilation efficiency
E_t	Thermal efficiency
EER	Energy efficiency ratio
EF	Energy factor
ERI	Energy rating index
F-factor	See Chapter 2—Definitions
FDD	Fault detection and diagnostics
FEI	Fan energy index
FL	Full load
ft^2	Square foot
gpm	Gallons per minute
HDD	Heating degree days
hp	Horsepower

ABBREVIATIONS AND NOTATIONS—continued	
HSPF	Heating seasonal performance factor
HVAC	Heating, ventilating and air conditioning
IEER	Integrated energy efficiency ratio
IPLV	Integrated Part Load Value
Kg/m^2	Kilograms per square meter
kW	Kilowatt
LPD	Light power density (lighting power allowance)
L/s	Liters per second
Ls	Liner system
m^2	Square meters
MERV	Minimum efficiency reporting value
NAECA	National Appliance Energy Conservation Act
NPLV	Nonstandard Part Load Value
Pa	Pascal
PF	Projection factor
pcf	Pounds per cubic foot
psf	Pounds per square foot
PTAC	Packaged terminal air conditioner
PTHP	Packaged terminal heat pump
R-value	See Chapter 2—Definitions
SCOP	Sensible coefficient of performance
SEER	Seasonal energy efficiency ratio
SHGC	Solar Heat Gain Coefficient
SPVAC	Single packaged vertical air conditioner
SPVHP	Single packaged vertical heat pump
SRI	Solar reflectance index
SWF	Service water heat recovery factor
U-factor	See Chapter 2—Definitions
VAV	Variable air volume
VRF	Variable refrigerant flow
VT	Visible transmittance
W	Watts
w.c.	Water column
w.g.	Water gauge

CONTENTS

IECC—COMMERCIAL PROVISIONS

CONTENTS

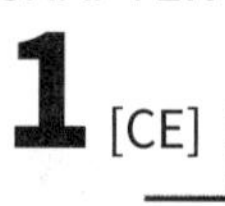
CHAPTER 1 [CE]

SCOPE AND ADMINISTRATION

User notes:

About this chapter: *Chapter 1 establishes the limits of applicability of the code and describes how the code is to be applied and enforced. Chapter 1 is in two parts: Part 1—Scope and Application and Part 2—Administration and Enforcement. Section C101 identifies what buildings, systems, appliances and equipment fall under its purview and references other I-Codes as applicable. Standards and codes are scoped to the extent referenced.*

The code is intended to be adopted as a legally enforceable document and it cannot be effective without adequate provisions for its administration and enforcement. The provisions of Chapter 1 establish the authority and duties of the code official appointed by the authority having jurisdiction and also establish the rights and privileges of the design professional, contractor and property owner.

QR code use: *A QR code is placed at the beginning of any section that has undergone technical revision. To see those revisions, scan the QR code with a smart device or enter the 7-digit code beneath the QR code at the end of the following URL: qr.iccsafe.org/ (see Formatting Changes to the 2024 International Codes for more information).*

PART 1—SCOPE AND APPLICATION

SECTION C101—SCOPE AND GENERAL REQUIREMENTS

C101.1 Title. This code shall be known as the *Energy Conservation Code* of **[NAME OF JURISDICTION]** and shall be cited as such. It is referred to herein as "this code."

C101.2 Scope. This code applies to the design and construction of buildings not covered by the scope of the IECC—Residential Provisions.

C101.2.1 Appendices. Provisions in the appendices shall not apply unless specifically adopted.

C101.3 Intent. The IECC—Commercial Provisions provide market-driven, enforceable requirements for the design and construction of *commercial buildings*, providing minimum efficiency requirements for buildings that result in the maximum level of energy efficiency that is safe, technologically feasible, and life cycle cost effective, considering economic feasibility, including potential costs and savings for consumers and building owners, and return on investment. Additionally, the code provides jurisdictions with supplemental requirements, including ASHRAE 90.1, and optional requirements that lead to achievement of zero energy buildings, presently, and through glidepaths that achieve zero energy buildings by 2030 and on additional timelines sought by governments, and achievement of additional policy goals as identified by the Energy and Carbon Advisory Council and approved by the Board of Directors. Requirements contained in the code will include, but not be limited to, prescriptive- and performance-based pathways. The code may include nonmandatory appendices incorporating additional energy efficiency and greenhouse gas reduction resources developed by the International Code Council and others. The code will aim to simplify code requirements to facilitate the code's use and compliance rate. The code is updated on a 3-year cycle with each subsequent edition providing increased energy savings over the prior edition. This code is intended to provide flexibility to permit the use of innovative approaches and techniques to achieve this intent. This code is not intended to abridge safety, health or environmental requirements contained in other applicable codes or ordinances.

C101.4 Compliance. *Residential buildings* shall meet the provisions of IECC—Residential Provisions. *Commercial buildings* shall meet the provisions of IECC—Commercial Provisions.

C101.4.1 Compliance materials. The *code official* shall be permitted to approve specific computer software, worksheets, compliance manuals and other similar materials that meet the intent of this code.

SECTION C102—APPLICABILITY

C102.1 Applicability. Where, in any specific case, different sections of this code specify different materials, methods of construction or other requirements, the most restrictive shall govern. Where there is a conflict between a general requirement and a specific requirement, the specific requirement shall govern.

C102.1.1 Mixed residential and commercial buildings. Where a *building* includes both *residential building* and *commercial building* portions, each portion shall be separately considered and meet the applicable provisions of IECC—Commercial Provisions or IECC—Residential Provisions.

C102.2 Other laws. The provisions of this code shall not be deemed to nullify any provisions of local, state or federal law.

C102.3 Applications of references. References to chapter or section numbers, or to provisions not specifically identified by number, shall be construed to refer to such chapter, section or provision of this code.

C102.4 Referenced codes and standards. The codes and standards referenced in this code shall be those listed in Chapter 6, and such codes and standards shall be considered as part of the requirements of this code to the prescribed extent of each such reference and as further regulated in Sections C102.4.1 and C102.4.2.

C102.4.1 Conflicts. Where conflicts occur between provisions of this code and referenced codes and standards, the provisions of this code shall apply.

C102.4.2 Provisions in referenced codes and standards. Where the extent of the reference to a referenced code or standard includes subject matter that is within the scope of this code, the provisions of this code, as applicable, shall take precedence over the provisions in the referenced code or standard.

C102.5 Partial invalidity. If a portion of this code is held to be illegal or void, such a decision shall not affect the validity of the remainder of this code.

PART 2—ADMINISTRATION AND ENFORCEMENT

SECTION C103—CODE COMPLIANCE AGENCY

C103.1 Creation of enforcement agency. The [INSERT NAME OF DEPARTMENT] is hereby created and the official in charge thereof shall be known as the authority having jurisdiction (AHJ). The function of the agency shall be the implementation, administration and enforcement of the provisions of this code.

C103.2 Appointment. The AHJ shall be appointed by the chief appointing authority of the jurisdiction.

C103.3 Deputies. In accordance with the prescribed procedures of this jurisdiction and with the concurrence of the appointing authority, the AHJ shall have the authority to appoint a deputy AHJ, other related technical officers, inspectors and other employees. Such employees shall have powers as delegated by the AHJ.

SECTION C104—ALTERNATIVE MATERIALS, DESIGN AND METHODS OF CONSTRUCTION AND EQUIPMENT

C104.1 General. The provisions of this code are not intended to prevent the installation of any material or to prohibit any design or method of construction not specifically prescribed by this code, provided that any such alternative has been *approved*. The *code official* shall have the authority to approve an alternative material, design or method of construction upon the written application of the *owner* or the owner's authorized agent. The *code official* shall first find that the proposed design is satisfactory and complies with the intent of the provisions of this code, and that the material, method or work offered is, for the purpose intended, not less than the equivalent of that prescribed in this code in quality, strength, effectiveness, *fire resistance*, durability, energy conservation and safety. The *code official* shall respond to the applicant, in writing, stating the reasons why the alternative was *approved* or was not *approved*.

C104.1.1 Above code programs. The *code official* or other authority having jurisdiction shall be permitted to deem a national, state or local energy efficiency program as exceeding the energy efficiency required by this code. Buildings *approved* in writing by such an energy efficiency program shall be considered to be in compliance with this code. The requirements identified in Table C407.2(1) shall be met.

SECTION C105—CONSTRUCTION DOCUMENTS

C105.1 General. *Construction documents* and other supporting data shall be submitted in one or more sets, or in a digital format where allowed by the *code official*, with each application for a permit. The *construction documents* shall be prepared by a *registered design professional* where required by the statutes of the jurisdiction in which the project is to be constructed. Where special conditions exist, the *code official* is authorized to require necessary *construction documents* to be prepared by a *registered design professional*.

Exception: The *code official* is authorized to waive the requirements for *construction documents* or other supporting data if the *code official* determines they are not necessary to confirm compliance with this code.

C105.2 Information on construction documents. *Construction documents* shall be drawn to scale on suitable material. Electronic media documents are permitted to be submitted where *approved* by the *code official*. *Construction documents* shall be of sufficient clarity to indicate the location, nature and extent of the work proposed, and show in sufficient detail pertinent data and features of the *building*, systems and equipment as herein governed. Details shall include, but are not limited to, the following as applicable:

1. Energy compliance path.
2. Insulation materials and their *R*-values.
3. Fenestration *U-factors* and solar heat gain coefficients (SHGC).
4. Area-weighted *U-factor* and *solar heat gain coefficient* (SHGC) calculations.
5. *Air barrier* and air sealing details, including the location of the *air barrier*.
6. *Thermal bridges* as identified in Section C402.7.
7. Mechanical system design criteria.
8. Mechanical and service water-heating systems and equipment types, sizes and efficiencies.
9. Economizer description.
10. Equipment and system controls.

11. Fan motor horsepower (hp) and controls.
12. Duct sealing, *duct* and pipe insulation and location.
13. Lighting fixture schedule with wattage and control narrative.
14. Location of *daylight* zones on floor plans.
15. Location of pathways for routing of raceways or cable from the on-site renewable energy system to the electrical distribution equipment.
16. Location reserved for inverters, metering equipment and energy storage systems (ESS), and a pathway reserved for routing of raceways or conduit from the renewable energy system to the point of interconnection with the electrical service and the ESS.
17. Location and layout of a designated area for ESS.
18. Rated energy capacity and rated power capacity of the installed or planned ESS.

C105.2.1 Building thermal envelope depiction. The *building thermal envelope* shall be represented on the construction drawings.

C105.3 Examination of documents. The *code official* shall examine or cause to be examined the accompanying *construction documents* and shall ascertain whether the construction indicated and described is in accordance with the requirements of this code and other pertinent laws or ordinances. The *code official* is authorized to utilize a *registered design professional*, or other *approved* entity not affiliated with the building design or construction, in conducting the review of the plans and specifications for compliance with the code.

C105.3.1 Approval of construction documents. When the *code official* issues a permit where *construction documents* are required, the *construction documents* shall be endorsed in writing and stamped "Reviewed for Code Compliance." Such *approved construction documents* shall not be changed, modified or altered without authorization from the *code official*. Work shall be done in accordance with the *approved construction documents*.

One set of *construction documents* so reviewed shall be retained by the *code official*. The other set shall be returned to the applicant, kept at the site of work and shall be open to inspection by the *code official* or a duly authorized representative.

C105.3.2 Previous approvals. This code shall not require changes in the *construction documents*, construction or designated occupancy of a structure for which a lawful permit has been heretofore issued or otherwise lawfully authorized, and the construction of which has been pursued in good faith within 180 days after the effective date of this code and has not been abandoned.

C105.3.3 Phased approval. The *code official* shall have the authority to issue a permit for the construction of part of an energy conservation system before the *construction documents* for the entire system have been submitted or *approved*, provided that adequate information and detailed statements have been filed complying with all pertinent requirements of this code. The holders of such permit shall proceed at their own risk without assurance that the permit for the entire energy conservation system will be granted.

C105.4 Amended construction documents. Changes made during construction that are not in compliance with the *approved construction documents* shall be resubmitted for approval as an amended set of *construction documents*.

C105.5 Retention of construction documents. One set of *approved construction documents* shall be retained by the *code official* for a period of not less than 180 days from date of completion of the permitted work, or as required by state or local laws.

C105.6 Building documentation and closeout submittal requirements. The *construction documents* shall specify that the documents described in this section be provided to the building *owner* or owner's authorized agent within 90 days of the date of receipt of the certificate of occupancy.

C105.6.1 Record documents. *Construction documents* shall be updated to convey a record of the completed work. Such updates shall include mechanical, electrical and control drawings that indicate all changes to size, type and location of components, equipment and assemblies.

C105.6.2 Compliance documentation. Energy code compliance documentation and supporting calculations shall be delivered in one document to the building *owner* as part of the project record documents or manuals, or as a standalone document. This document shall include the specific energy code edition utilized for compliance determination for each system, documentation demonstrating compliance with Section C303.1.3 for each fenestration product installed, and the interior lighting power compliance path, building area or space-by-space, used to calculate the lighting power allowance.

For projects complying with Item 2 of Section C401.2, the documentation shall include:

1. The envelope insulation compliance path.
2. All compliance calculations including those required by Sections C402.1.4, C403.8.1, C405.3 and C405.5.
3. A plan for annual energy use data gathering and disclosure as specified in Section C405.13.

For projects complying with Section C407, the documentation shall include that required by Sections C407.3.1 and C407.3.2.

C105.6.3 Systems operation control. Training shall be provided to those responsible for maintaining and operating equipment included in the manuals required by Section C105.6.2.

The training shall include:

1. Review of manuals and permanent certificate.

2. Hands-on demonstration of all normal maintenance procedures, normal operating modes, and all emergency shutdown and startup procedures.
3. Training completion report.

SECTION C106—FEES

C106.1 Fees. A permit shall not be valid until the fees prescribed by law have been paid. Nor shall an amendment to a permit be released until the additional fee, if any, has been paid.

C106.2 Schedule of permit fees. A fee for each permit shall be paid as required, in accordance with the schedule as established by the applicable governing authority.

C106.3 Valuation of work. The applicant for a permit shall provide an estimated value of the work for which the permit is being issued at the time of application. Such estimated valuations shall include the total value of the work, including materials and labor. Where, in the opinion of the *code official*, the valuation is underestimated, the permit shall be denied unless the applicant can show detailed estimates acceptable to the *code official*. The final valuation shall be *approved* by the *code official*.

C106.4 Work commencing before permit issuance. Any person who commences any work before obtaining the necessary permits shall be subject to a fee established by the *code official* that shall be in addition to the required permit fees.

C106.5 Related fees. The payment of the fee for the construction, *alteration*, removal or demolition of work done in connection to or concurrently with the work or activity authorized by a permit shall not relieve the applicant or holder of the permit from the payment of other fees that are prescribed by law.

C106.6 Refunds. The *code official* is authorized to establish a refund policy.

SECTION C107—INSPECTIONS

C107.1 General. Construction or work for which a permit is required shall be subject to inspection by the *code official*, his or her designated agent or an *approved agency*, and such construction or work shall remain visible and able to be accessed for inspection purposes until *approved*. Approval as a result of an inspection shall not be construed to be an approval of a violation of the provisions of this code or of other ordinances of the jurisdiction. Inspections presuming to give authority to violate or cancel the provisions of this code or of other ordinances of the jurisdiction shall not be valid. It shall be the duty of the permit applicant to cause the work to remain visible and able to be accessed for inspection purposes. Neither the *code official* nor the jurisdiction shall be liable for expense entailed in the removal or replacement of any material, product, system or building component required to allow inspection to validate compliance with this code.

C107.2 Required inspections. The *code official*, his or her designated agent or an *approved agency*, upon notification, shall make the inspections set forth in Sections C107.2.1 through C107.2.6.

C107.2.1 Footing and foundation insulation. Inspections shall verify the footing and foundation insulation *R-value*, location, thickness, depth of burial and protection of insulation as required by the code, *approved* plans and specifications.

C107.2.2 Building thermal envelope. Inspections shall verify the type of insulation, *R*-values, location of insulation, *thermal bridge* mitigation, *fenestration*, *U-factor*, SHGC and VT, and that *air leakage* controls are installed, as required by the code, *approved* plans and specifications.

C107.2.3 Plumbing system. Inspections shall verify the type of insulation, *R*-values, protection required, controls and *heat traps* as required by the code, *approved* plans and specifications.

C107.2.4 Mechanical system. Inspections shall verify the installed HVAC equipment for the type and size, controls, insulation, *R*-values, system and damper *air leakage*, minimum fan efficiency, energy recovery and economizer as required by the code, *approved* plans and specifications.

C107.2.5 Electrical system. Inspections shall verify lighting system controls, components and meters as required by the code, *approved* plans and specifications. Where an ESS area is required, inspections shall verify space availability and pathways to electrical service.

C107.2.6 Final inspection. The final inspection shall include verification of the installation and proper operation of all required building controls, and documentation verifying activities associated with required *building commissioning* have been conducted in accordance with Section C408.

C107.3 Reinspection. A *building* shall be reinspected where determined necessary by the *code official*.

C107.4 Approved inspection agencies. The *code official* is authorized to accept reports of third-party inspection agencies not affiliated with the building design or construction, provided that such agencies are *approved* as to qualifications and reliability relevant to the building components and systems that they are inspecting.

C107.5 Inspection requests. It shall be the duty of the holder of the permit or their duly authorized agent to notify the *code official* when work is ready for inspection. It shall be the duty of the permit holder to provide *access* to and means for inspections of such work that are required by this code.

C107.6 Reinspection and testing. Where any work or installation does not pass an initial test or inspection, the necessary corrections shall be made to achieve compliance with this code. The work or installation shall then be resubmitted to the *code official* for inspection and testing.

SECTION C108—NOTICE OF APPROVAL

C108.1 Approval. After the prescribed tests and inspections indicate that the work complies in all respects with this code, a notice of approval shall be issued by the *code official*.

C108.2 Revocation. The *code official* is authorized to suspend or revoke, in writing, a notice of approval issued under the provisions of this code wherever the certificate is issued in error, or on the basis of incorrect information supplied, or where it is determined that the *building* or structure, premise, or portion thereof is in violation of any ordinance or regulation or any of the provisions of this code.

SECTION C109—MEANS OF APPEALS

C109.1 General. In order to hear and decide appeals of orders, decisions or determinations made by the *code official* relative to the application and interpretation of this code, there shall be and is hereby created a board of appeals. The board of appeals shall be appointed by the governing authority and shall hold office at its pleasure. The board shall adopt rules of procedure for conducting its business, and shall render all decisions and findings in writing to the appellant with a duplicate copy to the *code official*.

C109.2 Limitations on authority. An application for appeal shall be based on a claim that the true intent of this code or the rules legally adopted thereunder have been incorrectly interpreted, the provisions of this code do not fully apply or an equally good or better form of construction is proposed. The board shall not have authority to waive requirements of this code.

C109.3 Qualifications. The board of appeals shall consist of members who are qualified by experience and training on matters pertaining to the provisions of this code and are not employees of the jurisdiction.

C109.4 Administration. The *code official* shall take action in accordance with the decisions of the board.

SECTION C110—STOP WORK ORDER

C110.1 Authority. Where the *code official* finds any work regulated by this code being performed in a manner contrary to the provisions of this code or in a dangerous or unsafe manner, the *code official* is authorized to issue a stop work order.

C110.2 Issuance. The stop work order shall be in writing and shall be given to the *owner* of the property, the owner's authorized agent or the person performing the work. Upon issuance of a stop work order, the cited work shall immediately cease. The stop work order shall state the reason for the order and the conditions under which the cited work is authorized to resume.

C110.3 Emergencies. Where an emergency exists, the *code official* shall not be required to give a written notice prior to stopping the work.

C110.4 Failure to comply. Any person who shall continue any work after having been served with a stop work order, except such work as that person is directed to perform to remove a violation or unsafe condition, shall be subject to fines established by the AHJ.

CHAPTER 2 [CE] DEFINITIONS

User notes:

About this chapter: *Codes, by their very nature, are technical documents. Every word, term and punctuation mark can add to or change the meaning of a technical requirement. It is necessary to maintain a consensus on the specific meaning of each term contained in the code. Chapter 2 performs this function by stating clearly what specific terms mean for the purposes of the code.*

SECTION C201—GENERAL

C201.1 Scope. Unless stated otherwise, the following words and terms in this code shall have the meanings indicated in this chapter.

C201.2 Interchangeability. Words used in the present tense include the future; words in the masculine gender include the feminine and neuter; the singular number includes the plural and the plural includes the singular.

C201.3 Terms defined in other codes. Terms thsat are not defined in this code but are defined in the *International Building Code*, *International Fire Code*, *International Fuel Gas Code*, *International Mechanical Code*, *International Plumbing Code* or the *International Residential Code* shall have the meanings ascribed to them in those codes.

C201.4 Terms not defined. Terms not defined by this chapter shall have ordinarily accepted meanings such as the context implies.

SECTION C202—GENERAL DEFINITIONS

ABOVE-GRADE WALL. See "*Wall, above-grade*."

ACCESS (TO). That which enables a device, appliance or equipment to be reached by *ready access* or by a means that first requires the removal or movement of a panel or similar obstruction.

ADDITION. An extension or increase in the *conditioned space* floor area, number of stories or height of a *building* or structure.

AIR BARRIER. One or more materials joined together in a continuous manner to restrict or prevent the passage of air through the *building thermal envelope* and its assemblies.

AIR CURTAIN UNIT. A device installed at the building entrance that generates and discharges a laminar airstream intended to prevent the *infiltration* of external, unconditioned air into the conditioned spaces or the loss of interior, conditioned air to the outside.

AIR LEAKAGE. The uncontrolled airflow through the *building thermal envelope* caused by pressure differences across the *building thermal envelope*. Air leakage can be inward (*infiltration*) or outward (exfiltration) through the *building thermal envelope*.

ALTERATION. Any construction, retrofit or renovation to an existing structure other than *repair* or *addition*. Also, a change in a building, electrical, gas, mechanical or plumbing system that involves an extension, *addition* or change to the arrangement, type or purpose of the original installation.

APPROVED. Acceptable to the *code official*.

APPROVED AGENCY. An established and recognized agency that is regularly engaged in conducting tests or furnishing inspection services, or furnishing product certification, where such agency has been *approved* by the *code official*.

APPROVED SOURCE. An independent person, firm or corporation *approved* by the code official, who is competent and experienced in the application of engineering principles to materials, methods or systems analyses.

AUTOMATIC. Self-acting, operating by its own mechanism when actuated by some impersonal influence, as, for example, a change in current strength, pressure, temperature or mechanical configuration (see "*Manual*").

BELOW-GRADE WALL. See "*Wall, below-grade*."

BEST EFFICIENCY POINT (BEP). The pump hydraulic power operating point (consisting of both flow and head conditions) that results in the maximum efficiency.

BIOGAS. A mixture of hydrocarbons that is a gas at 60°F (15.5°C) and 1 atmosphere of pressure that is produced through the anaerobic digestion of organic matter.

BIOMASS WASTE. Organic nonfossil material of biological origin that is a byproduct or a discarded product. *Biomass waste* includes municipal solid waste from biogenic sources; landfill gas; sludge waste; agricultural crop byproducts; straw; and other biomass solids, liquids and biogases, but excludes wood and wood-derived fuels (including black liquor), biofuel feedstock, biodiesel and fuel ethanol.

BOILER, MODULATING. A boiler that is capable of more than a single firing rate in response to a varying temperature or heating load.

BOILER SYSTEM. One or more boilers, their piping and controls that work together to supply steam or hot water to heat output devices remote from the boiler.

BUBBLE POINT. The refrigerant liquid saturation temperature at a specified pressure.

BUILDING. Any structure used or intended for supporting or sheltering any use or occupancy, including any mechanical systems, service water-heating systems and electric power and lighting systems located on the *building site* and supporting the *building*.

BUILDING COMMISSIONING. A process that verifies and documents that the selected building systems have been designed, installed and function according to the *owner*'s project requirements and *construction documents*, and to minimum code requirements.

BUILDING ENTRANCE. Any door, set of doors, doorway or other form of portal that is used to gain access to the *building* from the outside by the public.

BUILDING SITE. A contiguous area of land that is under the ownership or control of one entity.

BUILDING THERMAL ENVELOPE. The basement walls, *exterior walls*, floors, ceilings, roofs and any other building element assemblies that enclose *conditioned space* or provide a boundary between *conditioned space* and exempt or unconditioned space.

CAPTIVE KEY OVERRIDE. A lighting control that will not release the key that activates the override when the lighting is on.

CAVITY INSULATION. Insulating material located between framing members.

***C*-FACTOR (THERMAL CONDUCTANCE).** The coefficient of heat transmission (surface to surface) through a building component or assembly, equal to the time rate of heat flow per unit area and the unit temperature difference between the warm side and cold side surfaces (Btu/h × ft^2 × °F) [W/(m^2 × K)].

CHANGE OF OCCUPANCY. A change in the use of a *building* or a portion of a *building* that results in any of the following:

1. A *change of occupancy* classification.
2. A change from one group to another group within an occupancy classification.
3. Any change in use within a group for which there is a change in the application of the requirements of this code.

CHI-FACTOR (χ-FACTOR). The heat loss factor for a single *thermal bridge* characterized as a point element of a *building thermal envelope* (Btu/h × °F) [W/k].

CIRCULATING HOT WATER SYSTEM. A specifically designed water distribution system where one or more pumps are operated in the service hot water piping to circulate heated water from the water-heating equipment to the fixture supply and back to the water-heating equipment.

CLEAN WATER PUMP. A device that is designed for use in pumping water with a maximum nonabsorbent free solid content of 0.016 lb/ft^3 (0.256 kg/m^3) and with a maximum dissolved solid content of 3.1 lb/ft^3 (49.66 kg/m^3), provided that the total gas content of the water does not exceed the saturation volume, and disregarding any additives necessary to prevent the water from freezing at a minimum of 14°F (-10°C).

CLIMATE ZONE. A geographical region based on climatic criteria as specified in this code.

CODE OFFICIAL. The officer or other designated authority charged with the administration and enforcement of this code, or a duly authorized representative.

COEFFICIENT OF PERFORMANCE (COP) – COOLING. The ratio of the rate of heat removal to the rate of energy input, in consistent units, for a complete refrigerating system or some specific portion of that system under designated operating conditions.

COEFFICIENT OF PERFORMANCE (COP) – HEATING. The ratio of the rate of heat delivered to the rate of energy input, in consistent units, for a complete heat pump system, including the compressor and, if applicable, auxiliary heat, under designated operating conditions.

COMMERCIAL BUILDING. For this code, all buildings that are not included in the definition of "*Residential building*."

COMMON AREAS. All conditioned spaces within *Group R* occupancy buildings that are not *dwelling units* or *sleeping units*.

COMMUNITY RENEWABLE ENERGY FACILITY. A facility that produces energy harvested from *renewable energy resources* and is qualified as a community energy facility under applicable jurisdictional statutes and rules.

COMPUTER ROOM. A room whose primary function is to house equipment for the processing and storage of electronic data which has a design total *information technology equipment* (ITE) equipment power density less than or equal to 20 watts per square foot (20 watts per 0.092 m^2) of conditioned area or a design total ITE equipment load less than or equal to 10 kW.

CONDENSING UNIT. A factory-made assembly of refrigeration components designed to compress and liquefy a specific refrigerant. The unit consists of one or more refrigerant compressors, refrigerant condensers (air-cooled, evaporatively cooled or water-cooled), condenser fans and motors (where used) and factory-supplied accessories.

CONDITIONED FLOOR AREA. The horizontal projection of the floors associated with the *conditioned space*.

CONDITIONED SPACE. An area, room or space that is enclosed within the *building thermal envelope* and is directly or indirectly heated or cooled. Spaces are indirectly heated or cooled where they communicate through openings with *conditioned spaces*, where they are separated from *conditioned spaces* by uninsulated walls, floors or ceilings, or where they contain uninsulated *ducts*, piping or other sources of heating or cooling.

CONGREGATE LIVING FACILITIES. A *building* or part thereof that contains *sleeping units* where residents share bathroom or kitchen facilities, or both.

CONSTRUCTION DOCUMENTS. Written, graphic and pictorial documents prepared or assembled for describing the design, location and physical characteristics of the elements of a project necessary for obtaining a building permit.

CONTINUOUS INSULATION (ci). Insulating material that is continuous across all structural members without *thermal bridges* other than fasteners and service openings. It is installed on the interior or exterior or is integral to any opaque surface of the *building thermal envelope*.

CRAWL SPACE WALL. The opaque portion of a wall that encloses a crawl space and is partially or totally below grade.

CURTAIN WALL. Fenestration products used to create an external nonload-bearing wall that is designed to separate the exterior and interior environments.

DATA CENTER. A room or series of rooms that share *data center systems*, whose primary function is to house equipment for the processing and storage of electronic data and that has a design total ITE equipment power density exceeding 20 watts per square foot (20 watts per 0.092 m^2) of conditioned area and a total design ITE equipment load greater than 10 kW.

DATA CENTER SYSTEMS. HVAC systems and equipment, or portions thereof, used to provide cooling or *ventilation* in a *data center*.

DAYLIGHT RESPONSIVE CONTROL. A device or system that provides *automatic* control of electric light levels based on the amount of daylight in a space.

DAYLIGHT ZONE. That portion of a building's interior floor area that is illuminated by natural light.

DEDICATED OUTDOOR AIR SYSTEM (DOAS). A ventilation system that supplies 100 percent outdoor air primarily for the purpose of *ventilation* and that is a separate system from the *zone* space-conditioning system.

DEHUMIDIFIER. A self-contained, electrically operated and mechanically encased product with the sole purpose of dehumidifying the space consisting of the following:

1. A refrigerated surface (evaporator) that condenses moisture from the atmosphere.
2. A refrigerating system, including an electric motor.
3. An air-circulating fan.
4. A means for collecting or disposing of the condensate.

A dehumidifier does not include a portable air conditioner, room air conditioner or packaged terminal air conditioner.

DEMAND CONTROL KITCHEN VENTILATION (DCKV). A system that provides *automatic*, continuous control over exhaust hood and makeup air fan speed in response to temperature, optical or infrared (IR) sensors that monitor cooking activity or through direct communication with cooking appliances.

DEMAND CONTROL VENTILATION (DCV). A ventilation system capability that provides for the *automatic* reduction of outdoor air intake below design rates when the actual occupancy of spaces served by the system is less than design occupancy.

DEMAND RECIRCULATION WATER SYSTEM. A water distribution system where one or more pumps prime the service hot water piping with heated water upon a demand for hot water.

DEMAND RESPONSE SIGNAL. A signal that indicates a price or a request to modify electricity consumption for a limited time period.

DEMAND RESPONSIVE CONTROL. A control capable of receiving and automatically responding to a *demand response signal*.

DESSICANT DEHUMIDIFICATION SYSTEM. A mechanical dehumidification technology that uses a solid or liquid material to remove moisture from the air.

DIRECT DIGITAL CONTROL (DDC). A type of control where controlled and monitored analog or binary data, such as temperature and contact closures, are converted to digital format for manipulation and calculations by a digital computer or microprocessor, then converted back to analog or binary form to control physical devices.

DUCT. A tube or conduit utilized for conveying air. The air passages of self-contained systems are not to be construed as air *ducts*.

DUCT SYSTEM. A continuous passageway for the transmission of air that, in addition to *ducts*, includes duct fittings, dampers, plenums, fans and accessory air-handling equipment and appliances.

DWELLING UNIT. A single unit providing complete independent living facilities for one or more persons, including permanent provisions for living, sleeping, eating, cooking and sanitation.

DX-DEDICATED OUTDOOR AIR SYSTEM UNIT (DX-DOAS UNIT). A type of air-cooled, water-cooled or water source factory-assembled product that dehumidifies 100 percent outdoor air to a low dew point and includes reheat that is capable of controlling the supply dry-bulb temperature of the dehumidified air to the designated supply air temperature. It may precondition outdoor air with an *energy recovery ventilation system*.

DYNAMIC GLAZING. Any fenestration product that has the fully reversible ability to change its performance properties, including *U-factor*, *solar heat gain coefficient* (SHGC) or *visible transmittance* (VT).

EAST-ORIENTED. Facing within 45 degrees of true east to the south and within less than 22.5 degrees of true east to the north in the northern hemisphere or facing within 45 degrees of true east to the north and within less than 22.5 degrees of true east to the south in the southern hemisphere.

ECONOMIZER, AIR. A *duct* and damper arrangement and *automatic* control system that allows a cooling system to supply outside air to reduce or eliminate the need for mechanical cooling during mild or cold weather.

ECONOMIZER, WATER. A system where the supply air of a cooling system is cooled indirectly with water that is itself cooled by heat or mass transfer to the environment without the use of mechanical cooling.

EMITTANCE. The ratio of the radiant heat flux emitted by a specimen measured on a scale from 0 to 1, where a value of 1 indicates perfect release of thermal radiation.

ENCLOSED SPACE. A volume surrounded by solid surfaces such as walls, floors, roofs and openable devices, such as doors and operable windows.

ENERGY ANALYSIS. A method for estimating the annual energy use of the *proposed design* and *standard reference design* based on estimates of energy use.

ENERGY COST. The total estimated annual cost for *purchased energy* for the building functions regulated by this code, including applicable demand charges.

ENERGY RECOVERY, SERIES. A three-step process in which the first step is to remove energy from a single airstream without the use of mechanical cooling. In the second step, the airstream is mechanically cooled for the purpose of dehumidification. In the third step, the energy removed in the first step is reintroduced to the airstream.

ENERGY RECOVERY RATIO, SERIES (SERR). The difference between the dry-bulb air temperatures leaving the *series energy recovery* unit and leaving the dehumidifying coil divided by the difference between 75°F (24°C) and the dry-bulb temperature of the air leaving the dehumidifying cooling coil.

ENERGY RECOVERY VENTILATION SYSTEM. Systems that employ air-to-air heat exchangers to recover energy from exhaust air for the purpose of preheating, precooling, humidifying or dehumidifying outdoor *ventilation air* prior to supplying the air to a space, either directly or as part of an HVAC system.

ENERGY SIMULATION TOOL. An *approved* software program or calculation-based methodology that projects the annual energy use of a *building*.

ENERGY STORAGE SYSTEM (ESS). One or more devices, assembled together, capable of storing energy in order to supply electrical energy at a future time.

ENERGY USE INTENSITY (EUI). The metric indicating the total amount of energy consumed by a *building* in 1 year divided by the gross floor area of the *building*.

ENTHALPY RECOVERY RATIO (ERR). Change in the enthalpy of the outdoor air supply divided by the difference between the outdoor air and entering exhaust air enthalpy, expressed as a percentage.

ENTRANCE DOOR. A vertical *fenestration* product used for occupant ingress, egress and access in nonresidential buildings, including, but not limited to, exterior entrances utilizing latching hardware and *automatic* closers and containing over 50 percent glazing specifically designed to withstand heavy-duty usage.

EQUIPMENT ROOM. A space that contains either electrical equipment, mechanical equipment, machinery, water pumps or hydraulic pumps that are a function of the building's services.

EXTERIOR WALL. Walls including both *above-grade walls* and basement walls.

EXTERIOR WALL ENVELOPE. A system or assembly of exterior wall components, including exterior wall finish materials, that provides protection of the building structural members, including framing and sheathing materials, and conditioned interior space from the detrimental effects of the exterior environment.

FAN, EMBEDDED. A fan that is part of a manufactured assembly where the assembly includes functions other than air movement.

FAN ARRAY. Multiple fans in parallel between two plenum sections in an air distribution system.

FAN BRAKE HORSEPOWER (BHP). The horsepower delivered to the fan's shaft. Brake horsepower does not include the mechanical drive losses, such as that from belts and gears.

FAN ELECTRICAL INPUT POWER. The electrical input power in kilowatts required to operate an individual fan or *fan array* at design conditions. It includes the power consumption of motor controllers, where present.

FAN ENERGY INDEX (FEI). The ratio of the electric input power of a reference fan to the electric input power of the actual fan as calculated in accordance with AMCA 208.

FAN NAMEPLATE ELECTRICAL INPUT POWER. The nominal electrical input power rating stamped on a fan assembly nameplate.

FAN SYSTEM. All the fans that contribute to the movement of air serving spaces that pass through a point of a common *duct*, plenum or cabinet.

FAN SYSTEM, COMPLEX. A *fan system* that combines a *single-cabinet fan system* with other supply fans, exhaust fans or both.

FAN SYSTEM, EXHAUST OR RELIEF. A *fan system* dedicated to the removal of air from interior spaces to the outdoors.

FAN SYSTEM, RETURN. A *fan system* dedicated to removing air from the interior where some or all the air is to be recirculated except during economizer operation.

FAN SYSTEM, SINGLE-CABINET. A fan system that supplies air to a space and recirculates the air, wherein a single cabinet houses a single fan, a single *fan array*, a single set of fans operating in parallel or fans or fan arrays in series.

FAN SYSTEM, TRANSFER. A *fan system* that exclusively moves air from one occupied space to another.

FAN SYSTEM AIRFLOW. The sum of the airflow of all fans with *fan electrical input power* greater than 1 kW at *fan system design conditions*, excluding the airflow that passes through downstream fans with *fan electrical input power* less than 1 kW.

FAN SYSTEM BHP. The sum of the *fan brake horsepower* of all fans that are required to operate at *fan system design conditions* to supply air from the heating or cooling source to the *conditioned spaces* and return it to the source or exhaust it to the outdoors.

FAN SYSTEM DESIGN CONDITIONS. Operating conditions that can be expected to occur during normal system operation that result in the highest supply fan airflow rate to *conditioned spaces* served by the system, other than during *air economizer* operation.

FAN SYSTEM ELECTRICAL INPUT POWER. The sum of the fan electrical power of all fans that are required to operate at *fan system design conditions* to supply air from the heating or cooling source to the *conditioned spaces* and/or return it to the source or exhaust it to the outdoors.

FAN SYSTEM MOTOR NAMEPLATE HP. The sum of the motor *nameplate horsepower* of all fans that are required to operate at design conditions to supply air from the heating or cooling source to the *conditioned spaces* and return it to the source or exhaust it to the outdoors.

FAULT DETECTION AND DIAGNOSTICS (FDD) SYSTEM. A software platform that utilizes building analytic algorithms to convert data provided by sensors and devices to automatically identify faults in building systems and provide a prioritized list of actionable resolutions to those faults based on cost or energy avoidance, comfort and maintenance impact.

FENESTRATION. Products classified as either skylights or vertical *fenestration*.

FENESTRATION PRODUCT, FIELD-FABRICATED. A fenestration product whose frame is made at the construction site of standard dimensional lumber or other materials that were not previously cut or otherwise formed with the specific intention of being used to fabricate a fenestration product or exterior door. Field-fabricated does not include site-built fenestration.

FENESTRATION PRODUCT, SITE-BUILT. A *fenestration* designed to be made up of field-glazed or field-assembled units using specific factory cut or otherwise factory-formed framing and glazing units. Examples of site-built fenestration include *storefront* systems, *curtain walls* and atrium roof systems.

***F*-FACTOR.** The perimeter heat loss factor per unit perimeter length of slab-on-grade floors (Btu/h × ft × °F) [W/(m × K)].

FINANCIAL RENEWABLE ENERGY POWER PURCHASE AGREEMENT. A financial arrangement between a renewable electricty generator and a purchaser wherein the purchaser pays or guarantees a price to the generator for the project's renewable generation. Also known as a "financial power purchase agreement" and "virtual power purchase agreement."

FLOOR AREA, NET. The actual occupied area not including unoccupied accessory areas such as corridors, stairways, toilet rooms, mechanical rooms and closets.

GENERAL LIGHTING. Interior lighting that provides a substantially uniform level of illumination throughout a space.

GREEN RETAIL TARIFF. An electricity-rate structure qualified under applicable statutes or rules contracted by an electricity service provider to the building project owner to provide electricity generated with 100 percent *renewable energy resources* without the purchase of unbundled renewable energy certificates (RECs).

GREENHOUSE. A structure or a thermally isolated area of a *building* that maintains a specialized sunlit environment with a skylight roof ratio of 50 percent or more above the growing area exclusively used for, and essential to, the cultivation, protection or maintenance of plants. *Greenhouses* are those that are erected for a period of 180 days or more.

GROUP R. Buildings or portions of buildings that contain any of the following occupancies as established in the *International Building Code*:

1. Group R-1.
2. Group R-2 where located more than three stories in height above grade plane.
3. Group R-4 where located more than three stories in height above grade plane.

HEAT TRAP. An arrangement of piping and fittings, such as elbows, or a commercially available *heat trap* that prevents thermosyphoning of hot water during standby periods.

HEATED SLAB. Slab-on-grade construction in which the heating elements, hydronic tubing or hot air distribution system is in contact with, or placed within or under, the slab.

HIGH SPEED DOOR. A nonswinging door used primarily to facilitate vehicular access or material transportation, with a minimum opening rate of 32 inches (813 mm) per second, a minimum closing rate of 24 inches (610 mm) per second and that includes an automatic-closing device.

HIGH-CAPACITY GAS-FIRED WATER HEATER. Gas-fired instantaneous *water heaters* with a rated input greater than 200,000 Btu/h (58.6 kW) and not less than 4,000 Btu/h per gallon (310 W per liter) of stored water. Also, gas-fired storage *water heaters* with a rated input both greater than 105,000 Btu/h (30.8 kW) and less than 4,000 Btu/h per gallon (310 W per liter) of stored water.

HIGH-END TRIM. A lighting control setting that limits the maximum power to individual luminaries or groups of luminaries in a space.

HISTORIC BUILDING. Any *building* or structure that is one or more of the following:

1. Listed, or certified as eligible for listing, by the State Historic Preservation Officer or the Keeper of the National Register of Historic Places, in the National Register of Historic Places.
2. Designated as historic under an applicable state or local law.
3. Certified as a contributing resource within a National Register-listed, state-designated or locally designated historic district.

HORTICULTURAL LIGHTING. Electric lighting used for horticultural production, cultivation or maintenance.

HUMIDISTATIC CONTROLS. *Automatic* controls used to maintain humidity at a setpoint.

HVAC TOTAL SYSTEM PERFORMANCE RATIO (HVAC TSPR). The ratio of the sum of a building's annual heating and cooling load in thousands of Btu's to the sum of annual site energy consumption of the building HVAC systems in Btu.

IEC DESIGN H MOTOR. An electric motor that meets all of the following:

1. It is an induction motor designed for use with three-phase power.
2. It contains a cage rotor.
3. It is capable of direct-on-line starting.
4. It has four, six or eight poles.
5. It is rated from 0.4 kW to 1600 kW at a frequency of 60 hertz.

IEC DESIGN N MOTOR. An electric motor that meets all of the following:

1. It is an induction motor designed for use with three-phase power.
2. It contains a cage rotor.
3. It is capable of direct-on-line starting.
4. It has two, four, six or eight poles.
5. It is rated from 0.4 kW to 1600 kW at a frequency of 60 hertz.

INDOOR GROW. A space, other than a *greenhouse*, used exclusively for and essential to horticultural production, cultivation or maintenance.

INFILTRATION. The uncontrolled inward *air leakage* into a *building* caused by the pressure effects of wind or the effect of differences in the indoor and outdoor air density or both.

INFORMATION TECHNOLOGY EQUIPMENT (ITE). Items including computers, data storage devices, servers and network and communication equipment.

INTEGRATED HVAC SYSTEM. An HVAC system designed to handle both sensible and latent heat removal. *Integrated HVAC systems* include, but are not limited to, HVAC systems with a sensible heat ratio of 0.65 or less and the capability of providing cooling, dedicated outdoor air systems, single-package air conditioners with at least one refrigerant circuit providing hot gas reheat, and *dehumidifiers* modified to allow external heat rejection.

INTEGRATED PART LOAD VALUE (IPLV). A single-number figure of merit based on part-load EER, COP or kW/ton expressing part-load efficiency for air-conditioning and heat pump equipment on the basis of weighted operation at various load capacities for equipment.

INTERNAL CURTAIN SYSTEM. A system consisting of movable panels of fabric or plastic film used to cover and uncover the space enclosed in a *greenhouse* on a daily basis.

ISOLATION DEVICES. Devices that isolate HVAC *zones* so that they can be operated independently of one another. *Isolation devices* include separate systems, isolation dampers and controls providing shutoff at terminal boxes.

LABELED. Equipment, materials or products to which have been affixed a label, seal, symbol or other identifying mark of a nationally recognized testing laboratory, *approved agency* or other organization concerned with product evaluation that maintains periodic inspection of the production of the *labeled* items and whose labeling indicates either that the equipment, material or product meets identified standards or has been tested and found suitable for a specified purpose.

LARGE-DIAMETER CEILING FAN. A ceiling fan that is greater than or equal to $84^1/_2$ inches (2.15 m) in diameter. These fans are sometimes referred to as High-Volume, Low-Speed (HVLS) fans.

LINER SYSTEM (Ls). A system that includes the following:

1. A continuous vapor barrier liner membrane that is installed below the purlins and that is uninterrupted by framing members.
2. An uncompressed, unfaced insulation resting on top of the liner membrane and located between the purlins.

For multilayer installations, the last rated *R*-value of insulation is for unfaced insulation draped over purlins and then compressed when the metal roof panels are attached.

LISTED. Equipment, materials, products or services included in a list published by an organization acceptable to the *code official* and concerned with evaluation of products or services that maintains periodic inspection of production of *listed* equipment or materials or periodic evaluation of services and whose listing states either that the equipment, material, product or service meets identified standards or has been tested and found suitable for a specified purpose.

LOW SLOPE. A slope less than 2 units vertical in 12 units horizontal (17 percent slope) as applied to roofs.

LOW-VOLTAGE DRY-TYPE DISTRIBUTION TRANSFORMER. A transformer that is air-cooled, does not use oil as a coolant, has an input voltage less than or equal to 600 volts and is rated for operation at a frequency of 60 hertz.

LUMINAIRE-LEVEL LIGHTING CONTROLS. A lighting system consisting of one or more luminaires with embedded lighting control logic, occupancy and ambient light sensors, wireless networking capabilities and local override switching capability, where required.

MANUAL. Capable of being operated by personal intervention (see "*Automatic*").

NAMEPLATE HORSEPOWER. The nominal motor output power rating stamped on the motor nameplate.

NEMA DESIGN A MOTOR. A squirrel-cage motor that meets all of the following:

1. It is designed to withstand full-voltage starting and develop locked-rotor torque as shown in paragraph 12.38.1 of NEMA MG 1.
2. It has pull-up torque not less than the values shown in paragraph 12.40.1 of NEMA MG 1.

3. It has breakdown torque not less than the values shown in paragraph 12.39.1 of NEMA MG 1.
4. It has a locked-rotor current higher than the values shown in paragraph 12.35.1 of NEMA MG 1 for 60 hertz and paragraph 12.35.2 of NEMA MG 1 for 50 hertz.
5. It has a slip at rated load of less than 5 percent for motors with fewer than 10 poles.

NEMA DESIGN B MOTOR. A squirrel-cage motor that meets all of the following:

1. It is designed to withstand full-voltage starting.
2. It develops locked-rotor, breakdown and pull-up torques adequate for general application as specified in Sections 12.38, 12.39 and 12.40 of NEMA MG1.
3. It draws locked-rotor current not to exceed the values shown in Section 12.35.1 for 60 hertz and Section 12.35.2 for 50 hertz of NEMA MG1.
4. It has a slip at rated load of less than 5 percent for motors with fewer than 10 poles.

NEMA DESIGN C MOTOR. A squirrel-cage motor that meets all of the following:

1. Designed to withstand full-voltage starting and develop locked-rotor torque for high-torque applications up to the values shown in paragraph 12.38.2 of NEMA MG1 (incorporated by reference, see A§431.15).
2. It has pull-up torque not less than the values shown in paragraph 12.40.2 of NEMA MG1.
3. It has breakdown torque not less than the values shown in paragraph 12.39.2 of NEMA MG1.
4. It has a locked-rotor current not to exceed the values shown in paragraph 12.35.1 of NEMA MG1 for 60 hertz and paragraph 12.35.2 for 50 hertz.
5. It has a slip at rated load of less than 5 percent.

NETWORKED GUESTROOM CONTROL SYSTEM. A control system, with access from the front desk or other central location associated with a *Group R-1 building*, that is capable of identifying the rented and unrented status of each guestroom according to a timed schedule, and is capable of controlling HVAC in each hotel and motel guestroom separately.

NONSTANDARD PART LOAD VALUE (NPLV). A single-number part-load efficiency figure of merit calculated and referenced to conditions other than IPLV conditions, for units that are not designed to operate at AHRI standard rating conditions.

NORTH-ORIENTED. Facing within 67.5 degrees of true north in the northern hemisphere or facing within 67.5 degrees of true south in the southern hemisphere.

OCCUPANT SENSOR CONTROL. An *automatic* control device or system that detects the presence or absence of people within an area and causes lighting, equipment or appliances to be regulated accordingly.

OCCUPIED-STANDBY MODE. Mode of operation when an HVAC *zone* is scheduled to be occupied and an occupant sensor indicates no occupants are within the zone.

ON-SITE RENEWABLE ENERGY. Energy from *renewable energy resources* harvested at the building project site.

OPAQUE DOOR. A door that is not less than 50 percent opaque in surface area.

OWNER. Any person, agent, operator, entity, firm or corporation having any legal or equitable interest in the property; or recorded in the official records of the state, county or municipality as holding an interest or title to the property; or otherwise having possession or control of the property, including the guardian of the estate of any such person, and the executor or administrator of the estate of such person if ordered to take possession of real property by a court.

PARKING AREA, EXTERIOR. Parking spaces, drive aisles and ramps that are not located within a *building*, or that are located on a roof.

PARKING AREA, INTERIOR. Parking spaces, drive aisles and ramps located within a *building*.

PARKING GARAGE SECTION. A part of an enclosed parking garage that is separated from all other parts of the garage by full-height solid walls or operable openings that are intended to remain closed during normal operation and where vehicles cannot pass to other parts of the garage. A parking garage can have one or more *parking garage sections*, and *parking garage sections* can include multiple floors.

PHOTOSYNTHETIC PHOTON EFFICACY (PPE). Photosynthetic photon flux emitted by a light source divided by its electrical input power in units of micromoles per second per watt, or micromoles per joule (μmol/J) between 400–700nm as defined by ANSI/ASABE S640.

PHYSICAL RENEWABLE ENERGY POWER PURCHASE AGREEMENT. A contract for the purchase of renewable electricity from a specific renewable electricity generator to a purchaser of renewable electricity.

POWERED ROOF/WALL VENTILATORS. A fan consisting of a centrifugal or axial impeller with an integral driver in a weather-resistant housing and with a base designed to fit, usually by means of a curb, over a wall or roof opening.

PROCESS APPLICATION. A manufacturing, industrial or commercial procedure or activity where the primary purpose is other than conditioning spaces and maintaining comfort and amenities for the occupants of a building.

PROPOSED DESIGN. A description of the proposed building used to estimate annual energy use for determining compliance based on simulated *building* performance and HVAC total system performance ratio.

PSI-FACTOR (ψ-FACTOR). The heat loss factor per unit length of a *thermal bridge* characterized as a linear element of a *building thermal envelope* (Btu/h × ft × °F) [W/(m × K)].

PUMP ENERGY INDEX (PEI). The ratio of a pump's energy rating divided by the energy rating of a minimally compliant pump. For pumps with the constant load operating mode, the relevant PEI is PEI_{CL}. For pumps with the variable load operating mode, the relevant PEI is PEI_{VL}.

PURCHASED ENERGY. Energy or power purchased for consumption and delivered to the building site.

RADIANT HEATING SYSTEM. A heating system that transfers heat to objects and surfaces within a *conditioned space*, primarily by infrared radiation.

READY ACCESS (TO). That which enables a device, appliance or equipment to be directly reached without requiring the removal or movement of any panel or similar obstruction.

REFRIGERANT DEW POINT. The refrigerant vapor saturation temperature at a specified pressure.

REFRIGERATED WAREHOUSE COOLER. An enclosed storage space capable of being refrigerated to temperatures above 32°F (0°C) that can be walked into and has a total chilled storage area of not less than 3,000 square feet (279 m^2).

REFRIGERATED WAREHOUSE FREEZER. An enclosed storage space capable of being refrigerated to temperatures at or below 32°F (0°C) that can be walked into and has a total chilled storage area of not less than 3,000 square feet (279 m^2).

REFRIGERATION SYSTEM, LOW TEMPERATURE. Systems for maintaining food product in a frozen state in refrigeration applications.

REFRIGERATION SYSTEM, MEDIUM TEMPERATURE. Systems for maintaining food product above freezing in refrigeration applications.

REGISTERED DESIGN PROFESSIONAL. An individual who is registered or licensed to practice their respective design profession as defined by the statutory requirements of the professional registration laws of the state or jurisdiction in which the project is to be constructed.

RENEWABLE ENERGY CERTIFICATE (REC). A market-based instrument that represents and conveys the environmental, social and other nonpower attributes of 1 megawatt hour of renewable electricity generation and could be sold separately from the underlying physical electricity associated with *renewable energy resources*, also known as energy attribute and energy attribute certificate (EAC).

RENEWABLE ENERGY INVESTMENT FUND (REIF). A fund established by a jurisdiction to accept payment from building project owners to construct or acquire interests in qualifying renewable energy systems, together with their associated RECs, on the building project owners' behalf.

RENEWABLE ENERGY RESOURCES. Energy derived from solar radiation, wind, waves, tides, *biomass waste* or extracted from hot fluid or steam heated within the earth.

REPAIR. The reconstruction or renewal of any part of an existing building for the purpose of its maintenance or to correct damage.

REROOFING. The process of recovering or replacing an existing roof covering. See "*Roof recover*" and "*Roof replacement*."

RESIDENTIAL BUILDING. For this code, includes detached one- and two-family dwellings and multiple single-family dwellings (townhouses) and *Group R*-2, R-3 and R-4 buildings three stories or less in height above grade plane.

ROOF ASSEMBLY. A system designed to provide weather protection and resistance to design loads. The system consists of a roof covering and roof deck or a single component serving as both the roof covering and the roof deck. A *roof assembly* includes the roof covering, underlayment, roof deck, insulation, vapor retarder and interior finish.

ROOF RECOVER. The process of installing an additional roof covering over an existing roof covering without removing the existing roof covering.

ROOF REPAIR. Reconstruction or renewal of any part of an existing roof for the purpose of its maintenance.

ROOF REPLACEMENT. An *alteration* that includes the removal of all existing layers of *roof assembly* materials down to the roof deck and the installation of replacement materials above the existing roof deck.

ROOFTOP MONITOR. A raised section of a roof containing vertical *fenestration* along one or more sides.

***R*-VALUE (THERMAL RESISTANCE).** The inverse of the time rate of heat flow through a body from one of its bounding surfaces to the other surface for a unit temperature difference between the two surfaces, under steady state conditions, per unit area ($h \times ft^2 \times °F/Btu$) [$(m^2 \times K)/W$].

SATURATED CONDENSING TEMPERATURE. The saturation temperature corresponding to the measured refrigerant pressure at the condenser inlet for single component and azeotropic refrigerants, and the arithmetic average of the dew point and *bubble point* temperatures corresponding to the refrigerant pressure at the condenser entrance for zeotropic refrigerants.

SENSIBLE ENERGY RECOVERY RATIO. Change in the dry-bulb temperature of the outdoor air supply divided by the difference between the outdoor air and entering exhaust air dry-bulb temperatures, expressed as a percentage.

SERVICE WATER HEATING. Supply of hot water for purposes other than comfort heating.

SIMULATED BUILDING PERFORMANCE. A process in which the proposed building design is compared to a *standard reference design* for the purposes of estimating relative energy use against a baseline to determine code compliance.

SLEEPING UNIT. A room or space in which people sleep that can include permanent provisions for living, eating, and either sanitation or kitchen facilities but not both. Such rooms and spaces that are part of a *dwelling unit* are not *sleeping units*.

SMALL ELECTRIC MOTOR. A general purpose alternating-current single-speed induction motor.

SOLAR HEAT GAIN COEFFICIENT (SHGC). The ratio of the solar heat gain entering the space through the *fenestration* assembly to the incident solar radiation. Solar heat gain includes directly transmitted solar heat and absorbed solar radiation that is then reradiated, conducted or convected into the space.

SOUTH-ORIENTED. Facing within 45 degrees of true south in the northern hemisphere or facing within 45 degrees of true north in the southern hemisphere.

STANDARD REFERENCE DESIGN. A version of the *proposed design* that meets the minimum requirements of this code and is used to determine the maximum annual energy use requirement for compliance based on total simulated building performance and HVAC total system performance ratio.

STOREFRONT. A system of doors and windows mulled as a composite fenestration structure that has been designed to resist heavy use. *Storefront* systems include, but are not limited to, exterior fenestration systems that span from the floor level or above to the ceiling of the same story on *commercial buildings*, with or without mulled windows and doors.

SUBSTANTIAL IMPROVEMENT. Any *repair*, reconstruction, rehabilitation, *alteration*, *addition* or other improvement of a *building* or structure, the cost of which equals or is more than 50 percent of the market value of the structure before the improvement. Where the structure has sustained substantial damage, as defined in the *International Building Code*, any *repairs* are considered *substantial improvement* regardless of the actual *repair* work performed. *Substantial improvement* does not include the following:

1. Improvement of a *building* ordered by the code official to correct health, sanitary or safety code violations.
2. *Alteration* of a historic building where the *alteration* will not affect the designation as a historic building.

TESTING UNIT ENCLOSURE AREA. The area sum of all the boundary surfaces that define the *dwelling unit*, *sleeping unit* or *conditioned enclosed space*, including top/ceiling, bottom/floor and all side walls. This does not include interior partition walls within the *dwelling unit*, *sleeping unit* or *conditioned enclosed space*. Wall height shall be measured from the finished floor of the *conditioned space* to the finished floor or roof/ceiling *air barrier* above.

THERMAL BLOCK. A generic concept used in energy simulation. It can include one or more thermal zones. It represents a whole *building* or portion of a *building* with the same use type served by the same HVAC system type.

THERMAL BRIDGE. An element or interface of elements that has higher thermal conductivity than the surrounding *building thermal envelope*, which creates a path of least resistance for heat transfer.

THERMAL DISTRIBUTION EFFICIENCY (TDE). The resistance to changes in air heat as air is conveyed through a distance of air *duct*. TDE is a heat loss calculation evaluating the difference in the heat of the air between the air duct inlet and outlet caused by differences in temperatures between the air in the *duct* and the duct material. TDE is expressed as a percent difference between the inlet and outlet heat in the *duct*.

THERMOSTAT. An *automatic* control device used to maintain temperature at a fixed or adjustable setpoint.

TIME-SWITCH CONTROL. An *automatic* control device or system that controls lighting or other loads, including switching off, based on time schedules.

***U*-FACTOR (THERMAL TRANSMITTANCE).** The coefficient of heat transmission (air to air) through a building component or assembly, equal to the time rate of heat flow per unit area and unit temperature difference between the warm side and cold side air films (Btu/h × ft^2 × °F) [W/(m^2 × K)].

VARIABLE REFRIGERANT FLOW SYSTEM. An engineered direct-expansion (DX) refrigerant system that incorporates a common *condensing unit*, at least one variable-capacity compressor, a distributed refrigerant piping network to multiple indoor fan heating and cooling units each capable of individual zone temperature control, through integral zone temperature control devices and a common communications network. Variable refrigerant flow utilizes three or more steps of control on common interconnecting piping.

VEGETATIVE ROOF. An assembly of interacting components designed to waterproof a building's top surface that includes, by design, vegetation and related landscape elements.

VENTILATION. The natural or mechanical process of supplying conditioned or unconditioned air to, or removing such air from, any space.

VENTILATION AIR. That portion of supply air that comes from outside (outdoors) plus any recirculated air that has been treated to maintain the desired quality of air within a designated space.

VISIBLE TRANSMITTANCE (VT). The ratio of visible light entering the space through the fenestration product assembly to the incident visible light. Visible transmittance includes the effects of glazing material and frame and is expressed as a number between 0 and 1.

VISIBLE TRANSMITTANCE, ANNUAL (VT_{annual}). The ratio of visible light entering the space through the fenestration product assembly to the incident visible light during the course of a year, which includes the effects of glazing material, frame, and light well or tubular conduit, and is expressed as a number between 0 and 1.

VOLTAGE DROP. A decrease in voltage caused by losses in the wiring systems that connect the power source to the load.

WALK-IN COOLER. An enclosed storage space capable of being refrigerated to temperatures above 32°F (0°C) and less than 55°F (12.8°C) that can be walked into, has a ceiling height of not less than 7 feet (2134 mm) and has a total chilled storage area of less than 3,000 square feet (279 m^2).

WALK-IN FREEZER. An enclosed storage space capable of being refrigerated to temperatures at or below 32°F (0°C) that can be walked into, has a ceiling height of not less than 7 feet (2134 mm) and has a total chilled storage area of less than 3,000 square feet (279 m^2).

WALL, ABOVE-GRADE. A wall associated with the *building thermal envelope* that is more than 15 percent above grade and is on the exterior of the *building* or any wall that is associated with the *building thermal envelope* that is not on the exterior of the building. This includes, but is not limited to, between-floor spandrels, peripheral edges of floors, roof knee walls, dormer walls, gable end walls, walls enclosing a mansard roof, mechanical equipment penetrations and skylight shafts.

WALL, BELOW-GRADE. A wall associated with the basement or first story of the *building* that is part of the *building thermal envelope*, is not less than 85 percent below grade and is on the exterior of the building.

WATER HEATER. Any heating appliance or equipment that heats potable water and supplies such water to the potable hot water distribution system.

WEST-ORIENTED. Facing within 45 degrees of true west to the south and within less than 22.5 degrees of true west to the north in the northern hemisphere or facing within 45 degrees of true west to the north and within less than 22.5 degrees of true west to the south in the southern hemisphere.

WORK AREA. That portion or portions of a *building* consisting of all reconfigured spaces as indicated on the *construction documents*. *Work area* excludes other portions of the *building* where incidental work entailed by the intended work must be performed and portions of the building where work not initially intended by the *owner* is specifically required by this code.

ZONE. A space or group of spaces within a *building* with heating or cooling requirements that are sufficiently similar so that desired conditions can be maintained throughout using a single controlling device.

CHAPTER

3 [CE] GENERAL REQUIREMENTS

User notes:

About this chapter: *Chapter 3 addresses broadly applicable requirements that would not be at home in other chapters having more specific coverage of subject matter. This chapter establishes climate zone by US counties and territories and includes methodology for determining climate zones elsewhere. It also contains product rating, marking and installation requirements for materials such as insulation, windows, doors and siding.*

SECTION C301—CLIMATE ZONES

C301.1 General. *Climate zones* from Figure C301.1 or Table C301.1 shall be used for determining the applicable requirements from Chapter 4. Locations not indicated in Table C301.1 shall be assigned a *climate zone* in accordance with Section C301.3.

FIGURE C301.1—CLIMATE ZONES

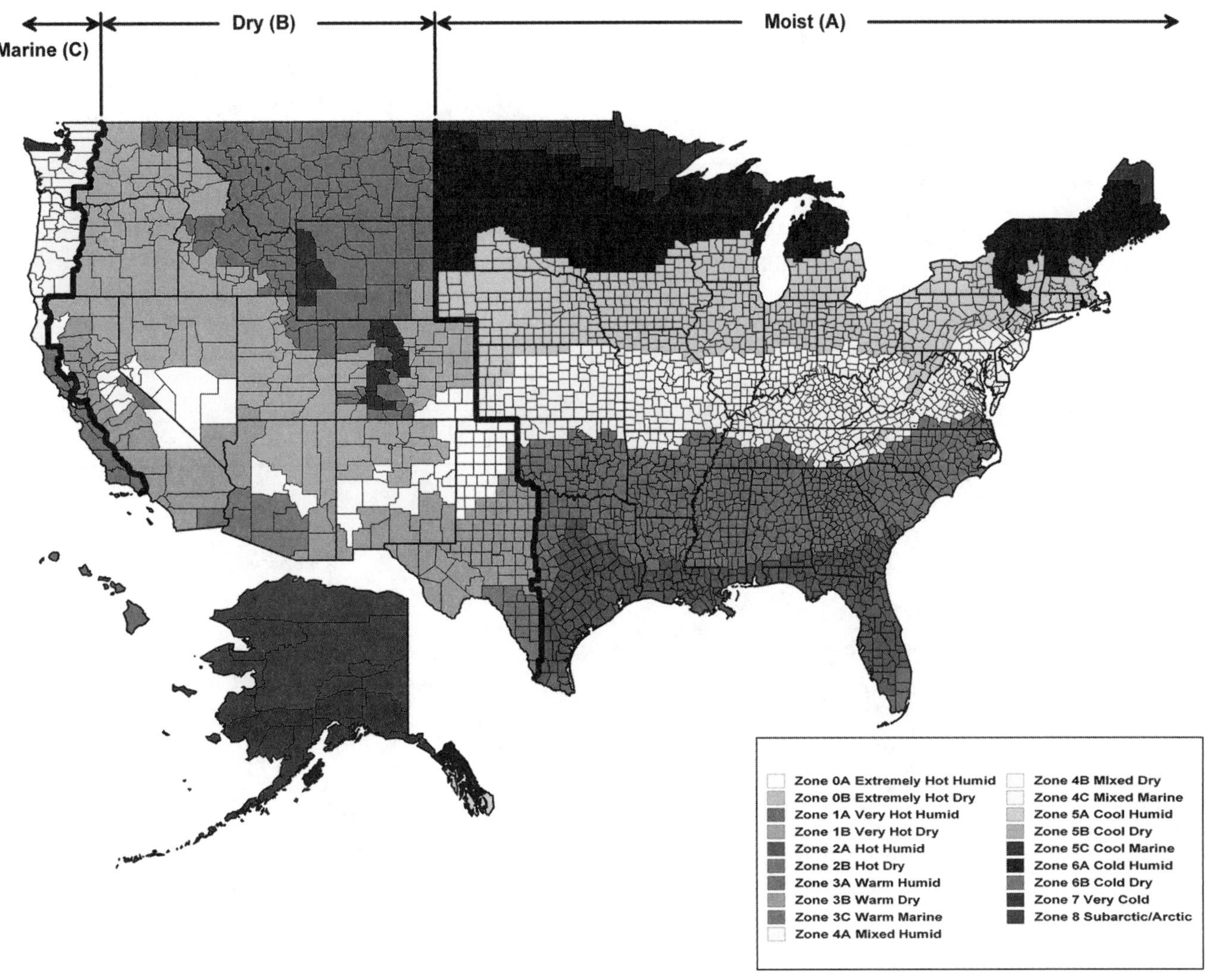

TABLE C301.1—CLIMATE ZONES, MOISTURE REGIMES AND WARM HUMID DESIGNATIONS BY STATE, COUNTY AND TERRITORY[a]

US STATES
ALABAMA
3A Autauga*
2A Baldwin*
3A Barbour*
3A Bibb
3A Blount
3A Bullock*
3A Butler*
3A Calhoun
3A Chambers
3A Cherokee
3A Chilton
3A Choctaw*
3A Clarke*
3A Clay
3A Cleburne
2A Coffee*
3A Colbert
3A Conecuh*
3A Coosa
2A Covington*
3A Crenshaw*
3A Cullman
2A Dale*
3A Dallas*
3A DeKalb
3A Elmore*
2A Escambia*
3A Etowah
3A Fayette
3A Franklin
2A Geneva*
3A Greene
3A Hale
2A Henry*
2A Houston*
3A Jackson
3A Jefferson
3A Lamar
3A Lauderdale
3A Lawrence
3A Lee
3A Limestone
3A Lowndes*
3A Macon*
3A Madison
3A Marengo*
3A Marion
3A Marshall
2A Mobile*
3A Monroe*
3A Montgomery*
3A Morgan
3A Perry*
3A Pickens
3A Pike*
3A Randolph
3A Russell*
3A Shelby
3A St. Clair
3A Sumter
3A Talladega
3A Tallapoosa
3A Tuscaloosa
3A Walker
3A Washington*
3A Wilcox*
3A Winston
ALASKA
7 Aleutians East
7 Aleutians West
7 Anchorage
7 Bethel
7 Bristol Bay
8 Denali
7 Dillingham
8 Fairbanks North Star
6A Haines
6A Juneau
7 Kenai Peninsula
5C Ketchikan Gateway
6A Kodiak Island
7 Lake and Peninsula
7 Matanuska-Susitna
8 Nome
8 North Slope
8 Northwest Arctic
5C Prince of Wales Outer Ketchikan
5C Sitka
6A Skagway-Hoonah-Angoon
8 Southeast Fairbanks

TABLE C301.1—CLIMATE ZONES, MOISTURE REGIMES AND WARM HUMID DESIGNATIONS BY STATE, COUNTY AND TERRITORY[a]—continued

US STATES—continued
ALASKA *(continued)*
7 Valdez-Cordova
8 Wade Hampton
6A Wrangell-Petersburg
7 Yakutat
8 Yukon-Koyukuk
ARIZONA
5B Apache
3B Cochise
5B Coconino
4B Gila
3B Graham
3B Greenlee
2B La Paz
2B Maricopa
3B Mohave
5B Navajo
2B Pima
2B Pinal
3B Santa Cruz
4B Yavapai
2B Yuma
ARKANSAS
3A Arkansas
3A Ashley
4A Baxter
4A Benton
4A Boone
3A Bradley
3A Calhoun
4A Carroll
3A Chicot
3A Clark
3A Clay
3A Cleburne
3A Cleveland
3A Columbia*
3A Conway
3A Craighead
3A Crawford
3A Crittenden
3A Cross
3A Dallas
3A Desha
3A Drew
3A Faulkner
3A Franklin
4A Fulton
3A Garland
3A Grant
3A Greene
3A Hempstead*
3A Hot Spring
3A Howard
3A Independence
4A Izard
3A Jackson
3A Jefferson
3A Johnson
3A Lafayette*
3A Lawrence
3A Lee
3A Lincoln
3A Little River*
3A Logan
3A Lonoke
4A Madison
4A Marion
3A Miller*
3A Mississippi
3A Monroe
3A Montgomery
3A Nevada
4A Newton
3A Ouachita
3A Perry
3A Phillips
3A Pike
3A Poinsett
3A Polk
3A Pope
3A Prairie
3A Pulaski
3A Randolph
3A Saline
3A Scott
4A Searcy
3A Sebastian
3A Sevier*
3A Sharp
3A St. Francis

TABLE C301.1—CLIMATE ZONES, MOISTURE REGIMES AND WARM HUMID DESIGNATIONS BY STATE, COUNTY AND TERRITORY[a]—continued

US STATES—continued
ARKANSAS *(continued)*
4A Stone
3A Union*
3A Van Buren
4A Washington
3A White
3A Woodruff
3A Yell
CALIFORNIA
3C Alameda
6B Alpine
4B Amador
3B Butte
4B Calaveras
3B Colusa
3B Contra Costa
4C Del Norte
4B El Dorado
3B Fresno
3B Glenn
4C Humboldt
2B Imperial
4B Inyo
3B Kern
3B Kings
4B Lake
5B Lassen
3B Los Angeles
3B Madera
3C Marin
4B Mariposa
3C Mendocino
3B Merced
5B Modoc
6B Mono
3C Monterey
3C Napa
5B Nevada
3B Orange
3B Placer
5B Plumas
3B Riverside
3B Sacramento
3C San Benito
3B San Bernardino
3B San Diego
3C San Francisco
3B San Joaquin
3C San Luis Obispo
3C San Mateo
3C Santa Barbara
3C Santa Clara
3C Santa Cruz
3B Shasta
5B Sierra
5B Siskiyou
3B Solano
3C Sonoma
3B Stanislaus
3B Sutter
3B Tehama
4B Trinity
3B Tulare
4B Tuolumne
3C Ventura
3B Yolo
3B Yuba
COLORADO
5B Adams
6B Alamosa
5B Arapahoe
6B Archuleta
4B Baca
4B Bent
5B Boulder
5B Broomfield
6B Chaffee
5B Cheyenne
7 Clear Creek
6B Conejos
6B Costilla
5B Crowley
5B Custer
5B Delta
5B Denver
6B Dolores
5B Douglas
6B Eagle
5B Elbert
5B El Paso
5B Fremont

TABLE C301.1—CLIMATE ZONES, MOISTURE REGIMES AND WARM HUMID DESIGNATIONS BY STATE, COUNTY AND TERRITORY[a]—continued

US STATES—continued
COLORADO *(continued)*
5B Garfield
5B Gilpin
7 Grand
7 Gunnison
7 Hinsdale
5B Huerfano
7 Jackson
5B Jefferson
5B Kiowa
5B Kit Carson
7 Lake
5B La Plata
5B Larimer
4B Las Animas
5B Lincoln
5B Logan
5B Mesa
7 Mineral
6B Moffat
5B Montezuma
5B Montrose
5B Morgan
4B Otero
6B Ouray
7 Park
5B Phillips
7 Pitkin
4B Prowers
5B Pueblo
6B Rio Blanco
7 Rio Grande
7 Routt
6B Saguache
7 San Juan
6B San Miguel
5B Sedgwick
7 Summit
5B Teller
5B Washington
5B Weld
5B Yuma
CONNECTICUT
5A (all)

DELAWARE
4A (all)
DISTRICT OF COLUMBIA
4A (all)
FLORIDA
2A Alachua*
2A Baker*
2A Bay*
2A Bradford*
2A Brevard*
1A Broward*
2A Calhoun*
2A Charlotte*
2A Citrus*
2A Clay*
2A Collier*
2A Columbia*
2A DeSoto*
2A Dixie*
2A Duval*
2A Escambia*
2A Flagler*
2A Franklin*
2A Gadsden*
2A Gilchrist*
2A Glades*
2A Gulf*
2A Hamilton*
2A Hardee*
2A Hendry*
2A Hernando*
2A Highlands*
2A Hillsborough*
2A Holmes*
2A Indian River*
2A Jackson*
2A Jefferson*
2A Lafayette*
2A Lake*
2A Lee*
2A Leon*
2A Levy*
2A Liberty*
2A Madison*
2A Manatee*

TABLE C301.1—CLIMATE ZONES, MOISTURE REGIMES AND WARM HUMID DESIGNATIONS BY STATE, COUNTY AND TERRITORY[a]—continued

US STATES—continued
FLORIDA *(continued)*
2A Marion*
2A Martin*
1A Miami-Dade*
1A Monroe*
2A Nassau*
2A Okaloosa*
2A Okeechobee*
2A Orange*
2A Osceola*
1A Palm Beach*
2A Pasco*
2A Pinellas*
2A Polk*
2A Putnam*
2A Santa Rosa*
2A Sarasota*
2A Seminole*
2A St. Johns*
2A St. Lucie*
2A Sumter*
2A Suwannee*
2A Taylor*
2A Union*
2A Volusia*
2A Wakulla*
2A Walton*
2A Washington*
GEORGIA
2A Appling*
2A Atkinson*
2A Bacon*
2A Baker*
3A Baldwin
3A Banks
3A Barrow
3A Bartow
3A Ben Hill*
2A Berrien*
3A Bibb
3A Bleckley*
2A Brantley*
2A Brooks*
2A Bryan*
3A Bulloch*
3A Burke
3A Butts
2A Calhoun*
2A Camden*
3A Candler*
3A Carroll
3A Catoosa
2A Charlton*
2A Chatham*
3A Chattahoochee*
3A Chattooga
3A Cherokee
3A Clarke
3A Clay*
3A Clayton
2A Clinch*
3A Cobb
2A Coffee*
2A Colquitt*
3A Columbia
2A Cook*
3A Coweta
3A Crawford
3A Crisp*
3A Dade
3A Dawson
2A Decatur*
3A DeKalb
3A Dodge*
3A Dooly*
2A Dougherty*
3A Douglas
2A Early*
2A Echols*
2A Effingham*
3A Elbert
3A Emanuel*
2A Evans*
3A Fannin
3A Fayette
3A Floyd
3A Forsyth
3A Franklin
3A Fulton
3A Gilmer
3A Glascock

TABLE C301.1—CLIMATE ZONES, MOISTURE REGIMES AND WARM HUMID DESIGNATIONS BY STATE, COUNTY AND TERRITORY[a]—continued

US STATES—continued
GEORGIA *(continued)*
2A Glynn*
3A Gordon
2A Grady*
3A Greene
3A Gwinnett
3A Habersham
3A Hall
3A Hancock
3A Haralson
3A Harris
3A Hart
3A Heard
3A Henry
3A Houston*
3A Irwin*
3A Jackson
3A Jasper
2A Jeff Davis*
3A Jefferson
3A Jenkins*
3A Johnson*
3A Jones
3A Lamar
2A Lanier*
3A Laurens*
3A Lee*
2A Liberty*
3A Lincoln
2A Long*
2A Lowndes*
3A Lumpkin
3A Macon*
3A Madison
3A Marion*
3A McDuffie
2A McIntosh*
3A Meriwether
2A Miller*
2A Mitchell*
3A Monroe
3A Montgomery*
3A Morgan
3A Murray
3A Muscogee
3A Newton
3A Oconee
3A Oglethorpe
3A Paulding
3A Peach*
3A Pickens
2A Pierce*
3A Pike
3A Polk
3A Pulaski*
3A Putnam
3A Quitman*
3A Rabun
3A Randolph*
3A Richmond
3A Rockdale
3A Schley*
3A Screven*
2A Seminole*
3A Spalding
3A Stephens
3A Stewart*
3A Sumter*
3A Talbot
3A Taliaferro
2A Tattnall*
3A Taylor*
3A Telfair*
3A Terrell*
2A Thomas*
2A Tift*
2A Toombs*
3A Towns
3A Treutlen*
3A Troup
3A Turner*
3A Twiggs*
3A Union
3A Upson
3A Walker
3A Walton
2A Ware*
3A Warren
3A Washington
2A Wayne*
3A Webster*

TABLE C301.1—CLIMATE ZONES, MOISTURE REGIMES AND WARM HUMID DESIGNATIONS BY STATE, COUNTY AND TERRITORY[a]—continued

US STATES—continued
GEORGIA *(continued)*
3A Wheeler*
3A White
3A Whitfield
3A Wilcox*
3A Wilkes
3A Wilkinson
2A Worth*
HAWAII
1A (all)*
IDAHO
5B Ada
6B Adams
6B Bannock
6B Bear Lake
5B Benewah
6B Bingham
6B Blaine
6B Boise
6B Bonner
6B Bonneville
6B Boundary
6B Butte
6B Camas
5B Canyon
6B Caribou
5B Cassia
6B Clark
5B Clearwater
6B Custer
5B Elmore
6B Franklin
6B Fremont
5B Gem
5B Gooding
5B Idaho
6B Jefferson
5B Jerome
5B Kootenai
5B Latah
6B Lemhi
5B Lewis
5B Lincoln
6B Madison
5B Minidoka
5B Nez Perce
6B Oneida
5B Owyhee
5B Payette
5B Power
5B Shoshone
6B Teton
5B Twin Falls
6B Valley
5B Washington
ILLINOIS
5A Adams
4A Alexander
4A Bond
5A Boone
5A Brown
5A Bureau
4A Calhoun
5A Carroll
5A Cass
5A Champaign
4A Christian
4A Clark
4A Clay
4A Clinton
4A Coles
5A Cook
4A Crawford
4A Cumberland
5A DeKalb
5A De Witt
5A Douglas
5A DuPage
5A Edgar
4A Edwards
4A Effingham
4A Fayette
5A Ford
4A Franklin
5A Fulton
4A Gallatin
4A Greene
5A Grundy
4A Hamilton
5A Hancock
4A Hardin

TABLE C301.1—CLIMATE ZONES, MOISTURE REGIMES AND WARM HUMID DESIGNATIONS BY STATE, COUNTY AND TERRITORY[a]—continued

US STATES—continued
ILLINOIS *(continued)*
5A Henderson
5A Henry
5A Iroquois
4A Jackson
4A Jasper
4A Jefferson
4A Jersey
5A Jo Daviess
4A Johnson
5A Kane
5A Kankakee
5A Kendall
5A Knox
5A Lake
5A La Salle
4A Lawrence
5A Lee
5A Livingston
5A Logan
5A Macon
4A Macoupin
4A Madison
4A Marion
5A Marshall
5A Mason
4A Massac
5A McDonough
5A McHenry
5A McLean
5A Menard
5A Mercer
4A Monroe
4A Montgomery
5A Morgan
5A Moultrie
5A Ogle
5A Peoria
4A Perry
5A Piatt
5A Pike
4A Pope
4A Pulaski
5A Putnam
4A Randolph
4A Richland
5A Rock Island
4A Saline
5A Sangamon
5A Schuyler
5A Scott
4A Shelby
5A Stark
4A St. Clair
5A Stephenson
5A Tazewell
4A Union
5A Vermilion
4A Wabash
5A Warren
4A Washington
4A Wayne
4A White
5A Whiteside
5A Will
4A Williamson
5A Winnebago
5A Woodford
INDIANA
5A Adams
5A Allen
4A Bartholomew
5A Benton
5A Blackford
5A Boone
4A Brown
5A Carroll
5A Cass
4A Clark
4A Clay
5A Clinton
4A Crawford
4A Daviess
4A Dearborn
4A Decatur
5A De Kalb
5A Delaware
4A Dubois
5A Elkhart
4A Fayette
4A Floyd

TABLE C301.1—CLIMATE ZONES, MOISTURE REGIMES AND WARM HUMID DESIGNATIONS BY STATE, COUNTY AND TERRITORY[a]—continued

US STATES—continued
INDIANA *(continued)*
5A Fountain
4A Franklin
5A Fulton
4A Gibson
5A Grant
4A Greene
5A Hamilton
5A Hancock
4A Harrison
4A Hendricks
5A Henry
5A Howard
5A Huntington
4A Jackson
5A Jasper
5A Jay
4A Jefferson
4A Jennings
4A Johnson
4A Knox
5A Kosciusko
5A LaGrange
5A Lake
5A LaPorte
4A Lawrence
5A Madison
4A Marion
5A Marshall
4A Martin
5A Miami
4A Monroe
5A Montgomery
4A Morgan
5A Newton
5A Noble
4A Ohio
4A Orange
4A Owen
5A Parke
4A Perry
4A Pike
5A Porter
4A Posey
5A Pulaski
4A Putnam
5A Randolph
4A Ripley
4A Rush
4A Scott
4A Shelby
4A Spencer
5A Starke
5A Steuben
5A St. Joseph
4A Sullivan
4A Switzerland
5A Tippecanoe
5A Tipton
4A Union
4A Vanderburgh
5A Vermillion
4A Vigo
5A Wabash
5A Warren
4A Warrick
4A Washington
5A Wayne
5A Wells
5A White
5A Whitley
IOWA
5A Adair
5A Adams
5A Allamakee
5A Appanoose
5A Audubon
5A Benton
6A Black Hawk
5A Boone
5A Bremer
5A Buchanan
5A Buena Vista
5A Butler
5A Calhoun
5A Carroll
5A Cass
5A Cedar
6A Cerro Gordo
5A Cherokee
5A Chickasaw

TABLE C301.1—CLIMATE ZONES, MOISTURE REGIMES AND WARM HUMID DESIGNATIONS BY STATE, COUNTY AND TERRITORY[a]—continued

US STATES—continued
IOWA *(continued)*
5A Clarke
6A Clay
5A Clayton
5A Clinton
5A Crawford
5A Dallas
5A Davis
5A Decatur
5A Delaware
5A Des Moines
6A Dickinson
5A Dubuque
6A Emmet
5A Fayette
5A Floyd
5A Franklin
5A Fremont
5A Greene
5A Grundy
5A Guthrie
5A Hamilton
6A Hancock
5A Hardin
5A Harrison
5A Henry
5A Howard
5A Humboldt
5A Ida
5A Iowa
5A Jackson
5A Jasper
5A Jefferson
5A Johnson
5A Jones
5A Keokuk
6A Kossuth
5A Lee
5A Linn
5A Louisa
5A Lucas
6A Lyon
5A Madison
5A Mahaska
5A Marion
5A Marshall
5A Mills
6A Mitchell
5A Monona
5A Monroe
5A Montgomery
5A Muscatine
6A O'Brien
6A Osceola
5A Page
6A Palo Alto
5A Plymouth
5A Pocahontas
5A Polk
5A Pottawattamie
5A Poweshiek
5A Ringgold
5A Sac
5A Scott
5A Shelby
6A Sioux
5A Story
5A Tama
5A Taylor
5A Union
5A Van Buren
5A Wapello
5A Warren
5A Washington
5A Wayne
5A Webster
6A Winnebago
5A Winneshiek
5A Woodbury
6A Worth
5A Wright
KANSAS
4A Allen
4A Anderson
4A Atchison
4A Barber
4A Barton
4A Bourbon
4A Brown
4A Butler
4A Chase

TABLE C301.1—CLIMATE ZONES, MOISTURE REGIMES AND WARM HUMID DESIGNATIONS BY STATE, COUNTY AND TERRITORY[a]—continued

US STATES—continued
KANSAS *(continued)*
4A Chautauqua
4A Cherokee
5A Cheyenne
4A Clark
4A Clay
4A Cloud
4A Coffey
4A Comanche
4A Cowley
4A Crawford
5A Decatur
4A Dickinson
4A Doniphan
4A Douglas
4A Edwards
4A Elk
4A Ellis
4A Ellsworth
4A Finney
4A Ford
4A Franklin
4A Geary
5A Gove
4A Graham
4A Grant
4A Gray
5A Greeley
4A Greenwood
4A Hamilton
4A Harper
4A Harvey
4A Haskell
4A Hodgeman
4A Jackson
4A Jefferson
5A Jewell
4A Johnson
4A Kearny
4A Kingman
4A Kiowa
4A Labette
4A Lane
4A Leavenworth
4A Lincoln
4A Linn
5A Logan
4A Lyon
4A Marion
4A Marshall
4A McPherson
4A Meade
4A Miami
4A Mitchell
4A Montgomery
4A Morris
4A Morton
4A Nemaha
4A Neosho
4A Ness
5A Norton
4A Osage
4A Osborne
4A Ottawa
4A Pawnee
5A Phillips
4A Pottawatomie
4A Pratt
5A Rawlins
4A Reno
5A Republic
4A Rice
4A Riley
4A Rooks
4A Rush
4A Russell
4A Saline
5A Scott
4A Sedgwick
4A Seward
4A Shawnee
5A Sheridan
5A Sherman
5A Smith
4A Stafford
4A Stanton
4A Stevens
4A Sumner
5A Thomas
4A Trego
4A Wabaunsee

TABLE C301.1—CLIMATE ZONES, MOISTURE REGIMES AND WARM HUMID DESIGNATIONS BY STATE, COUNTY AND TERRITORY[a]—continued

US STATES—continued
KANSAS *(continued)*
5A Wallace
4A Washington
5A Wichita
4A Wilson
4A Woodson
4A Wyandotte
KENTUCKY
4A (all)
LOUISIANA
2A Acadia*
2A Allen*
2A Ascension*
2A Assumption*
2A Avoyelles*
2A Beauregard*
3A Bienville*
3A Bossier*
3A Caddo*
2A Calcasieu*
3A Caldwell*
2A Cameron*
3A Catahoula*
3A Claiborne*
3A Concordia*
3A De Soto*
2A East Baton Rouge*
3A East Carroll
2A East Feliciana*
2A Evangeline*
3A Franklin*
3A Grant*
2A Iberia*
2A Iberville*
3A Jackson*
2A Jefferson*
2A Jefferson Davis*
2A Lafayette*
2A Lafourche*
3A La Salle*
3A Lincoln*
2A Livingston*
3A Madison*
3A Morehouse
3A Natchitoches*
2A Orleans*
3A Ouachita*
2A Plaquemines*
2A Pointe Coupee*
2A Rapides*
3A Red River*
3A Richland*
3A Sabine*
2A St. Bernard*
2A St. Charles*
2A St. Helena*
2A St. James*
2A St. John the Baptist*
2A St. Landry*
2A St. Martin*
2A St. Mary*
2A St. Tammany*
2A Tangipahoa*
3A Tensas*
2A Terrebonne*
3A Union*
2A Vermilion*
3A Vernon*
2A Washington*
3A Webster*
2A West Baton Rouge*
3A West Carroll
2A West Feliciana*
3A Winn*
MAINE
6A Androscoggin
7 Aroostook
6A Cumberland
6A Franklin
6A Hancock
6A Kennebec
6A Knox
6A Lincoln
6A Oxford
6A Penobscot
6A Piscataquis
6A Sagadahoc
6A Somerset
6A Waldo
6A Washington
6A York

TABLE C301.1—CLIMATE ZONES, MOISTURE REGIMES AND WARM HUMID DESIGNATIONS BY STATE, COUNTY AND TERRITORY[a]—continued

US STATES—continued
MARYLAND
5A Allegany
4A Anne Arundel
4A Baltimore
4A Baltimore (city)
4A Calvert
4A Caroline
4A Carroll
4A Cecil
4A Charles
4A Dorchester
4A Frederick
5A Garrett
4A Harford
4A Howard
4A Kent
4A Montgomery
4A Prince George's
4A Queen Anne's
4A Somerset
4A St. Mary's
4A Talbot
4A Washington
4A Wicomico
4A Worcester
MASSACHUSETTS
5A (all)
MICHIGAN
6A Alcona
6A Alger
5A Allegan
6A Alpena
6A Antrim
6A Arenac
6A Baraga
5A Barry
5A Bay
6A Benzie
5A Berrien
5A Branch
5A Calhoun
5A Cass
6A Charlevoix
6A Cheboygan
6A Chippewa
6A Clare
5A Clinton
6A Crawford
6A Delta
6A Dickinson
5A Eaton
6A Emmet
5A Genesee
6A Gladwin
6A Gogebic
6A Grand Traverse
5A Gratiot
5A Hillsdale
6A Houghton
5A Huron
5A Ingham
5A Ionia
6A Iosco
6A Iron
6A Isabella
5A Jackson
5A Kalamazoo
6A Kalkaska
5A Kent
7 Keweenaw
6A Lake
5A Lapeer
6A Leelanau
5A Lenawee
5A Livingston
6A Luce
6A Mackinac
5A Macomb
6A Manistee
7 Marquette
6A Mason
6A Mecosta
6A Menominee
5A Midland
6A Missaukee
5A Monroe
5A Montcalm
6A Montmorency
5A Muskegon
6A Newaygo
5A Oakland

TABLE C301.1—CLIMATE ZONES, MOISTURE REGIMES AND WARM HUMID DESIGNATIONS BY STATE, COUNTY AND TERRITORY[a]—continued

US STATES—continued
MICHIGAN *(continued)*
6A Oceana
6A Ogemaw
6A Ontonagon
6A Osceola
6A Oscoda
6A Otsego
5A Ottawa
6A Presque Isle
6A Roscommon
5A Saginaw
5A Sanilac
6A Schoolcraft
5A Shiawassee
5A St. Clair
5A St. Joseph
5A Tuscola
5A Van Buren
5A Washtenaw
5A Wayne
6A Wexford
MINNESOTA
7 Aitkin
6A Anoka
6A Becker
7 Beltrami
6A Benton
6A Big Stone
6A Blue Earth
6A Brown
7 Carlton
6A Carver
7 Cass
6A Chippewa
6A Chisago
6A Clay
7 Clearwater
7 Cook
6A Cottonwood
7 Crow Wing
6A Dakota
6A Dodge
6A Douglas
6A Faribault
5A Fillmore
6A Freeborn
6A Goodhue
6A Grant
6A Hennepin
5A Houston
7 Hubbard
6A Isanti
7 Itasca
6A Jackson
6A Kanabec
6A Kandiyohi
7 Kittson
7 Koochiching
6A Lac qui Parle
7 Lake
7 Lake of the Woods
6A Le Sueur
6A Lincoln
6A Lyon
7 Mahnomen
7 Marshall
6A Martin
6A McLeod
6A Meeker
6A Mille Lacs
6A Morrison
6A Mower
6A Murray
6A Nicollet
6A Nobles
7 Norman
6A Olmsted
6A Otter Tail
7 Pennington
7 Pine
6A Pipestone
7 Polk
6A Pope
6A Ramsey
7 Red Lake
6A Redwood
6A Renville
6A Rice
6A Rock
7 Roseau
6A Scott

TABLE C301.1—CLIMATE ZONES, MOISTURE REGIMES AND WARM HUMID DESIGNATIONS BY STATE, COUNTY AND TERRITORY[a]—continued

US STATES—continued
MINNESOTA *(continued)*
6A Sherburne
6A Sibley
6A Stearns
6A Steele
6A Stevens
7 St. Louis
6A Swift
6A Todd
6A Traverse
6A Wabasha
7 Wadena
6A Waseca
6A Washington
6A Watonwan
6A Wilkin
5A Winona
6A Wright
6A Yellow Medicine
MISSISSIPPI
3A Adams*
3A Alcorn
3A Amite*
3A Attala
3A Benton
3A Bolivar
3A Calhoun
3A Carroll
3A Chickasaw
3A Choctaw
3A Claiborne*
3A Clarke
3A Clay
3A Coahoma
3A Copiah*
3A Covington*
3A DeSoto
3A Forrest*
3A Franklin*
2A George*
3A Greene*
3A Grenada
2A Hancock*
2A Harrison*
3A Hinds*
3A Holmes
3A Humphreys
3A Issaquena
3A Itawamba
2A Jackson*
3A Jasper
3A Jefferson*
3A Jefferson Davis*
3A Jones*
3A Kemper
3A Lafayette
3A Lamar*
3A Lauderdale
3A Lawrence*
3A Leake
3A Lee
3A Leflore
3A Lincoln*
3A Lowndes
3A Madison
3A Marion*
3A Marshall
3A Monroe
3A Montgomery
3A Neshoba
3A Newton
3A Noxubee
3A Oktibbeha
3A Panola
2A Pearl River*
3A Perry*
3A Pike*
3A Pontotoc
3A Prentiss
3A Quitman
3A Rankin*
3A Scott
3A Sharkey
3A Simpson*
3A Smith*
2A Stone*
3A Sunflower
3A Tallahatchie
3A Tate
3A Tippah
3A Tishomingo

TABLE C301.1—CLIMATE ZONES, MOISTURE REGIMES AND WARM HUMID DESIGNATIONS BY STATE, COUNTY AND TERRITORY[a]—continued

US STATES—continued
MISSISSIPPI *(continued)*
3A Tunica
3A Union
3A Walthall*
3A Warren*
3A Washington
3A Wayne*
3A Webster
3A Wilkinson*
3A Winston
3A Yalobusha
3A Yazoo
MISSOURI
5A Adair
5A Andrew
5A Atchison
4A Audrain
4A Barry
4A Barton
4A Bates
4A Benton
4A Bollinger
4A Boone
4A Buchanan
4A Butler
4A Caldwell
4A Callaway
4A Camden
4A Cape Girardeau
4A Carroll
4A Carter
4A Cass
4A Cedar
4A Chariton
4A Christian
5A Clark
4A Clay
4A Clinton
4A Cole
4A Cooper
4A Crawford
4A Dade
4A Dallas
5A Daviess
5A DeKalb
4A Dent
4A Douglas
3A Dunklin
4A Franklin
4A Gasconade
5A Gentry
4A Greene
5A Grundy
5A Harrison
4A Henry
4A Hickory
5A Holt
4A Howard
4A Howell
4A Iron
4A Jackson
4A Jasper
4A Jefferson
4A Johnson
5A Knox
4A Laclede
4A Lafayette
4A Lawrence
5A Lewis
4A Lincoln
5A Linn
5A Livingston
5A Macon
4A Madison
4A Maries
5A Marion
4A McDonald
5A Mercer
4A Miller
4A Mississippi
4A Moniteau
4A Monroe
4A Montgomery
4A Morgan
4A New Madrid
4A Newton
5A Nodaway
4A Oregon
4A Osage
4A Ozark
3A Pemiscot

TABLE C301.1—CLIMATE ZONES, MOISTURE REGIMES AND WARM HUMID DESIGNATIONS BY STATE, COUNTY AND TERRITORY[a]—continued

US STATES—continued
MISSOURI *(continued)*
4A Perry
4A Pettis
4A Phelps
5A Pike
4A Platte
4A Polk
4A Pulaski
5A Putnam
5A Ralls
4A Randolph
4A Ray
4A Reynolds
4A Ripley
4A Saline
5A Schuyler
5A Scotland
4A Scott
4A Shannon
5A Shelby
4A St. Charles
4A St. Clair
4A St. Francois
4A St. Louis
4A St. Louis (city)
4A Ste. Genevieve
4A Stoddard
4A Stone
5A Sullivan
4A Taney
4A Texas
4A Vernon
4A Warren
4A Washington
4A Wayne
4A Webster
5A Worth
4A Wright
MONTANA
6B (all)
NEBRASKA
5A (all)
NEVADA
4B Carson City (city)
5B Churchill
3B Clark
4B Douglas
5B Elko
4B Esmeralda
5B Eureka
5B Humboldt
5B Lander
4B Lincoln
4B Lyon
4B Mineral
4B Nye
5B Pershing
5B Storey
5B Washoe
5B White Pine
NEW HAMPSHIRE
6A Belknap
6A Carroll
5A Cheshire
6A Coos
6A Grafton
5A Hillsborough
5A Merrimack
5A Rockingham
5A Strafford
6A Sullivan
NEW JERSEY
4A Atlantic
5A Bergen
4A Burlington
4A Camden
4A Cape May
4A Cumberland
4A Essex
4A Gloucester
4A Hudson
5A Hunterdon
4A Mercer
4A Middlesex
4A Monmouth
5A Morris
4A Ocean
5A Passaic
4A Salem
5A Somerset
5A Sussex

TABLE C301.1—CLIMATE ZONES, MOISTURE REGIMES AND WARM HUMID DESIGNATIONS BY STATE, COUNTY AND TERRITORY[a]—continued

US STATES—continued
NEW JERSEY *(continued)*
4A Union
5A Warren
NEW MEXICO
4B Bernalillo
4B Catron
3B Chaves
4B Cibola
5B Colfax
4B Curry
4B DeBaca
3B Doña Ana
3B Eddy
4B Grant
4B Guadalupe
5B Harding
3B Hidalgo
3B Lea
4B Lincoln
5B Los Alamos
3B Luna
5B McKinley
5B Mora
3B Otero
4B Quay
5B Rio Arriba
4B Roosevelt
5B Sandoval
5B San Juan
5B San Miguel
5B Santa Fe
3B Sierra
4B Socorro
5B Taos
5B Torrance
4B Union
4B Valencia
NEW YORK
5A Albany
5A Allegany
4A Bronx
5A Broome
5A Cattaraugus
5A Cayuga
5A Chautauqua
5A Chemung
6A Chenango
6A Clinton
5A Columbia
5A Cortland
6A Delaware
5A Dutchess
5A Erie
6A Essex
6A Franklin
6A Fulton
5A Genesee
5A Greene
6A Hamilton
6A Herkimer
6A Jefferson
4A Kings
6A Lewis
5A Livingston
6A Madison
5A Monroe
6A Montgomery
4A Nassau
4A New York
5A Niagara
6A Oneida
5A Onondaga
5A Ontario
5A Orange
5A Orleans
5A Oswego
6A Otsego
5A Putnam
4A Queens
5A Rensselaer
4A Richmond
5A Rockland
5A Saratoga
5A Schenectady
5A Schoharie
5A Schuyler
5A Seneca
5A Steuben
6A St. Lawrence
4A Suffolk
6A Sullivan

TABLE C301.1—CLIMATE ZONES, MOISTURE REGIMES AND WARM HUMID DESIGNATIONS BY STATE, COUNTY AND TERRITORY[a]—continued

US STATES—continued
NEW YORK *(continued)*
5A Tioga
5A Tompkins
6A Ulster
6A Warren
5A Washington
5A Wayne
4A Westchester
5A Wyoming
5A Yates
NORTH CAROLINA
3A Alamance
3A Alexander
5A Alleghany
3A Anson
5A Ashe
5A Avery
3A Beaufort
3A Bertie
3A Bladen
3A Brunswick*
4A Buncombe
4A Burke
3A Cabarrus
4A Caldwell
3A Camden
3A Carteret*
3A Caswell
3A Catawba
3A Chatham
3A Cherokee
3A Chowan
3A Clay
3A Cleveland
3A Columbus*
3A Craven
3A Cumberland
3A Currituck
3A Dare
3A Davidson
3A Davie
3A Duplin
3A Durham
3A Edgecombe
3A Forsyth
3A Franklin
3A Gaston
3A Gates
4A Graham
3A Granville
3A Greene
3A Guilford
3A Halifax
3A Harnett
4A Haywood
4A Henderson
3A Hertford
3A Hoke
3A Hyde
3A Iredell
4A Jackson
3A Johnston
3A Jones
3A Lee
3A Lenoir
3A Lincoln
4A Macon
4A Madison
3A Martin
4A McDowell
3A Mecklenburg
4A Mitchell
3A Montgomery
3A Moore
3A Nash
3A New Hanover*
3A Northampton
3A Onslow*
3A Orange
3A Pamlico
3A Pasquotank
3A Pender*
3A Perquimans
3A Person
3A Pitt
3A Polk
3A Randolph
3A Richmond
3A Robeson
3A Rockingham
3A Rowan

TABLE C301.1—CLIMATE ZONES, MOISTURE REGIMES AND WARM HUMID DESIGNATIONS BY STATE, COUNTY AND TERRITORY[a]—continued

US STATES—continued
NORTH CAROLINA *(continued)*
3A Rutherford
3A Sampson
3A Scotland
3A Stanly
4A Stokes
4A Surry
4A Swain
4A Transylvania
3A Tyrrell
3A Union
3A Vance
3A Wake
3A Warren
3A Washington
5A Watauga
3A Wayne
3A Wilkes
3A Wilson
4A Yadkin
5A Yancey
NORTH DAKOTA
6A Adams
6A Barnes
7 Benson
6A Billings
7 Bottineau
6A Bowman
7 Burke
6A Burleigh
6A Cass
7 Cavalier
6A Dickey
7 Divide
6A Dunn
6A Eddy
6A Emmons
6A Foster
6A Golden Valley
7 Grand Forks
6A Grant
6A Griggs
6A Hettinger
6A Kidder
6A LaMoure
6A Logan
7 McHenry
6A McIntosh
6A McKenzie
6A McLean
6A Mercer
6A Morton
6A Mountrail
7 Nelson
6A Oliver
7 Pembina
7 Pierce
7 Ramsey
6A Ransom
7 Renville
6A Richland
7 Rolette
6A Sargent
6A Sheridan
6A Sioux
6A Slope
6A Stark
6A Steele
6A Stutsman
7 Towner
6A Traill
7 Walsh
7 Ward
6A Wells
6A Williams
OHIO
4A Adams
5A Allen
5A Ashland
5A Ashtabula
4A Athens
5A Auglaize
5A Belmont
4A Brown
4A Butler
5A Carroll
5A Champaign
5A Clark
4A Clermont
4A Clinton
5A Columbiana

TABLE C301.1—CLIMATE ZONES, MOISTURE REGIMES AND WARM HUMID DESIGNATIONS BY STATE, COUNTY AND TERRITORY[a]—continued

US STATES—continued
OHIO *(continued)*
5A Coshocton
5A Crawford
5A Cuyahoga
5A Darke
5A Defiance
5A Delaware
5A Erie
5A Fairfield
4A Fayette
4A Franklin
5A Fulton
4A Gallia
5A Geauga
4A Greene
5A Guernsey
4A Hamilton
5A Hancock
5A Hardin
5A Harrison
5A Henry
4A Highland
4A Hocking
5A Holmes
5A Huron
4A Jackson
5A Jefferson
5A Knox
5A Lake
4A Lawrence
5A Licking
5A Logan
5A Lorain
5A Lucas
4A Madison
5A Mahoning
5A Marion
5A Medina
4A Meigs
5A Mercer
5A Miami
5A Monroe
5A Montgomery
5A Morgan
5A Morrow
5A Muskingum
5A Noble
5A Ottawa
5A Paulding
5A Perry
4A Pickaway
4A Pike
5A Portage
5A Preble
5A Putnam
5A Richland
4A Ross
5A Sandusky
4A Scioto
5A Seneca
5A Shelby
5A Stark
5A Summit
5A Trumbull
5A Tuscarawas
5A Union
5A Van Wert
4A Vinton
4A Warren
4A Washington
5A Wayne
5A Williams
5A Wood
5A Wyandot
OKLAHOMA
3A Adair
4A Alfalfa
3A Atoka
4B Beaver
3A Beckham
3A Blaine
3A Bryan
3A Caddo
3A Canadian
3A Carter
3A Cherokee
3A Choctaw
4B Cimarron
3A Cleveland
3A Coal
3A Comanche

TABLE C301.1—CLIMATE ZONES, MOISTURE REGIMES AND WARM HUMID DESIGNATIONS BY STATE, COUNTY AND TERRITORY[a]—continued

US STATES—continued
OKLAHOMA *(continued)*
3A Cotton
4A Craig
3A Creek
3A Custer
4A Delaware
3A Dewey
4A Ellis
4A Garfield
3A Garvin
3A Grady
4A Grant
3A Greer
3A Harmon
4A Harper
3A Haskell
3A Hughes
3A Jackson
3A Jefferson
3A Johnston
4A Kay
3A Kingfisher
3A Kiowa
3A Latimer
3A Le Flore
3A Lincoln
3A Logan
3A Love
4A Major
3A Marshall
3A Mayes
3A McClain
3A McCurtain
3A McIntosh
3A Murray
3A Muskogee
3A Noble
4A Nowata
3A Okfuskee
3A Oklahoma
3A Okmulgee
4A Osage
4A Ottawa
3A Pawnee
3A Payne
3A Pittsburg
3A Pontotoc
3A Pottawatomie
3A Pushmataha
3A Roger Mills
3A Rogers
3A Seminole
3A Sequoyah
3A Stephens
4B Texas
3A Tillman
3A Tulsa
3A Wagoner
4A Washington
3A Washita
4A Woods
4A Woodward
OREGON
5B Baker
4C Benton
4C Clackamas
4C Clatsop
4C Columbia
4C Coos
5B Crook
4C Curry
5B Deschutes
4C Douglas
5B Gilliam
5B Grant
5B Harney
5B Hood River
4C Jackson
5B Jefferson
4C Josephine
5B Klamath
5B Lake
4C Lane
4C Lincoln
4C Linn
5B Malheur
4C Marion
5B Morrow
4C Multnomah
4C Polk
5B Sherman

TABLE C301.1—CLIMATE ZONES, MOISTURE REGIMES AND WARM HUMID DESIGNATIONS BY STATE, COUNTY AND TERRITORY[a]—continued

US STATES—continued
OREGON *(continued)*
4C Tillamook
5B Umatilla
5B Union
5B Wallowa
5B Wasco
4C Washington
5B Wheeler
4C Yamhill
PENNSYLVANIA
4A Adams
5A Allegheny
5A Armstrong
5A Beaver
5A Bedford
4A Berks
5A Blair
5A Bradford
4A Bucks
5A Butler
5A Cambria
5A Cameron
5A Carbon
5A Centre
4A Chester
5A Clarion
5A Clearfield
5A Clinton
5A Columbia
5A Crawford
4A Cumberland
4A Dauphin
4A Delaware
5A Elk
5A Erie
5A Fayette
5A Forest
4A Franklin
5A Fulton
5A Greene
5A Huntingdon
5A Indiana
5A Jefferson
5A Juniata
5A Lackawanna
4A Lancaster
5A Lawrence
4A Lebanon
5A Lehigh
5A Luzerne
5A Lycoming
5A McKean
5A Mercer
5A Mifflin
5A Monroe
4A Montgomery
5A Montour
5A Northampton
5A Northumberland
4A Perry
4A Philadelphia
5A Pike
5A Potter
5A Schuylkill
5A Snyder
5A Somerset
5A Sullivan
5A Susquehanna
5A Tioga
5A Union
5A Venango
5A Warren
5A Washington
5A Wayne
5A Westmoreland
5A Wyoming
4A York
RHODE ISLAND
5A (all)
SOUTH CAROLINA
3A Abbeville
3A Aiken
3A Allendale*
3A Anderson
3A Bamberg*
3A Barnwell*
2A Beaufort*
3A Berkeley*
3A Calhoun
3A Charleston*
3A Cherokee

TABLE C301.1—CLIMATE ZONES, MOISTURE REGIMES AND WARM HUMID DESIGNATIONS BY STATE, COUNTY AND TERRITORY[a]—continued

US STATES—continued
SOUTH CAROLINA *(continued)*
3A Chester
3A Chesterfield
3A Clarendon
3A Colleton*
3A Darlington
3A Dillon
3A Dorchester*
3A Edgefield
3A Fairfield
3A Florence
3A Georgetown*
3A Greenville
3A Greenwood
3A Hampton*
3A Horry*
2A Jasper*
3A Kershaw
3A Lancaster
3A Laurens
3A Lee
3A Lexington
3A Marion
3A Marlboro
3A McCormick
3A Newberry
3A Oconee
3A Orangeburg
3A Pickens
3A Richland
3A Saluda
3A Spartanburg
3A Sumter
3A Union
3A Williamsburg
3A York
SOUTH DAKOTA
6A Aurora
6A Beadle
5A Bennett
5A Bon Homme
6A Brookings
6A Brown
5A Brule
6A Buffalo
6A Butte
6A Campbell
5A Charles Mix
6A Clark
5A Clay
6A Codington
6A Corson
6A Custer
6A Davison
6A Day
6A Deuel
6A Dewey
5A Douglas
6A Edmunds
6A Fall River
6A Faulk
6A Grant
5A Gregory
5A Haakon
6A Hamlin
6A Hand
6A Hanson
6A Harding
6A Hughes
5A Hutchinson
6A Hyde
5A Jackson
6A Jerauld
5A Jones
6A Kingsbury
6A Lake
6A Lawrence
6A Lincoln
5A Lyman
6A Marshall
6A McCook
6A McPherson
6A Meade
5A Mellette
6A Miner
6A Minnehaha
6A Moody
6A Pennington
6A Perkins
6A Potter
6A Roberts

TABLE C301.1—CLIMATE ZONES, MOISTURE REGIMES AND WARM HUMID DESIGNATIONS BY STATE, COUNTY AND TERRITORY[a]—continued

US STATES—continued
SOUTH DAKOTA *(continued)*
6A Sanborn
6A Shannon
6A Spink
5A Stanley
6A Sully
5A Todd
5A Tripp
6A Turner
5A Union
6A Walworth
5A Yankton
6A Ziebach
TENNESSEE
4A Anderson
3A Bedford
4A Benton
4A Bledsoe
4A Blount
4A Bradley
4A Campbell
4A Cannon
4A Carroll
4A Carter
4A Cheatham
3A Chester
4A Claiborne
4A Clay
4A Cocke
3A Coffee
3A Crockett
4A Cumberland
3A Davidson
3A Decatur
4A DeKalb
4A Dickson
3A Dyer
3A Fayette
4A Fentress
3A Franklin
3A Gibson
3A Giles
4A Grainger
4A Greene
3A Grundy
4A Hamblen
3A Hamilton
4A Hancock
3A Hardeman
3A Hardin
4A Hawkins
3A Haywood
3A Henderson
4A Henry
3A Hickman
4A Houston
4A Humphreys
4A Jackson
4A Jefferson
4A Johnson
4A Knox
4A Lake
3A Lauderdale
3A Lawrence
3A Lewis
3A Lincoln
4A Loudon
4A Macon
3A Madison
3A Marion
3A Marshall
3A Maury
4A McMinn
3A McNairy
4A Meigs
4A Monroe
4A Montgomery
3A Moore
4A Morgan
4A Obion
4A Overton
3A Perry
4A Pickett
4A Polk
4A Putnam
4A Rhea
4A Roane
4A Robertson
3A Rutherford
4A Scott
4A Sequatchie

TABLE C301.1—CLIMATE ZONES, MOISTURE REGIMES AND WARM HUMID DESIGNATIONS BY STATE, COUNTY AND TERRITORY[a]—continued

US STATES—continued
TENNESSEE *(continued)*
4A Sevier
3A Shelby
4A Smith
4A Stewart
4A Sullivan
4A Sumner
3A Tipton
4A Trousdale
4A Unicoi
4A Union
4A Van Buren
4A Warren
4A Washington
3A Wayne
4A Weakley
4A White
3A Williamson
4A Wilson
TEXAS
2A Anderson*
3B Andrews
2A Angelina*
2A Aransas*
3A Archer
4B Armstrong
2A Atascosa*
2A Austin*
4B Bailey
2B Bandera
2A Bastrop*
3B Baylor
2A Bee*
2A Bell*
2A Bexar*
3A Blanco*
3B Borden
2A Bosque*
3A Bowie*
2A Brazoria*
2A Brazos*
3B Brewster
4B Briscoe
2A Brooks*
3A Brown*
2A Burleson*
3A Burnet*
2A Caldwell*
2A Calhoun*
3B Callahan
1A Cameron*
3A Camp*
4B Carson
3A Cass*
4B Castro
2A Chambers*
2A Cherokee*
3B Childress
3A Clay
4B Cochran
3B Coke
3B Coleman
3A Collin*
3B Collingsworth
2A Colorado*
2A Comal*
3A Comanche*
3B Concho
3A Cooke
2A Coryell*
3B Cottle
3B Crane
3B Crockett
3B Crosby
3B Culberson
4B Dallam
2A Dallas*
3B Dawson
4B Deaf Smith
3A Delta
3A Denton*
2A DeWitt*
3B Dickens
2B Dimmit
4B Donley
2A Duval*
3A Eastland
3B Ector
2B Edwards
2A Ellis*
3B El Paso

TABLE C301.1—CLIMATE ZONES, MOISTURE REGIMES AND WARM HUMID DESIGNATIONS BY STATE, COUNTY AND TERRITORY[a]—continued

US STATES—continued
TEXAS *(continued)*
3A Erath*
2A Falls*
3A Fannin
2A Fayette*
3B Fisher
4B Floyd
3B Foard
2A Fort Bend*
3A Franklin*
2A Freestone*
2B Frio
3B Gaines
2A Galveston*
3B Garza
3A Gillespie*
3B Glasscock
2A Goliad*
2A Gonzales*
4B Gray
3A Grayson
3A Gregg*
2A Grimes*
2A Guadalupe*
4B Hale
3B Hall
3A Hamilton*
4B Hansford
3B Hardeman
2A Hardin*
2A Harris*
3A Harrison*
4B Hartley
3B Haskell
2A Hays*
3B Hemphill
3A Henderson*
1A Hidalgo*
2A Hill*
4B Hockley
3A Hood*
3A Hopkins*
2A Houston*
3B Howard
3B Hudspeth
3A Hunt*
4B Hutchinson
3B Irion
3A Jack
2A Jackson*
2A Jasper*
3B Jeff Davis
2A Jefferson*
2A Jim Hogg*
2A Jim Wells*
2A Johnson*
3B Jones
2A Karnes*
3A Kaufman*
3A Kendall*
2A Kenedy*
3B Kent
3B Kerr
3B Kimble
3B King
2B Kinney
2A Kleberg*
3B Knox
3A Lamar*
4B Lamb
3A Lampasas*
2B La Salle
2A Lavaca*
2A Lee*
2A Leon*
2A Liberty*
2A Limestone*
4B Lipscomb
2A Live Oak*
3A Llano*
3B Loving
3B Lubbock
3B Lynn
2A Madison*
3A Marion*
3B Martin
3B Mason
2A Matagorda*
2B Maverick
3B McCulloch
2A McLennan*

TABLE C301.1—CLIMATE ZONES, MOISTURE REGIMES AND WARM HUMID DESIGNATIONS BY STATE, COUNTY AND TERRITORY[a]—continued

US STATES—continued
TEXAS *(continued)*
2A McMullen*
2B Medina
3B Menard
3B Midland
2A Milam*
3A Mills*
3B Mitchell
3A Montague
2A Montgomery*
4B Moore
3A Morris*
3B Motley
3A Nacogdoches*
2A Navarro*
2A Newton*
3B Nolan
2A Nueces*
4B Ochiltree
4B Oldham
2A Orange*
3A Palo Pinto*
3A Panola*
3A Parker*
4B Parmer
3B Pecos
2A Polk*
4B Potter
3B Presidio
3A Rains*
4B Randall
3B Reagan
2B Real
3A Red River*
3B Reeves
2A Refugio*
4B Roberts
2A Robertson*
3A Rockwall*
3B Runnels
3A Rusk*
3A Sabine*
3A San Augustine*
2A San Jacinto*
2A San Patricio*
3A San Saba*
3B Schleicher
3B Scurry
3B Shackelford
3A Shelby*
4B Sherman
3A Smith*
3A Somervell*
2A Starr*
3A Stephens
3B Sterling
3B Stonewall
3B Sutton
4B Swisher
2A Tarrant*
3B Taylor
3B Terrell
3B Terry
3B Throckmorton
3A Titus*
3B Tom Green
2A Travis*
2A Trinity*
2A Tyler*
3A Upshur*
3B Upton
2B Uvalde
2B Val Verde
3A Van Zandt*
2A Victoria*
2A Walker*
2A Waller*
3B Ward
2A Washington*
2B Webb
2A Wharton*
3B Wheeler
3A Wichita
3B Wilbarger
1A Willacy*
2A Williamson*
2A Wilson*
3B Winkler
3A Wise
3A Wood*
4B Yoakum

TABLE C301.1—CLIMATE ZONES, MOISTURE REGIMES AND WARM HUMID DESIGNATIONS BY STATE, COUNTY AND TERRITORY[a]—continued

US STATES—continued
TEXAS *(continued)*
3A Young
2B Zapata
2B Zavala
UTAH
5B Beaver
5B Box Elder
5B Cache
5B Carbon
6B Daggett
5B Davis
6B Duchesne
5B Emery
5B Garfield
5B Grand
5B Iron
5B Juab
5B Kane
5B Millard
6B Morgan
5B Piute
6B Rich
5B Salt Lake
5B San Juan
5B Sanpete
5B Sevier
6B Summit
5B Tooele
6B Uintah
5B Utah
6B Wasatch
3B Washington
5B Wayne
5B Weber
VERMONT
6A (all)
VIRGINIA
4A (all except as follows:)
5A Alleghany
5A Bath
3A Brunswick
3A Chesapeake
5A Clifton Forge
5A Covington
3A Emporia
3A Franklin
3A Greensville
3A Halifax
3A Hampton
5A Highland
3A Isle of Wight
3A Mecklenburg
3A Newport News
3A Norfolk
3A Pittsylvania
3A Portsmouth
3A South Boston
3A Southampton
3A Suffolk
3A Surry
3A Sussex
3A Virginia Beach
WASHINGTON
5B Adams
5B Asotin
5B Benton
5B Chelan
5C Clallam
4C Clark
5B Columbia
4C Cowlitz
5B Douglas
6B Ferry
5B Franklin
5B Garfield
5B Grant
4C Grays Harbor
5C Island
4C Jefferson
4C King
5C Kitsap
5B Kittitas
5B Klickitat
4C Lewis
5B Lincoln
4C Mason
5B Okanogan
4C Pacific
6B Pend Oreille
4C Pierce
5C San Juan

TABLE C301.1—CLIMATE ZONES, MOISTURE REGIMES AND WARM HUMID DESIGNATIONS BY STATE, COUNTY AND TERRITORY[a]—continued

US STATES—continued
WASHINGTON *(continued)*
4C Skagit
5B Skamania
4C Snohomish
5B Spokane
6B Stevens
4C Thurston
4C Wahkiakum
5B Walla Walla
4C Whatcom
5B Whitman
5B Yakima
WEST VIRGINIA
5A Barbour
4A Berkeley
4A Boone
4A Braxton
5A Brooke
4A Cabell
4A Calhoun
4A Clay
4A Doddridge
4A Fayette
4A Gilmer
5A Grant
4A Greenbrier
5A Hampshire
5A Hancock
5A Hardy
5A Harrison
4A Jackson
4A Jefferson
4A Kanawha
4A Lewis
4A Lincoln
4A Logan
5A Marion
5A Marshall
4A Mason
4A McDowell
4A Mercer
5A Mineral
4A Mingo
5A Monongalia
4A Monroe
4A Morgan
4A Nicholas
5A Ohio
5A Pendleton
4A Pleasants
5A Pocahontas
5A Preston
4A Putnam
4A Raleigh
5A Randolph
4A Ritchie
4A Roane
4A Summers
5A Taylor
5A Tucker
4A Tyler
4A Upshur
4A Wayne
4A Webster
5A Wetzel
4A Wirt
4A Wood
4A Wyoming
WISCONSIN
5A Adams
6A Ashland
6A Barron
6A Bayfield
6A Brown
6A Buffalo
6A Burnett
5A Calumet
6A Chippewa
6A Clark
5A Columbia
5A Crawford
5A Dane
5A Dodge
6A Door
6A Douglas
6A Dunn
6A Eau Claire
6A Florence
5A Fond du Lac
6A Forest
5A Grant

TABLE C301.1—CLIMATE ZONES, MOISTURE REGIMES AND WARM HUMID DESIGNATIONS BY STATE, COUNTY AND TERRITORY[a]—continued

US STATES—continued
WISCONSIN *(continued)*
5A Green
5A Green Lake
5A Iowa
6A Iron
6A Jackson
5A Jefferson
5A Juneau
5A Kenosha
6A Kewaunee
5A La Crosse
5A Lafayette
6A Langlade
6A Lincoln
6A Manitowoc
6A Marathon
6A Marinette
6A Marquette
6A Menominee
5A Milwaukee
5A Monroe
6A Oconto
6A Oneida
5A Outagamie
5A Ozaukee
6A Pepin
6A Pierce
6A Polk
6A Portage
6A Price
5A Racine
5A Richland
5A Rock
6A Rusk
5A Sauk
6A Sawyer
6A Shawano
6A Sheboygan
6A St. Croix
6A Taylor
6A Trempealeau
5A Vernon
6AVilas
5A Walworth
6A Washburn
5A Washington
5A Waukesha
6A Waupaca
5A Waushara
5A Winnebago
6A Wood
WYOMING
6B Albany
6B Big Horn
6B Campbell
6B Carbon
6B Converse
6B Crook
6B Fremont
5B Goshen
6B Hot Springs
6B Johnson
5B Laramie
7 Lincoln
6B Natrona
6B Niobrara
6B Park
5B Platte
6B Sheridan
7 Sublette
6B Sweetwater
7 Teton
6B Uinta
6B Washakie
6B Weston
US TERRITORIES
AMERICAN SAMOA
1A (all)*
GUAM
1A (all)*
NORTHERN MARIANA ISLANDS
1A (all)*
PUERTO RICO
1A (all except as follows:)*
2B Barraquitas
2B Cayey
VIRGIN ISLANDS
1A (all)*

a. Key: A – Moist, B – Dry, C – Marine. Absence of moisture designation indicates moisture regime is irrelevant. Asterisk (*) indicates a Warm Humid location.

C301.2 Warm Humid counties. In Table C301.1, Warm Humid counties are identified by an asterisk.

C301.3 Climate zone definitions. To determine the *climate zones* for locations not listed in this code, use the following information to determine *climate zone* numbers and letters in accordance with Items 1 through 5.

1. Determine the thermal climate zone, 0 through 8, from Table C301.3 using the heating (HDD) and cooling degree-days (CDD) for the location.
2. Determine the moisture zone (Marine, Dry or Humid) in accordance with Items 2.1 through 2.3.
   - 2.1. If monthly average temperature and precipitation data are available, use the Marine, Dry and Humid definitions to determine the moisture zone (C, B or A).
   - 2.2. If annual average temperature information (including degree-days) and annual precipitation (i.e., annual mean) are available, use Items 2.2.1 through 2.2.3 to determine the moisture zone. If the moisture zone is not Marine, then use the Dry definition to determine whether Dry or Humid.
     - 2.2.1. If thermal climate zone is 3 and CDD50°F ≤ 4,500 (CDD10°C ≤ 2500), climate zone is Marine (3C).
     - 2.2.2. If thermal climate zone is 4 and CDD50°F ≤ 2,700 (CDD10°C ≤ 1500), climate zone is Marine (4C).
     - 2.2.3. If thermal climate zone is 5 and CDD50°F ≤ 1,800 (CDD10°C ≤ 1000), climate zone is Marine (5C).
   - 2.3. If only degree-day information is available, use Items 2.3.1 through 2.3.3 to determine the moisture zone. If the moisture zone is not Marine, then it is not possible to assign Humid or Dry moisture zone for this location.
     - 2.3.1. If thermal climate zone is 3 and CDD50°F ≤ 4,500 (CDD10°C ≤ 2500), climate zone is Marine (3C).
     - 2.3.2. If thermal climate zone is 4 and CDD50°F ≤ 2,700 (CDD10°C ≤ 1500), climate zone is Marine (4C).
     - 2.3.3. If thermal climate zone is 5 and CDD50°F ≤ 1,800 (CDD10°C ≤ 1000), climate zone is Marine (5C).
3. Marine (C) Zone definition: Locations meeting all the criteria in Items 3.1 through 3.4.
   - 3.1. Mean temperature of coldest month between 27°F (-3°C) and 65°F (18°C).
   - 3.2. Warmest month mean < 72°F (22°C).
   - 3.3. Not fewer than four months with mean temperatures over 50°F (10°C).
   - 3.4. Dry season in summer. The month with the heaviest precipitation in the cold season has at least three times as much precipitation as the month with the least precipitation in the rest of the year. The cold season is October through March in the Northern Hemisphere and April through September in the Southern Hemisphere.
4. Dry (B) definition: Locations meeting the criteria in Items 4.1 through 4.4.
   - 4.1. Not Marine (C).
   - 4.2. If 70 percent or more of the precipitation, *P*, occurs during the high sun period, defined as April through September in the Northern Hemisphere and October through March in the Southern Hemisphere, then the dry/humid threshold is in accordance with Equation 3-1.

     Equation 3-1 $P < 0.44 \times (T - 7)$
     $[P < 20.0 \times (T + 14) \text{ in SI units}]$

     where:

     P = Annual precipitation, inches (mm).

     T = Annual mean temperature, °F (°C).
   - 4.3. If between 30 and 70 percent of the precipitation, *P*, occurs during the high sun period, defined as April through September in the Northern Hemisphere and October through March in the Southern Hemisphere, then the dry/humid threshold is in accordance with Equation 3-2.

     Equation 3-2 $P < 0.44 \times (T - 19.5)$
     $[P < 20.0 \times (T + 7) \text{ in SI units}]$

     where:

     P = Annual precipitation, inches (mm).

     T = Annual mean temperature, °F (°C).
   - 4.4. If 30 percent or less of the precipitation, *P*, occurs during the high sun period, defined as April through September in the Northern Hemisphere and October through March in the Southern Hemisphere, then the dry/humid threshold is in accordance with Equation 3-3.

     Equation 3-3 $P < 0.44 \times (T - 32)$
     $[P < 20.0 \times T \text{ in SI units}]$

     where:

     P = Annual precipitation, inches (mm).

     T = Annual mean temperature, °F (°C).
5. Humid (A) definition: Locations that are not Marine (C) or Dry (B).

TABLE C301.3—THERMAL CLIMATE ZONE DEFINITIONS		
ZONE NUMBER	THERMAL CRITERIA	
	IP Units	SI Units
0	10,800 < CDD50°F	6000 < CDD10°C
1	9,000 < CDD50°F < 10,800	5000 < CDD10°C < 6000
2	6,300 < CDD50°F ≤ 9,000	3500 < CDD10°C ≤ 5000
3	CDD50°F ≤ 6,300 AND HDD65°F ≤ 3,600	CDD10°C < 3500 AND HDD18°C ≤ 2000
4	CDD50°F ≤ 6,300 AND 3,600 < HDD65°F ≤ 5,400	CDD10°C < 3500 AND 2000 < HDD18°C ≤ 3000
5	CDD50°F < 6,300 AND 5,400 < HDD65°F ≤ 7,200	CDD10°C < 3500 AND 3000 < HDD18°C ≤ 4000
6	7,200 < HDD65°F ≤ 9,000	4000 < HDD18°C ≤ 5000
7	9,000 < HDD65°F ≤ 12,600	5000 < HDD18°C ≤ 7000
8	12,600 < HDD65°F	7000 < HDD18°C

For SI: °C = [(°F) – 32]/1.8.

C301.4 Tropical climate region. The tropical climate region shall be defined as:

1. Hawaii, Puerto Rico, Guam, American Samoa, US Virgin Islands, Commonwealth of Northern Mariana Islands; and
2. Islands in the area between the Tropic of Cancer and the Tropic of Capricorn.

SECTION C302—DESIGN CONDITIONS

C302.1 Interior design conditions. The interior design temperatures used for heating and cooling load calculations shall be a maximum of 72°F (22°C) for heating and minimum of 75°F (24°C) for cooling.

SECTION C303—MATERIALS, SYSTEMS AND EQUIPMENT

C303.1 Identification. Materials, systems and equipment shall be identified in a manner that will allow a determination of compliance with the applicable provisions of this code.

C303.1.1 Building thermal envelope insulation. An *R-value* identification mark shall be applied by the manufacturer to each piece of *building thermal envelope* insulation 12 inches (305 mm) or greater in width. Alternatively, the insulation installers shall provide a certification listing the type, manufacturer and *R-value* of insulation installed in each element of the *building thermal envelope*. For blown-in or sprayed fiberglass and cellulose insulation, the initial installed thickness, settled thickness, settled *R-value*, installed density, coverage area and number of bags installed shall be indicated on the certification. For sprayed polyurethane foam (SPF) insulation, the installed thickness of the areas covered and *R-value* of installed thickness shall be indicated on the certification. For insulated siding, the *R-value* shall be *labeled* on the product's package and shall be indicated on the certification. The insulation installer shall sign, date and post the certification in a conspicuous location on the job site.

Exception: For roof insulation installed above the deck, the *R-value* shall be *labeled* as required by the material standards specified in Table 1508.2 of the *International Building Code*.

C303.1.1.1 Blown-in or sprayed roof/ceiling insulation. The thickness of blown-in or sprayed fiberglass and cellulose roof/ceiling insulation shall be written in inches (mm) on markers and one or more of such markers shall be installed for every 300 square feet (28 m^2) of attic area throughout the attic space. The markers shall be affixed to the trusses or joists and marked with the minimum initial installed thickness with numbers not less than 1 inch (25 mm) in height. Each marker shall face the attic *access* opening. Spray polyurethane foam thickness and installed *R-value* shall be indicated on certification provided by the insulation installer.

C303.1.2 Insulation mark installation. Insulating materials shall be installed such that the manufacturer's *R-value* mark is readily observable upon inspection. For insulation materials that are installed without an observable manufacturer's *R-value* mark, such as blown or draped products, an insulation certificate complying with Section C303.1.1 shall be left immediately after installation by the installer, in a conspicuous location within the *building*, to certify the installed *R-value* of the insulation material.

Exception: For roof insulation installed above the deck, the *R-value* shall be *labeled* as specified by the material standards in Table 1508.2 of the *International Building Code*.

C303.1.3 Fenestration product rating. *U-factors*, *solar heat gain coefficient* (SHGC) and visible transmittance (VT) of fenestration products shall be determined as follows:

1. For windows, doors and skylights, *U-factor*, SHGC and VT ratings shall be determined in accordance with NFRC 100 and NFRC 200.
2. Where required for garage doors and rolling doors, *U-factor* ratings shall be determined in accordance with either NFRC 100 or ANSI/DASMA 105.

U-factors, SHGC and VT shall be determined by an accredited, independent laboratory, and *labeled* and certified by the manufacturer with a label affixed to the product or a label certificate specific to the products in the project.

Products lacking such a *labeled U-factor* shall be assigned a default *U-factor* from Table C303.1.3(1) or Table C303.1.3(2). Products lacking such a *labeled* SHGC or VT shall be assigned a default SHGC or VT from Table C303.1.3(3). For Tubular Daylighting Devices, VT_{annual} shall be measured and rated in accordance with NFRC 203.

TABLE C303.1.3(1)—DEFAULT GLAZED WINDOW, GLASS DOOR AND SKYLIGHT *U*-FACTORS

FRAME TYPE	WINDOW AND GLASS DOOR		SKYLIGHT	
	Single	Double	Single	Double
Metal	1.20	0.80	2.00	1.30
Metal with thermal break	1.10	0.65	1.90	1.10
Nonmetal or metal clad	0.95	0.55	1.75	1.05
Glass block	0.60			

TABLE C303.1.3(2)—DEFAULT OPAQUE DOOR *U*-FACTORS

DOOR TYPE	OPAQUE *U*-FACTOR
Uninsulated metal	1.20
Insulated metal (rolling)	0.90
Insulated metal (other)	0.60
Wood	0.50
Insulated, nonmetal edge, max 45% glazing, any glazing double pane	0.35

TABLE C303.1.3(3)—DEFAULT GLAZED FENESTRATION SHGC AND VT

	SINGLE GLAZED		DOUBLE GLAZED		GLAZED BLOCK
	Clear	Tinted	Clear	Tinted	
SHGC	0.8	0.7	0.7	0.6	0.6
VT	0.6	0.3	0.6	0.3	0.6

C303.1.4 Insulation product rating. The thermal resistance (*R-value*) of insulation shall be determined in accordance with the US Federal Trade Commission *R*-value rule (CFR Title 16, Part 460) in units of $h \times ft^2 \times °F/Btu$ at a mean temperature of 75°F (24°C).

C303.1.4.1 Insulated siding. The thermal resistance (*R-value*) of insulated siding shall be determined in accordance with ASTM C1363. Installation for testing shall be in accordance with the manufacturer's instructions.

C303.2 Installation. Materials, systems and equipment shall be installed in accordance with the manufacturer's instructions and the *International Building Code*.

C303.2.1 Protection of exposed foundation insulation. Insulation applied to the exterior of basement walls, *crawl space walls* and the perimeter of slab-on-grade floors shall have a rigid, opaque and weather-resistant protective covering to prevent the degradation of the insulation's thermal performance. The protective covering shall cover the exposed exterior insulation and extend not less than 6 inches (153 mm) below grade.

C303.2.2 Multiple layers of continuous insulation board. Where two or more layers of continuous insulation board are used in a construction assembly, the continuous insulation boards shall be installed in accordance with Section C303.2. Where the continuous insulation board manufacturer's instructions do not address installation of two or more layers, the edge joints between each layer of continuous insulation boards shall be staggered.

CHAPTER 4 [CE] COMMERCIAL ENERGY EFFICIENCY

User notes:

About this chapter: *Chapter 4 presents the paths and options for compliance with the energy efficiency provisions. Chapter 4 contains energy efficiency provisions for the building envelope, mechanical and water-heating systems, lighting and additional efficiency requirements. A performance alternative is also provided to allow for energy code compliance other than by the prescriptive method.*

SECTION C401—GENERAL

C401.1 Scope. The provisions in this chapter are applicable to *commercial buildings* and their *building sites*.

C401.2 Application. *Commercial buildings* shall comply with Section C401.2.1 or C401.2.2.

C401.2.1 International Energy Conservation Code. *Commercial buildings* shall comply with one of the following:

1. Prescriptive Compliance. The Prescriptive Compliance option requires compliance with Sections C402 through C406 and Section C408. *Dwelling units* and *sleeping units* in Group R-2 buildings shall be deemed to be in compliance with this chapter, provided that they comply with Section R406.
2. *Simulated Building Performance*. The *Simulated Building Performance* option requires compliance with Section C407.

Exception: *Additions*, *alterations*, *repairs* and changes of occupancy to existing buildings complying with Chapter 5.

C401.2.2 ASHRAE 90.1. *Commercial buildings* shall comply with the requirements of ANSI/ASHRAE/IES 90.1.

C401.3 Building thermal envelope certificate. A permanent *building thermal envelope* certificate shall be completed by an *approved* party. Such certificate shall be posted on a wall in the space where the space conditioning equipment is located, a utility room or other *approved* location. If located on an electrical panel, the certificate shall not cover or obstruct the visibility of the circuit directory label, service disconnect label or other required labels. A copy of the certificate shall also be included in the construction files for the project. The certificate shall include the following:

1. *R*-values of insulation installed in or on ceilings, roofs, walls, foundations and slabs, *basement walls*, *crawl space walls* and floors and *ducts* outside *conditioned spaces*.
2. *U-factors* and *solar heat gain coefficients* (SHGC) of *fenestrations*.
3. Results from any *building thermal envelope air leakage* testing performed on the *building*.

Where there is more than one value for any component of the *building thermal envelope*, the certificate shall indicate the area-weighted average value where available. If the area-weighted average is not available, the certificate shall list each value that applies to 10 percent or more of the total component area.

SECTION C402—BUILDING THERMAL ENVELOPE REQUIREMENTS

C402.1 General. *Building thermal envelope* assemblies for buildings that are intended to comply with the code on a prescriptive basis in accordance with the compliance path described in Item 1 of Section C401.2.1 shall comply with the following:

1. The opaque portions of the *building thermal envelope* shall comply with the specific insulation requirements of Section C402.2 and the thermal requirements of Section C402.1.2, C402.1.3 or C402.1.4. Where the total area of through penetrations of mechanical equipment is greater than 1 percent of the opaque *above-grade wall* area, the *building thermal envelope* shall comply with Section C402.1.2.1.8.
2. Wall solar reflectance and thermal *emittance* shall comply with Section C402.3.
3. Roof solar reflectance and thermal *emittance* shall comply with Section C402.4.
4. *Fenestration* in the *building thermal envelope* shall comply with Section C402.5. Where *buildings* have a vertical *fenestration* area or skylight area greater than that allowed in Section C402.5, the *building* and *building thermal envelope* shall comply with Item 2 of Section C401.2.1, C401.2.2 or C402.1.4.
5. *Air leakage* of *building thermal envelope* shall comply with Section C402.6.
6. *Thermal bridges* in *above-grade walls* shall comply with Section C402.7.
7. *Walk-in coolers*, *walk-in freezers*, *refrigerated warehouse coolers* and *refrigerated warehouse freezers* shall comply with Section C403.12.

C402.1.1 Low-energy buildings and greenhouses. The following low-energy buildings, or portions thereof separated from the remainder of the *building* by *building thermal envelope* assemblies complying with this section, shall be exempt from the *building thermal envelope* provisions of Section C402.

C402.1.1.1 Low-energy buildings. *Buildings* that comply with either of the following:

1. Those with a peak design rate of energy usage less than 3.4 Btu/h × ft^2 (10.7 W/m^2) or 1.0 watt per square foot (10.7 W/m^2) of floor area for space conditioning purposes.
2. Those that do not contain *conditioned space*.

C402.1.1.2 Greenhouses. *Greenhouse* structures or areas that are mechanically heated or cooled and that comply with all of the following shall be exempt from the *building thermal envelope* requirements of this code:

1. Exterior opaque envelope assemblies comply with Sections C402.2 and C402.5.5.

   Exception: Low energy greenhouses that comply with Section C402.1.1.
2. Interior partition *building thermal envelope* assemblies that separate the *greenhouse* from *conditioned space* comply with Sections C402.2, C402.5.3 and C402.5.5.
3. *Fenestration* assemblies that comply with the *building thermal envelope* requirements in Table C402.1.1.2. The *U-factor* for a roof shall be for the *roof assembly* or a roof that includes the assembly and an *internal curtain system*.

   Exception: Unconditioned greenhouses.

TABLE C402.1.1.2—FENESTRATION BUILDING THERMAL ENVELOPE MAXIMUM REQUIREMENTS

COMPONENT	*U*-FACTOR (Btu/h × ft^2 × °F)
Skylight	0.5
Vertical fenestration	0.7

C402.1.1.3 Equipment buildings. *Buildings* that comply with the following shall be exempt from the *building thermal envelope* provisions of this code:

1. Are separate *buildings* with floor area not more than 1,200 square feet (111 m^2).
2. Are intended to house electric equipment with installed equipment power totaling not less than 7 watts per square foot (75 W/m^2) and not intended for human occupancy.
3. Have a heating system capacity not greater than 20,000 Btu/h (6 kW) and a heating *thermostat* setpoint that is restricted to not more than 50°F (10°C).
4. Have an average wall and roof *U-factor* less than 0.2 in Climate Zones 1 through 5 and less than 0.12 in Climate Zones 6 through 8.
5. Comply with the roof solar reflectance and thermal *emittance* provisions for Climate Zone 1.

C402.1.2 Assembly *U*-factor, *C*-factor or *F*-factor method. *Building thermal envelope* opaque assemblies shall have a *U*-, *C*- or *F-factor* not greater than that specified in Table C402.1.2. *Commercial buildings* or portions of *commercial buildings* enclosing *Group R* occupancies shall use the *U*-, *C*- or *F-factor* from the "*Group R*" column of Table C402.1.2. *Commercial buildings* or portions of *commercial buildings* enclosing occupancies other than *Group R* shall use the *U*-, *C*- or *F*-factor from the "All other" column of Table C402.1.2

TABLE C402.1.2—OPAQUE BUILDING THERMAL ENVELOPE ASSEMBLY MAXIMUM REQUIREMENTS, *U*-FACTOR METHOD[a, b]

CLIMATE ZONE	0 AND 1		2		3		4 EXCEPT MARINE		5 AND MARINE 4		6		7		8	
	All Other	Group R	All Other	Group R	All Other	Group R	All Other	Group R	All Other	Group R	All Other	Group R	All Other	Group R	All Other	Group R
Roofs																
Insulation entirely above roof deck	U-0.048	U-0.039	U-0.039	U-0.039	U-0.039	U-0.039	U-0.032	U-0.032	U-0.032	U-0.032	U-0.032	U-0.032	U-0.028	U-0.028	U-0.028	U-0.028
Metal buildings	U-0.035	U-0.035	U-0.035	U-0.035	U-0.035	U-0.035	U-0.035	U-0.035	U-0.035	U-0.035	U-0.031	U-0.029	U-0.029	U-0.029	U-0.026	U-0.026
Attic and other	U-0.027	U-0.027	U-0.027	U-0.027	U-0.027	U-0.027	U-0.021	U-0.021	U-0.021	U-0.021	U-0.021	U-0.021	U-0.017	U-0.017	U-0.017	U-0.017
Walls, above grade																
Mass[f]	U-0.151	U-0.151	U-0.151	U-0.123	U-0.123	U-0.104	U-0.104	U-0.090	U-0.090	U-0.080	U-0.080	U-0.071	U-0.071	U-0.071	U-0.037	U-0.037
Metal building	U-0.079	U-0.079	U-0.079	U-0.079	U-0.079	U-0.052	U-0.052	U-0.050	U-0.050	U-0.050	U-0.050	U-0.050	U-0.044	U-0.039	U-0.039	U-0.039
Metal framed	U-0.077	U-0.077	U-0.077	U-0.064	U-0.064	U-0.064	U-0.064	U-0.064	U-0.055	U-0.055	U-0.049	U-0.049	U-0.049	U-0.042	U-0.037	U-0.037
Wood framed and other[c]	U-0.064	U-0.064	U-0.064	U-0.064	U-0.064	U-0.064	U-0.064	U-0.064	U-0.051	U-0.051	U-0.051	U-0.051	U-0.051	U-0.051	U-0.032	U-0.032
Walls, below grade																
Below-grade wall[c]	C-1.140[e]	C-1.140[e]	C-1.140[e]	C-1.140[e]	C-1.140[e]	C-1.140[e]	C-0.119	C-0.092	C-0.119	C-0.092	C-0.092	C-0.063	C-0.063	C-0.063	C-0.063	C-0.063
Floors																
Mass[d]	U-0.322[e]	U-0.322[e]	U-0.107	U-0.087	U-0.074	U-0.074	U-0.057	U-0.051	U-0.057	U-0.051	U-0.051	U-0.051	U-0.042	U-0.042	U-0.038	U-0.038
Joist/framing	U-0.066[e]	U-0.066[e]	U-0.033	U-0.033	U-0.033	U-0.033	U-0.033	U-0.033	U-0.033	U-0.033	U-0.027	U-0.027	U-0.027	U-0.027	U-0.027	U-0.027
Slab-on-grade floors																
Unheated slabs	F-0.73[e]	F-0.73[e]	F-0.73[e]	F-0.73[e]	F-0.73[e]	F-0.54	F-0.52	F-0.52	F-0.52	F-0.51	F-0.51	F-0.434	F-0.51	F-0.434	F-0.434	F-0.424
Heated slabs	F-0.69	F-0.69	F-0.69	F-0.69	F-0.66	F-0.66	F-0.62	F-0.62	F-0.62	F-0.62	F-0.62	F-0.602	F-0.602	F-0.602	F-0.602	F-0.602
Opaque doors																
Nonswinging door	U-0.31	U-0.31	U-0.31	U-0.31	U-0.31	U-0.31	U-0.31	U-0.31	U-0.31	U-0.31	U-0.31	U-0.31	U-0.31	U-0.31	U-0.31	U-0.31
Swinging door[g]	U-0.37	U-0.37	U-0.37	U-0.37	U-0.37	U-0.37	U-0.37	U-0.37	U-0.37	U-0.37	U-0.37	U-0.37	U-0.37	U-0.37	U-0.37	U-0.37
Garage door < 14% glazing[h]	U-0.31	U-0.31	U-0.31	U-0.31	U-0.31	U-0.31	U-0.31	U-0.31	U-0.31	U-0.31	U-0.31	U-0.31	U-0.31	U-0.31	U-0.31	U-0.31

For SI: 1 pound per square foot = 4.88 kg/m², 1 pound per cubic foot = 16 kg/m³.

a. Where assembly *U*-factors, *C*-factors and *F*-factors are established in ANSI/ASHRAE/IES 90.1 Appendix A, such opaque assemblies shall be a compliance alternative where those values meet the criteria of this table, and provided that the construction, excluding the cladding system on walls, complies with the appropriate construction details from ANSI/ASHRAE/IES 90.1 Appendix A.

b. Where *U*-factors have been established by testing in accordance with ASTM C1363, such opaque assemblies shall be a compliance alternative where those values meet the criteria of this table. The *R*-value of continuous insulation shall be permitted to be added to or subtracted from the original tested design.

c. Where heated slabs are below grade, below-grade walls shall comply with the *U*-factor requirements for above-grade mass walls.

d. "Mass floors" shall be in accordance with Section C402.1.3.4.

e. These *C*-, *F*- and *U*-factors are based on assemblies that are not required to contain insulation.

f. "Mass walls" shall be in accordance with Section C402.1.3.4.

g. Swinging door *U*-factors shall be determined in accordance with NFRC-100.

h. Garage doors having a single row of fenestration shall have an assembly *U*-factor less than or equal to 0.44 in Climate Zones 0 through 6 and less than or equal to 0.36 in Climate Zones 7 and 8, provided that the fenestration area is not less than 14 percent and not more than 25 percent of the total door area.

C402.1.2.1 Methods of determining *U*-, *C*- and *F*-factors. Where assembly *U-factors*, *C*-factors and *F*-factors and calculation procedures are established in ANSI/ASHRAE/IES 90.1 Appendix A for opaque assemblies, such opaque assemblies shall be a compliance alternative provided they meet the criteria of Table C402.1.2 and the construction, excluding cladding system on walls, complies with the applicable construction details from ANSI/ASHRAE/IES 90.1 Appendix A. Where *U-factors* have been established by testing in accordance with ASTM C1363, such opaque assemblies shall be a compliance alternative provided they meet the criteria of Table C402.1.4. The *R-value* of *continuous insulation* shall be permitted to be added to or subtracted from the original tested design. Airspaces used for assembly evaluations shall comply with Section C402.2.7.

C402.1.2.1.1 Tapered, above-deck insulation based on thickness. For tapered, above-deck roof insulation, area-weighted *U-factors* of non-uniform insulation thickness shall be determined by an *approved* method.

Exception: The area-weighted *U*-factor shall be permitted to be determined by using the inverse of the average *R*-value determined in accordance with the exception to Section C402.1.3.2.

C402.1.2.1.2 Suspended ceilings. Insulation installed on suspended ceilings having removable ceiling tiles shall not be considered part of the assembly *U-factor* of the roof-ceiling construction.

C402.1.2.1.3 Concrete masonry units, integral insulation. In determining compliance with Table C402.1.2, the *U-factor* of concrete masonry units with integral insulation shall be permitted to be used.

C402.1.2.1.4 Mass walls and floors. Compliance with required maximum *U-factors* for mass walls and mass floors in accordance with Table C402.1.2 shall be permitted for assemblies complying with Section C402.1.3.4.

C402.1.2.1.5 Area-weighted averaging of above-grade wall *U*-factors. Where *above-grade walls* include more than one assembly type or a penetration of the opaque wall area, the area-weighted *U-factor* of the *above-grade wall* is permitted to be determined by an *approved* method.

C402.1.2.1.6 Cold-formed steel assemblies. *U*-factors for *building thermal envelopes* containing cold-formed steel-framed ceilings and walls shall be permitted to be determined in accordance with AISI S250 as modified herein.

1. Where the steel-framed wall contains no *cavity insulation*, and uses *continuous insulation* to satisfy the *U-factor* maximum, the steel-framed wall member spacing is permitted to be installed at any on-center spacing.
2. Where the steel-framed wall contains framing at 24 inches (610 mm) on center with a 23 percent framing factor or framing at 16 inches (406 mm) on center with a 25 percent framing factor, the next lower framing member spacing input values shall be used when calculating using AISI S250.
3. Where the steel-framed wall contains less than 23 percent framing factors, the AISI S250 shall be used without any modifications.
4. Where the steel-framed wall contains other than standard C-shape framing members, the AISI S250 calculation option for other than standard C-shape framing is permitted to be used.

C402.1.2.1.7 Spandrel panels. *U-factors* of opaque assemblies within *fenestration* framing systems shall be determined in accordance with the default values in Table C402.1.2.1.7, ASTM C1363 or ANSI/NFRC 100.

TABLE C402.1.2.1.7—EFFECTIVE *U*-FACTORS FOR SPANDREL PANELS[a]

RATED *R*-VALUE OF INSULATION BETWEEN FRAMING MEMBERS		R-4	R-7	R-10	R-15	R-20	R-25	R-30
Frame Type	Spandrel Panel	Default *U*-Factor						
Aluminum without thermal break[b]	Single glass pane, stone, or metal panel	0.285	0.259	0.247	0.236	0.230	0.226	0.224
	Double glazing with no low-e coatings	0.273	0.254	0.244	0.234	0.229	0.226	0.223
	Triple glazing or double glazing with low-e glass	0.263	0.249	0.241	0.233	0.228	0.225	0.223
Aluminum with thermal break[c]	Single glass pane, stone, or metal panel	0.243	0.212	0.197	0.184	0.176	0.172	0.169
	Double glazing with no low-e coatings	0.228	0.205	0.193	0.182	0.175	0.171	0.168
	Triple glazing or double glazing with low-e glass	0.217	0.199	0.189	0.180	0.174	0.170	0.167
Structural glazing[d]	Single glass pane, stone, or metal panel	0.217	0.180	0.161	0.145	0.136	0.130	0.126
	Double glazing with no low-e coatings	0.199	0.172	0.157	0.143	0.135	0.129	0.126
	Triple glazing or double glazing with low-e glass	0.186	0.165	0.152	0.140	0.133	0.128	0.125
No framing or insulation is continuous[e]	Single glass pane, stone, or metal panel	0.160	0.108	0.082	0.058	0.045	0.037	0.031
	Double glazing with no low-e coatings	0.147	0.102	0.078	0.056	0.044	0.036	0.030
	Triple glazing or double glazing with low-e glass	0.139	0.098	0.076	0.055	0.043	0.035	0.030

a. Extrapolation outside of the table shall not be permitted. Assemblies with distance between framing less than 30 inches, or not included in the default table, shall have a *U*-factor determined by testing in compliance with ASTM C1363 or modeling in compliance with ANSI/NFRC 100. Spandrel panel assemblies in the table do not include metal backpans. For designs with metal backpans, multiply the *U*-factor by 1.2.
b. This frame type shall be used for systems that do not contain a nonmetallic element separating the metal exposed to the exterior from the metal exposed to the interior condition.
c. This frame type shall be used for systems where a nonmetallic element separates the metal exposed to the exterior from the metal that is exposed to the interior condition.
d. This frame type shall be used for systems that have no exposed mullion on the exterior.
e. This frame type shall be used for systems where there is no framing or the insulation is continuous and uninterrupted between framing.

C402.1.2.1.8 Mechanical equipment penetrations. Where the total area of through penetrations of mechanical equipment is greater than 1 percent of the opaque above-grade wall area, such area shall be calculated as a separate wall assembly, in accordance with either Section C402.1.2.1.5 or Section C402.1.4 using a published and *approved U-factor* for that equipment or a default *U-factor* of 0.5.

C402.1.3 Insulation component *R*-value method. For opaque portions of the *building thermal envelope* the *R*-values for *cavity insulation* and *continuous insulation* shall be not less than that specified in Table C402.1.3. *Group R* occupancy buildings or portions of *commercial buildings* enclosing *Group R* occupancies shall use the *R*-values from the "*Group R*" column of Table C402.1.3. *Commercial buildings* or portions of *commercial buildings* enclosing occupancies other than *Group R* shall use the *R*-values from the "All other" column of Table C402.1.3.

TABLE C402.1.3—OPAQUE BUILDING THERMAL ENVELOPE INSULATION COMPONENT MINIMUM REQUIREMENTS, *R*-VALUE METHOD[a]

CLIMATE ZONE	0 AND 1		2		3		4 EXCEPT MARINE		5 AND MARINE 4		6		7		8	
	All Other	Group R	All Other	Group R	All Other	Group R	All Other	Group R	All Other	Group R	All Other	Group R	All Other	Group R	All Other	Group R
Roofs																
Insulation entirely above roof deck	R-20ci	R-25ci	R-25ci	R-25ci	R-25ci	R-25ci	R-30ci	R-30ci	R-30ci	R-30ci	R-30ci	R-30ci	R-35ci	R-35ci	R-35ci	R-35ci
Metal buildings[b]	R-19 + R-11 LS	R-19 + R-11 LS	R-19 + R11 LS	R-19 + R-11 LS	R-19 + R-11 LS	R-19 + R-11 LS	R-19 + R-11 LS	R-19 + R-11 LS	R-19 + R-11 LS	R-19 + R-11 LS	R-25 + R-11 LS	R-30 + R-11 LS	R-30 + R-11 LS	R-30 + R-11 LS	R-25 + R-11 + R-11 LS	R-25 + R-11 + R-11 LS
Attic and other	R-38	R-38	R-38	R-38	R-38	R-38	R-49	R-49	R-49	R-49	R-49	R-49	R-60	R-60	R-60	R-60
Walls, above grade																
Mass[f]	R-5.7ci[c]	R-5.7ci[c]	R-5.7ci[c]	R-7.6ci	R-7.6ci	R-9.5ci	R-9.5ci	R-11.4ci	R-11.4ci	R-13.3ci	R-13.3ci	R-15.2ci	R-15.2ci	R-15.2ci	R-25ci	R-25ci
Metal building	R-13 + R-6.5ci	R-13 + R-6.5ci	R13 + R-6.5ci	R-13 + R-13ci	R-13 + R-6.5ci	R-13 + R-13ci	R-13 + R-13ci	R-13 + R-14ci	R-13 + R-14ci	R-13 + R-14ci	R-13 + R-14ci	R-13 + R-14ci	R-13 + R-17ci	R-13 + R-19.5ci	R-13 + R-19.5ci	R-13 + R-19.5ci
Metal framed[h, i]	R-0 + R-10ci or R-13 + R-5ci or R-20 + R-3.8ci	R-0 + R-10ci or R-13 + R-5ci or R-20 + R-3.8ci	R-0 + R-10ci or R-13 + R-5ci or R-20 + R-3.8ci	R-0 + R-12.6ci or R-13 + R-7.5ci or R-20 + R-6.3ci	R-0 + R-12.6ci or R-13 + R-7.5ci or R-20 + R-6.3ci	R-0 + R-12.6ci or R-13 + R-7.5ci or R-20 + R-6.3ci	R-0 + R-12.6ci or R-13 + R-7.5ci or R-20 + R-6.3ci	R-0 + R-12.6ci or R-13 + R-7.5ci or R-20 + R-6.3ci	R-0 + R-15.2ci or R-13 + R-10ci or R-20 + R-9ci	R-0 + R-15.2ci or R-13 + R-10ci or R-20 + R-9ci	R-0 + R-17.3ci or R-13 + R-12.5ci or R-20 + R-11ci	R-0 + R-17.3ci or R-13 + R-12.5ci or R-20 + R-11ci	R-0 + R-17.3ci or R-13 + R-12.5ci or R-20 + R-11ci	R-0 + R-21ci or R-13 + R-15.6ci or R-20 + R-14.3ci	R-0 + R-24ci or R-13 + R-18.8ci or R-20 + R-17.5ci	R-0 + R-24ci or R-13 + R-18.8ci or R-20 + R-17.5ci
Wood framed and other[h, i]	R-0 + R-12ci or R-13 + R-3.8ci or R-20	R-0 + R-12ci or R-13 + R-3.8ci or R-20	R-0 + R-12ci or R-13 + R-3.8ci or R-20	R-0 + R-12ci or R-13 + R-3.8ci or R-20	R-0 + R-12ci or R-13 + R-3.8ci or R-20	R-0 + R-12ci or R-13 + R-3.8ci or R-20	R-0 + R-12ci or R-13 + R-3.8ci or R-20	R-0 + R-12ci or R-13 + R-3.8ci or R-20	R-0 + R-16ci or R-13 + R-7.5ci or R20 + R3.8ci or R-27	R-0 + R-16ci or R-13 + R-7.5ci or R-20 + R-3.8ci or R-27	R-0 + R-16ci or R-13 + R-7.5ci or R-20 + R-3.8ci or R-27	R-0 + R-16ci or R-13 + R-7.5ci or R-20 + R-3.8ci or R-27	R-0 + R-16ci or R-13 + R-7.5ci or R-20 + R-3.8ci or R-27	R-0 + R-16ci or R-13 + R-7.5ci or R-20 + R-3.8ci or R-27	R-0 + R-27.5ci or R-13 + R-18.8ci or R-20 + R-14ci	R-0 + R-27.5ci or R-13 + R-18.8ci or R-20 + R-14ci
Walls, below grade																
Below-grade wall[d]	NR	NR	NR	NR	NR	NR	R-7.5ci	R-10ci	R-7.5ci	R-10ci	R-10ci	R-15ci	R-15ci	R-15ci	R-15ci	R-15ci
Floors																
Mass[e]	NR	NR	R-6.3ci	R-8.3ci	R-10ci	R-10ci	R-14.6ci	R-16.7ci	R-14.6ci	R-16.7ci	R-16.7ci	R-16.7ci	R-20.9ci	R-20.9ci	R-23ci	R-23ci
Joist/framing	R-13	R-13	R-30	R-30	R-30	R-30	R-30	R-30	R-30	R-30	R-38	R-38	R-38	R-38	R-38	R-38
Slab-on-grade floors																
Unheated slabs	NR	NR	NR	NR	NR	R-10 for 24" below	R-15 for 24" below	R-15 for 24" below	R-15 for 24" below	R-20 for 24" below	R-20 for 24" below	R-20 for 48" below	R-20 for 24" below	R-20 for 48" below	R-20 for 48" below	R-25 for 48" below
Heated slabs[g]	R-7.5 for 12" below + R-5 full slab	R-7.5 for 12" below + R-5 full slab	R-7.5 for 12" below + R-5 full slab	R-7.5 for 12" below + R-5 full slab	R-10 for 24" below + R-5 full slab	R-10 for 24" below + R-5 full slab	R-15 for 24" below + R-5 full slab	R-15 for 24" below + R-5 full slab	R-15 for 36" below + R-5 full slab	R-15 for 36" below + R-5 full slab	R-15 for 36" below + R-5 full slab	R-20 for 48" below + R-5 full slab	R-20 for 48" below + R-5 full slab	R-20 for 48" below + R-5 full slab	R-20 for 48" below + R-5 full slab	R-20 for 48" below + R-5 full slab

TABLE C402.1.3—OPAQUE THERMAL ENVELOPE INSULATION COMPONENT MINIMUM REQUIREMENTS, *R*-VALUE METHOD[a]—continued

For SI: 1 inch = 25.4 mm, 1 pound per square foot = 4.88 kg/m^2, 1 pound per cubic foot = 16 kg/m^3.

ci = Continuous Insulation, NR = No Requirement, LS = Liner System.

a. Assembly descriptions can be found in ANSI/ASHRAE/IES 90.1 Appendix A.

b. Where using *R*-value compliance method, a thermal spacer block shall be provided, otherwise use the *U*-factor compliance method in Table C402.1.2.

c. R-5.7ci is allowed to be substituted with concrete block walls complying with ASTM C90, ungrouted or partially grouted not less than 32 inches on center vertically and not less than 48 inches on center horizontally, with ungrouted cores filled with materials having a maximum thermal conductivity of 0.44 Btu-in/h-ft^2°F.

d. Where heated slabs are below grade, below-grade walls shall comply with the *R*-value requirements for above-grade mass walls.

e. "Mass floors" shall be in accordance with Section C402.1.3.4.

f. "Mass walls" shall be in accordance with Section C402.1.3.4.

g. The first value is for perimeter insulation and the second value is for full, under-slab insulation. Perimeter insulation and full-slab insulation components shall be installed in accordance with Section C402.2.4.

h. The first value is cavity insulation; the second value is continuous insulation. Therefore, "R-0 + R-12ci" means R-12 continuous insulation and no cavity insulation; "R-13 + R-3.8ci" means R-13 cavity insulation and R-3.8 continuous insulation; "R-20" means R-20 cavity insulation and no continuous insulation. R-13, R-20 and R-27 cavity insulation, as used in this table, apply to a nominal 4-inch, 6-inch and 8-inch-deep wood or cold-formed steel stud cavities, respectively.

i. Where the required *R*-value in Table C402.1.3 is met by using continuous insulation such that cavity insulation is not required, the *R*-value is applicable to any wall framing spacing.

C402.1.3.1 *R*-value of multi-layered insulation components. Where *cavity insulation* is installed in multiple layers, the *cavity insulation R*-values shall be summed to determine compliance with the *cavity insulation R-value* requirements. Where *continuous insulation* is installed in multiple layers, the *continuous insulation R*-values shall be summed to determine compliance with the *continuous insulation R-value* requirements. *Cavity insulation R*-values shall not be used to determine compliance with the *continuous insulation R-value* requirements in Table C402.1.3.

C402.1.3.2 Area-weighted averaging of *R*-values. Area-weighted averaging shall not be permitted for *R-value* compliance.

Exception: For tapered above-deck roof insulation, compliance with the *R*-values required in Table C402.1.3 shall be permitted to be demonstrated by multiplying the rated *R-value* per inch of the insulation material by the average thickness of the roof insulation. The average thickness of the roof insulation shall equal the total volume of the roof insulation divided by the area of the roof.

C402.1.3.3 Suspended ceilings. Insulation installed on suspended ceilings having removable ceiling tiles shall not be considered part of the minimum thermal resistance (*R-value*) of roof insulation in roof-ceiling construction.

C402.1.3.4 Mass walls and mass floors. Compliance with required maximum *U-factors* for mass walls and mass floors in accordance with Table C402.1.2 and minimum *R*-values for insulation components applied to mass walls and mass floors in accordance with Table C402.1.3 shall be permitted for assemblies complying with the following:

1. Where used as a component of the *building thermal envelope*, mass walls shall comply with one of the following:
   1.1. Weigh not less that 35 pounds per square foot (171 kg/m^2) of wall surface area.
   1.2. Weigh not less than 25 pounds per square foot (122 kg/m^2) of wall surface area where the material weight is not more than 120 pounds per cubic foot (pcf) (1922 kg/m^3).
   1.3. Have a heat capacity exceeding 7 Btu/ft^2 × °F (144 kJ/m^2 × K).
   1.4. Have a heat capacity exceeding 5 Btu/ft^2 × °F (103 kJ/m^2 × K) where the material weight is not more than 120 pcf (1922 kg/m^3).
2. Where used as a component of the *building thermal envelope*, the minimum weight of mass floors shall comply with one of the following:
   2.1. Thirty-five pounds per square foot (171 kg/m^2) of floor surface area.
   2.2. Twenty-five pounds per square foot (122 kg/m^2) of floor surface area where the material weight is not more than 120 pcf (1922 kg/m^3).

C402.1.4 Component performance method. *Building thermal envelope* values and *fenestration* areas determined in accordance with Equation 4-1 shall be an alternative to compliance with the *U*-, *F*-, psi-, chi-, and *C*-factors in Tables C402.1.2, C402.1.2.1.7, C402.1.4 and C402.5 and the maximum allowable *fenestration* areas in Section C402.5.1. *Fenestration* shall meet the applicable SHGC requirements of Section C402.5.3.

Equation 4-1 $A_P + B_P + C_P + T_P \leq A_T + B_T + C_T + T_T - V_F - V_S$

where:

A_P = Sum of the (area × *U*-factor) for each proposed building thermal envelope assembly, other than slab-on-grade or below-grade wall assemblies.

B_P = Sum of the (length × *F*-factor) for each proposed slab-on-grade edge condition.

C_P = Sum of the (area × *C*-factor) for each proposed below-grade wall assembly.

T_P = Sum of the (ψLP) and (χNP) values for each type of thermal bridge condition of the building thermal envelope as identified in Section C402.7 in the proposed building. For the purposes of this section, the (ψLP) and (χNP) values for thermal bridges caused by materials with a thermal conductivity less than or equal to 3.0 Btu × in/h × ft^2 × °F shall be assigned as zero. For buildings or structures located in Climate Zones 0 through 3, the value of T_P shall be assigned as zero.

ψLP = Psi-factor × length of the thermal bridge elements in the proposed building thermal envelope.

χNP = Chi-factor × number of the thermal bridge point elements other than fasteners, ties or brackets in the proposed building thermal envelope.

A_T = Sum of the (area × *U*-factor permitted by Tables C402.1.2 and C402.5) for each proposed building thermal envelope assembly, other than slab-on-grade or below-grade wall assemblies.

B_T = Sum of the (length × *F*-factor permitted by Table C402.1.2) for each proposed slab-on-grade edge condition.

C_T = Sum of the (area × *C*-factor permitted by Table C402.1.2) for each proposed below-grade wall assembly.

T_T = Sum of the (ψLT) and (χNT) values for each type of thermal bridge condition in the proposed building thermal envelope as identified in Section C402.7 with values specified as "compliant" in Table C402.1.4. For the purposes of this section, the (ψLT) and (χNT) values for thermal bridges caused by materials with a thermal conductivity less than or equal to 3.0 Btu × in/h × ft^2 × °F shall be assigned as zero. For buildings or structures located in Climate Zones 0 through 3, the value of T_T shall be assigned as zero.

ψLT = (Psi-factor specified as "compliant" in Table C402.1.4) × length of the thermal bridge elements in the proposed building thermal envelope.

χNT = (Chi-factor specified as "compliant" in Table C402.1.4) × number of the thermal bridge point elements other than fasteners, ties or brackets in the proposed building thermal envelope.

P_F = Maximum vertical fenestration area allowable by Section C402.5.1, C402.5.1.1 or C402.5.1.2.

Q_F = Proposed vertical fenestration area.

$R_F = Q_F - P_F$, but not less than zero (excess vertical fenestration area).

S_F = Area-weighted average *U*-factor permitted by Table C402.5 of all vertical fenestration assemblies.

T_F = Area-weighted average *U*-factor permitted by Table C402.1.2 of all exterior opaque wall assemblies.

$U_F = S_F - T_F$ (excess *U*-factor for excess vertical fenestration area).

$V_F = R_F \times U_F$ (excess $U \times A$ due to excess vertical fenestration area).

P_S = Maximum skylight area allowable by Section C402.1.2.

Q_S = Actual skylight area.

$R_S = Q_S - P_S$, but not less than zero (excess skylight area).

S_S = Area-weighted average *U*-factor permitted by Table C402.5 of all skylights.

T_S = Area-weighted average *U*-factor permitted by Table C402.1.2 of all opaque roof assemblies.

$U_S = S_S - T_S$ (excess *U*-factor for excess skylight area).

$V_S = R_S \times U_S$ (excess $U \times A$ due to excess skylight area).

A proposed psi- or *chi-factor* for each thermal bridge shall comply with one of the following, as applicable:

1. Where the proposed mitigation of a thermal bridge is compliant with the requirements of Section C402.7, the "compliant" values in Table C402.1.4 shall be used for the proposed psi- or chi-factors.
2. Where a thermal bridge is not mitigated in a manner at least equivalent to Section C402.7, the "noncompliant" values in Table C402.1.4 shall be used for the proposed psi- or chi-factors.
3. Where the proposed mitigation of a thermal bridge provides a psi- or chi-factor less than the "compliant" values in Table C402.1.4, the proposed psi- or chi-factor shall be determined by thermal analysis, testing or other approved sources.

TABLE C402.1.4—PSI- and CHI-FACTORS TO DETERMINE THERMAL BRIDGES FOR THE COMPONENT PERFORMANCE METHOD

THERMAL BRIDGE PER SECTION C402.7	THERMAL BRIDGE COMPLIANT WITH SECTION C402.7		THERMAL BRIDGE NONCOMPLIANT WITH SECTION C402.7	
	Psi-Factor (Btu/h × ft × °F)	Chi-Factor (Btu/h × °F)	Psi-Factor (Btu/h × ft × °F)	Chi-Factor (Btu/h × °F)
C402.7.1 Balconies and floor decks	0.2	N/A	0.5	N/A
C402.7.2 Cladding supports	0.2	N/A	0.3	N/A
C402.7.3 Structural beams and columns	N/A	1.0 carbon steel 0.3 concrete	N/A	2.0 carbon steel 1.0 concrete
C402.7.4 Vertical fenestration	0.15	N/A	0.3	N/A
C402.7.5 Parapets	0.2	N/A	0.4	N/A

For SI: 1 W/m × K = 0.578 Btu/h × ft × °F, 1 W/K = 1.9 Btu/h × °F.
N/A = Not Applicable.

C402.1.5 Rooms containing fuel-burning appliances. In Climate Zones 3 through 8, where combustion air is supplied through openings in an *exterior wall* to a room or space containing a space-conditioning fuel-burning appliance, one of the following shall apply:

1. The room or space containing the appliance shall be located outside of the *building thermal envelope*.
2. The room or space containing the appliance shall be enclosed and isolated from *conditioned spaces* inside the *building thermal envelope*. Such rooms shall comply with all of the following:
   - 2.1. The walls, floors and ceilings that separate the enclosed room or space from *conditioned spaces* shall be insulated to be not less than equivalent to the insulation requirement of *below-grade walls* as specified in Table C402.1.3 or Table C402.1.2.
   - 2.2. The walls, floors and ceilings that separate the enclosed room or space from *conditioned spaces* shall be sealed in accordance with Section C402.6.1.2.
   - 2.3. The doors into the enclosed room or space shall be fully gasketed.
   - 2.4. Piping serving as part of a heating or cooling system and *ducts* in the enclosed room or space shall be insulated in accordance with Section C403. Service water piping shall be insulated in accordance with Section C404.
   - 2.5. Where an air *duct* supplying combustion air to the enclosed room or space passes through *conditioned space*, the *duct* shall be insulated to an *R-value* of not less than R-8.

Exception: Fireplaces and stoves complying with Sections 901 through 905 of the *International Mechanical Code*, and Section 2111.14 of the *International Building Code*.

C402.2 Specific insulation and installation requirements. Insulation in *building thermal envelope* opaque assemblies shall be installed in accordance with Section C303.2 and Sections C402.2.1 through C402.2.7, or an *approved* design.

C402.2.1 Roof-ceiling construction. Insulation materials in the roof-ceiling construction shall be installed between the roof or ceiling framing, continuously below the ceiling framing, continuously above, below, or within the roof deck or in any *approved* combination thereof. Insulation installed above the roof deck shall comply with Sections C402.2.1.1 through C402.2.1.3.

C402.2.1.1 Joints staggered. Continuous, above-deck insulation board located above the roof deck shall be installed in not less than two layers and the edge joints between each layer of insulation shall be staggered, except where insulation tapers to the roof deck at a gutter edge, roof drain or scupper.

C402.2.1.2 Skylight curbs. Skylight curbs shall be insulated to the level of the above-deck roof insulation or R-5, whichever is less.

Exception: Unit skylight curbs included as a component of a skylight *listed* and *labeled* in accordance with NFRC 100 shall not be required to be insulated.

C402.2.1.3 Minimum thickness of tapered insulation. The thickness of tapered above-deck roof insulation at its lowest point, gutter edge, roof drain or scupper, shall be not less than 1 inch (25 mm).

C402.2.2 Above-grade walls. *Above-grade wall* insulation materials shall be installed between the wall framing, be integral to the wall assembly, be continuous on the wall assembly, or be any combination of these insulation methods. Where *continuous insulation* is layered on the exterior side of a wall assembly, the joints shall be staggered.

C402.2.3 Floors over outdoor air or unconditioned space. Floor insulation shall be installed between floor framing, be integral to the floor assembly, be continuous on the floor assembly, or be any combination of these insulation methods. Where *continuous insulation* is layered on the exterior side of a floor assembly, the joints shall be staggered. Floor framing *cavity insulation* or structural slab insulation shall be installed to maintain permanent contact with the underside of the subfloor decking or structural slabs.

Exceptions:

1. The floor framing *cavity insulation* or structural slab insulation shall be permitted to be installed in contact with the top side of sheathing or *continuous insulation* installed on the bottom side of floor assemblies. Floor framing or structural slab members at the perimeter of the floor assembly shall be insulated vertically for their full depth with insulation equivalent to that required for the *above-grade wall* construction.
2. Insulation applied to the underside of concrete floor slabs shall be permitted an airspace of not more than 1 inch (25 mm) where it turns up and is in contact with the underside of the floor under walls associated with the *building thermal envelope*.

C402.2.4 Slabs-on-grade. Where installed, the perimeter insulation for slab-on-grade shall be placed on the outside of the foundation or on the inside of the foundation wall. For installations complying with Table C402.1.3, the perimeter insulation shall extend downward from the top of the slab for the minimum distance shown in the table or to the top of the footing, whichever is less, or downward to not less than the bottom of the slab and then horizontally to the interior or exterior for the total distance shown in the table. Where installed, full slab insulation shall be continuous under the entire area of the slab-on-grade floor, except at structural column locations and service penetrations. Insulation required at the *heated slab* perimeter shall not be required to extend below the bottom of the heated slab and shall be continuous with the full slab insulation.

Exception: Where the slab-on-grade floor is greater than 24 inches (610 mm) below the finished exterior grade, perimeter insulation is not required.

C402.2.5 Below-grade walls. Below-grade wall insulation shall be installed between framing members, be integral to the wall assembly, be continuous on the wall assembly, or be any combination of these insulation methods. For installations complying with Section C401.2.1, insulation shall extend to a depth of not less than 10 feet (3048 mm) below the outside finished ground level or to the level of the lowest floor of the *conditioned space* enclosed by the *below-grade wall*, whichever is less.

C402.2.6 Insulation of radiant heating system panels. *Radiant heating system* panels, and their associated components that are installed in interior or exterior assemblies, shall be insulated to an *R*-value of not less than R-3.5 on all surfaces not facing the space being heated. *Radiant heating system* panels that are installed in the *building thermal envelope* shall be separated from the exterior of the building or unconditioned or exempt spaces by not less than the *R*-value of insulation installed in the opaque assembly in which they are installed or the assembly shall comply with Section C402.1.2.

C402.2.7 Airspaces. Where the *R-value* of an airspace is used for compliance in accordance with Section C402.1, the airspace shall be enclosed in a cavity bounded on all sides by building components and constructed to minimize airflow into and out of the enclosed airspace. Airflow shall be deemed minimized where one of the following conditions occur:

1. The enclosed airspace is unventilated.
2. The enclosed airspace is bounded on at least one side by an anchored masonry veneer, constructed in accordance with Chapter 14 of the *International Building Code* and vented by veneer weep holes located only at the bottom of the airspace and spaced not less than 15 inches (381 mm) on center with top of the cavity airspace closed.

Exception: For ventilated cavities, the effect of the *ventilation* of airspaces located on the exterior side of the continuous *air barrier* and adjacent to and behind the exterior wall-covering material shall be determined in accordance with ASTM C1363 modified with an airflow entering the bottom and exiting the top of the airspace at an air movement rate of not less than 70 mm/second.

C402.3 Above-grade wall solar reflectance. For Climate Zone 0, above-grade *east-oriented*, *south-oriented* and *west-oriented* walls shall comply with either of the following:

1. Not less than 75 percent of the opaque *above-grade wall* area shall have an area-weighted initial solar reflectance of not less than 0.30 where tested in accordance with ASTM C1549 with AM1.5GV output or ASTM E903 with AM1.5GV output, or determined in accordance with an *approved source*. This *above-grade wall* area shall have an *emittance* or emissivity of not less than 0.75 where tested in accordance with ASTM C835, ASTM C1371, ASTM E408 or determined in accordance with an *approved source*. For the portion of the *above-grade wall* that is glass spandrel area, a solar reflectance of not less than 0.29, as determined in accordance with NFRC 300 or ISO 9050, shall be permitted. Area-weighted averaging is permitted using only *south-*, *east-* and *west-oriented* walls enclosing the same occupancy classification.
2. Not less than 30 percent of the opaque *above-grade wall* area shall be shaded by manmade structures, existing buildings, hillsides, permanent building projections, *on-site renewable energy* systems or a combination of these. Shade coverage shall be calculated by projecting the shading surface downward on the *above-grade wall* at an angle of 45 degrees.

Exception: *Above-grade walls* of low-energy buildings complying with Section C402.1.1.1, greenhouses complying with Section C402.1.1.2 and equipment buildings complying with Section C402.1.1.3.

C402.4 Roof solar reflectance and thermal emittance. *Low slope* roofs directly above cooled *conditioned spaces* in Climate Zones 0 through 3 shall comply with one or more of the options in Table C402.4.

Exceptions: The following roofs and portions of roofs are exempt from the requirements of Table C402.4:

1. Portions of the roof that include or are covered by the following:
   - 1.1. Photovoltaic systems or components.
   - 1.2. Solar air or water-heating systems or components.
   - 1.3. *Vegetative roofs* or landscaped roofs.
   - 1.4. Above-roof decks or walkways.
   - 1.5. Skylights.
   - 1.6. HVAC systems and components, and other opaque objects mounted above the roof.
2. Portions of the roof shaded during the peak sun angle on the summer solstice by permanent features of the *building* or by permanent features of adjacent buildings.
3. Portions of roofs that are ballasted with a minimum stone ballast of 17 pounds per square foot (74 kg/m^2) or 23 psf (117 kg/m^2) pavers.
4. Roofs where not less than 75 percent of the roof area complies with one or more of the exceptions to this section.

TABLE C402.4—MINIMUM ROOF REFLECTANCE AND EMITTANCE OPTIONS[a]
Three-year-aged solar reflectance[b] of 0.55 and 3-year aged thermal emittance[c] of 0.75
Three-year-aged solar reflectance index[d] of 64

a. The use of area-weighted averages to comply with these requirements shall be permitted. Materials lacking 3-year-aged tested values for either solar reflectance or thermal emittance shall be assigned both a 3-year-aged solar reflectance in accordance with Section C402.4.1 and a 3-year-aged thermal emittance of 0.90.
b. Aged solar reflectance tested in accordance with ASTM C1549, ASTM E903 or ASTM E1918 or CRRC-S100.
c. Aged thermal emittance tested in accordance with ASTM C1371 or ASTM E408 or CRRC-S100.
d. Solar reflectance index (SRI) shall be determined in accordance with ASTM E1980 using a convection coefficient of 2.1 Btu/h × ft^2 × °F (12 W/m^2 × K). Calculation of aged SRI shall be based on aged tested values of solar reflectance and thermal emittance.

C402.4.1 Aged roof solar reflectance. Where an aged solar reflectance required by Section C402.4 is not available, it shall be determined in accordance with Equation 4-2.

Equation 4-2 $R_{aged} = [0.2 + 0.7(R_{initial} - 0.2)]$

where:

R_{aged} = The aged solar reflectance.

$R_{initial}$ = The initial solar reflectance determined in accordance with CRRC-S100.

C402.5 Fenestration. *Fenestration* shall comply with Sections C402.5.1 through C402.5.5 and Table C402.5. *Daylight responsive controls* shall comply with this section and Section C405.2.4.

TABLE C402.5—BUILDING THERMAL ENVELOPE FENESTRATION MAXIMUM *U*-FACTOR AND SHGC REQUIREMENTS

CLIMATE ZONE	0 AND 1		2		3		4 EXCEPT MARINE		5 AND MARINE 4		6		7		8	
Vertical fenestration																
***U*-factor**																
Fixed fenestration	0.50		0.45		0.38		0.34		0.34		0.34		0.28		0.25	
Operable fenestration	0.62		0.60		0.54		0.45		0.45		0.42		0.36		0.32	
Entrance doors	0.83		0.77		0.68		0.63		0.63		0.63		0.63		0.63	
SHGC																
	Fixed	Operable	Fixed	Operable	Fixed	Operable	Fixed	Operable	Fixed	Operable	Fixed	Operable	Fixed	Operable	Fixed	Operable
PF < 0.2	0.23	0.21	0.25	0.23	0.25	0.23	0.36	0.33	0.38	0.33	0.38	0.34	0.40	0.36	0.40	0.36
0.2 ≤ PF < 0.5	0.28	0.25	0.30	0.28	0.30	0.28	0.43	0.40	0.46	0.40	0.46	0.41	0.48	0.43	0.48	0.43
PF ≥ 0.5	0.37	0.34	0.40	0.37	0.40	0.37	0.58	0.53	0.61	0.53	0.61	0.54	0.64	0.58	0.64	0.58
Skylights																
U-factor	0.70		0.65		0.55		0.50		0.50		0.50		0.44		0.41	
SHGC	0.30		0.30		0.30		0.40		0.40		0.40		NR		NR	

NR = No Requirement, PF = Projection Factor.

C402.5.1 Maximum area. The vertical *fenestration* area, not including *opaque doors* and opaque spandrel panels, shall be not greater than 30 percent of the gross *above-grade wall* area. The skylight area shall be not greater than 3 percent of the gross roof area.

C402.5.1.1 Increased vertical fenestration area with daylight responsive controls. In Climate Zones 0 through 6, not more than 40 percent of the gross *above-grade wall* area shall be vertical *fenestration,* provided that all of the following requirements are met:

1. In buildings not greater than two stories above grade, not less than 50 percent of the *net floor area* is within a primary sidelit *daylight zone* or a toplit *daylight zone*.
2. In buildings three or more stories above grade, not less than 25 percent of the *net floor area* is within a primary sidelit *daylight zone* or a toplit *daylight zone.*
3. *Daylight responsive controls* are installed in *daylight zones.*
4. *Visible transmittance* (VT) of vertical *fenestration* is not less than 1.1 times *solar heat gain coefficient* (SHGC).

   Exception: *Fenestration* that is outside the scope of NFRC 200 is not required to comply with Item 4.

C402.5.1.2 Increased skylight area with daylight responsive controls. The skylight area shall be not more than 6 percent of the roof area provided that *daylight responsive controls* are installed in toplit *daylight zones.*

C402.5.2 Minimum skylight fenestration area. Skylights shall be provided in *enclosed spaces* greater than 2,500 square feet (232 m^2) in floor area, directly under a roof with not less than 75 percent of the ceiling area with a ceiling height greater than 15 feet (4572 mm), and used as an office, lobby, atrium, concourse, corridor, storage space, gymnasium/exercise center, convention center, automotive service area, space where manufacturing occurs, nonrefrigerated warehouse, retail store, distribution/sorting area, transportation depot or workshop. The total *toplit daylight zone* shall be not less than half the floor area and shall comply with one of the following:

1. A minimum skylight area to toplit *daylight zone* of not less than 3 percent where all skylights have a VT of not less than 0.40, or VT_{annual} of not less than 0.26, as determined in accordance with Section C303.1.3.
2. A minimum skylight effective aperture, determined in accordance with Equation 4-3, of:
   - 2.1. Not less than 1 percent using a skylight's VT rating; or
   - 2.2. Not less than 0.66 percent using a Tubular Daylight Device's VT_{annual} rating.

Equation 4-3

$$\text{Skylight Effective Aperture} = \frac{0.85 \times \text{Skylight Area} \times \text{Skylight VT} \times \text{WF}}{\text{Toplit Zone}}$$

where:

Skylight area = Total fenestration area of skylights.

Skylight VT = Area-weighted average visible transmittance of skylights.

WF = Area-weighted average well factor, where well factor is 0.9 if light well depth is less than 2 feet (610 mm), or 0.7 if light well depth is 2 feet (610 mm) or greater, or 1.0 for Tubular Daylighting Devices with VT_{annual} ratings.

Light well depth = Measure vertically from the underside of the lowest point of the skylight glazing to the ceiling plane under the skylight.

Exceptions: Skylights above *daylight zones* of *enclosed spaces* are not required in:

1. Buildings in *Climate Zones* 6 through 8.
2. Spaces where the designed *general lighting* power densities are less than 0.5 W/ft^2 (5.4 W/m^2).
3. Areas where it is documented that existing structures or natural objects block direct beam sunlight on not less than half of the roof over the enclosed area for more than 1,500 daytime hours per year between 8 a.m. and 4 p.m.
4. Spaces where the *daylight zone* under rooftop monitors is greater than 50 percent of the *enclosed space* floor area.
5. Spaces where the total area minus the area of *sidelit daylight zones* is less than 2,500 square feet (232 m^2), and where the lighting is controlled in accordance with Section C405.2.3.
6. Spaces designed as storm shelters complying with ICC 500.

C402.5.2.1 Lighting controls in toplit daylight zones. *Daylight responsive controls* shall be provided in toplit *daylight zones*.

C402.5.2.2 Haze factor. Skylights in office, storage, automotive service, manufacturing, nonrefrigerated warehouse, retail store and distribution/sorting area spaces shall have a glazing material or diffuser with a haze factor greater than 90 percent when tested in accordance with ASTM D1003.

Exception: Skylights and tubular daylighting devices designed and installed to exclude direct sunlight entering the occupied space by the use of fixed or automated baffles, the geometry of skylight and light well or the use of optical diffuser components.

C402.5.3 Maximum *U*-factor and SHGC. The maximum *U-factor* and *solar heat gain coefficient* (SHGC) for *fenestration* shall be as specified in Table C402.5.

The window projection factor shall be determined in accordance with Equation 4-4.

Equation 4-4 $PF = A/B$

where:

PF = Projection factor (decimal).

A = Distance measured horizontally from the farthest continuous extremity of any overhang, eave or permanently attached shading device to the vertical surface of the glazing.

B = Distance measured vertically from the bottom of the glazing to the underside of the overhang, eave or permanently attached shading device.

Where different windows or glass doors have different *PF* values, they shall each be evaluated separately.

C402.5.3.1 Increased skylight SHGC. In Climate Zones 0 through 6, skylights shall be permitted a maximum SHGC of 0.60 where located above *daylight zones* provided with *daylight responsive controls*.

C402.5.3.2 Increased skylight *U*-factor. Where skylights are installed above *daylight zones* provided with *daylight responsive controls*, a maximum *U-factor* of 0.9 shall be permitted in Climate Zones 0 through 3 and a maximum *U-factor* of 0.75 shall be permitted in *Climate Zones* 4 through 8.

C402.5.3.3 Dynamic glazing. Where *dynamic glazing* is intended to satisfy the SHGC and VT requirements of Table C402.5, the ratio of the higher to lower *labeled* SHGC shall be greater than or equal to 2.4, and the *dynamic glazing* shall be automatically controlled to modulate the amount of solar gain into the space in multiple steps. *Dynamic glazing* shall be considered separately from other *fenestration*, and area-weighted averaging with other *fenestration* that is not *dynamic glazing* shall not be permitted.

Exception: *Dynamic glazing* is not required to comply with this section where both the lower and higher *labeled* SHGC already comply with the requirements of Table C402.5.

C402.5.3.4 Area-weighted *U*-factor. An area-weighted average shall be permitted to satisfy the *U-factor* requirements for each *fenestration* product category listed in Table C402.5. Individual *fenestration* products from different *fenestration* product categories listed in Table C402.5 shall not be combined in calculating area-weighted average *U-factor*.

C402.5.4 Daylight zones. *Daylight zones* referenced in Sections C402.5.1.1 through C402.5.3.2 shall comply with Sections C405.2.4.2 and C405.2.4.3, as applicable. *Daylight zones* shall include toplit *daylight zones* and sidelit *daylight* zones.

C402.5.5 Doors. Opaque swinging doors shall comply with Table C402.1.2. Opaque nonswinging doors shall comply with Table C402.1.2. *Opaque doors* shall be considered as part of the gross area of *above-grade walls* that are part of the *building thermal envelope*. *Opaque doors* shall comply with Section C402.5.5.1 or C402.5.5.2. Other doors shall comply with the provisions of Section C402.5.3 for vertical *fenestration*.

C402.5.5.1 Opaque swinging doors. Opaque swinging doors shall comply with Table C402.1.2.

C402.5.5.2 Nonswinging doors. Opaque nonswinging doors that are horizontally hinged sectional doors with a single row of *fenestration* shall have an assembly *U-factor* less than or equal to 0.440 in Climate Zones 0 through 6 and less than or equal to 0.360 in Climate Zones 7 and 8, provided that the *fenestration* area is not less than 14 percent and not more than 25 percent of the total door area.

Exception: Other doors shall comply with the provisions of Section C402.5.3 for vertical *fenestration*.

C402.6 Air leakage—building thermal envelope. The *building thermal envelope* shall comply with Sections C402.6.1 through C402.6.7.

C402.6.1 Air barriers. A continuous *air barrier* shall be provided throughout the *building thermal envelope*. The *air barrier* is permitted to be located at any combination of inside, outside or within the *building thermal envelope*. The *air barrier* shall comply with Sections C402.6.1.1 and C402.6.1.2. The *air leakage* performance of the *air barrier* shall be verified in accordance with Section C402.6.2.

Exception: *Air barriers* are not required in buildings located in *Climate Zone* 2B.

C402.6.1.1 Air barrier design and documentation requirements. Design of the continuous *air barrier* shall be documented as follows:

1. Components comprising the continuous *air barrier* and their position within each *building thermal envelope* assembly shall be identified.
2. Joints, interconnections and penetrations of the continuous *air barrier* components shall be detailed.
3. The continuity of the *air barrier* building element assemblies that enclose *conditioned space* or provide a boundary between *conditioned space* and unconditioned space shall be identified.
4. Documentation of the continuous *air barrier* shall detail methods of sealing the *air barrier,* such as wrapping, caulking, gasketing, taping or other *approved* methods at the following locations:
    - 4.1. Joints around *fenestration* and door frames.
    - 4.2. Joints between walls and floors; between walls at building corners; between walls and roofs, including parapets and copings; where *above-grade walls* meet foundations; and at similar intersections.
    - 4.3. Penetrations or attachments through the continuous *air barrier*.
    - 4.4. Building assemblies used as *ducts* or plenums.
    - 4.5. Changes in continuous *air barrier* materials and assemblies.
5. Identify where testing will or will not be performed in accordance with Section C402.6.2. Where testing will not be performed, a plan for field inspections required by Section C402.6.2.3 shall be provided that includes the following:
    - 5.1. A schedule for periodic inspection.
    - 5.2. The continuous *air barrier* scope of work.
    - 5.3. A list of critical inspection items.
    - 5.4. Inspection documentation requirements.
    - 5.5. Provisions for corrective actions where needed.

C402.6.1.2 Air barrier construction. The *continuous air barrier* shall be constructed to comply with the following:

1. The *air barrier* shall be continuous for all assemblies that compromise the *building thermal envelope* and across the joints and assemblies.
2. *Air barrier* joints and seams shall be sealed, including sealing transitions in places and changes in materials. The joints and seals shall be securely installed in or on the joint for its entire length so as not to dislodge, loosen or otherwise impair its ability to resist positive and negative pressure differentials such as those from wind, stack effect and mechanical ventilation.
3. Penetrations of the *air barrier* shall be caulked, gasketed or otherwise sealed in a manner compatible with the construction materials and location. Sealing shall allow for expansion, contraction and mechanical vibration. Sealing materials shall be securely installed around the penetration so as not to dislodge, loosen or otherwise impair the penetrations' ability to resist positive and negative pressure. Sealing of concealed fire sprinklers, where required, shall be in a manner that is recommended by the fire sprinkler manufacturer. Caulking or other adhesive sealants shall not be used to fill voids between fire sprinkler cover plates and walls or ceilings.
4. Recessed lighting fixtures shall comply with Section C402.6.1.2.1. Where similar objects are installed that penetrate the *air barrier*, provisions shall be made to maintain the integrity of the *air barrier*.
5. Electrical and communication boxes shall comply with Section C402.6.1.2.2.

C402.6.1.2.1 Recessed lighting. Recessed luminaires installed in the *building thermal envelope* shall be all of the following:

1. IC-rated.
2. *Labeled* as having an *air leakage* rate of not greater than 2.0 cfm (0.944 L/s) when tested in accordance with ASTM E283 at a 1.57 psf (75 Pa) pressure differential.
3. Sealed with a gasket or caulk between the housing and interior wall or ceiling covering.

C402.6.1.2.2 Electrical and communication boxes. Electrical and communication boxes that penetrate the *air barrier* of the *building thermal envelope*, and that do not comply with Section C402.6.1.2.2.1, shall be caulked, taped, gasketed or otherwise sealed to the *air barrier* element being penetrated. All openings on the concealed portion of the box shall be sealed. Where present, insulation shall rest against all concealed portions of the box.

C402.6.1.2.2.1 Air-sealed boxes. Where air-sealed boxes are installed, they shall be marked in accordance with NEMA OS 4. Air-sealed boxes shall be installed in accordance with the manufacturer's instructions.

C402.6.2 Air leakage compliance. *Air leakage* of the *building thermal envelope* shall be tested by an *approved* third party in accordance with Section C402.6.2.1. The measured *air leakage* shall not be greater than 0.35 cubic feet per minute per square foot (1.8 L/s x m^2) of the *building thermal envelope* area at a pressure differential of 0.3 inch water gauge (75 Pa) with the calculated *building thermal envelope* surface area being the sum of the above- and below-grade *building thermal envelope*.

Exceptions:

1. Where the measured *air leakage* rate is greater than 0.35 cfm/ft^2 (1.8 L/s × m^2) but is not greater than 0.45 cfm/ft^2 (2.3 L/s × m^2), the *approved* third party shall perform a diagnostic evaluation using a smoke tracer or infrared imaging. The evaluation shall be conducted while the building is pressurized or depressurized along with a visual inspection of the *air barrier* in accordance with ASTM E1186. All identified leaks shall be sealed where such sealing can be made without damaging existing building components. A report specifying the corrective actions taken to seal leaks shall be deemed to establish compliance with the requirements of this section where submitted to the *code official* and the *building* owner. Where the measured *air leakage* rate is greater than 0.45 cfm/ft^2 (2.3 L/s × m^2), corrective actions must be made to the *building* and an additional test completed for which the results are 0.45 cfm/ft^2 (2.3 L/s × m^2) or less.
2. Buildings in Climate Zone 2B.
3. Buildings larger than 25,000 square feet (2323 m^2) floor area in Climate Zones 0 through 4, other than Group I and R occupancies, that comply with Section C402.6.2.3.
4. As an alternative, *buildings* or portions of *buildings* containing Group I-1 and R-2 occupancies shall be permitted to be tested by an *approved* third party in accordance with Section C402.6.2.2. The reported *air leakage* of the *building thermal envelope* shall not be greater than 0.27 cfm/ft^2 (1.4 L/s × m^2) of the *testing unit enclosure area* at a pressure differential of 0.2 inch water gauge (50 Pa).

C402.6.2.1 Whole building test method and reporting. The *building thermal envelope* shall be tested by an *approved* third party in accordance with ASTM E3158 or an equivalent *approved* method. A report that includes the tested surface area, floor area, air by volume, stories above grade, and *air leakage* rates shall be submitted to the *code official* and the building *owner*.

Exceptions:

1. For *buildings* less than 10,000 square feet (929 m^2), the entire *building thermal envelope* shall be permitted to be tested in accordance with ASTM E779, ASTM E3158, ASTM E1827 or an equivalent *approved* method.
2. For *buildings* greater than 50,000 square feet (4645 m^2), portions of the *building* shall be permitted to be tested and the measured *air leakage* shall be area weighted by the surface areas of the *building thermal envelope* in each portion. The weighted-average tested *air leakage* shall not be greater than the whole building *air leakage* limit. The following portions of the *building* shall be tested:
    2.1. The entire *building thermal envelope* area of stories that have any *conditioned spaces* directly under a roof.

2.2. The entire *building thermal envelope* area of stories that have a *building entrance*, have a floor over unconditioned space, have a loading dock or that are below grade.

2.3. Representative above-grade portions of the building totaling not less than 25 percent of the wall area enclosing the remaining *conditioned space*.

C402.6.2.2 Dwelling and sleeping unit enclosure method and reporting. The *building thermal envelope* shall be tested for *air leakage* in accordance with ASTM E779, ANSI/RESNET/ICC 380, ASTM E1827 or an equivalent *approved* method. Where multiple *dwelling units* or *sleeping units* or other spaces are contained within one *building thermal envelope*, each shall be considered an individual testing unit, and the *building air leakage* shall be the weighted average of all tested unit results, weighted by each *testing unit enclosure area*. Units shall be tested without simultaneously testing adjacent units and shall be separately tested as follows:

1. Where buildings have less than eight total dwelling or *sleeping units*, each testing unit shall be tested.
2. Where buildings have eight or more dwelling or *sleeping units*, the greater of seven units or 20 percent of the units in the building shall be tested, including a top floor unit, a middle floor unit, a ground floor unit and a unit with the largest *testing unit enclosure area*. For each tested unit that exceeds the maximum *air leakage* rate, an additional three units shall be tested, including a mixture of testing unit types and locations.
3. *Enclosed spaces* with not less than one *exterior wall* in the *building thermal envelope* shall be tested in accordance with Section C402.6.2.1.

   Exception: Corridors, stairwells, and *enclosed spaces* having a *conditioned floor area* not greater than 1,500 square feet (139 m^2) shall be permitted to comply with Section C402.6.2.3 and either Section C402.6.2.3.1 or Section C402.6.2.3.2.

C402.6.2.3 Building thermal envelope design and construction verification criteria. Where Section C402.6.2.1 and C402.6.2.2 are not applicable the installation of the continuous *air barrier* shall be verified by the *code official*, a *registered design professional* or *approved* agency in accordance with the following:

1. A review of the *construction documents* and other supporting data shall be conducted to assess compliance with the requirements in Section C402.6.1.
2. Inspection of continuous *air barrier* components and assemblies shall be conducted during construction to verify compliance with the requirements of Sections C402.6.2.3.1 and C402.6.2.3.2. The *air barrier* shall be provided with access for inspection and repair.
3. A final inspection report shall be provided for inspections completed by the *registered design professional* or *approved* agency. The inspection report shall be provided to the *building owner* or *owner's* authorized agent and the *code official*. The report shall identify deficiencies found during inspection and details of corrective measures taken.

C402.6.2.3.1 Materials. Materials with an air permeability not greater than 0.004 cfm/ft^2(0.02 L/s × m ^{2}) under a pressure differential of 0.3 inch water gauge (75 Pa) when tested in accordance with ASTM E2178 shall comply with this section. Materials in Items 1 through 16 shall be deemed to comply with this section, provided that joints are sealed and materials are installed as *air barriers* in accordance with the manufacturer's instructions.

1. Plywood with a thickness of not less than $^3/_8$ inch (10 mm).
2. Oriented strand board having a thickness of not less than $^3/_8$ inch (10 mm).
3. Extruded polystyrene insulation board having a thickness of not less than $^1/_2$ inch (12.7 mm).
4. Foil-back polyisocyanurate insulation board having a thickness of not less than $^1/_2$ inch (12.7 mm).
5. Closed-cell spray foam having a minimum density of 1.5 pcf (2.4 kg/m^3) and having a thickness of not less than $1^1/_2$ inches (38 mm).
6. Open-cell spray foam with a density between 0.4 and 1.5 pcf (0.6 and 2.4 kg/m^3) and having a thickness of not less than 4.5 inches (113 mm).
7. Exterior or interior gypsum board having a thickness of not less than $^1/_2$ inch (12.7 mm).
8. Cement board having a thickness of not less than $^1/_2$ inch (12.7 mm).
9. Built-up roofing membrane.
10. Modified bituminous roof membrane.
11. Single-ply roof membrane.
12. A Portland cement/sand parge, or gypsum plaster having a thickness of not less than $^5/_8$ inch (15.9 mm).
13. Cast-in-place and precast concrete.
14. Fully grouted concrete block masonry.
15. Sheet steel or aluminum.
16. Solid or hollow masonry constructed of clay or shale masonry units.

C402.6.2.3.2 Assemblies. Assemblies of materials and components with an average *air leakage* not greater than 0.04 cfm/ft^2 (0.2 L/s × m^2) under a pressure differential of 0.3 inch of water gauge (75 Pa) where tested in accordance with ASTM E2357,

ASTM E1677, ASTM D8052 or ASTM E283 shall comply with this section. Assemblies listed in Items 1 through 3 below shall be deemed to comply, provided that joints are sealed and the requirements of Section C402.6.1.2 are met.

1. Concrete masonry walls coated with either one application of block filler or two applications of a paint or sealer coating.
2. Masonry walls constructed of clay or shale masonry units with a nominal width greater than or equal to 4 inches (102 mm).
3. A Portland cement/sand parge, stucco or plaster not less than $^1/_2$ inch (12.7 mm) in thickness.

C402.6.3 Air leakage of fenestration and opaque doors. The *air leakage* of *fenestration* and opaque door assemblies shall comply with Table C402.6.3. Testing shall be conducted by an accredited, independent testing laboratory in accordance with applicable reference test standards in Table C402.6.3 and *labeled* by the manufacturer.

Exceptions:

1. Field-fabricated *fenestration* assemblies that are sealed in accordance with Section C402.6.1.
2. *Fenestration* in *buildings* that is tested in accordance with Section C402.6.2 is not required to meet the *air leakage* requirements in Table C402.6.3.

TABLE C402.6.3—MAXIMUM AIR LEAKAGE RATE FOR FENESTRATION ASSEMBLIES

FENESTRATION ASSEMBLY	MAXIMUM RATE (cfm/ft^2)	TEST PROCEDURE
Windows	0.20[a]	AAMA/WDMA/CSA101/I.S.2/A440 or NFRC 400
Sliding doors	0.20[a]	
Swinging doors	0.20[a]	
Skylights—with condensation weepage openings	0.30	
Skylights—all other	0.20[a]	
Curtain walls	0.06	NFRC 400 or ASTM E283 at 1.57 psf (75 Pa)
Storefront glazing	0.06	
Commercial glazed swinging entrance doors	1.00	
Power-operated sliding doors and power operated folding doors	1.00	
Revolving doors	1.00	
Garage doors	0.40	ANSI/DASMA 105, NFRC 400, or ASTM E283 at 1.57 psf (75 Pa)
Rolling doors	1.00	
High-speed doors	1.30	

For SI: 1 cubic foot per minute = 0.47 L/s, 1 square foot = 0.093 m^2.

a. The maximum rate for windows, sliding and swinging doors, and skylights is permitted to be 0.3 cfm per square foot of fenestration or door area when tested in accordance with AAMA/WDMA/CSA101/I.S.2/A440 at 6.24 psf (300 Pa).

C402.6.4 Doors and access openings to shafts, chutes, stairways and elevator lobbies. Doors and *access* openings from *conditioned space* to shafts, chutes, stairways and elevator lobbies not within the scope of the *fenestration* assemblies covered by Section C402.6.3 shall be gasketed, weather-stripped or sealed.

Exceptions:

1. Door openings required to comply with Section 716 of the *International Building Code*.
2. Doors and door openings required by the *International Building Code* to comply with UL 1784.

C402.6.5 Air intakes, exhaust openings, stairways and shafts. Stairway enclosures, elevator shaft vents and other outdoor air intakes and exhaust openings integral to the *building thermal envelope* shall be provided with dampers in accordance with Section C403.7.7.

C402.6.6 Vestibules. *Building entrances* shall be protected with an enclosed vestibule. Doors opening into and out of the vestibule equipped with self-closing devices. Vestibules shall be designed so that in passing through the vestibule it is not necessary for the interior and exterior doors to open at the same time. The installation of one or more revolving doors in the *building entrance* shall not eliminate the requirement that a vestibule be provided on any doors adjacent to revolving doors.

Exceptions: Vestibules are not required for the following:

1. Buildings in Climate Zones 0 through 2.
2. Doors not intended to be used by the public, such as doors to mechanical or electrical *equipment rooms*, or intended solely for employee use.
3. Doors opening directly from a *sleeping unit* or *dwelling unit*.
4. Doors that open directly from a space less than 3,000 square feet (298 m^2) in area.

5. Revolving doors.
6. Doors used primarily to facilitate vehicular movement or material handling and adjacent personnel doors.
7. Doors that have an air curtain unit with a velocity of not less than 6.56 feet per second (2 m/s) at 6 inches (152 mm) above the floor that has been tested in accordance with ANSI/AMCA 220 or ISO 27327-1 and installed in accordance with the manufacturer's instructions. *Manual* or *automatic* controls shall be provided that will operate the *air curtain unit* with the opening and closing of the door and comply with Section C403.4.1.5. *Air curtain units* and their controls shall comply with Section C408.2.3.

C402.6.7 Loading dock weather seals. Cargo door openings and loading door openings shall be equipped with weather seals that restrict *air leakage* and provide direct contact along the top and sides of vehicles that are parked in the doorway.

C402.7 Thermal bridges in above-grade walls. *Thermal bridges* in *above-grade walls* shall comply with this section or an *approved* design.

Exceptions:

1. *Buildings* and structures located in Climate Zones 0 through 3.
2. Any *thermal bridge* with a material thermal conductivity not greater than 3.0 Btu/h × ft × °F (5.19 W/m × K).
3. Blocking, coping, flashing and other similar materials for attachment of roof coverings.
4. *Thermal bridges* accounted for in the *U-factor* or *C-factor* for a *building thermal envelope*.

C402.7.1 Balconies and floor decks. Balconies and concrete floor decks shall not penetrate the *building thermal envelope*. Such assemblies shall be separately supported or shall be supported by structural attachments or elements that minimize thermal bridging through the *building thermal envelope*.

Exceptions: Balconies and concrete floor decks shall be permitted to penetrate the *building thermal envelope* where one of the following applies:

1. An area-weighted *U-factor* is used for *above-grade wall* compliance that includes a *U-factor* of 0.8 Btu/h × °F × ft^2 (1.38 W/m × K) for the area of the *above-grade wall* penetrated by the concrete floor deck in accordance with Section C402.1.2.1.5.
2. An *approved* thermal break device with not less than R-10 insulation material is installed in accordance with the manufacturer's instructions.
3. An *approved* design where the *above-grade wall U-factor* used for compliance accounts for all balcony and concrete floor deck *thermal bridges*.

C402.7.2 Cladding supports. Linear elements supporting opaque cladding shall be offset from the structure with attachments that allow the *continuous insulation*, where present, to pass behind the cladding support element except at the point of attachment.

Exceptions:

1. An *approved* design where the *above-grade wall U-factor* used for compliance accounts for the cladding support element *thermal bridge*.
2. Anchoring for *curtain wall* and window wall systems where *curtain wall* and window wall systems comply with Section C402.7.4.

C402.7.3 Structural beams and columns. Structural steel and concrete beams and columns that project through the *building thermal envelope* shall be covered with not less than R-5 insulation for not less than 2 feet (610 mm) beyond the interior or exterior surface of an insulation component within the *building thermal envelope*.

Exceptions:

1. Where an *approved* thermal break device is installed in accordance with the manufacturer's instructions.
2. An *approved* design where the *above-grade wall U-factor* used to demonstrate compliance accounts for the beam or column *thermal bridge*.

C402.7.4 Vertical fenestration. Vertical *fenestration* intersections with *above-grade walls* shall comply with one or more of the following:

1. Where *above-grade walls* include *continuous insulation*, the plane of the exterior glazing layer or, for metal frame *fenestration*, a nonmetal thermal break in the frame shall be positioned within 2 inches (610 mm) of the interior or exterior surface of the *continuous insulation*.
2. Where *above-grade walls* do not include *continuous insulation*, the plane of the exterior glazing layer or, for metal frame *fenestration*, a nonmetal thermal break in the frame shall be positioned within the thickness of the integral or *cavity insulation*.
3. The surface of the rough opening, not covered by the *fenestration* frame, shall be insulated with insulation of not less than R-3 material or covered with a wood buck that is not less than 1.5 inches (38 mm) thick.
4. For the intersection between vertical *fenestration* and opaque spandrel in a shared framing system, manufacturer's data for the spandrel *U-factor* shall account for *thermal bridges*.

Exceptions:

1. Where an *approved* design for the *above-grade wall U-factor* used for compliance accounts for *thermal bridges* at the intersection with the vertical *fenestration*.
2. Doors.

C402.7.5 Parapets. Parapets shall comply with one or more of the following as applicable:

1. Where *continuous insulation* is installed on the exterior side of the *above-grade wall* and the roof is insulated with insulation entirely above deck, the *continuous insulation* shall extend up both sides of the parapet not less than 2 feet (610 mm) above the roof covering or to the top of the parapet, whichever is less. Parapets that are an integral part of a fire-resistance rated wall, and the exterior continuous insulation applied to the parapet, shall comply with the fire-resistance ratings of the *International Building Code*.
2. Where *continuous insulation* is installed on the exterior side of the *above-grade wall* and the roof insulation is below the roof deck, the *continuous insulation* shall extend up the exterior side of the parapet to not less than the height of the top surface of the *roof assembly*.
3. Where *continuous insulation* is not installed on the exterior side of the *above-grade wall* and the roof is insulated with insulation entirely above deck, the wall cavity or integral insulation shall extend into the parapet up to the exterior face of the roof insulation or equivalent *R-value* insulation shall be installed not less than 2 feet (610 mm) horizontally inward on the underside of the roof deck.
4. Where *continuous insulation* is not installed on the exterior side of the *above-grade wall* and the roof insulation is below the roof deck, the wall and roof insulation components shall be adjacent to each other at the roof-ceiling-wall intersection.
5. Where a thermal break device with not less than R-10 insulation material aligned with the *above-grade wall* and roof insulation is installed in accordance with the manufacturer's instructions.

Exception: An *approved design* where the *above-grade wall U-factor* used for compliance accounts for the parapet *thermal bridge*.

SECTION C403—BUILDING MECHANICAL SYSTEMS

C403.1 General. Mechanical systems and equipment serving the building heating, cooling, ventilating or refrigerating needs shall comply with one of the following:

1. Section C403.1.1 and Sections C403.2 through C403.17.
2. *Data Centers* shall comply with Section C403.1.1, Section C403.1.2 and Sections C403.6 through C403.17.
3. Section C409.

C403.1.1 Calculation of heating and cooling loads. Design loads associated with heating, ventilating and air conditioning of the *building* shall be determined in accordance with ANSI/ASHRAE/ACCA Standard 183 or by an *approved* equivalent computational procedure using the design parameters specified in Chapter 3. Heating and cooling loads shall be adjusted to account for load reductions that are achieved where energy recovery systems are utilized in the HVAC system in accordance with the ASHRAE HVAC Systems and Equipment Handbook by an *approved* equivalent computational procedure.

C403.1.2 Data centers. *Data center systems* shall comply with Sections 6 and 8 of ASHRAE 90.4.

C403.2 System design. Mechanical systems shall be designed to comply with Sections C403.2.1 through C403.2.3. Where elements of a building's mechanical systems are addressed in Sections C403.3 through C403.14, such elements shall comply with the applicable provisions of those sections.

C403.2.1 Zone isolation required. HVAC systems serving *zones* that are over 25,000 square feet (2323 m^2) in floor area or that span more than one floor and are designed to operate or be occupied nonsimultaneously shall be divided into isolation areas. Each isolation area shall be equipped with *isolation devices* and controls configured to automatically shut off the supply of conditioned air and outdoor air to and exhaust air from the isolation area. Each isolation area shall be controlled independently by a device meeting the requirements of Section C403.4.2.2. Central systems and plants shall be provided with controls and devices that will allow system and equipment operation for any length of time while serving only the smallest isolation area served by the system or plant.

Exceptions:

1. Exhaust air and outdoor air connections to isolation areas where the *fan system* to which they connect is not greater than 5,000 cfm (2360 L/s).
2. Exhaust airflow from a single isolation area of less than 10 percent of the design airflow of the exhaust system to which it connects.
3. Isolation areas intended to operate continuously or intended to be inoperative only when all other isolation areas in a *zone* are inoperative.

C403.2.2 Ventilation. *Ventilation*, either natural or mechanical, shall be provided in accordance with Chapter 4 of the *International Mechanical Code*. Where mechanical ventilation is provided, the system shall provide the capability to reduce the outdoor air supply to the minimum required by Chapter 4 of the *International Mechanical Code*.

C403.2.3 Fault detection and diagnostics. Buildings with a gross *conditioned floor area* of not less than 100,000 square feet (9290 m^2) served by one or more HVAC systems that are controlled by a *direct digital control* (DDC) system shall include a fault detection and diagnostics (FDD) system to monitor the HVAC system's performance and automatically identify faults. The *FDD system* shall:

1. Include permanently installed sensors and devices to monitor HVAC system performance.
2. Sample HVAC system performance at least once every 15 minutes.

3. Automatically identify and report HVAC system faults.
4. Automatically notify authorized personnel of identified HVAC system faults.
5. Automatically provide prioritized recommendations for *repair* of identified faults based on analysis of data collected from the sampling of HVAC system performance.
6. Be capable of transmitting the prioritized fault repair recommendations to remotely located authorized personnel.

Exception: R-1 and R-2 occupancies.

C403.3 Heating and cooling equipment efficiencies. Heating and cooling equipment installed in mechanical systems shall be sized in accordance with Section C403.3.1 and shall be not less efficient in the use of energy than as specified in Section C403.3.2.

C403.3.1 Equipment sizing. The output capacity of heating and cooling equipment shall be not greater than that of the smallest available equipment size that exceeds the loads calculated in accordance with Section C403.1.1. A single piece of equipment providing both heating and cooling shall satisfy this provision for one function with the capacity for the other function as small as possible, within available equipment options.

Exceptions:

1. Required standby equipment and systems provided with controls and devices that allow such systems or equipment to operate automatically only when the primary equipment is not operating.
2. Multiple units of the same equipment type with combined capacities exceeding the design load and provided with controls that are configured to sequence the operation of each unit based on load.

C403.3.2 HVAC equipment performance requirements. Equipment shall meet the minimum efficiency requirements of Tables C403.3.2(1) through C403.3.2(16) when tested and rated in accordance with the applicable test procedure. Plate-type liquid-to-liquid heat exchangers shall meet the minimum requirements of AHRI 400. The efficiency shall be verified through certification under an *approved* certification program or, where a certification program does not exist, the equipment efficiency ratings shall be supported by data furnished by the manufacturer. Where multiple rating conditions or performance requirements are provided, the equipment shall satisfy all stated requirements. Where components, such as indoor or outdoor coils, from different manufacturers are used, calculations and supporting data shall be furnished by the designer that demonstrates that the combined efficiency of the specified components meets the requirements herein.

TABLE C403.3.2(1)—ELECTRICALLY OPERATED UNITARY AIR CONDITIONERS AND CONDENSING UNITS—MINIMUM EFFICIENCY REQUIREMENTS[c]

EQUIPMENT TYPE	SIZE CATEGORY	HEADING SECTION TYPE	SUBCATEGORY OR RATING CONDITION	MINIMUM EFFICIENCY	TEST PROCEDURE[a]
Air conditioners, air cooled	< 65,000 Btu/h[b]	All	Split system, three phase and applications outside US single phase[b]	13.4 SEER2	AHRI 210/240—2023
			Single-package, three phase and applications outside US single phase[b]	13.4 SEER2	
Space constrained, air cooled	≤ 30,000 Btu/h[b]	All	Split system, three phase and applications outside US single phase[b]	11.7 SEER2	AHRI 210/240—2023
			Single package, three phase and applications outside US single phase[b]	11.7 SEER2	
Small duct, high velocity, air cooled	< 65,000 Btu/h[b]	All	Split system, three phase and applications outside US single phase[b]	12. 0 SEER2	AHRI 210/240—2023
Air conditioners, air cooled	≥ 65,000 Btu/h and < 135,000 Btu/h	Electric resistance (or none)	Split system and single package	14.8 IEER	AHRI 340/360
		All other		14.6 IEER	
	≥ 135,000 Btu/h and < 240,000 Btu/h	Electric resistance (or none)		14.2 IEER	
		All other		14.0 IEER	
	≥ 240,000 Btu/h and < 760,000 Btu/h	Electric resistance (or none)	Split system and single package	13.2 IEER	AHRI 340/360
		All other		13.0 IEER	
	≥ 760,000 Btu/h	Electric resistance (or none)		12.5 IEER	
		All other		12.3 IEER	

TABLE C403.3.2(1)—ELECTRICALLY OPERATED UNITARY AIR CONDITIONERS AND CONDENSING UNITS—MINIMUM EFFICIENCY REQUIREMENTS[c]—continued

EQUIPMENT TYPE	SIZE CATEGORY	HEADING SECTION TYPE	SUBCATEGORY OR RATING CONDITION	MINIMUM EFFICIENCY	TEST PROCEDURE[a]
Air conditioners, water cooled	< 65,000 Btu/h	All	Split system and single package	12.1 EER 12.3 IEER	AHRI 210/240
	≥ 65,000 Btu/h and < 135,000 Btu/h	Electric resistance (or none)		12.1 EER 13.9 IEER	AHRI 340/360
		All other		11.9 EER 13.7 IEER	
	≥ 135,000 Btu/h and < 240,000 Btu/h	Electric resistance (or none)		12.5 EER 13.9 IEER	
		All other		12.3 EER 13.7 IEER	
	≥ 240,000 Btu/h and < 760,000 Btu/h	Electric resistance (or none)		12.4 EER 13.6 IEER	
		All other		12.2 EER 13.4 IEER	
	≥ 760,000 Btu/h	Electric resistance (or none)		12.2 EER 13.5 IEER	
		All other		12.0 EER 13.3 IEER	
Air conditioners, evaporatively cooled	< 65,000 Btu/h[b]	All	Split systemand single package	12.1 EER 12.3 IEER	AHRI 210/240
	≥ 65,000 Btu/h and < 135,000 Btu/h	Electric resistance (or none)		12.1 EER 12.3 IEER	AHRI 340/360
		All other		11.9 EER 12.1 IEER	
	≥ 135,000 Btu/h and < 240,000 Btu/h	Electric resistance (or none)		12.0 EER 12.2 IEER	
		All other		11.8 EER 12.0 IEER	
	≥ 240,000 Btu/h and < 760,000 Btu/h	Electric resistance (or none)		11.9 EER 12.1 IEER	
		All other		11.7 EER 11.9 IEER	
	≥ 760,000 Btu/h	Electric resistance (or none)		11.7 EER 11.9 IEER	
		All other		11.5 EER 11.7 IEER	
Condensing units, air cooled	≥ 135,000 Btu/h	—	—	10.5 EER 11.8 IEER	AHRI 365
Condensing units, water cooled	≥ 135,000 Btu/h	—	—	13.5 EER 14.0 IEER	AHRI 365
Condensing units, evaporatively cooled	≥ 135,000 Btu/h	—	—	13.5 EER 14.0 IEER	AHRI 365

For SI: 1 British thermal unit per hour = 0.2931 W.

a. Chapter 6 contains a complete specification of the referenced standards, which include test procedures, including the reference year version of the test procedure.

b. Single-phase, US air-cooled air conditioners less than 65,000 Btu/h are regulated as consumer products by the US Department of Energy Code of Federal Regulations DOE 10 CFR 430. SEER and SEER2 values for single-phase products are set by the US Department of Energy.

c. DOE 10 CFR 430 Subpart B Appendix M1 includes the test procedure updates effective January 1, 2023, documented in AHRI 210/240—2023.

TABLE C403.3.2(2)—ELECTRICALLY OPERATED AIR-COOLED UNITARY HEAT PUMPS—MINIMUM EFFICIENCY REQUIREMENTS[c]					
EQUIPMENT TYPE	**SIZE CATEGORY**	**HEADING SECTION TYPE**	**SUBCATEGORY OR RATING CONDITION**	**MINIMUM EFFICIENCY**	**TEST PROCEDURE[a]**
Air cooled (cooling mode)	< 65,000 Btu/h	All	Split system, three phase and applications outside US single phase[b]	14.3 SEER2	AHRI 210/240—2023
			Single package, three phase and applications outside US single phase[b]	13.4 SEER2	
Space constrained, air cooled (cooling mode)	≤ 30,000 Btu/h	All	Split system, three phase and applications outside US single phase[b]	11.7 SEER2	AHRI 210/240—2023
			Single package, three phase and applications outside US single phase[b]	11.7 SEER2	
Small duct, high velocity, air cooled (cooling mode)	< 65,000 Btu/h	All	Split system, three phase and applications outside US single phase[b]	12.0 SEER2	AHRI 210/240—2023
Air cooled (cooling mode)	≥ 65,000 Btu/h and < 135,000 Btu/h	Electric resistance (or none)	Split system and single package	14.1 IEER	AHRI 340/360
		All other		13.9 IEER	
	≥ 135,000 Btu/h and < 240,000 Btu/h	Electric resistance (or none)		13.5 IEER	
		All other		13.3 IEER	
	≥ 240,000 Btu/h	Electric resistance (or none)		12.5 IEER	
		All other		12.3 IEER	
Air cooled (heating mode)	< 65,000 Btu/h (cooling capacity)	—	Split system, three phase and applications outside US single phase[b]	7.5 HSPF2	AHRI 210/240—2023
			Single package, three phase and applications outside US single phase[b]	6.7 HSPF2	
Space constrained, air cooled (heating mode)	≤ 30,000 Btu/h (cooling capacity)	—	Split system, three phase and applications outside US single phase[b]	6.3 HSPF2	AHRI 210/240—2023
			Single package, three phase and applications outside US single phase[b]	6.3 HSPF2	
Small duct high velocity, air cooled (heating mode)	< 65,000 Btu/h	—	Split system, three phase and applications outside US single phase[b]	6.1 HSPF2	AHRI 210/240—2023
Air cooled (heating mode)	≥ 65,000 Btu/h and < 135,000 Btu/h (cooling capacity)	—	47°F db/43°F wb outdoor air	3.40 COP_H	AHRI 340/360
			17°F db/15°F wb outdoor air	2.25 COP_H	
	≥ 135,000 Btu/h and < 240,000 Btu/h (cooling capacity)		47°F db/43°F wb outdoor air	3.30 SOP_H	
			17°F db/15°F wb outdoor air	2.05 COP_H	
	≥ 240,000 Btu/h (cooling capacity)		47°F db/43°F wb outdoor air	3.20 COP_H	
			17°F db/15°F wb outdoor air	2.05 COP_H	

For SI: 1 British thermal unit per hour = 0.2931 W, °C = (°F – 32)/1.8, wb = wet bulb, db = dry bulb.

a. Chapter 6 contains a complete specification of the referenced standards, which include test procedures, including the reference year version of the test procedure.

b. Single-phase, US air-cooled heat pumps less than 65,000 Btu/h are regulated as consumer products by the US Department of Energy Code of Federal Regulations DOE 10 CFR 430. SEER, SEER2 and HSPF values for single-phase products are set by the US Department of Energy.

c. DOE 10 CFR 430 Subpart B Appendix M1 includes the test procedure updates effective January 1, 2023, documented in AHRI 210/240—2023.

TABLE C403.3.2(3)—LIQUID-CHILLING PACKAGES—MINIMUM EFFICIENCY REQUIREMENTS[a, b, e]

EQUIPMENT TYPE	SIZE CATEGORY	UNITS	PATH A	PATH B	TEST PROCEDURE[c]
Air cooled	< 150 tons	EER (Btu/Wh)	≥ 10.100 FL	≥ 9.700 FL	AHRI 550/590
			≥ 13.700 IPLV.IP	≥ 15.800 IPLV.IP	
	≥ 150 tons		≥ 10.100 FL	≥ 9.700FL	
			≥ 14.000 IPLV.IP	≥ 16.100 IPLV.IP	
Air cooled without condenser, electrically operated	All capacities	EER (Btu/Wh)	Air-cooled without condenser must be rated with matching condensers and comply with air-cooled chiller efficiency requirements		AHRI 550/590
Liquid-cooled, electrically operated positive displacement	< 75 tons	kW/ton	≤ 0.750 FL	≤ 0.780 FL	AHRI 550/590
			≤ 0.600 IPLV.IP	≤ 0.500 IPLV.IP	
	≥ 75 tons and < 150 tons		≤ 0.720 FL	≤ 0.750 FL	
			≤ 0.560 IPLV.IP	≤ 0.490 IPLV.IP	
	≥ 150 tons and < 300 tons		≤ 0.660 FL	≤ 0.680 FL	
			≤ 0.540 IPLV.IP	≤ 0.440 IPLV.IP	
	≥ 300 tons and < 600 tons		≤ 0.610 FL	≤ 0.625 FL	
			≤ 0.520 IPLV.IP	≤ 0.410 IPLV.IP	
	≥ 600 tons		≤ 0.560 FL	≤ 0.585 FL	
			≤ 0.500 IPLV.IP	≤ 0.380 IPLV.IP	
Liquid-cooled, electrically operated centrifugal	< 150 tons	kW/ton	≤ 0.610 FL	≤ 0.695 FL	AHRI 550/590
			≤ 0.550 IPLV.IP	≤ 0.440 IPLV.IP	
	≥150 tons and <300 tons		≤ 0.610 FL	≤ 0.635 FL	
			≤ 0.550 IPLV.IP	≤ 0.400 IPLV.IP	
	≥ 300 tons and < 400 tons		≤ 0.560 FL	≤ 0.595 FL	
			≤ 0.520 IPLV.IP	≤ 0.390 IPLV.IP	
	≥ 400 tons and < 600 tons		≤ 0.560 FL	≤ 0.585 FL	
			≤ 0.500 IPLV.IP	≤ 0.380 IPLV.IP	
	≥ 600 tons		≤ 0.560 FL	≤ 0.585 FL	
			≤ 0.500 IPLV.IP	≤ 0.380 IPLV.IP	
Air cooled absorption, single effect	All capacities	COP (W/W)	≥ 0.600 FL	NA[d]	AHRI 560
Liquid-cooled absorption, single effect	All capacities	COP (W/W)	≥ 0.700 FL	NA[d]	AHRI 560
Absorption double effect, indirect fired	All capacities	COP (W/W)	≥ 1.000 FL	NA[d]	AHRI 560
			≥ 0.150 IPLV.IP		
Absorption double effect, direct fired	All capacities	COP (W/W)	≥ 1.000 FL	NA[d]	AHRI 560
			≥ 1.000 IPLV		

a. Chapter 6 contains a complete specification of the referenced standards, which include test procedures, including the reference year version of the test procedure.
b. The requirements for centrifugal chillers shall be adjusted for nonstandard rating conditions per Section C403.3.2.1 and are applicable only for the range of conditions listed there. The requirements for air-cooled, water-cooled positive displacement and absorption chillers are at standard rating conditions defined in the reference test procedure.
c. Both the full-load and IPLV.IP requirements must be met or exceeded to comply with this standard. When there is a Path B, compliance can be with either Path A or Path B for any application.
d. NA means the requirements are not applicable for Path B, and only Path A can be used for compliance.
e. FL is the full-load performance requirements, and IPLV.IP is for the part-load performance requirements.

TABLE C403.3.2(4)—ELECTRICALLY OPERATED PACKAGED TERMINAL AIR CONDITIONERS, PACKAGED TERMINAL HEAT PUMPS, SINGLE-PACKAGE VERTICAL AIR CONDITIONERS, SINGLE-PACKAGE VERTICAL HEAT PUMPS, ROOM AIR CONDITIONERS AND ROOM AIR-CONDITIONER HEAT PUMPS—MINIMUM EFFICIENCY REQUIREMENTS[e]				
EQUIPMENT TYPE	**SIZE CATEGORY (INPUT)**	**SUBCATEGORY OR RATING CONDITION**	**MINIMUM EFFICIENCY[d]**	**TEST PROCEDURE[a]**
PTAC (cooling mode) standard size	< 7,000 Btu/h	95°F db/75°F wb outdoor air[c]	11.9 EER	AHRI 310/380
	≥ 7,000 Btu/h and ≤ 15,000 Btu/h		14.0 – (0.300 × Cap/1,000) EER[d]	
	> 15,000 Btu/h		9.5 EER	
PTAC (cooling mode) nonstandard size[a]	< 7,000 Btu/h	95°F db/75°F wb outdoor air[c]	9.4 EER	AHRI 310/380
	≥ 7,000 Btu/h and ≤ 15,000 Btu/h		10.9 – (0.213 × Cap/1,000) EER[d]	
	> 15,000 Btu/h		7.7 EER	
PTHP (cooling mode) standard size	< 7,000 Btu/h	95°F db/75°F wb outdoor air[c]	11.9 EER	AHRI 310/380
	≥ 7,000 Btu/h and ≤ 15,000 Btu/h		14.0 – (0.300 × Cap/1,000) EER[d]	
	> 15,000 Btu/h		9.5 EER	
PTHP (cooling mode) nonstandard size[b]	< 7,000 Btu/h	95°F db/75°F wb outdoor air[c]	9.3 EER	AHRI 310/380
	≥ 7,000 Btu/h and ≤ 15,000 Btu/h		10.8 – (0.213 × Cap/1,000) EER[d]	
	> 15,000 Btu/h		7.6 EER	
PTHP (heating mode) standard size	< 7,000 Btu/h	47°F db/43°F wb outdoor air	3.3 COP_H	AHRI 310/380
	≥ 7,000 Btu/h and ≤ 15,000 Btu/h		3.7 – (0.052 × Cap/1,000) COP_H[d]	
	> 15,000 Btu/h		2.90 COP_H	
PTHP (heating mode) nonstandard size[b]	< 7,000 Btu/h	47°F db/43°F wb outdoor air	2.7 COP_H	AHRI 310/380
	≥ 7,000 Btu/h and ≤ 15,000 Btu/h		2.9 – (0.026 × Cap/1000) COP_H[d]	
	> 15,000 Btu/h		2.5 COP_H	
SPVAC (cooling mode) single and three phase	< 65,000 Btu/h	95°F db/75°F wb outdoor air[c]	11.0 EER	AHRI 390
	≥ 65,000 Btu/h and ≤ 135,000 Btu/h		10.0 EER	
	≥ 135,000 Btu/h and ≤ 240,000 Btu/h		10.0 EER	
SPVHP (cooling mode)	< 65,000 Btu/h	95°F db/75°F wb outdoor air[c]	11.0 EER	AHRI 390
	≥ 65,000 Btu/h and ≤ 135,000 Btu/h		10.0 EER	
	≥ 135,000 Btu/h and ≤ 240,000 Btu/h		10. 0 EER	
SPVHP (heating mode)	< 65,000 Btu/h	47°F db/43°F wb outdoor air	3.3 COP_H	AHRI 390
	≥ 65,000 Btu/h and ≤ 135,000 Btu/h		3.0 COP_H	
	≥ 135,000 Btu/h and ≤ 240,000 Btu/h		3.0 COP_H	
Room air conditioners without reverse cycle with louvered sides for applications outside US[d]	< 6,000 Btu/h	—	11.0 CEER	ANSI/AHAM RAC-1
	≥ 6,000 Btu/h and < 8,000 Btu/h	—	11.0 CEER	
	≥ 8,000 Btu/h and < 14,000 Btu/h	—	10.9 CEER	
	≥ 14,000 Btu/h and < 20,000 Btu/h	—	10.7 CEER	
	≥ 20,000 Btu/h and < 28,000 Btu/h	—	9.4 CEER	
	≥ 28,000 Btu/h	—	9.0 CEER	

TABLE C403.3.2(4)—ELECTRICALLY OPERATED PACKAGED TERMINAL AIR CONDITIONERS, PACKAGED TERMINAL HEAT PUMPS, SINGLE-PACKAGE VERTICAL AIR CONDITIONERS, SINGLE-PACKAGE VERTICAL HEAT PUMPS, ROOM AIR CONDITIONERS AND ROOM AIR-CONDITIONER HEAT PUMPS—MINIMUM EFFICIENCY REQUIREMENTS[e]—continued

EQUIPMENT TYPE	SIZE CATEGORY (INPUT)	SUBCATEGORY OR RATING CONDITION	MINIMUM EFFICIENCY[d]	TEST PROCEDURE[a]
Room air conditioners without louvered sides	< 6,000 Btu/h	—	10.0 CEER	ANSI/AHAM RAC-1
	≥ 6,000 Btu/h and < 8,000 Btu/h	—	10.0 CEER	
	≥ 8,000 Btu/h and < 11,000 Btu/h	—	9.6 CEER	
	≥ 11,000 Btu/h and < 14,000 Btu/h	—	9.5 CEER	
	≥ 14,000 Btu/h and < 20,000 Btu/h	—	9.3 CEER	
	≥ 20,000 Btu/h	—	9.4 CEER	
Room air conditioners with reverse cycle, with louvered sides for applications outside US[d]	< 20,000 Btu/h	—	9.8 CEER	ANSI/AHAM RAC-1
	≥ 20,000 Btu/h	—	9.3 CEER	
Room air conditioners with reverse cycle without louvered sides for applications outside US[d]	< 14,000 Btu/h	—	9.3 CEER	ANSI/AHAM RAC-1
	≥ 14,000 Btu/h	—	8.7 CEER	
Room air conditioners, casement only for applications outside US[d]	All	—	9.5 CEER	ANSI/AHAM RAC-1
Room air conditioners, casement slider for applications outside US[d]	All	—	10.4 CEER	ANSI/AHAM RAC-1

For SI: 1 British thermal unit per hour = 0.2931 W, °C = (°F – 32)/1.8, wb = wet bulb, db = dry bulb.

"Cap" = The rated cooling capacity of the project in Btu/h. Where the unit's capacity is less than 7,000 Btu/h, use 7,000 Btu/h in the calculation. Where the unit's capacity is greater than 15,000 Btu/h, use 15,000 Btu/h in the calculations.

a. Chapter 6 contains a complete specification of the referenced standards, which include test procedures, including the reference year version of the test procedure.

b. Nonstandard size units must be factory labeled as follows: "MANUFACTURED FOR NONSTANDARD SIZE APPLICATIONS ONLY; NOT TO BE INSTALLED IN NEW STANDARD PROJECTS." Nonstandard size efficiencies apply only to units being installed in existing sleeves having an external wall opening of less than 16 inches high or less than 42 inches wide and having a cross-sectional area less than 670 square inches.

c. The cooling-mode wet bulb temperature requirement only applies for units that reject condensate to the condenser coil.

d. Room air conditioners are regulated as consumer products by 10 CFR 430. For US applications of room air conditioners, refer to Informative Appendix F, Table F-3, for the US DOE minimum efficiency requirements for US applications.

e. "Cap" in EER and COP_H equations for PTACs and PTHPs means cooling capacity in Btu/h at 95°F outdoor dry-bulb temperature.

TABLE C403.3.2(5)—WARM-AIR FURNACES AND COMBINATION WARM-AIR FURNACES/ AIR-CONDITIONING UNITS, WARM-AIR DUCT FURNACES AND UNIT HEATERS—MINIMUM EFFICIENCY REQUIREMENTS[g]

DESCRIPTION	FUEL	ELECTRIC POWER PHASE	APPLICATION LOCATION	HEATING CAPACITY (INPUT), Btu/h[b]	COMBO-UNIT COOLING CAPACITY, Btu/h	SUBTYPE	MINIMUM EFFICIENCY	TEST PROCEDURE[a]
Warm-air furnace	Gas	1	Inside US	< 225,000	< 65,000	See Informative Appendix F, Table F-4[f]		
Warm-air furnace	Gas	1	Inside US	< 225,000	≥ 65,000	Nonweatherized	80% AFUE	Appendix N[g]
						Weatherized	81% AFUE or 80% E_t[c]	Appendix N[g]
								ANSI Z21.47
Warm-air furnace	Gas	1	Outside US	< 225,000	All	Nonweatherized	80% AFUE	Appendix N[g]
						Weatherized	81% AFUE or 80% E_t[c]	Appendix N[g]
								ANSI Z21.47
Warm-air furnace	Gas	3	All	< 225,000	All	Nonweatherized	80% AFUE	Appendix N[g]
						Weatherized	81% AFUE or 80% E_t[c]	Appendix N[g]
								ANSI Z21.47
Warm-air furnace	Gas	All	All	≥ 225,000 and ≤ 400,000	All	All	81% E_t[c]	ANSI Z21.47
Warm-air furnace	Gas	All	Inside US	> 400,000	All	All	80%E_t[c] before 1/1/2023 81% E_t[c] after 1/1/2023	ANSI Z21.47
Warm-air furnace	Gas	All	Outside US	> 400,000	All	All	80%Etc before 1/1/2023 81% Etc after 1/1/2023	ANSI Z21.47 or ANSI Z83.8
Warm-air furnace	Oil	1	Inside US	< 225,000	< 65,000	See Informative Appendix F, Table F-4[f]		
Warm-air furnace	Oil	1	Inside US	< 225,000	≥ 65,000	Nonweatherized	83% AFUE	Appendix N[g]
						Weatherized	78% AFUE or 80% E_t[d]	Appendix N[g]
								Section 42 UL 727
Warm-air furnace	Oil	1	Outside US	< 225,000	All	Nonweatherized	83% AFUE	Appendix N[g]
						Weatherized	78% AFUE or 80% E_t[d]	Appendix N[g]
								Section 42 UL 727
Warm-air furnace	Oil	3	All	<225,000	All	Nonweatherized	83% AFUE	Appendix N[g]
						Weatherized	78% AFUE or 80%E_t[d]	Appendix N[g]
								Section 42 UL 727
Warm-air furnace	Oil	All	All	≥ 225,000	All	All	82% E_t[d]	Section 42 UL 727
Warm-air furnace	Electric	1	Inside US	< 225,000	< 65,000	See Informative Appendix F, Table F-4[f]		
Warm-air furnace	Electric	1	Inside US	< 225,000	≥ 65,000	All	96% AFUE	Appendix N[g]
Warm-air furnace	Electric	1	Outside US	< 225,000	All	All	96% AFUE	Appendix N[g]
Warm-air furnace	Electric	3	All	< 225,000	All	All	96% AFUE	Appendix N[g]
Warm-air duct furnaces	Gas	All	All	All	All	All	80% E_c[d]	ANSI Z83.8

TABLE C403.3.2(5)—WARM-AIR FURNACES AND COMBINATION WARM-AIR FURNACES/ AIR-CONDITIONING UNITS, WARM-AIR DUCT FURNACES AND UNIT HEATERS—MINIMUM EFFICIENCY REQUIREMENTS[g]—continued

DESCRIPTION	FUEL	ELECTRIC POWER PHASE	APPLICATION LOCATION	HEATING CAPACITY (INPUT), Btu/h[b]	COMBO-UNIT COOLING CAPACITY, Btu/h	SUBTYPE	MINIMUM EFFICIENCY	TEST PROCEDURE[a]
Warm-air unit heaters	Gas	All	All	All	All	All	80% E_c [d, e]	ANSI Z83.8
Warm-air unit heaters	Oil	All	All	All	All	All	80% E_c [d, e]	Section 40 UL 731

For SI: 1 British thermal unit per hour = 0.2931 W.

a. Chapter 6 contains a complete specification of the referenced standards, which include test procedures, including the reference year version of the test procedure. For this table, the following applies:
- Appendix N = 10 CFR 430 Appendix N
- ANSI Z21.47 = Section 2.39, Thermal Efficiency, ANSI Z21.47
- ANSI Z83.3 = Section 2.10, Efficiency, ANSI Z83.3
- UL 727 = Section 42, Combustion, UL 727
- UL 731 = Section 40, Combustion, UL 731

b. Compliance of multiple firing rate units shall be at the maximum firing rate.

c. E_t = thermal efficiency. Units must also include an interrupted or intermittent ignition device (IID), have jacket losses not exceeding 0.75 percent of the input rating, and have either power venting or a flue damper. A vent damper is an acceptable alternative to a flue damper for those furnaces where combustion air is drawn from the conditioned space.

d. E_c = combustion efficiency (100 percent less flue losses). See test procedure for detailed discussion.

e. Units must also include an interrupted or intermittent ignition device (IID) and have either power venting or an *automatic* flue damper.

f. Includes combination units with cooling capacity < 65,000 Btu/h. For US applications of federally covered < 225,000 Btu/h products, see Informative Appendix F, Table F-4.

g. 10 CFR 430 is limited to-single phase equipment that is not contained within the same cabinet with a central air conditioner whose rated cooling capacity is above 65,000 Btu/h but for the test and rating procedures are not impacted for three-phase and can be used for AFUE ratings for ASHRAE/IES Standard 90.1 three-phase products and single-phase products with a cooling capacity greater than 65,000 Btu/h.

TABLE C403.3.2(6)—GAS- AND OIL-FIRED BOILERS—MINIMUM EFFICIENCY REQUIREMENTS

EQUIPMENT TYPE[b]	SUBCATEGORY OR RATING CONDITION	SIZE CATEGORY (INPUT)	MINIMUM EFFICIENCY	TEST PROCEDURE[a]
Boilers, hot water	Gas fired	< 300,000 Btu/h[g, h] for applications outside US	84% AFUE	DOE 10 CFR 430 Appendix N
		≥ 300,000 Btu/h and ≤ 2,500,000 Btu/h[e]	84% E_t[d]	DOE 10 CFR 431.86
		> 2,500,000 Btu/h[b] and ≤ 10,000,000 Btu/h[b]	82% E_c[c]	
		> 10,000,000 Btu/h[b]	82% E_c[c]	
	Oil fired[f]	< 300,000 Btu/h[g, h] for applications outside US	86% AFUE	DOE 10 CFR 430 Appendix N
		≥ 300,000 Btu/h and ≤ 2,500,000 Btu/h[e]	82% E_t[d]	DOE 10 CFR 431.86
		> 2,500,000 Btu/h[b] and ≤ 10,000,000 Btu/h[b]	84% E_c[c]	
		> 10,000,000 Btu/h[b]	84% E_c[c]	
Boilers, steam	Gas fired	< 300,000 Btu/h[g] for applications outside US	82% AFUE	DOE 10 CFR 430 Appendix N
	Gas fired—all, except natural draft	≥ 300,000 Btu/h and ≤ 2,500,000 Btu/h[e]	79% E_t[d]	DOE 10 CFR 431.86
		> 2,500,000 Btu/h[b] and ≤ 10,000,000 Btu/h[b]	79% E_t[d]	
		> 10,000,000 Btu/h[b]	79% E_t[d]	
	Gas fired—natural draft	≥ 300,000 Btu/h and ≤ 2,500,000 Btu/h[e]	79% E_t[d]	
		> 2,500,000 Btu/h[b]	79% E_t[d]	
	Oil fired[f]	< 300,000 Btu/h[g] for applications outside US	82% AFUE	DOE 10 CFR 430 Appendix N
		≥ 300,000 Btu/h and ≤ 2,500,000 Btu/h[e]	84% E_t[d]	DOE 10 CFR 431.86
		> 2,500,000 Btu/h[b] and ≤ 10,000,000 Btu/h[b]	81% E_t[d]	
		> 10,000,000 Btu/h[b]	81% E_t[d]	

For SI: 1 British thermal unit per hour = 0.2931 W.

a. Chapter 6 contains a complete specification of the referenced standards, which include test procedures, including the reference year version of the test procedure.

b. These requirements apply to boilers with rated input of 8,000,000 Btu/h or less that are not packaged boilers and to all packaged boilers. Minimum efficiency requirements for boilers cover all capacities of packaged boilers.

c. E_c = Combustion efficiency (100 percent less flue losses).

d. E_t = Thermal efficiency.

e. Maximum capacity—minimum and maximum ratings as provided for and allowed by the unit's controls.

f. Includes oil-fired (residual).

g. Boilers shall not be equipped with a constant burning pilot light.

h. A boiler not equipped with a tankless domestic water-heating coil shall be equipped with an automatic means for adjusting the temperature of the water such that an incremental change in inferred heat load produces a corresponding incremental change in the temperature of the water supplied.

TABLE C403.3.2(7)—PERFORMANCE REQUIREMENTS FOR HEAT REJECTION EQUIPMENT—MINIMUM EFFICIENCY REQUIREMENTS

EQUIPMENT TYPE	TOTAL SYSTEM HEAT-REJECTION CAPACITY AT RATED CONDITIONS	SUBCATEGORY OR RATING CONDITION[h]	PERFORMANCE REQUIRED[a, b, c, f, g]	TEST PROCEDURE[d, e]
Propeller or axial fan open-circuit cooling towers	All	95°F entering water 85°F leaving water 75°F entering wb	≥ 40.2 gpm/hp	CTI ATC-105 and CTI STD-201 RS
Centrifugal fan open-circuit cooling towers	All	95°F entering water 85°F leaving water 75°F entering wb	≥ 20.0 gpm/hp	CTI ATC-105 and CTI STD-201 RS
Propeller or axial fan closed-circuit cooling towers	All	102°F entering water 90°F leaving water 75°F entering wb	≥ 16.1 gpm/hp	CTI ATC-105S and CTI STD-201 RS
Centrifugal fan closed-circuit cooling towers	All	102°F entering water 90°F leaving water 75°F entering wb	≥ 7.0 gpm/hp	CTI ATC-105S and CTI STD-201 RS
Propeller or axial fan dry coolers (air-cooled fluid coolers)	All	115°F entering water 105°F leaving water 95°F entering wb	≥ 4.5 gpm/hp	CTI ATC-105DS
Propeller or axial fan evaporative condensers	All	R-448A test fluid 165°F entering gas temperature 105°F condensing temperature 75°F entering wb	≥ 160,000 Btu/h × hp	CTI ATC-106
Propeller or axial fan evaporative condensers	All	Ammonia test fluid 140°F entering gas temperature 96.3°F condensing temperature 75°F entering wb	≥ 134,000 Btu/h × hp	CTI ATC-106
Centrifugal fan evaporative condensers	All	R-448A test fluid 165°F entering gas temperature 105°F condensing temperature 75°F entering wb	≥ 137,000 Btu/h × hp	CTI ATC-106
Centrifugal fan evaporative condensers	All	Ammonia test fluid 140°F entering gas temperature 96.3°F condensing temperature 75°F entering wb	≥ 110,000 Btu/h × hp	CTI ATC-106
Air-cooled condensers	All	125°F condensing temperature 190°F entering gas temperature 15°F subcooling 95°F entering db	≥ 176,000 Btu/h × hp	AHRI 460

For SI: °C = (°F – 32)/1.8, L/s × kW = (gpm/hp)/(11.83), COP = (Btu/h × hp)/(2550.7), db = dry bulb temperature, wb = wet bulb temperature.

a. For purposes of this table, open-circuit cooling tower performance is defined as the water-flow rating of the tower at the thermal rating condition listed in the table divided by the fan motor nameplate power.

b. For purposes of this table, closed-circuit cooling tower performance is defined as the process water-flow rating of the tower at the thermal rating condition listed in the table divided by the sum of the fan motor nameplate power and the integral spray pump motor nameplate power.

c. For purposes of this table, dry-cooler performance is defined as the process water-flow rating of the unit at the thermal rating condition listed in the table divided by the total fan motor nameplate power of the unit, and air-cooled condenser performance is defined as the heat rejected from the refrigerant divided by the total fan motor nameplate power of the unit.

d. ASHRAE 90.1 Section 13 contains a complete specification of the referenced test procedure, including the referenced year version of the test procedure.

e. The efficiencies and test procedures for both open- and closed-circuit cooling towers are not applicable to hybrid cooling towers that contain a combination of separate wet and dry heat exchange sections. The certification requirements do not apply to field-erected cooling towers.

f. All cooling towers shall comply with the minimum efficiency listed in the table for that specific type of tower with the capacity effect of any project-specific accessories and/or options included in the capacity of the cooling tower.

g. For purposes of this table, evaporative condenser performance is defined as the heat rejected at the specified rating condition in the table, divided by the sum of the fan motor nameplate power and the integral spray pump nameplate power.

h. Requirements for evaporative condensers are listed with ammonia R-717 and R-448A as test fluids in the table. Evaporative condensers intended for use with halocarbon refrigerants other than R-448A must meet the minimum efficiency requirements listed with R-448A as the test fluid. For ammonia, the condensing temperature is defined as the saturation temperature corresponding to the refrigerant pressure at the condenser entrance. For R-448A, which is a zeotropic refrigerant, the condensing temperature is defined as the arithmetic average of the dew point and the bubble point temperatures corresponding to the refrigerant pressure at the condenser entrance.

TABLE C403.3.2(8)—ELECTRICALLY OPERATED VARIABLE-REFRIGERANT-FLOW AIR CONDITIONERS—MINIMUM EFFICIENCY REQUIREMENTS

EQUIPMENT TYPE	SIZE CATEGORY	HEATING SECTION TYPE	SUBCATEGORY OR RATING CONDITION	MINIMUM EFFICIENCY	TEST PROCEDURE[a]
VRF air conditioners, air cooled	< 65,000 Btu/h three-phase for applications in the US and single- and three-phase for applications outside the US	All	VRF multisplit system	13.0 SEER	AHRI 210/240
	≥ 65,000 Btu/h and < 135,000 Btu/h	Electric resistance (or none)	VRF multisplit system	10.5 EER 15.5 IEER	AHRI 1230
	≥ 135,000 Btu/h and < 240,000 Btu/h	Electric resistance (or none)	VRF multisplit system	10.3 EER 14.9 IEER	
	≥ 240,000 Btu/h	Electric resistance (or none)	VRF multisplit system	9.5 EER 13.9 IEER	

For SI: 1 British thermal unit per hour = 0.2931 W.

a. Chapter 6 contains a complete specification of the referenced standards, which include test procedures, including the reference year version of the test procedure.

TABLE C403.3.2(9)—ELECTRICALLY OPERATED VARIABLE-REFRIGERANT-FLOW AND APPLIED HEAT PUMPS—MINIMUM EFFICIENCY REQUIREMENTS

EQUIPMENT TYPE	SIZE CATEGORY	HEATING SECTION TYPE	SUBCATEGORY OR RATING CONDITION	MINIMUM EFFICIENCY	TEST PROCEDURE[a]
VRF air cooled (cooling mode)	< 65,000 Btu/h three-phase for applications in the US and single- and three- phase for applications outside the US	All	VRF multisplit system	SEER2 = 13.4	AHRI 210/240
	≥ 65,000 Btu/h and < 135,000 Btu/h	Electric resistance (or none)		10.3 EER 14.6 IEER	AHRI 1230
			VRF multisplit system with heat recovery	10.1 EER 14.4 IEER	
	≥ 135,000 Btu/h and < 240,000 Btu/h		VRF multisplit system	9.9 EER 14.4 IEER	
			VRF multisplit system with heat recovery	9.7 EER 13.9 IEER	
	≥ 240,000 Btu/h		VRF multisplit system	9.1 EER 12.7 IEER	
			VRF multisplit system with heat recovery	8.9 EER 12.5 IEER	
VRF water source (cooling mode)	< 65,000 Btu/h	All	VRF multisplit systems 86°F entering water	12.0 EER 16.0 IEER	AHRI 1230
			VRF multisplit systems with heat recovery 86°F entering water	11.8 EER 15.8 IEER	
	≥ 65,000 Btu/h and < 135,000 Btu/h		VRF multisplit system 86°F entering water	12.0 EER 16.0 IEER	
			VRF multisplit system with heat recovery 86°F entering water	11.8 EER 15.8 IEER	
	≥ 135,000 Btu/h and < 240,000 Btu/h		VRF multisplit system 86°F entering water	10.0 EER 14.0 IEER	
			VRF multisplit system with heat recovery 86°F entering water	9.8 EER 13.8 IEER	
	≥ 240,000 Btu/h		VRF multisplit system 86°F entering water	10.0 EER 12.0 IEER	
			VRF multisplit system with heat recovery 86°F entering water	9.8 EER 11.8 IEER	

TABLE C403.3.2(9)—ELECTRICALLY OPERATED VARIABLE-REFRIGERANT-FLOW AND APPLIED HEAT PUMPS—MINIMUM EFFICIENCY REQUIREMENTS—continued

EQUIPMENT TYPE	SIZE CATEGORY	HEATING SECTION TYPE	SUBCATEGORY OR RATING CONDITION	MINIMUM EFFICIENCY	TEST PROCEDURE[a]
VRF groundwater source (cooling mode)	< 135,000 Btu/h	All	VRF multisplit system 59°F entering water	16.2 EER	AHRI 1230
			VRF multisplit system with heat recovery 59°F entering water	16.0 EER	
	≥ 135,000 Btu/h		VRF multisplit system 59°F entering water	13.8 EER	
			VRF multisplit system with heat recovery 59°F entering water	13.6 EER	
VRF ground source (cooling mode)	< 135,000 Btu/h	All	VRF multisplit system 77°F entering water	13.4 EER	AHRI 1230
			VRF multisplit system with heat recovery 77°F entering water	13.2 EER	
	≥ 135,000 Btu/h		VRF multisplit system 77°F entering water	11.0 EER	
			VRF multisplit system with heat recovery 77°F entering water	10.8 EER	
VRF air cooled (heating mode)	< 65,000 Btu/h (cooling capacity) three-phase for applications in the US and single- and three-phase for applications outside the US		VRF multisplit system	HSPF2 = 7.5	AHRI 210/240
	≥ 65,000 Btu/h and < 135,000 Btu/h (cooling capacity)		VRF multisplit system 47°F db/43°F wb outdoor air	3.3 COP_H	AHRI 1230
			17°F db/15°F wb outdoor air	2.25 COP_H	
	≥ 135,000 Btu/h (cooling capacity)		VRF multisplit system 47°F db/43°F wb outdoor air	3.2 COP_H	
			17°F db/15°F wb outdoor air	2.05 COP_H	
VRF water source (heating mode)	< 65,000 Btu/h (cooling capacity)		VRF multisplit system 68°F entering water	4.3 COP_H	AHRI 1230
	≥ 65,000 Btu/h and < 135,000 Btu/h (cooling capacity)		VRF multisplit system 68°F entering water	4.3 COP_H	
	≥ 135,000 Btu/h and < 240,000 Btu/h (cooling capacity)	All	VRF multisplit system 68°F entering water	4.0 COP_H	
	≥ 240,000 Btu/h (cooling capacity)		VRF multisplit system 68°F entering water	3.9 COP_H	
VRF groundwater source (heating mode)	< 135,000 Btu/h (cooling capacity)		VRF multisplit system 50°F entering water	3.6 COP_H	AHRI 1230
	≥ 135,000 Btu/h (cooling capacity)		VRF multisplit system 50°F entering water	3.3 COP_H	
VRF ground source (heating mode)	< 135,000 Btu/h (cooling capacity)		VRF multisplit system 32°F entering water	3.1 COP_H	AHRI 1230
	≥ 135,000 Btu/h (cooling capacity)		VRF multisplit system 32°F entering water	2.8 COP_H	

For SI: °C = (°F – 32)/1.8, 1 British thermal unit per hour = 0.2931 W, db = dry bulb temperature, wb = wet bulb temperature.

a. Chapter 6 contains a complete specification of the referenced standards, which include test procedures, including the reference year version of the test procedure.

TABLE C403.3.2(10)—FLOOR-MOUNTED AIR CONDITIONERS AND CONDENSING UNITS SERVING COMPUTER ROOMS—MINIMUM EFFICIENCY REQUIREMENTS

EQUIPMENT TYPE	STANDARD MODEL	NET SENSIBLE COOLING CAPACITY	MINIMUM NET SENSIBLE COP	RATING CONDITIONS RETURN AIR (dry bulb/dew point)	TEST PROCEDURE
Air cooled	Downflow	< 80,000 Btu/h	2.70	85°F/52°F (Class 2)	AHRI 1360
		≥ 80,000 Btu/h and < 295,000 Btu/h	2.58		
		≥ 295,000 Btu/h	2.36		
	Upflow—ducted	< 80,000 Btu/h	2.67		
		≥ 80,000 Btu/h and < 295,000 Btu/h	2.55		
		≥ 295,000 Btu/h	2.33		
	Upflow—nonducted	< 65,000 Btu/h	2.16	75°F/52°F (Class 1)	
		≥ 65,000 Btu/h and < 240,000 Btu/h	2.04		
		≥ 240,000 Btu/h	1.89		
	Horizontal	< 65,000 Btu/h	2.65	95°F/52°F (Class 3)	
		≥ 65,000 Btu/h and < 240,000 Btu/h	2.55		
		≥ 240,000 Btu/h	2.47		
Air cooled with fluid economizer	Downflow	< 80,000 Btu/h	2.70	85°F/52°F (Class 1)	AHRI 1360
		≥ 80,000 Btu/h and < 295,000 Btu/h	2.58		
		≥ 295,000 Btu/h	2.36		
	Upflow—ducted	< 80,000 Btu/h	2.67		
		≥ 80,000 Btu/h and < 295,000 Btu/h	2.55		
		≥ 295,000 Btu/h	2.33		
	Upflow—nonducted	< 65,000 Btu/h	2.09	75°F/52°F (Class 1)	
		≥ 65,000 Btu/h and < 240,000 Btu/h	1.99		
		≥ 240,000 Btu/h	1.81		
	Horizontal	< 65,000 Btu/h	2.65	95°F/52°F (Class 3)	
		≥ 65,000 Btu/h and < 240,000 Btu/h	2.55		
		≥ 240,000 Btu/h	2.47		
Water cooled	Downflow	< 80,000 Btu/h	2.82	85°F/52°F (Class 1)	AHRI 1360
		≥ 80,000 Btu/h and < 295,000 Btu/h	2.73		
		≥ 295,000 Btu/h	2.67		
	Upflow—ducted	< 80,000 Btu/h	2.79		
		≥ 80,000 Btu/h and < 295,000 Btu/h	2.70		
		≥ 295,000 Btu/h	2.64		
	Upflow—nonducted	< 65,000 Btu/h	2.43	75°F/52°F (Class 1)	
		≥ 65,000 Btu/h and < 240,000 Btu/h	2.32		
		≥ 240,000 Btu/h	2.20		
	Horizontal	< 65,000 Btu/h	2.79	95°F/52°F (Class 3)	
		≥ 65,000 Btu/h and < 240,000 Btu/h	2.68		
		≥ 240,000 Btu/h	2.60		

TABLE C403.3.2(10)—FLOOR-MOUNTED AIR CONDITIONERS AND CONDENSING UNITS SERVING COMPUTER ROOMS—MINIMUM EFFICIENCY REQUIREMENTS—continued

EQUIPMENT TYPE	STANDARD MODEL	NET SENSIBLE COOLING CAPACITY	MINIMUM NET SENSIBLE COP	RATING CONDITIONS RETURN AIR (dry bulb/dew point)	TEST PROCEDURE
Water cooled with fluid economizer	Downflow	< 80,000 Btu/h	2.77	85°F/52°F (Class 1)	AHRI 1360
		≥ 80,000 Btu/h and < 295,000 Btu/h	2.68		
		≥ 295,000 Btu/h	2.61		
	Upflow—ducted	< 80,000 Btu/h	2.74		
		≥ 80,000 Btu/h and < 295,000 Btu/h	2.65		
		≥ 295,000 Btu/h	2.58		
	Upflow—nonducted	< 65,000 Btu/h	2.35	75°F/52°F (Class 1)	
		≥ 65,000 Btu/h and < 240,000 Btu/h	2.24		
		≥ 240,000 Btu/h	2.12		
	Horizontal	< 65,000 Btu/h	2.71	95°F/52°F (Class 3)	
		≥ 65,000 Btu/h and < 240,000 Btu/h	2.60		
		≥ 240,000 Btu/h	2.54		
Glycol cooled	Downflow	< 80,000 Btu/h	2.56	85°F/52°F (Class 1)	AHRI 1360
		≥ 80,000 Btu/h and < 295,000 Btu/h	2.24		
		≥ 295,000 Btu/h	2.21		
	Upflow—ducted	< 80,000 Btu/h	2.53		
		≥ 80,000 Btu/h and < 295,000 Btu/h	2.21		
		≥ 295,000 Btu/h	2.18		
	Upflow, nonducted	< 65,000 Btu/h	2.08	75°F/52°F (Class 1)	
		≥ 65,000 Btu/h and < 240,000 Btu/h	1.90		
		≥ 240,000 Btu/h	1.81		
	Horizontal	< 65,000 Btu/h	2.48	95°F/52°F (Class 3)	
		≥ 65,000 Btu/h and < 240,000 Btu/h	2.18		
		≥ 240,000 Btu/h	2.18		
Glycol cooled with fluid economizer	Downflow	< 80,000 Btu/h	2.51	85°F/52°F (Class 1)	AHRI 1360
		≥ 80,000 Btu/h and < 295,000 Btu/h	2.19		
		≥ 295,000 Btu/h	2.15		
	Upflow—ducted	< 80,000 Btu/h	2.48		
		≥ 80,000 Btu/h and < 295,000 Btu/h	2.16		
		≥ 295,000 Btu/h	2.12		
	Upflow—nonducted	< 65,000 Btu/h	2.00	75°F/52°F (Class 1)	
		≥ 65,000 Btu/h and < 240,000 Btu/h	1.82		
		≥ 240,000 Btu/h	1.73		
	Horizontal	< 65,000 Btu/h	2.44	95°F/52°F (Class 3)	
		≥ 65,000 Btu/h and < 240,000 Btu/h	2.10		
		≥ 240,000 Btu/h	2.10		

For SI: 1 British thermal unit per hour = 0.2931 W, °C = (°F – 32)/1.8, COP = (Btu/h × hp)/(2,550.7).

TABLE C403.3.2(11)—VAPOR-COMPRESSION-BASED INDOOR POOL DEHUMIDIFIERS—MINIMUM EFFICIENCY REQUIREMENTS

EQUIPMENT TYPE	SUBCATEGORY OR RATING CONDITION	MINIMUM EFFICIENCY	TEST PROCEDURE
Single package indoor (with or without economizer)	Rating conditions: A or C	3.5 MRE	AHRI 910
Single package indoor water-cooled (with or without economizer)	Rating conditions: A, B or C	3.5 MRE	
Single package indoor air-cooled (with or without economizer)	Rating conditions: A, B or C	3.5 MRE	
Split system indoor air-cooled (with or without economizer)	Rating conditions: A, B or C	3.5 MRE	

TABLE C403.3.2(12)—ELECTRICALLY OPERATED DX-DOAS UNITS, SINGLE-PACKAGE AND REMOTE CONDENSER, WITHOUT ENERGY RECOVERY—MINIMUM EFFICIENCY REQUIREMENTS

EQUIPMENT TYPE	SUBCATEGORY OR RATING CONDITION	MINIMUM EFFICIENCY	TEST PROCEDURE[a]
Air cooled (dehumidification mode)	—	3.8 ISMRE2	AHRI 920
Air-source heat pumps (dehumidification mode)	—	3.8 ISMRE2	AHRI 920
Water cooled (dehumidification mode)	Cooling tower condenser water	4.7 ISMRE2	AHRI 920
Air-source heat pump (heating mode)	—	2.05 ISCOP2	AHRI 920
Water-source heat pump (dehumidification mode)	Ground source, closed and open loop[b]	4.6 ISMRE2	AHRI 920
	Water source	3.8 ISMRE2	
Water-source heat pump (heating mode)	Ground source, closed and open loop[b]	2.13 ISCOP2	AHRI 920
	Water source	2.13 ISCOP2	

a. Chapter 6 contains a complete specification of the referenced standards, which include test procedures, including the reference year version of the test procedure.
b. Open-loop systems are rated using closed-loop test conditions.

TABLE C403.3.2(13)—ELECTRICALLY OPERATED DX-DOAS UNITS, SINGLE-PACKAGE AND REMOTE CONDENSER, WITH ENERGY RECOVERY—MINIMUM EFFICIENCY REQUIREMENTS

EQUIPMENT TYPE	SUBCATEGORY OR RATING CONDITION	MINIMUM EFFICIENCY	TEST PROCEDURE[a]
Air cooled (dehumidification mode)	—	5.0 ISMRE2	AHRI 920
Air-source heat pumps (dehumidification mode)	—	5.0 ISMRE2	AHRI 920
Water cooled (dehumidification mode)	Cooling tower condenser water	5.1 ISMRE2	AHRI 920
Air-source heat pump (heating mode)	—	3.2 ISCOP2	AHRI 920
Water-source heat pump (dehumidification mode)	Ground source, closed and open loop[b]	5.0 ISMRE2	AHRI 920
	Water source	4.6 ISMRE2	
Water-source heat pump (heating mode)	Ground source, closed and open loop[b]	3.5 ISCOP2	AHRI 920
	Water source	4.04 ISCOP2	

a. Chapter 6 contains a complete specification of the referenced standards, which include test procedures, including the reference year version of the test procedure.
b. Open-loop systems are rated using closed-loop test conditions.

TABLE C403.3.2(14)—ELECTRICALLY OPERATED WATER-SOURCE HEAT PUMPS—MINIMUM EFFICIENCY REQUIREMENTS[b]

EQUIPMENT TYPE	SIZE CATEGORY	HEATING SECTION TYPE	SUBCATEGORY OR RATING CONDITION	MINIMUM EFFICIENCY	TEST PROCEDURE[a]
Water-to-air, water loop (cooling mode)	< 17,000 Btu/h	All	86°F entering water	12.2 EER	ISO 13256-1
	≥ 17,000 Btu/h and < 65,000 Btu/h			13.0 EER	
	≥ 65,000 Btu/h and < 135,000 Btu/h			13.0 EER	
Water-to-air, ground water (cooling mode)	< 135,000 Btu/h	All	59°F entering water	18.0 EER	ISO 13256-1
Brine-to-air, ground loop (cooling mode)	< 135,000 Btu/h	All	77°F entering water	14.1 EER	ISO 13256-1
Water-to-water, water loop (cooling mode)	< 135,000 Btu/h	All	86°F entering water	10.6 EER	ISO 13256-2
Water-to-water, ground water (cooling mode)	< 135,000 Btu/h	All	59°F entering water	16.3 EER	ISO 13256-2
Brine-to-water, ground loop (cooling mode)	< 135,000 Btu/h	All	77°F entering water	12.1 EER	ISO 13256-2
Water-to-water, water loop (heating mode)	< 135,000 Btu/h (cooling capacity)	—	68°F entering water	4.3 COP_H	ISO 13256-1
Water-to-air, ground water (heating mode)	< 135,000 Btu/h (cooling capacity)	—	50°F entering water	3.7 COP_H	ISO 13256-1
Brine-to-air, ground loop (heating mode)	< 135,000 Btu/h (cooling capacity)	—	32°F entering water	3.2 COP_H	ISO 13256-1
Water-to-water, water loop (heating mode)	< 135,000 Btu/h (cooling capacity)	—	68°F entering water	3.7 COP_H	ISO 13256-1
Water-to-water, ground water (heating mode)	< 135,000 Btu/h (cooling capacity)	—	50°F entering water	3.1 COP_H	ISO 13256-2
Brine-to-water, ground loop (heating mode)	< 135,000 Btu/h (cooling capacity)	—	32°F entering water	2.5 COP_H	ISO 13256-2

For SI: 1 British thermal unit per hour = 0.2931 W, °C = (°F – 32)/1.8.

a. Chapter 6 contains a complete specification of the referenced standards, which include test procedures, including the reference year version of the test procedure.

b. Single-phase, US air-cooled heat pumps < 65,000 Btu/h are regulated as consumer products by 10 CFR 430. SEER, SEER2, HPSF and HPSF2 values for single-phase products are set by the US DOE. Informative Note: See ASHRAE 90.1 Informative Appendix F for the US DOE minimum.

TABLE C403.3.2(15)—HEAT-PUMP AND HEAT RECOVERY CHILLER PACKAGES—MINIMUM EFFICIENCY REQUIREMENTS[g, o]

HEATING OPERATION EFFICIENCY[b, e, j]																	
EQUIPMENT TYPE	SIZE CATEGORY REFRIGERATING CAPACITY[n], ton_R	COOLING OPERATION EFFICIENCY[a, d, e, j] AIR-SOURCE EER (FL/IPLV), Btu/W × h LIQUID-SOURCE POWER INPUT PER CAPACITY (FL/IPLV), kW/ton_R		HEATING SOURCE CONDITIONS (leaving liquid) OR OUTDOOR AIR TEMPERATURE (db/wb), °F	HEAT-PUMP HEATING FULL-LOAD HEATING EFFICIENCY (COP_H)[f, h, k], W/W				SIMULTANEOUS COOLING AND HEATING FULL-LOAD EFFICIENCY (COP_{SHC})[b, i], W/W				HEAT RECOVERY HEATING FULL-LOAD EFFICIENCY (COP_{HR})[c, j], W/W				Test Procedure[a]
					Entering/Leaving Heating Liquid Temperature				Entering/Leaving Heating Liquid Temperature				Entering/Leaving Heating Liquid Temperature				
		Path A	Path B		Low	Medium	High	Boost	Low	Medium	High	Boost	Low	Medium	Hot-Water 1	Hot-Water 2	
					95°F/ 105°F	105°F/ 120°F	120°F/ 140°F	120°F/ 140°F	95°F/ 105°F	105°F/ 120°F	120°F/ 140°F	120°F/ 140°F	95°F/ 105°F	105°F/ 120°F	90°F/ 140°F	120°F/ 140°F	
Air source	< 150.0	≥ 9.595 FL ≥ 13.02 IPLV.IP	≥ 9.215 FL ≥ 15.01 IPLV.IP	47 db 43 wb[l]	≥ 3.290	≥ 2.770	≥ 2.310	NA	NA	NA	NA	NA	NA	NA	NA	NA	AHRI 550/590
				17 db 15 wb[l]	≥ 2.029	≥ 1.775	≥ 1.483	NA	NA	NA	NA	NA	NA	NA	NA	NA	
	≥ 150.0	≥ 9.595 FL ≥ 13.30 IPLV.IP	≥ 9.215 FL ≥ 15.30 IPLV.IP	47 db 43 wb[l]	≥ 3.290	≥ 2.770	≥ 2.310	NA	NA	NA	NA	NA	NA	NA	NA	NA	
				17 db 15 wb[l]	≥ 2.029	≥ 1.775	≥ 1.483	NA	NA	NA	NA	NA	NA	NA	NA	NA	
Liquid-source electrically operated positive displace-ment	≥ 11.25[p] and < 150	≤ 0.7895 FL ≤ 0.6316 IPLV.IP	≤ 0.8211 FL ≤ 0.5263 IPLV.IP	44[m]	≥ 4.640	≥ 3.680	≥ 2.680	NA	≥ 8.330	≥ 6.410	≥ 4.420	NA	≥ 8.330	≥ 6.410	≥ 4.862	≥ 4.420	AHRI 550/590
				65[m]	NA	NA	NA	≥ 3.550	NA	NA	NA	≥ 6.150	NA	NA	NA	NA	
	≥ 150 and < 300	≤ 0.7579 FL ≤ 0.5895 IPLV.IP	≤ 0.7895 FL ≤ 0.5158 IPLV.IP	44[m]	≥ 4.640	≥ 3.680	≥ 2.680	NA	≥ 8.330	≥ 6.410	≥ 4.420	NA	≥ 8.330	≥ 6.410	≥ 4.862	≥ 4.420	
				65[m]	NA	NA	NA	≥ 3.550	NA	NA	NA	≥ 6.150	NA	NA	NA	NA	
	≥ 300 and < 400	≤ 0.6947 FL ≤ 0.5684 IPLV.IP	≤ 0.7158 FL ≤ 0.4632 IPLV.IP	44[m]	≥ 4.640	≥ 3.680	≥ 2.680	NA	≥ 8.330	≥ 6.410	≥ 4.420	NA	≥ 8.330	≥ 6.410	≥ 4.862	≥ 4.420	
				65[m]	NA	NA	NA	≥ 3.550	NA	NA	NA	≥ 6.150	NA	NA	NA	NA	
	≥ 400 and < 600	≤ 0.6421 FL ≤ 0.5474 IPLV.IP	≤ 0.6579 FL ≤ 0.4316 IPLV.IP	44[m]	≥ 4.930	≥ 3.960	≥ 2.970	NA	≥ 8.900	≥ 6.980	≥ 5.000	NA	≥ 8.900	≥ 6.980	≥ 5.500	≥ 5.000	
				65[m]	NA	NA	NA	≥ 3.900	NA	NA	NA	≥ 6.850	NA	NA	NA	NA	
	≥ 600	≤ 0.5895 FL ≤ 0.5263 IPLV.IP	≤ 0.6158 FL ≤ 0.4000 IPLV.IP	44[m]	≥ 4.930	≥ 3.960	≥ 2.970	NA	≥ 8.900	≥ 6.980	≥ 5.000	NA	≥ 8.900	≥ 6.980	≥ 5.500	≥ 5.000	
				65[m]	NA	NA	NA	≥ 3.900	NA	NA	NA	≥ 6.850	NA	NA	NA	NA	
Liquid-source electrically operated centrifugal	≥11.25[p] and <150	≤ 0.6421 FL ≤ 0.5789 IPLV.IP	≤ 0.7316 FL ≤ 0.4632 IPLV.IP	44[m]	≥ 4.640	≥ 3.680	≥ 2.680	NA	≥ 8.330	≥ 6.410	≥ 4.420	NA	≥ 8.330	≥ 6.410	≥ 4.862	≥ 4.420	AHRI 550/590
				65[m]	NA	NA	NA	≥ 3.550	NA	NA	NA	≥ 6.150	NA	NA	NA	NA	
	≥ 150 and < 300	≤ 0.6190 FL ≤ 0.5748 IPLV.IP	≤ 0.6684 FL ≤ 0.4211 IPLV.IP	44[m]	≥ 4.640	≥ 3.680	≥ 2.680	NA	≥ 8.330	≥ 6.410	≥ 4.420	NA	≥ 8.330	≥ 6.410	≥ 4.862	≥ 4.420	
				65[m]	NA	NA	NA	≥ 3.550	NA	NA	NA	≥ 6.150	NA	NA	NA	NA	
	≥ 300 and < 400	≤ 0.5895 FL ≤ 0.5526 IPLV.IP	≤ 0.6263 FL ≤ 0.4105 IPLV.IP	44[m]	≥ 4.640	≥ 3.680	≥ 2.680	NA	≥ 8.330	≥ 6.410	≥ 4.420	NA	≥ 8.330	≥ 6.410	≥ 4.862	≥ 4.420	
				65[m]	NA	NA	NA	≥ 3.550	NA	NA	NA	≥ 6.150	NA	NA	NA	NA	
	≥ 400 and < 600	≤ 0.5895 FL ≤ 0.5263 IPLV.IP	≤ 0.6158 FL ≤ 0.4000 IPLV.IP	44[m]	≥ 4.930	≥ 3.960	≥ 2.970	NA	≥ 8.900	≥ 6.980	≥ 5.000	NA	≥ 8.900	≥ 6.980	≥ 5.500	≥ 5.000	
				65[m]	NA	NA	NA	≥ 3.900	NA	NA	NA	≥ 6.850	NA	NA	NA	NA	
	≥ 60	≤ 0.5895 FL ≤ 0.5263 IPLV.IP	≤ 0.6158 FL ≤ 0.4000 IPLV.IP	44[m]	≥ 4.930	≥ 3.960	≥ 2.970	NA	≥ 8.900	≥ 6.980	≥ 5.000	NA	≥ 8.900	≥ 6.980	≥ 5.500	≥ 5.000	
				65[m]	NA	NA	NA	≥ 3.900	NA	NA	NA	≥ 6.850	NA	NA	NA	NA	

TABLE C403.3.2(15)—HEAT-PUMP AND HEAT RECOVERY CHILLER PACKAGES—MINIMUM EFFICIENCY REQUIREMENTS[g, o]—continued

For SI: °C = (°F – 32)/1.8.

NA = Not Applicable.

a. Cooling rating conditions are standard rating conditions defined in AHRI 550/590 (I-P), Table 4, except for liquid-cooled centrifugal chilling packages which can adjust cooling efficiency for nonstandard rating conditions using K_{adj} procedure in accordance with ASHRAE 90.1 Section 6.4.1.2.1.

b. Heating full-load rating conditions are at standard rating conditions defined in AHRI 550/590 (I-P), Table 4; includes the impact of defrost for air source heating ratings.

c. For liquid-source heat recovery chilling packages that have capabilities for heat rejection to a heat recovery condenser and a tower condenser the COP_{HR} applies to operation at full load with 100 percent heat recovery (no tower rejection). Units that only have capabilities for partial heat recovery shall meet the requirements of ASHRAE 90.1 Table 6.8.1-3.

d. For cooling operation, compliance with both the FL and IPLV is required, but only compliance with Path A or Path B cooling efficiency is required.

e. For units that operate in both cooling and heating, compliance with both the cooling and heating efficiency is required.

f. For applications where the chilling package is installed to operate only in heating, compliance only with the heating performance COP_H is required at only one of the heating AHRI 550/590 (I-P) standard rating conditions of Low, Medium, High, or Boost. Compliance with cooling performance is not required.

g. For air source heat pumps, compliance with both the 47°F and 17°F heating source outdoor air temperature (OAT) rating efficiency is required for heating.

h. For heat-pump chilling package applications where the cooling capacity is not being used for conditioning, compliance with the heating performance COP_H is only required at one of the four heating AHRI 550/590 standard ratings conditions of Low, Medium, High, or Boost. Compliance with the cooling performance is required as defined in notes a and d, except as noted in note f.

i. For simultaneous cooling and heating chillers applications where there is simultaneous cooling and heating, compliance with the simultaneous cooling performance heat recovery COP_{SHC} is only required at one of the four simultaneous cooling and heating AHRI 550/590 (I-P) standard ratings conditions of Low, Medium, High, or Boost. Compliance with the cooling only performance is required as defined in notes a and d.

j. For heat recovery heating chilling package applications where there is simultaneous cooling and heating, compliance with the heating performance heat recovery COP_{HR} is only required at one of the four heating AHRI 550/590 (I-P) standard ratings conditions of Low, Medium, Hot-Water 1, or Hot-Water 2. Compliance with the cooling only performance is required as defined in notes a and d.

k. Chilling packages employing a freeze-protection liquid in accordance with ASHRAE 90.1 Section 6.4.1.2.2 shall be tested or rated with water for the purpose of compliance with the requirements of this table.

l. Outdoor air entering dry-bulb (db) temperature and wet-bulb (wb) temperature.

m. Source-leaving liquid temperature.
   - The cooling evaporator liquid flow rate used for the heating rating for a reverse cycle air-to-water heat pump shall be the flow rate determined during the full-load cooling rating.
   - The cooling evaporator liquid flow rate for the simultaneous cooling and heating and heat recovery liquid cooled chilling packages rating shall be the liquid flow rates from the cooling operation full load rating.
   - For heating-only fluid-to-fluid chiller packages, the evaporator flow rate obtained with an entering liquid temperature of 54°F and a leaving liquid temperature of 44°F shall be used.

n. The size category is the full-load net refrigerating cooling mode capacity, which is the capacity of the evaporator available for cooling of the thermal load external to the chilling package.

o. A heat recovery condenser at its maximum load point must remove enough heat from the refrigerant to cool the refrigerant to remove all superheat energy and begin condensation of the refrigerant. A heat recovery system where only the superheat is reduced is not covered by ASHRAE 90.1 Table 6.8.1-16 and is considered a desuperheater, and the chiller package must comply with ASHRAE 90.1 Table 6.8.1-3.

p. Water-to-water heat pumps with a capacity less than 135,000 Btu/h are covered by ASHRAE 90.1 Table 6.8.1-15.

**TABLE C403.3.2(16)—
CEILING-MOUNTED COMPUTER ROOM AIR CONDITIONERS—MINIMUM EFFICIENCY REQUIREMENTS**

EQUIPMENT TYPE	STANDARD MODEL	NET SENSIBLE COOLING CAPACITY	MINIMUM NET SENSIBLE COP	RATING CONDITIONS RETURN AIR (dry bulb/dew point)	TEST PROCEDURE[a]
Air cooled with free air discharge condenser	Ducted	< 29,000 Btu/h	2.05	75°F/52°F (Class 1)	AHRI 1360
		≥ 29,000 Btu/h and < 65,000 Btu/h	2.02		
		≥ 65,000 Btu/h	1.92		
	Nonducted	< 29,000 Btu/h	2.08		
		≥ 29,000 Btu/h and < 65,000 Btu/h	2.05		
		≥ 65,000 Btu/h	1.94		
Air cooled with free air discharge condenser with fluid economizer	Ducted	< 29,000 Btu/h	2.01	75°F/52°F (Class 1)	AHRI 1360
		≥ 29,000 Btu/h and < 65,000 Btu/h	1.97		
		≥ 65,000 Btu/h	1.87		
	Nonducted	< 29,000 Btu/h	2.04		
		≥ 29,000 Btu/h and < 65,000 Btu/h	2.00		
		≥ 65,000 Btu/h	1.89		
Air cooled with ducted condenser	Ducted	< 29,000 Btu/h	1.86	75°F/52°F (Class 1)	AHRI 1360
		≥ 29,000 Btu/h and < 65,000 Btu/h	1.83		
		≥ 65,000 Btu/h	1.73		
	Nonducted	< 29,000 Btu/h	1.89		
		≥ 29,000 Btu/h and < 65,000 Btu/h	1.86		
		≥ 65,000 Btu/h	1.75		
Air cooled with fluid economizer and ducted condenser	Ducted	< 29,000 Btu/h	1.82	75°F/52°F (Class 1)	AHRI 1360
		≥ 29,000 Btu/h and < 65,000 Btu/h	1.78		
		≥ 65,000 Btu/h	1.68		
	Nonducted	< 29,000 Btu/h	1.85		
		≥ 29,000 Btu/h and < 65,000 Btu/h	1.81		
		≥ 65,000 Btu/h	1.70		
Water cooled	Ducted	< 29,000 Btu/h	2.38	75°F/52°F (Class 1)	AHRI 1360
		≥ 29,000 Btu/h and < 65,000 Btu/h	2.28		
		≥ 65,000 Btu/h	2.18		
	Nonducted	< 29,000 Btu/h	2.41		
		≥ 29,000 Btu/h and < 65,000 Btu/h	2.31		
		≥ 65,000 Btu/h	2.20		

**TABLE C403.3.2(16)—
CEILING-MOUNTED COMPUTER ROOM AIR CONDITIONERS—MINIMUM EFFICIENCY REQUIREMENTS—continued**

EQUIPMENT TYPE	STANDARD MODEL	NET SENSIBLE COOLING CAPACITY	MINIMUM NET SENSIBLE COP	RATING CONDITIONS RETURN AIR (dry bulb/dew point)	TEST PROCEDURE[a]
Water cooled with fluid economizer	Ducted	< 29,000 Btu/h	2.33	75°F/52°F (Class 1)	AHRI 1360
		≥ 29,000 Btu/h and < 65,000 Btu/h	2.23		
		≥ 65,000 Btu/h	2.13		
	Nonducted	< 29,000 Btu/h	2.36		
		≥ 29,000 Btu/h and < 65,000 Btu/h	2.26		
		≥ 65,000 Btu/h	2.16		
Glycol cooled	Ducted	< 29,000 Btu/h	1.97	75°F/52°F (Class 1)	AHRI 1360
		≥ 29,000 Btu/h and < 65,000 Btu/h	1.93		
		≥ 65,000 Btu/h	1.78		
	Nonducted	< 29,000 Btu/h	2.00		
		≥ 29,000 Btu/h and < 65,000 Btu/h	1.98		
		≥ 65,000 Btu/h	1.81		
Glycol cooled with fluid economizer	Ducted	< 29,000 Btu/h	1.92	75°F/52°F (Class 1)	AHRI 1360
		≥ 29,000 Btu/h and < 65,000 Btu/h	1.88		
		≥ 65,000 Btu/h	1.73		
	Nonducted	< 29,000 Btu/h	1.95		
		≥ 29,000 Btu/h and < 65,000 Btu/h	1.93		
		≥ 65,000 Btu/h	1.76		

For SI: 1 British thermal unit per hour = 0.2931 W, °C = (°F – 32)/1.8, COP = (Btu/h × hp)/(2,550.7).

a. Chapter 6 contains a complete specification of the referenced standards, which include test procedures, including the reference year version of the test procedure.

C403.3.2.1 Water-cooled centrifugal chilling packages. Equipment not designed for operation at AHRI Standard 550/590 test conditions of 44.00°F leaving and 54.00°F entering chilled-fluid temperatures, and with 85.00°F entering and 94.30°F leaving condenser-fluid temperatures, shall have maximum full-load kW/ton (FL) and part-load rating requirements adjusted using the following equations:

Equation 4-5 $FL_{adj} = FL/K_{adj}$

Equation 4-6 $PLV_{adj} = IPLV.IP/K_{adj}$

where:

$K_{adj} = A \times B$

FL = Full-load kW/ton value from Table C403.3.2(3).

FL_{adj} = Maximum full-load kW/ton rating, adjusted for nonstandard conditions.

$IPLV.IP$ = $IPLV.IP$ value from Table C403.3.2(3).

PLV_{adj} = Maximum *NPLV* rating, adjusted for nonstandard conditions.

$A = 0.00000014592 \times (LIFT)^4 - 0.0000346496 \times (LIFT)^3 + 0.00314196 \times (LIFT)^2 - 0.147199 \times (LIFT) + 3.93073$

$B = 0.0015 \times L_{vg}E_{vap} + 0.934$

$LIFT = L_{vg}Cond - L_{vg}E_{vap}$

$L_{vg}Cond$ = Full-load condenser leaving fluid temperature (°F).

$L_{vg}E_{vap}$ = Full-load evaporator leaving temperature (°F).

The FL_{adj} and PLV_{adj} values are applicable only for centrifugal chillers meeting all of the following full-load design ranges:

- $36.00°F \le L_{vg}E_{vap} \le 60.00°F$

- $L_{vg}Cond \leq 115.00°F$
- $20.00°F \leq LIFT \leq 80.00°F$

Manufacturers shall calculate the FL_{adj} and PLV_{adj} before determining whether to label the chiller. Centrifugal chillers designed to operate outside of these ranges are not covered by this code.

C403.3.2.2 Positive displacement (air- and water-cooled) chilling packages. Equipment with a leaving fluid temperature higher than 32°F (0°C) and water-cooled positive displacement chilling packages with a condenser leaving fluid temperature below 115°F (46°C) shall meet the requirements of the tables in Section C403.3.2 when tested or certified with water at standard rating conditions, in accordance with the referenced test procedure.

C403.3.3 Hot gas bypass limitation. Cooling systems shall not use hot gas bypass or other evaporator pressure control systems unless the system is designed with multiple steps of unloading or continuous capacity modulation. The capacity of the hot gas bypass shall be limited as indicated in Table C403.3.3, as limited by Section C403.5.1.

TABLE C403.3.3—MAXIMUM HOT GAS BYPASS CAPACITY

RATED CAPACITY	MAXIMUM HOT GAS BYPASS CAPACITY (% of total capacity)
≤ 240,000 Btu/h	50
> 240,000 Btu/h	25

For SI: 1 British thermal unit per hour = 0.2931 W.

C403.3.4 Boilers. Boiler systems shall comply with the following:

1. Combustion air positive shutoff shall be provided on all newly installed boiler systems that meet one or more of the following conditions:
   1.1. The total input capacity is not less than 2,500,000 Btu/h (733 kW) and one or more of the boilers are designed to operate with a nonpositive vent static pressure.
   1.2. Any stack serving the *boiler system* is connected to two or more boilers with a total combined input capacity of not less than 2,500,000 Btu/h (733 kW).
2. Newly installed boilers or boiler systems with a combustion air fan motor *nameplate horsepower* rating of 10 horsepower (7.46 kW) or more shall comply with one of the following:
   2.1. The fan motor shall be variable speed.
   2.2. The fan motor shall include controls that modulate fan airflow as a function of the load to a speed 50 percent or less of design air volume.

C403.3.4.1 Boiler oxygen concentration controls. Newly installed boilers with an input capacity of 5,000,000 Btu/h (1465 kW) and steady state full-load less than 90 percent shall maintain stack-gas oxygen concentrations not greater than the values specified in Table C403.3.4.1. Combustion air volume shall be controlled with respect to measured flue gas oxygen concentration. The use of a common gas and combustion air control linkage or jack shaft is not permitted.

Exception: These concentration limits do not apply where 50 percent or more of the *boiler system* capacity serves Group R-2 occupancies.

TABLE C403.3.4.1—BOILER OXYGEN CONCENTRATIONS

BOILER APPLICATION	MAXIMUM STACK-GAS OXYGEN CONCENTRATION[a]
Commercial boilers or where ≤ 10% of the boiler system capacity is used for process applications at design conditions	5%
Process boilers	3%

a. Concentration levels measured by volume on a dry basis over firing rates of 20 to 100 percent.
Exception: These concentration limits do not apply where 50 percent or more of the boiler system capacity serves Group R-2 occupancies.

C403.3.4.2 Boiler turndown. *Boiler systems* with design input of greater than 1,000,000 Btu/h (293 kW) shall comply with the turndown ratio specified in Table C403.3.4.2.

The system turndown requirement shall be met through the use of multiple single-input boilers, one or more *modulating boilers* or a combination of single-input and *modulating boilers*.

TABLE C403.3.4.2—BOILER TURNDOWN	
BOILER SYSTEM DESIGN INPUT (Btu/h)	MINIMUM TURNDOWN RATIO
≥ 1,000,000 and ≤ 5,000,000	3 to 1
> 5,000,000 and ≤ 10,000,000	4 to 1
> 10,000,000	5 to 1
For SI: 1 British thermal unit per hour = 0.2931 W.	

C403.4 Heating and cooling system controls. Heating and cooling system shall be provided with controls in accordance with Sections C403.4.1 through C403.4.8.

C403.4.1 Thermostatic controls. The supply of heating and cooling energy to each *zone* shall be controlled by individual thermostatic controls capable of responding to temperature within the *zone*. Where humidification or dehumidification or both is provided, not fewer than one humidity control device shall be provided for each humidity control system.

Exception: Independent perimeter systems that are designed to offset only *building thermal envelope* heat losses, gains or both serving one or more perimeter *zones* also served by an interior system provided that both of the following conditions are met:

1. The perimeter system includes not fewer than one thermostatic control *zone* for each *building* exposure having *exterior walls* facing only one orientation (within ±45 degrees) (0.8 rad) for more than 50 contiguous feet (15 240 mm).
2. The perimeter system heating and cooling supply is controlled by *thermostats* located within the *zones* served by the system.

C403.4.1.1 Heat pump supplementary heat. Heat pumps having supplementary electric resistance heat shall have controls that limit supplemental heat operation to only those times when one of the following applies:

1. The vapor compression cycle cannot provide the necessary heating energy to satisfy the *thermostat* setting.
2. The heat pump is operating in defrost mode.
3. The vapor compression cycle malfunctions.
4. The *thermostat* malfunctions.

C403.4.1.2 Deadband. Where used to control both heating and cooling, *zone* thermostatic controls shall:

1. Have separate setpoints for heating and cooling, each individually adjustable.
2. Be capable of and initially configured to provide a temperature range or deadband between the two setpoints of not less than 5°F (3°C) within which the supply of heating and cooling energy to the zone is shut off or reduced to a minimum.
3. Have a minimum deadband of not less than 1°F (0.56°C) when setpoints are adjusted.

Exceptions:

1. *Thermostats* requiring *manual* changeover between heating and cooling modes.
2. Occupancies or applications where applicable codes or accreditation standards requiring precision in indoor temperature control shall be permitted to be initially configured to not less than 1°F (0.56°C) deadband.

C403.4.1.3 Setpoint adjustment and display. Where thermostatic control setpoints are capable of being adjusted by occupants or HVAC system operators, the adjustment shall be independent for the heating setpoint and the cooling setpoint; when one setpoint is changed, the other shall not change except as needed to maintain the minimum deadband required by Section C403.4.1.2. For thermostatic controls that display setpoints, both the heating and cooling setpoints shall be displayed simultaneously, or the setpoint of the currently active mode (heating or cooling) shall be displayed along with an indication of that mode.

C403.4.1.4 Setpoint overlap restriction. Where heating and cooling to a *zone* are controlled by separate *zone* thermostatic controls located within the *zone*, mechanical or software means shall be provided to prevent the heating setpoint from exceeding the cooling setpoint, minus the deadband required by Section C403.4.1.2.

C403.4.1.5 Heated or cooled vestibules. The heating system for heated vestibules and air curtains with integral heating shall be provided with controls configured to shut off the source of heating when the outdoor air temperature is greater than 45°F (7°C). Vestibule heating and cooling systems shall be controlled by a *thermostat* located in the vestibule configured to limit heating to a temperature not greater than 60°F (16°C) and cooling to a temperature not less than 85°F (29°C).

Exception: Control of heating or cooling provided by site-recovered energy or transfer air that would otherwise be exhausted.

C403.4.1.6 Hot water boiler outdoor temperature setback control. Hot water boilers that supply heat to the *building* through one- or two-pipe heating systems shall have an outdoor setback control that lowers the boiler water temperature based on the outdoor temperature.

C403.4.2 Off-hour controls. Each *zone* shall be provided with thermostatic setback controls that are controlled by either an *automatic* time clock or programmable control system.

Exceptions:

1. *Zones* that will be operated continuously.
2. *Zones* with a full HVAC load demand not exceeding 6,800 Btu/h (2 kW) and having a *manual* shutoff switch located with *ready access*.

C403.4.2.1 Thermostatic setback. Thermostatic setback controls shall be configured to set back or temporarily operate the system to maintain *zone* temperatures down to 55°F (13°C) or up to 85°F (29°C).

C403.4.2.2 Automatic setback and shutdown. *Automatic* time clock or programmable controls shall be capable of starting and stopping the system for seven different daily schedules per week and retaining their programming and time setting during a loss of power for not fewer than 10 hours. Additionally, the controls shall have a *manual* override that allows temporary operation of the system for up to 2 hours; a manually operated timer configured to operate the system for up to 2 hours; or an occupancy sensor.

C403.4.2.3 Optimum start and stop. Optimum start and stop controls shall be provided for each heating and cooling system with direct control of individual *zones*. The optimum start controls shall be configured to automatically adjust the daily start time of the heating and cooling system in order to bring each space to the desired occupied temperature immediately prior to scheduled occupancy. The optimum stop controls shall be configured to reduce the heating and cooling system's heating temperature setpoint and increase the cooling temperature setpoint by not less than 2°F (1.11°C) before scheduled unoccupied periods based on the thermal lag and acceptable drift in space temperature that is within comfort limits.

Exception: *Dwelling units* and *sleeping units* are not required to have optimum start controls.

C403.4.3 Hydronic systems controls. The heating of fluids that have been previously mechanically cooled and the cooling of fluids that have been previously mechanically heated shall be limited in accordance with Sections C403.4.3.1 through C403.4.3.3. Hydronic heating systems comprised of multiple-packaged boilers and designed to deliver conditioned water or steam into a common distribution system shall include *automatic* controls configured to sequence operation of the boilers. Hydronic heating systems composed of a single boiler and greater than 500,000 Btu/h (146.5 kW) input design capacity shall include either a multistaged or modulating burner.

C403.4.3.1 Three-pipe system. Hydronic systems that use a common return system for both hot water and chilled water are prohibited.

C403.4.3.2 Two-pipe changeover system. Systems that use a common distribution system to supply both heated and chilled water shall be designed to allow a deadband between changeover from one mode to the other of not less than 15°F (8.3°C) outside air temperatures; be designed to and provided with controls that will allow operation in one mode for not less than 4 hours before changing over to the other mode; and be provided with controls that allow heating and cooling supply temperatures at the changeover point to be not more than 30°F (16.7°C) apart.

C403.4.3.3 Hydronic (water loop) heat pump systems. Hydronic heat pump systems shall comply with Sections C403.4.3.3.1 through C403.4.3.3.3.

C403.4.3.3.1 Temperature deadband. Hydronic heat pumps connected to a common heat pump water loop with central devices for heat rejection and heat addition shall have controls that are configured to provide a heat pump water supply temperature deadband of not less than 20°F (11°C) between initiation of heat rejection and heat addition by the central devices.

Exception: Where a system loop temperature optimization controller is installed and can determine the most efficient operating temperature based on real-time conditions of demand and capacity, deadbands of less than 20°F (11°C) shall be permitted.

C403.4.3.3.2 Heat rejection. The following shall apply to hydronic water loop heat pump systems in Climate Zones 3 through 8:

1. Where a closed-circuit cooling tower is used directly in the heat pump loop, either an *automatic* valve shall be installed to bypass the flow of water around the closed-circuit cooling tower, except for any flow necessary for freeze protection, or low-leakage positive-closure dampers shall be provided.
2. Where an open-circuit cooling tower is used directly in the heat pump loop, an *automatic* valve shall be installed to bypass all heat pump water flow around the open-circuit cooling tower.
3. Where an open-circuit or closed-circuit cooling tower is used in conjunction with a separate heat exchanger to isolate the open-circuit cooling tower from the heat pump loop, heat loss shall be controlled by shutting down the circulation pump on the cooling tower loop.

Exception: Where it can be demonstrated that a heat pump system will be required to reject heat throughout the year.

C403.4.3.3.3 Two-position valve. Each hydronic heat pump on the hydronic system having a total pump system power exceeding 10 hp (7.5 kW) shall have a two-position *automatic* valve interlocked to shut off the water flow when the compressor is off.

C403.4.4 Part-load controls. Hydronic systems greater than or equal to 300,000 Btu/h (87.9 kW) in design output capacity supplying heated or chilled water to comfort conditioning systems shall include controls that are configured to do all of the following:

1. Automatically reset the supply-water temperatures in response to varying *building* heating and cooling demand using coil valve position, zone-return water temperature, building-return water temperature or outside air temperature. The temperature shall be reset by not less than 25 percent of the design supply-to-return water temperature difference.
2. Automatically vary fluid flow for hydronic systems with a combined pump motor capacity of 2 hp (1.5 kW) or larger with three or more control valves or other devices by reducing the system design flow rate by not less than 50 percent or the maximum reduction allowed by the equipment manufacturer for proper operation of equipment by valves that modulate or step open and close, or pumps that modulate or turn on and off as a function of load.
3. Automatically vary pump flow on heating-water systems, chilled-water systems and heat rejection loops serving water-cooled unitary air conditioners as follows:
   3.1. Where pumps operate continuously or operate based on a time schedule, pumps with nominal output motor power of 2 hp or more shall have a variable speed drive.
   3.2. Where pumps have *automatic direct digital control* configured to operate pumps only when zone heating or cooling is required, a variable speed drive shall be provided for pumps with motors having the same or greater nominal output power indicated in Table C403.4.4 based on the *climate zone* and system served.
4. Where a variable speed drive is required by Item 3 of this section, pump motor power input shall be not more than 30 percent of design wattage at 50 percent of the design water flow. Pump flow shall be controlled to maintain one control valve nearly wide open or to satisfy the minimum differential pressure.

Exceptions:

1. Supply-water temperature reset is not required for chilled-water systems supplied by off-site district chilled water or chilled water from ice storage systems.
2. Variable pump flow is not required on dedicated coil circulation pumps where needed for freeze protection.
3. Variable pump flow is not required on dedicated equipment circulation pumps where configured in primary/secondary design to provide the minimum flow requirements of the equipment manufacturer for proper operation of equipment.
4. Variable speed drives are not required on heating water pumps where more than 50 percent of annual heat is generated by an electric boiler.

TABLE C403.4.4—VARIABLE SPEED DRIVE (VSD) REQUIREMENTS FOR DEMAND-CONTROLLED PUMPS

CHILLED WATER AND HEAT REJECTION LOOP PUMPS IN THESE CLIMATE ZONES	HEATING WATER PUMPS IN THESE CLIMATE ZONES	VSD REQUIRED FOR MOTORS WITH RATED OUTPUT OF:
0A, 0B, 1A, 1B, 2B	—	≥ 2 hp
2A, 3B	—	≥ 3 hp
3A, 3C, 4A, 4B	7, 8	≥ 5 hp
4C, 5A, 5B, 5C, 6A, 6B	3C, 5A, 5C, 6A, 6B	≥ 7.5 hp
—	4A, 4C, 5B	≥ 10 hp
7, 8	4B	≥ 15 hp
—	2A, 2B, 3A, 3B	≥ 25 hp
—	0B, 1B	≥ 100 hp
—	0A, 1A	≥ 200 hp

For SI: 1 hp = 0.746 kW.

C403.4.5 Pump isolation. Chilled water plants including more than one chiller shall be capable of and configured to reduce flow automatically through the chiller plant when a chiller is shut down. Chillers piped in series for the purpose of increased temperature differential shall be considered as one chiller.

Boiler systems including more than one boiler shall be capable of and configured to reduce flow automatically through the *boiler system* when a boiler is shut down.

C403.4.6 Reserved.

C403.4.7 Heating and cooling system controls for operable openings to the outdoors. All doors from a *conditioned space* to the outdoors and all other operable openings from a *conditioned space* to the outdoors that are larger than 40 square feet (3.7 m^2) when fully open shall have *automatic* controls interlocked with the heating and cooling system. The controls shall be configured to do the following within 5 minutes of opening:

1. Disable mechanical heating to the zone or reset the space heating temperature setpoint to 55°F (12.5°C) or less.

2. Disable mechanical cooling to the zone or reset the space cooling temperature setpoint to 90°F (32°C) or more. Mechanical cooling can remain enabled if the outdoor air temperature is below the space temperature.

Exceptions:

1. *Building entrances* with *automatic* closing devices.
2. Emergency exits with an *automatic* alarm that sounds when open.
3. Operable openings and doors serving *enclosed spaces* without a *thermostat* or heating or cooling temperature sensor.
4. Separately zoned areas associated with the preparation of food that contain appliances that contribute to the heating or cooling loads of a restaurant or similar type of occupancy.
5. Warehouses that utilize operable openings for the function of the occupancy, where *approved* by the *code official*.
6. The first *entrance doors* where located in the *exterior wall* and are part of a vestibule system.
7. Operable openings into spaces served by radiant heating and cooling systems.
8. *Alterations* where walls would have to be opened solely for the purpose of meeting this requirement and where *approved*.
9. Doors served by air curtains meeting the requirements of Section C402.6.6.

C403.4.8 Humidification and dehumidification controls. Humidification and dehumidification controls shall be in accordance with this section.

C403.4.8.1 Dehumidification. *Humidistatic controls* shall not use mechanical cooling to reduce the humidity below the lower of a dew point of 55°F (13°C) or relative humidity of 60 percent in the coldest *zone* served by the system. Lower humidity shall be permitted where mechanical cooling is being used for temperature control.

Exceptions:

1. Where approved, systems serving zones where specific humidity levels are required, such as museums and hospitals, and where *humidistatic controls* are capable of and configured to maintain a dead band of at least 10 percent relative humidity where no active humidification or dehumidification takes place.
2. Systems serving *zones* where humidity levels are required to be maintained with precision of not more than ±5 percent relative humidity to comply with applicable codes or accreditation standards or as *approved* by the authority having jurisdiction.

C403.4.8.2 Humidification. *Humidistatic controls* shall not use fossil fuels or electricity to produce relative humidity above 30 percent in the warmest *zone* served by the system.

Exceptions:

1. Where *approved*, systems serving *zones* where specific humidity levels are required, such as museums and hospitals, and where *humidistatic controls* are capable of and configured to maintain a deadband of at least 10 percent relative humidity where no active humidification or dehumidification takes place.
2. Systems serving *zones* where humidity levels are required to be maintained with precision of not more than ±5 percent relative humidity to comply with applicable codes or accreditation standards or as *approved* by the authority having jurisdiction.

C403.4.8.3 Control interlock. Where a *zone* is served by a system or systems with both humidification and dehumidification capability, means such as limit switches, mechanical stops, or for DDC systems, software programming, shall be provided capable of and configured to prevent simultaneous operation of humidification and dehumidification equipment.

Exception: Systems serving *zones* where humidity levels are required to be maintained with precision of not more than ±5 percent relative humidity to comply with applicable codes or accreditation standards or as *approved* by the authority having jurisdiction.

C403.5 Economizers. Economizers shall comply with Sections C403.5.1 through C403.5.5.

An air or *water economizer* shall be provided for the following cooling systems:

1. Chilled water systems with a total cooling capacity, less cooling capacity provided with air economizers, as specified in Table C403.5(1).
2. Individual *fan systems* with cooling capacity greater than or equal to 54,000 Btu/h (15.8 kW) in *buildings* having other than a *Group R* occupancy.

   The total supply capacity of all fan cooling units not provided with economizers shall not exceed 20 percent of the total supply capacity of all fan cooling units in the *building* or 300,000 Btu/h (88 kW), whichever is greater.
3. Individual *fan systems* with cooling capacity greater than or equal to 270,000 Btu/h (79.1 kW) in *buildings* having a *Group R* occupancy.

   The total supply capacity of all fan cooling units not provided with economizers shall not exceed 20 percent of the total supply capacity of all fan cooling units in the *building* or 1,500,000 Btu/h (440 kW), whichever is greater.

Exceptions: Economizers are not required for the following systems.

1. Individual *fan systems* not served by chilled water for *buildings* located in Climate Zones 0A, 0B, 1A and 1B.
2. Where more than 25 percent of the air designed to be supplied by the system is to spaces that are designed to be humidified above 35°F (1.7°C) dew-point temperature to satisfy process needs.

3. Systems expected to operate less than 20 hours per week.
4. Systems serving supermarket areas with open refrigerated casework.
5. Where the cooling efficiency is greater than or equal to the efficiency requirements in Table C403.5(2).
6. Systems that include a heat recovery system in accordance with Section C403.11.5.
7. Direct-expansion fan coils or unitary equipment with a capacity less than 54,000 Btu/h (15.8 kW) and multiple stages of compressor capacity installed with a dedicated outdoor air system.

TABLE C403.5(1)—MINIMUM CHILLED-WATER SYSTEM COOLING CAPACITY FOR DETERMINING ECONOMIZER COOLING REQUIREMENTS

CLIMATE ZONES (COOLING)	TOTAL CHILLED-WATER SYSTEM CAPACITY LESS CAPACITY OF COOLING UNITS WITH AIR ECONOMIZERS	
	Local Water-Cooled Chilled-Water Systems	Air-Cooled Chilled-Water Systems or District Chilled-Water Systems
0A, 1A	Economizer not required	Economizer not required
0B, 1B, 2A, 2B	960,000 Btu/h	1,250,000 Btu/h
3A, 3B, 3C, 4A, 4B, 4C	720,000 Btu/h	940,000 Btu/h
5A, 5B, 5C, 6A, 6B, 7, 8	1,320,000 Btu/h	1,720,000 Btu/h

For SI: 1 British thermal unit per hour = 0.2931 W.

TABLE C403.5(2)—EQUIPMENT EFFICIENCY PERFORMANCE EXCEPTION FOR ECONOMIZERS

CLIMATE ZONES	COOLING EQUIPMENT PERFORMANCE IMPROVEMENT (EER OR IPLV)
2A, 2B	10% efficiency improvement
3A, 3B	15% efficiency improvement
4A, 4B	20% efficiency improvement

C403.5.1 Integrated economizer control. Economizer systems shall be integrated with the mechanical cooling system and be configured to provide partial cooling even where additional mechanical cooling is required to provide the remainder of the cooling load. Controls shall not be capable of creating a false load in the mechanical cooling systems by limiting or disabling the economizer or any other means, such as hot gas bypass, except at the lowest stage of mechanical cooling.

Units that include an *air economizer* shall comply with the following:

1. Unit controls shall have the mechanical cooling capacity control interlocked with the *air economizer* controls such that the outdoor air damper is at the 100 percent open position when mechanical cooling is on and the outdoor air damper does not begin to close to prevent coil freezing due to minimum compressor run time until the leaving air temperature is less than 45°F (7°C).
2. Direct expansion (DX) units that control 75,000 Btu/h (22 kW) or greater of rated capacity of the capacity of the mechanical cooling directly based on occupied space temperature shall have not fewer than two stages of mechanical cooling capacity.
3. Other DX units, including those that control space temperature by modulating the airflow to the space, shall be in accordance with Table C403.5.1.

TABLE C403.5.1—DX COOLING STAGE REQUIREMENTS FOR MODULATING AIRFLOW UNITS

RATING CAPACITY	MINIMUM NUMBER OF MECHANICAL COOLING STAGES	MINIMUM COMPRESSOR DISPLACEMENT[a]
≥ 65,000 Btu/h and < 240,000 Btu/h	3 stages	≤ 35% of full load
≥ 240,000 Btu/h	4 stages	≤ 25% full load

For SI: 1 British thermal unit per hour = 0.2931 W.

a. For mechanical cooling stage control that does not use variable compressor displacement, the percent displacement shall be equivalent to the mechanical cooling capacity reduction evaluated at the full load rating conditions for the compressor.

C403.5.2 Economizer heating system impact. HVAC system design and economizer controls shall be such that economizer operation does not increase *building* heating energy use during normal operation.

Exception: Economizers on variable air volume (VAV) systems that cause *zone* level heating to increase because of a reduction in supply air temperature.

C403.5.3 Air economizers. Where economizers are required by Section C403.5, air economizers shall comply with Sections C403.5.3.1 through C403.5.3.5.

C403.5.3.1 Design capacity. *Air economizer* systems shall be configured to modulate outdoor air and return air dampers to provide up to 100 percent of the design supply air quantity as outdoor air for cooling.

C403.5.3.2 Control signal. Economizer controls and dampers shall be configured to sequence the dampers with the mechanical cooling equipment and shall not be controlled by only mixed-air temperature.

Exception: The use of mixed-air temperature limit control shall be permitted for systems controlled from space temperature (such as single-*zone* systems).

C403.5.3.3 High-limit shutoff. Air economizers shall be configured to automatically reduce outdoor air intake to the design minimum outdoor air quantity when outdoor air intake will not reduce cooling energy usage. High-limit shutoff control types for specific climates shall be chosen from Table C403.5.3.3. High-limit shutoff control settings for these control types shall be those specified in Table C403.5.3.3.

TABLE C403.5.3.3—HIGH-LIMIT SHUTOFF CONTROL SETTING FOR AIR ECONOMIZERS[b]

DEVICE TYPE	CLIMATE ZONE	REQUIRED HIGH LIMIT (ECONOMIZER OFF WHEN):	
		Equation	Description
Fixed dry bulb	0B, 1B, 2B, 3B, 3C, 4B, 4C, 5B, 5C, 6B, 7, 8	$T_{OA} > 75°F$	Outdoor air temperature exceeds 75°F
	5A, 6A	$T_{OA} > 70°F$	Outdoor air temperature exceeds 70°F
	0A, 1A, 2A, 3A, 4A	$T_{OA} > 65°F$	Outdoor air temperature exceeds 65°F
Differential dry bulb	0B, 1B, 2B, 3B, 3C, 4B, 4C, 5A, 5B, 5C, 6A, 6B, 7, 8	$T_{OA} > T_{RA}$	Outdoor air temperature exceeds return air temperature
Fixed enthalpy with fixed dry-bulb temperatures	All	$h_{OA} > 28$ Btu/lb[a] or $T_{OA} > 75°F$	Outdoor air enthalpy exceeds 28 Btu/lb of dry air[a] or
			Outdoor air temperature exceeds 75°F
Differential enthalpy with fixed dry-bulb temperature	All	$h_{OA} > h_{RA}$ or $T_{OA} > 75°F$	Outdoor air enthalpy exceeds return air enthalpy or
			Outdoor air temperature exceeds 75°F

For SI: °C = (°F – 32)/1.8, 1 Btu/lb = 2.33 kJ/kg.

a. At altitudes substantially different than sea level, the fixed enthalpy limit shall be set to the enthalpy value at 75°F and 50 percent relative humidity. As an example, at approximately 6,000 feet elevation, the fixed enthalpy limit is approximately 30.7 Btu/lb.

b. Devices with selectable setpoints shall be capable of being set to within 2°F and 2 Btu/lb of the setpoint listed.

C403.5.3.4 Relief of excess outdoor air. Systems shall provide one of the following means to relieve excess outdoor air during *air economizer* operation to prevent overpressurizing the *building*.

1. Return or relief fan(s) meeting the requirements of Section C403.11.1.
2. A barometric or motorized damper relief path with a total pressure drop at a design relief airflow rate less than 0.10 inches water column (25 Pa) from the occupied space to the outdoors. Design relief airflow rate shall be the design supply airflow rate minus any continuous exhaust flows, such as toilet exhaust fans, whose makeup is provided by the economizer system.

The relief air outlet shall be located to avoid recirculation into the *building*.

C403.5.3.5 Economizer dampers. Return, exhaust/relief and outdoor air dampers used in economizers shall comply with Section C403.7.7.

C403.5.4 Water-side economizers. Where economizers are required by Section C403.5, water-side economizers shall comply with Sections C403.5.4.1 and C403.5.4.2.

C403.5.4.1 Design capacity. *Water economizer* systems shall be configured to cool supply air by indirect evaporation and providing up to 100 percent of the expected system cooling load at outdoor air temperatures of not greater than 50°F (10°C) dry bulb/45°F (7°C) wet bulb.

Exceptions:

1. Systems primarily serving *computer rooms* in which 100 percent of the expected system cooling load at 40°F (4°C) dry bulb/35°F (1.7°C) wet bulb is met with evaporative water economizers.
2. Systems primarily serving *computer rooms* with dry cooler water economizers that satisfy 100 percent of the expected system cooling load at 35°F (1.7°C) dry bulb.
3. Systems where dehumidification requirements cannot be met using outdoor air temperatures of 50°F (10°C) dry bulb/45°F (7°C) wet bulb and where 100 percent of the expected system cooling load at 45°F (7°C) dry bulb/40°F (4°C) wet bulb is met with evaporative water economizers.

C403.5.4.2 Maximum pressure drop. Precooling coils and water-to-water heat exchangers used as part of a *water economizer* system shall either have a water-side pressure drop of less than 15 feet (45 kPa) of water or a secondary loop shall be created so that the coil or heat exchanger pressure drop is not seen by the circulating pumps when the system is in the normal cooling (noneconomizer) mode.

C403.5.5 Economizer fault detection and diagnostics. Air-cooled unitary direct-expansion units listed in the tables in Section C403.3.2 and variable refrigerant flow (VRF) units that are equipped with an economizer in accordance with Sections C403.5 through C403.5.4 shall include a fault detection and diagnostics system complying with the following:

1. The following temperature sensors shall be permanently installed to monitor system operation:
   - 1.1. Outside air.
   - 1.2. Supply air.
   - 1.3. Return air.
2. Temperature sensors shall have an accuracy of ±2°F (1.1°C) over the range of 40°F to 80°F (4°C to 26.7°C).
3. Refrigerant pressure sensors, where used, shall have an accuracy of ±3 percent of full scale.
4. The unit controller shall be configured to provide system status by indicating the following:
   - 4.1. Free cooling available.
   - 4.2. Economizer enabled.
   - 4.3. Compressor enabled.
   - 4.4. Heating enabled.
   - 4.5. Mixed air low limit cycle active.
   - 4.6. The current value of each sensor.
5. The unit controller shall be capable of manually initiating each operating mode so that the operation of compressors, economizers, fans and the heating system can be independently tested and verified.
6. The unit shall be configured to report faults to a fault management application available for *access* by day-to-day operating or service personnel, or annunciated locally on *zone thermostats*.
7. The fault detection and diagnostics system shall be configured to detect the following faults:
   - 7.1. Air temperature sensor failure/fault.
   - 7.2. Not economizing when the unit should be economizing.
   - 7.3. Economizing when the unit should not be economizing.
   - 7.4. Damper not modulating.
   - 7.5. Excess outdoor air.

C403.6 Requirements for mechanical systems serving multiple zones. Sections C403.6.1 through C403.6.9 shall apply to mechanical systems serving multiple *zones*.

C403.6.1 Variable air volume and multiple-zone systems. Supply air systems serving multiple *zones* shall be variable air volume (VAV) systems that have *zone* controls configured to reduce the volume of air that is reheated, recooled or mixed in each *zone* to one of the following:

1. Thirty percent of the *zone* design peak supply for systems with *direct digital control* (DDC).
2. Systems with DDC where all of the following apply:
   - 2.1. The airflow rate in the deadband between heating and cooling does not exceed the highest of the allowed rates under Items 3, 4, 5 or 6 of this section.
   - 2.2. The first stage of heating modulates the *zone* supply air temperature setpoint up to a maximum setpoint while the airflow is maintained at the deadband flow rate.
   - 2.3. The second stage of heating modulates the airflow rate from the deadband flow rate up to the heating maximum flow rate that is less than 50 percent of the *zone* design peak supply rate.
3. The outdoor airflow rate required to meet the minimum *ventilation* requirements of Chapter 4 of the *International Mechanical Code*.
4. The minimum primary airflow rate required to meet the Simplified Procedure *ventilation* requirements of ASHRAE 62.1 for the zone and is permitted to be the average airflow rate as allowed by ASHRAE 62.1.
5. Any higher rate that can be demonstrated to reduce overall system annual energy use by offsetting reheat/recool energy losses through a reduction in outdoor air intake for the system as *approved* by the *code official*.
6. The airflow rate required to comply with applicable codes or accreditation standards such as pressure relationships or minimum air change rates.

Exception: The following individual *zones* or entire air distribution systems are exempted from the requirement for VAV control:

1. *Zones* or supply air systems where not less than 75 percent of the energy for reheating or for providing warm air in mixing systems is provided from a site-recovered, including condenser heat, or site-solar energy source.
2. Systems that prevent reheating, recooling, mixing or simultaneous supply of air that has been previously cooled, either mechanically or through the use of economizer systems, and air that has been previously mechanically heated.

C403.6.2 Single-duct VAV systems, terminal devices. Single-duct VAV systems shall use terminal devices capable of and configured to reduce the supply of primary supply air before reheating or recooling takes place.

C403.6.3 Dual-duct and mixing VAV systems, terminal devices. Systems that have one warm air *duct* and one cool air *duct* shall use terminal devices that are configured to reduce the flow from one *duct* to a minimum before mixing of air from the other *duct* takes place.

C403.6.4 Single-fan dual-duct and mixing VAV systems, economizers. Individual dual-duct or mixing heating and cooling systems with a single fan and with total capacities greater than 90,000 Btu/h [(26.4 kW) 7.5 tons] shall not be equipped with air economizers.

C403.6.5 Supply-air temperature reset controls. Multiple-zone HVAC systems shall include controls that are capable of and configured to automatically reset the supply-air temperature in response to representative *building* loads, or to outdoor air temperature. The controls shall be configured to reset the supply air temperature not less than 25 percent of the difference between the design supply-air temperature and the design room air temperature. Controls that adjust the reset based on *zone* humidity are allowed in Climate Zones 0B, 1B, 2B, 3B, 3C and 4 through 8. HVAC *zones* that are expected to experience relatively constant loads shall have maximum airflow designed to accommodate the fully reset supply-air temperature.

Exceptions:

1. Systems that prevent reheating, recooling or mixing of heated and cooled supply air.
2. Seventy-five percent of the energy for reheating is from site-recovered or site-solar energy sources.
3. Systems in Climate Zones 0A, 1A and 3A with less than 3,000 cfm (1500 L/s) of design outside air.
4. Systems in Climate Zone 2A with less than 10,000 cfm (5000 L/s) of design outside air.
5. Systems in Climate Zones 0A, 1A, 2A and 3A with not less than 80 percent outside air and employing exhaust air energy recovery complying with Section C403.7.4.

C403.6.5.1 Dehumidification control interaction. In Climate Zones 0A, 1A, 2A and 3A, the system design shall allow supply-air temperature reset while dehumidification is provided. When dehumidification control is active, air economizers shall be locked out.

C403.6.6 Multiple-zone VAV system ventilation optimization control. Multiple-zone VAV systems with *direct digital control* of individual *zone* boxes reporting to a central control panel shall have *automatic* controls configured to reduce outdoor air intake flow below design rates in response to changes in system *ventilation* efficiency (E_v) as defined by the *International Mechanical Code*.

Exceptions:

1. VAV systems with zonal transfer fans that recirculate air from other *zones* without directly mixing it with outdoor air, dual-duct dual-fan VAV systems, and VAV systems with fan-powered terminal units.
2. Systems where total design exhaust airflow is more than 70 percent of total design outdoor air intake flow requirements.

C403.6.7 Parallel-flow fan-powered VAV air terminal control. Parallel-flow fan-powered VAV air terminals shall have *automatic* controls configured to:

1. Turn off the terminal fan except when space heating is required or where required for *ventilation*.
2. Turn on the terminal fan as the first stage of heating before the heating coil is activated.
3. During heating for warmup or setback temperature control, either:
    3.1. Operate the terminal fan and heating coil without primary air.
    3.2. Reverse the terminal damper logic and provide heating from the central air handler by primary air.

C403.6.8 Setpoints for direct digital control. For systems with *direct digital control* of individual *zones* reporting to the central control panel, the static pressure setpoint shall be reset based on the *zone* requiring the most pressure. In such case, the setpoint is reset lower until one *zone* damper is nearly wide open. The *direct digital controls* shall be capable of monitoring zone damper positions or shall have an alternative method of indicating the need for static pressure that is configured to provide all of the following:

1. *Automatic* detection of any *zone* that excessively drives the reset logic.
2. Generation of an alarm to the system operational location.
3. Allowance for an operator to readily remove one or more *zones* from the reset algorithm.

C403.6.9 Static pressure sensor location. Static pressure sensors used to control VAV fans shall be located such that the controller setpoint is not greater than 1.2 inches w.c. (299 Pa). Where this results in one or more sensors being located downstream of major duct splits, not less than one sensor shall be located on each major branch to ensure that static pressure can be maintained in each branch.

C403.7 Ventilation and exhaust systems. In addition to other requirements of Section C403 applicable to the provision of *ventilation* air or the exhaust of air, *ventilation* and exhaust systems shall be in accordance with Sections C403.7.1 through C403.7.9.

C403.7.1 Demand control ventilation. Demand control ventilation (DCV) shall be provided for the following:

1. Spaces with *ventilation* provided by single-zone systems where an air-side economizer is provided in accordance with Section C403.5.

2. Spaces larger than 250 square feet (23 m^2) in Climate Zones 5A, 6, 7, and 8 and spaces larger than 500 square feet (46.5 m^2) in other *climate zones* that have a design occupant load of 15 people or greater per 1,000 square feet (93 m^2) of floor area, as established in Table 403.3.1.1 of the *International Mechanical Code*, and are served by systems with one or more of the following:
   2.1. An air-side economizer.
   2.2. *Automatic* modulating control of the outdoor air damper.
   2.3. A design outdoor airflow greater than 3,000 cfm (1416 L/s).

Exceptions:

1. Spaces served by systems with energy recovery in accordance with Section C403.7.4.2 and that have a floor area less than:
   1.1. 6,000 square feet (557 m^2) in Climate Zone 3C.
   1.2. 2,000 square feet (186 m^2) in Climate Zones 1A, 3B and 4B.
   1.3. 1,000 square feet (93 m^2) in Climate Zones 2A, 2B, 3A, 4A, 4C, 5 and 6.
   1.4. 400 square feet (37 m^2) in Climate Zones 7 and 8.
2. Multiple-zone systems without *direct digital control* of individual zones communicating with a central control panel.
3. Spaces served by multiple-zone systems with a design outdoor airflow less than 750 cfm (354 L/s).
4. Spaces where more than 75 percent of the space design outdoor airflow is required for makeup air that is exhausted from the space or transfer air that is required for makeup air that is exhausted from other spaces.
5. Spaces with one of the following occupancy classifications as defined in Table 403.3.1.1 of the *International Mechanical Code*: correctional cells, education laboratories, barber, beauty and nail salons, and bowling alley seating areas.
6. Spaces where the *registered design professional* demonstrates an engineered ventilation system design that:
   6.1. Prevents the maximum concentration of contaminants from being more than that obtainable by the required rate of outdoor air *ventilation*.
   6.2. Allows the required minimum design rate of outdoor air to be reduced by not less than 15 percent.

C403.7.2 Parking garage ventilation controls. Ventilation systems employed in enclosed parking garages shall comply with Section 404.1 of the *International Mechanical Code* and the following:

1. Separate ventilation systems and control systems shall be provided for each *parking garage section*.
2. Control systems for each *parking garage section* shall be capable of and configured to reduce fan airflow to not less than 0.05 cfm per square foot [0.00025 $m^3/(s \times m^2)$] of the floor area served and not more than 20 percent of the design capacity.
3. The ventilation system for each *parking garage section* shall have controls and devices that result in fan motor demand of not more than 30 percent of design wattage at 50 percent of the design airflow.

Exception: Garage ventilation systems serving a single *parking garage section* having a total ventilation system motor *nameplate horsepower* (ventilation system motor nameplate kilowatt) not exceeding 5 hp (3.7 kW) at *fan system design conditions* and where the *parking garage section* has no mechanical cooling or mechanical heating.

Nothing in this section shall be construed to require more than one *parking garage section* in any parking structure.

C403.7.3 Ventilation air heating control. Units that provide *ventilation* air to multiple *zones* and operate in conjunction with *zone* heating and cooling systems shall not use heating or heat recovery to warm supply air to a temperature greater than 60°F (16°C) when representative *building* loads or outdoor air temperatures indicate that the majority of *zones* require cooling.

Exception: Units that heat the airstream using only *series energy recovery* when representative *building* loads or outdoor air temperature indicates that the majority of *zones* require cooling in Climate Zones 0A, 1A, 2A, 3A and 4A.

C403.7.4 Energy recovery systems. Energy recovery ventilation systems shall be provided as specified in either Section C403.7.4.1 or C403.7.4.2, as applicable.

C403.7.4.1 Nontransient dwelling units. Nontransient dwelling units shall be provided with outdoor air energy recovery ventilation systems complying with not less than one of the following:

1. The system shall have an enthalpy recovery ratio of not less than 50 percent at cooling design condition and not less than 60 percent at heating design condition.
2. The system shall have a sensible recovery efficiency (SRE) that is not less than 65 percent at 32°F (0°C) and in Climate Zones 0A, 1A, 2A and 3A shall have a net moisture transfer (NMT) that is not less than 40 percent at 95°F (35°C). SRE and NMT shall be determined from a listed value or from interpolation of listed values at an airflow not less than the design airflow, based on testing in accordance with CAN/CSA C439.

Exceptions:

1. Nontransient dwelling units in Climate Zone 3C.
2. Nontransient dwelling units with not more than 500 square feet (46 m^2) of *conditioned floor area* in Climate Zones 0, 1, 2, 3, 4C and 5C.
3. *Enthalpy recovery ratio* requirements at heating design condition in Climate Zones 0, 1 and 2.
4. *Enthalpy recovery ratio* requirements at cooling design condition in Climate Zones 4, 5, 6, 7 and 8.

C403.7.4.2 Spaces other than nontransient dwelling units. Where the supply airflow rate of a *fan system* serving a space other than a nontransient dwelling unit exceeds the values specified in Tables C403.7.4.2(1) and C403.7.4.2(2), the system shall include an energy recovery system. The energy recovery system shall provide an *enthalpy recovery ratio* of not less than 50 percent at design conditions. Where an *air economizer* is required, the energy recovery system shall include a bypass or controls that permit operation of the economizer as required by Section C403.5.

Exception: An *energy recovery ventilation system* shall not be required in any of the following conditions:

1. Where energy recovery systems are prohibited by the *International Mechanical Code*.
2. Laboratory fume hood systems that include not fewer than one of the following features:
   2.1. Variable-air-volume hood exhaust and room supply systems configured to reduce exhaust and makeup air volume to 50 percent or less of design values.
   2.2. Direct makeup (auxiliary) air supply equal to or greater than 75 percent of the exhaust rate, heated not warmer than 2°F (1.1°C) above room setpoint, cooled to not cooler than 3°F (1.7°C) below room setpoint, with no humidification added, and no simultaneous heating and cooling used for dehumidification control.
3. Systems serving spaces that are heated to less than 60°F (15.5°C) and that are not cooled.
4. Heating energy recovery where more than 60 percent of the outdoor heating energy is provided from site-recovered or site-solar energy in Climate Zones 5 through 8.
5. *Enthalpy recovery ratio* requirements at heating design condition in Climate Zones 0, 1 and 2.
6. *Enthalpy recovery ratio* requirements at cooling design condition in Climate Zones 3C, 4C, 5B, 5C, 6B, 7 and 8.
7. Systems in Climate Zones 0 through 4 requiring dehumidification that employ series energy recovery land have a minimum SERR of 0.40.
8. Where the largest source of air exhausted at a single location at the *building* exterior is less than 75 percent of the design outdoor airflow rate.
9. Systems expected to operate less than 20 hours per week at the *outdoor air* percentage covered by Table C403.7.4.2(1).
10. Systems exhausting toxic, flammable, paint or corrosive fumes or dust.
11. Commercial kitchen hoods used for collecting and removing grease vapors and smoke.

TABLE C403.7.4.2(1)—ENERGY RECOVERY REQUIREMENT (Ventilation systems operating less than 8,000 hours per year)

CLIMATE ZONE	PERCENT (%) OUTDOOR AIR AT FULL DESIGN AIRFLOW RATE							
	≥ 10% and < 20%	≥ 20% and < 30%	≥ 30% and < 40%	≥ 40% and < 50%	≥ 50% and < 60%	≥ 60% and < 70%	≥ 70% and < 80%	≥ 80%
	Design Supply Fan Airflow Rate (cfm)							
3B, 3C, 4B, 4C, 5B	NR	NR	NR	NR	NR	NR	NR	NR
0B, 1B, 2B, 5C	NR	NR	NR	NR	≥ 26,000	≥ 12,000	≥ 5,000	≥ 4,000
6B	≥ 28,000	≥ 26,500	≥ 11,000	≥ 5,500	≥ 4,500	≥ 3,500	≥ 2,500	≥ 1,500
0A, 1A, 2A, 3A, 4A, 5A, 6A	≥ 26,000	≥ 16,000	≥ 5,500	≥ 4,500	≥ 3,500	≥ 2,000	≥ 1,000	> 120
7, 8	≥ 4,500	≥ 4,000	≥ 2,500	≥ 1,000	> 140	> 120	> 100	> 80

For SI: 1 cfm = 0.4719 L/s.
NR = Not Required.

TABLE C403.7.4.2(2)—ENERGY RECOVERY REQUIREMENT (Ventilation systems operating not less than 8,000 hours per year)

CLIMATE ZONE	PERCENT (%) OUTDOOR AIR AT FULL DESIGN AIRFLOW RATE							
	≥ 10% and < 20%	≥ 20% and < 30%	≥ 30% and < 40%	≥ 40% and < 50%	≥ 50% and < 60%	≥ 60% and < 70%	≥ 70% and < 80%	≥ 80%
	Design Supply Fan Airflow Rate (cfm)							
3C	NR	NR	NR	NR	NR	NR	NR	NR
0B, 1B, 2B, 3B, 4C, 5C	NR	≥ 19,500	≥ 9,000	≥ 5,000	≥ 4,000	≥ 3,000	≥ 1,500	≥ 120
0A, 1A, 2A, 3A, 4B, 5B	≥ 2,500	≥ 2,000	≥ 1,000	≥ 500	≥ 140	≥ 120	≥ 100	≥ 80
4A, 5A, 6A, 6B, 7, 8	≥ 200	≥ 130	≥ 100	≥ 80	≥ 70	≥ 60	≥ 50	≥ 40

For SI: 1 cfm = 0.4719 L/s.
NR = Not Required.

C403.7.5 Kitchen exhaust systems. Replacement air introduced directly into the exhaust hood cavity shall not be greater than 10 percent of the hood exhaust airflow rate. Conditioned supply air delivered to any space shall not exceed the greater of the following:

1. The ventilation rate required to meet the space heating or cooling load.
2. The hood exhaust flow minus the available transfer air from adjacent space where available transfer air is considered to be that portion of outdoor *ventilation* air not required to satisfy other exhaust needs, such as restrooms, and not required to maintain pressurization of adjacent spaces.

Kitchen exhaust hood systems serving Type I exhaust hoods shall be provided with *demand control kitchen ventilation* (DCKV) controls where a kitchen or kitchen/dining facility has a total Type I kitchen hood exhaust airflow rate greater than 5,000 cubic feet per minute (2360 L/s). DCKV systems shall be configured to provide a minimum of 50 percent reduction in exhaust and replacement air system airflow rates. Systems shall include controls necessary to modulate exhaust and replacement air system airflows in response to appliance operation and to maintain full capture and containment of smoke, effluent and combustion products during cooking and idle operation. Each hood shall be a factory-built commercial exhaust hood *listed* by a nationally recognized testing laboratory and shall have a maximum exhaust rate as specified in Table C403.7.5.

Where a single hood, or hood section, is installed over appliances with different duty ratings, the maximum allowable flow rate for the hood or hood section shall be based on the requirements for the highest appliance duty rating under the hood or hood section.

Exceptions:

1. UL 710 listed exhaust hoods that have a design maximum exhaust flow rate not greater than 250 cubic feet per minute (118 L/s) per linear foot (305 mm) of hood that serve kitchen or kitchen/dining facilities with a total kitchen hood exhaust airflow rate less than 5,000 cfm (2360 L/s).
2. Where allowed by the *International Mechanical Code*, an *energy recovery ventilation system* is installed on the kitchen exhaust with a sensible heat recovery effectiveness of not less than 40 percent on not less than 50 percent of the total exhaust hood airflow.

TABLE C403.7.5—MAXIMUM NET EXHAUST FLOW RATE, CFM PER LINEAR FOOT OF HOOD LENGTH

TYPE OF HOOD	LIGHT-DUTY EQUIPMENT	MEDIUM-DUTY EQUIPMENT	HEAVY-DUTY EQUIPMENT	EXTRA-HEAVY-DUTY EQUIPMENT
Wall-mounted canopy	140	210	280	385
Single island	280	350	420	490
Double island (per side)	175	210	280	385
Eyebrow	175	175	NA	NA
Backshelf/Pass-over	210	210	280	NA

For SI: 1 cfm = 0.4719 L/s; 1 foot = 304.8 mm.
NA = Not Allowed.

C403.7.6 Automatic control of HVAC systems serving guestrooms. In Group R-1 *buildings* containing more than 50 guestrooms, each guestroom shall be provided with controls complying with the provisions of Sections C403.7.6.1 and C403.7.6.2.

C403.7.6.1 Temperature setpoint controls. Controls shall be provided on each HVAC system that are capable of and configured with three modes of temperature control.

1. When the guestroom is rented but unoccupied, the controls shall automatically raise the cooling setpoint and lower the heating setpoint by not less than 4°F (2°C) from the occupant setpoint within 30 minutes after the occupants have left the guestroom.
2. When the guestroom is unrented and unoccupied, the controls shall automatically raise the cooling setpoint to not lower than 80°F (27°C) and lower the heating setpoint to not higher than 60°F (16°C). Unrented and unoccupied guestroom mode shall be initiated within 16 hours of the guestroom being continuously occupied or where a *networked guestroom control system* indicates that the guestroom is unrented and the guestroom is unoccupied for more than 20 minutes. A *networked guestroom control system* that is capable of returning the *thermostat* setpoints to default occupied setpoints 60 minutes prior to the time a guestroom is scheduled to be occupied is not precluded by this section. Cooling that is capable of limiting relative humidity with a setpoint not lower than 65 percent relative humidity during unoccupied periods is not precluded by this section.
3. When the guestroom is occupied, HVAC setpoints shall return to their occupied setpoints once occupancy is sensed.

C403.7.6.2 Ventilation controls. Controls shall be provided on each HVAC system that are capable of and configured to automatically turn off the *ventilation* and exhaust fans within 20 minutes of the occupants leaving the guestroom, or *isolation* *devices* shall be provided to each guestroom that are capable of automatically shutting off the supply of outdoor air to and exhaust air from the guestroom.

Exception: Guestroom ventilation systems are not precluded from having an *automatic* daily pre-occupancy purge cycle that provides daily outdoor air *ventilation* during unrented periods at the design ventilation rate for 60 minutes, or at a rate and duration equivalent to one air change.

C403.7.7 Shutoff dampers. Outdoor air intake and exhaust openings and stairway and shaft vents shall be provided with Class I motorized dampers. The dampers shall have an *air leakage* rate not greater than 4 cfm/ft ^{2}(20.3 L/s × m^2) of damper surface area at 1.0 inch water gauge (249 Pa) and shall be *labeled* by an *approved agency* when tested in accordance with AMCA 500D for such purpose.

Outdoor air intake and exhaust dampers shall be installed with *automatic* controls configured to close when the systems or spaces served are not in use or during unoccupied period warm-up and setback operation, unless the systems served require outdoor or exhaust air in accordance with the *International Mechanical Code* or the dampers are opened to provide intentional economizer cooling.

Stairway and elevator shaft vent dampers shall be installed with *automatic* controls configured to open upon the activation of any fire alarm initiating device of the *building's* fire alarm system, the interruption of power to the damper, or by thermostatic control systems.

Exception: Nonmotorized gravity dampers shall be an alternative to motorized dampers for exhaust and relief openings as follows:

1. In *buildings* less than three stories in height above grade plane.
2. In *buildings* of any height located in Climate Zones 0, 1, 2 or 3.
3. Where the design exhaust capacity is not greater than 300 cfm (142 L/s).

Nonmotorized gravity dampers shall have an *air leakage* rate not greater than 20 cfm/ft^2(101.6 L/s × m^2) where not less than 24 inches (610 mm) in either dimension and 40 cfm/ft^2(203.2 L/s × m^2) where less than 24 inches (610 mm) in either dimension. The rate of *air leakage* shall be determined at 1.0 inch water gauge (249 Pa) when tested in accordance with AMCA 500D for such purpose. The dampers shall be *labeled* by an *approved agency*.

C403.7.8 Occupied standby controls. The following spaces shall be equipped with occupied standby controls in accordance with Section C403.7.8.1 for each ventilation zone:

1. Postsecondary classrooms, lecture rooms and training rooms.
2. Conference/meeting/multipurpose rooms.
3. Lounges/breakrooms.
4. Enclosed offices.
5. Open-plan office areas.
6. Corridors.

Exception: Zones that are part of a multiple-zone system without automatic zone flow control dampers.

C403.7.8.1 Occupied-standby zone controls. Within 5 minutes of all spaces in that *zone* entering *occupied-standby mode*, the *zone* control shall operate as follows:

1. The active heating setpoint shall be set back by not less than 1°F (0.55°C).
2. The active cooling setpoint shall be set up by not less than 1°F (0.55°C).
3. All airflow supplied to the *zone* shall be shut off whenever the space temperature is between the active heating and cooling setpoints.
4. Multiple-zone systems shall comply with Section C403.7.8.1.1.

C403.7.8.1.1 Multiple-zone system controls. Multiple-zone systems required to automatically reset the effective minimum outdoor air setpoint, per Section C403.6.6, shall reset the effective minimum outdoor air setpoint based on a *zone* outdoor air requirement of zero for all *zones* in *occupied-standby mode*. Sequences of operation for system outside air reset shall comply with an *approved* method.

C403.7.9 Dwelling unit ventilation system. A fan that is the air mover for a heating or cooling system that serves an individual *dwelling unit* shall not be used to provide outdoor air.

Exception: Where the fan efficacy is not less than 1.2 cubic feet per minute (0.56 L/s) of outdoor airflow per watt when there is no demand for heating or cooling.

C403.8 Fans and fan controls. Fans in HVAC systems shall comply with Sections C403.8.1 through C403.8.6.1.

C403.8.1 Allowable fan horsepower. Where the summed fan system motor *nameplate horsepower* on an HVAC *fan system* is greater than 5 hp (3.7 kW) at *fan system design conditions*, it shall not be greater than the allowable total *fan system motor nameplate hp* (Option 1) or *fan system bhp* (Option 2), as specified in Table C403.8.1(1). Such summed HVAC fan system motor *nameplate horsepower* shall include supply fans, exhaust fans, return or relief fans, and fan-powered terminal units associated with systems providing heating or cooling capability. Single-zone variable air volume systems shall comply with the constant volume fan power limitation.

Exceptions:

1. Hospital, vivarium and laboratory systems that utilize flow control devices on exhaust or return to maintain space pressure relationships necessary for occupant health and safety or environmental control shall be permitted to use variable volume fan power limitation.
2. Individual exhaust fans with motor *nameplate horsepower* of 1 hp (0.746 kW) or less are exempt from the allowable fan horsepower requirement.

TABLE C403.8.1(1)—FAN POWER LIMITATION

	LIMIT	CONSTANT VOLUME	VARIABLE VOLUME
Option 1: Fan system motor nameplate hp	Allowable nameplate motor hp	$hp \le CFM_S \times 0.0011$	$hp \le CFM_S \times 0.0015$
Option 2: Fan system bhp	Allowable fan system bhp	$bhp \le CFM_S \times 0.00094 + A$	$bhp \le CFM_S \times 0.0013 + A$

For SI: 1 bhp = 735.5 W, 1 hp = 745.5 W, 1 cfm = 0.4719 L/s.

where:

CFM_S = The maximum design supply airflow rate to conditioned spaces served by the system in cubic feet per minute.
hp = The maximum combined motor nameplate horsepower.
bhp = The maximum combined fan brake horsepower.
A = Sum of $[PD \times CFM_D / 4131]$.

where:

PD = Each applicable pressure drop adjustment from Table C403.8.1(2) in. w.c.
CFM_D = The design airflow through each applicable device from Table C403.8.1(2) in cubic feet per minute.

TABLE C403.8.1(2)—FAN POWER LIMITATION PRESSURE DROP ADJUSTMENT

DEVICE	ADJUSTMENT
Credits	
Return air or exhaust systems required by code or accreditation standards to be fully ducted, or systems required to maintain air pressure differentials between adjacent rooms	0.5 inch w.c. (2.15 inches w.c. for laboratory and vivarium systems)
Return and exhaust airflow control devices	0.5 inch w.c.
Exhaust filters, scrubbers or other exhaust treatment	The pressure drop of device calculated at fan system design condition
Particulate filtration credit: MERV 9 thru 12	0.5 inch w.c.
Particulate filtration credit: MERV 13 thru 15	0.9 inch w.c.
Particulate filtration credit: MERV 16 and greater and electronically enhanced filters	Pressure drop calculated at 2 times the clean filter pressure drop at fan system design condition.
Carbon and other gas-phase air cleaners	Clean filter pressure drop at fan system design condition.
Biosafety cabinet	Pressure drop of device at fan system design condition.
Energy recovery device, other than coil runaround loop	For each airstream, (2.2 × energy recovery effectiveness - 0.5) inch w.c.
Coil runaround loop	0.6 inch w.c. for each airstream.
Evaporative humidifier/cooler in series with another cooling coil	Pressure drop of device at fan system design conditions.
Sound attenuation section (fans serving spaces with design background noise goals below NC35)	0.15 inch w.c.
Exhaust system serving fume hoods	0.35 inch w.c.
Laboratory and vivarium exhaust systems in high-rise buildings	0.25 inch w.c./100 feet of vertical duct exceeding 75 feet.
Deductions	
Systems without central cooling device	- 0.6 inch w.c.
Systems without central heating device	- 0.3 inch w.c.
Systems with central electric resistance heat	- 0.2 inch w.c.

For SI: 1 inch w.c. = 249 Pa, 1 inch = 25.4 mm, 1 foot = 304.8 mm.
w.c. = Water Column, NC = Noise Criterion.

C403.8.2 Motor nameplate horsepower. For each fan, the *fan brake horsepower* (bhp) shall be indicated on the *construction documents* and the selected motor shall be not larger than the first available motor size greater than the following:

1. For fans less than 6 bhp (4476 W), 1.5 times the *fan brake horsepower*.
2. For fans 6 bhp (4476 W) and larger, 1.3 times the *fan brake horsepower*.

Exceptions:

1. Fans equipped with electronic speed control devices to vary the fan airflow as a function of load.
2. Fans with a *fan nameplate electrical input power* of less than 0.89 kW.
3. Systems complying with Section C403.8.1 *fan system motor nameplate hp* (Option 1).
4. Fans with motor *nameplate horsepower* less than 1 hp (746 W).

C403.8.3 Fan efficiency. Each fan and *fan array* shall have a *fan energy index* (FEI) of not less than 1.00 at the design point of operation, as determined in accordance with AMCA 208 by an *approved* independent testing laboratory and *labeled* by the manufacturer. Each fan and *fan array* used for a variable-air-volume system shall have an FEI of not less than 0.95 at the design point of operation, as determined in accordance with AMCA 208 by an *approved* independent testing laboratory and *labeled* by the manufacturer. The FEI for fan arrays shall be calculated in accordance with AMCA 208 Annex C.

Exceptions: The following fans are not required to have a *fan energy index*:

1. Fans that are not *embedded fans* with motor *nameplate horsepower* of less than 1.0 hp (0.75 kW) or with a nameplate electrical input power of less than 0.89 kW.
2. *Embedded fans* that have a motor *nameplate horsepower* of 5 hp (3.7 kW) or less, or with a *fan system electrical input power* of 4.1 kW or less.
3. Multiple fans operated in series or parallel as the functional equivalent of a single fan that have a combined motor *nameplate horsepower* of 5 hp (3.7 kW) or less or with a *fan system electrical input power* of 4.1 kW or less.
4. Fans that are part of equipment covered in Section C403.3.2.
5. Fans included in an equipment package certified by an *approved agency* for air or energy performance.
6. Ceiling fans, which are defined as nonportable devices suspended from a ceiling or overhead structure for circulating air via the rotation of the blades.
7. Fans used for moving gases at temperatures above 482°F (250°C).
8. Fans used for operation in explosive atmospheres.
9. Reversible fans used for tunnel ventilation.
10. Fans that are intended to operate only during emergency conditions.
11. Fans outside the scope of AMCA 208.

C403.8.4 Fractional hp fan motors. Motors for fans that are not less than $^1/_{12}$ hp (0.062 kW) and are less than 1 hp (0.746 kW) shall be electronically commutated motors or shall have a minimum motor efficiency of 70 percent, rated in accordance with DOE 10 CFR 431. These motors shall have the means to adjust motor speed for either balancing or remote control. The use of belt-driven fans to sheave adjustments for airflow balancing instead of a varying motor speed shall be permitted.

Exceptions: The following motors are not required to comply with this section

1. Motors in the airstream within fan coils and terminal units that only provide heating to the space served.
2. Motors in space-conditioning equipment that comply with Section C403.3.2 or Sections C403.8.1. through C403.8.3.
3. Motors that comply with Section C405.8.

C403.8.5 Low-capacity ventilation fans. Mechanical ventilation system fans with motors less than $^1/_{12}$ hp (0.062 kW) in capacity shall meet the efficacy requirements of Table C403.8.5 at one or more rating points. Airflow shall be tested in accordance with the test procedure referenced in Table C403.8.5 and *listed*. The airflow shall be reported in the product listing or on the label. Fan efficacy shall be reported in the product listing or shall be derived from the input power and airflow values reported in the product listing or on the label. Fan efficacy for fully ducted HRV, ERV, balanced and in-line fans shall be determined at a static pressure not less than 0.2 inch w.c. (49.8 Pa). Fan efficacy for ducted range hoods, bathroom and utility room fans shall be determined at a static pressure not less than 0.1 inch w.c. (24.9 Pa).

Exceptions:

1. Where ventilation fans are a component of a *listed* heating or cooling appliance.
2. Dryer exhaust duct power ventilators, domestic range hoods and domestic range booster fans that operate intermittently.
3. Fans in radon mitigation systems.
4. Fans not covered within the scope of the test methods referenced in Table C403.8.5.
5. Ceiling fans regulated under 10 CFR 430, Appendix U.

TABLE C403.8.5—LOW-CAPACITY VENTILATION FAN EFFICACY[a]

SYSTEM TYPE	AIRFLOW RATE (CFM)	MINIMUM EFFICACY (CFM/WATT)	TEST PROCEDURE
Balanced ventilation system without heat or energy recovery	Any	1.2[a]	ASHRAE Standard 51 (ANSI/AMCA Standard 210)
HRV, ERV	Any	1.2	CAN/CSA 439
Range hood	Any	2.8	ASHRAE 51 (ANSI/AMCA Standard 210)
In-line supply or exhaust fan	Any	3.8	
Other exhaust fan	≤ 90	2.8	
	≥ 90 and < 200	3.5	
	≥ 200	4.0	

For SI: 1 cfm/ft = 0.47 L/s.

a. For balanced systems, HRVs and ERVs, determine the efficacy as the outdoor airflow divided by the total fan power.

C403.8.6 Fan control. Controls shall be provided for fans in accordance with Section C403.8.6.1 and as required for specific systems provided in Section C403.

C403.8.6.1 Fan airflow control. Each cooling system listed in Table C403.8.6.1 shall be designed to vary the indoor fan airflow as a function of load and shall comply with the following requirements:

1. Direct expansion (DX) and chilled water cooling units that control the capacity of the mechanical cooling directly based on space temperature shall have not fewer than two stages of fan control. Low or minimum speed shall not be greater than 66 percent of full speed. At low or minimum speed, the *fan system* shall draw not more than 40 percent of the fan power at full fan speed. Low or minimum speed shall be used during periods of low cooling load and ventilation-only operation.
2. Other units including DX cooling units and chilled water units that control the space temperature by modulating the airflow to the space shall have modulating fan control. Minimum speed shall be not greater than 50 percent of full speed. At minimum speed the *fan system* shall draw not more than 30 percent of the power at full fan speed. Low or minimum speed shall be used during periods of low cooling load and ventilation-only operation.
3. Units that include an air-side economizer in accordance with Section C403.5 shall have not fewer than two speeds of fan control during economizer operation.

Exceptions:

1. Modulating fan control is not required for chilled water and evaporative cooling units with fan motors of less than 1 hp (0.746 kW) where the units are not used to provide *ventilation air* and the indoor fan cycles with the load.
2. Where the volume of outdoor air required to comply with the ventilation requirements of the *International Mechanical Code* at low speed exceeds the air that would be delivered at the speed defined in Section C403.8.6, the minimum speed shall be selected to provide the required *ventilation air*.

TABLE C403.8.6.1—COOLING SYSTEMS

COOLING SYSTEM TYPE	FAN MOTOR SIZE	MECHANICAL COOLING CAPACITY
DX cooling	Any	≥ 65,000 Btu/h
Chilled water and evaporative cooling	≥ $^{1}/_{4}$ hp	Any

For SI: 1 British thermal unit per hour = 0.2931 W; 1 hp = 0.746 kW.

C403.8.6.2 Intermittent exhaust control for bathrooms and toilet rooms. Where an exhaust system serving a bathroom or toilet room is designed for intermittent operation, the exhaust system shall be provided with *manual* on capability and one or more of the following controls:

1. A timer control that has a minimum setpoint not greater than 30 minutes.
2. An occupant sensor control that automatically turns off exhaust fans within 30 minutes after all occupants have left the space.
3. A humidity control capable of manual or automatic adjustment from a minimum setpoint not greater than 50 percent to a maximum setpoint not greater than 80 percent relative humidity.
4. A contaminant control that responds to a particle or gaseous concentration.

Exception: Bathroom and toilet room exhaust systems serving as an integral component of an outdoor air ventilation system in Group R-2, R-3 and R-4 occupancies shall not be required to provide controls other than manual on capability.

An off setpoint shall not be used to comply with a minimum setpoint requirement.

C403.9 Large-diameter ceiling fans. Where provided, *large-diameter ceiling fans* shall be tested and *labeled* in accordance with AMCA 230 and shall meet the efficiency requirements of Table C403.9 and Section C403.9.1.

TABLE C403.9—CEILING FAN EFFICIENCY REQUIREMENTS[a]

EQUIPMENT TYPE	MINIMUM EFFICIENCY[b, c]	TEST PROCEDURE
Large-diameter ceiling fan for applications outside the US[c]	CFEI ≥ 1.00 at high (maximum) speed CFEI ≥ 1.31 at 40% of high speed or the nearest speed that is not less than 40% of high speed	10 CFR 430, Appendix U or AMCA 230 and AMCA 208 (for FEI calculations)
Large-diameter ceiling fan	CFEI ≥ 1.00 at high (maximum) speed; and CFEI ≥ 1.31 at 40% of high speed or the nearest speed that is not less than 40% of high speed	10 CFR 430, Appendix U

a. The minimum efficiency requirements at both high speed and 40% of maximum speed shall be met or exceeded to comply with this code.
b. Ceiling fans are regulated as consumer products by 10 CFR 430.
c. Chapter 6 contains a complete specification of the referenced test procedure, including the referenced year version of the test procedure.

C403.9.1 Ceiling Fan Energy Index (CFEI). The Ceiling Fan Energy Index shall be calculated as the ratio of the electric input power of a reference *large-diameter ceiling fan* to the electric input power of the actual *large-diameter ceiling fan* as calculated in accordance with AMCA 208 with the following modifications to the calculations for the reference fan: using an airflow constant (Q) of 26,500 cfm (12.5 m^3/s), a pressure constant (P) of 0.0027 inch of water (0.6719 Pa), and fan efficiency constant (η) of 42 percent.

C403.10 Buildings with high-capacity space-heating gas boiler systems. Gas hot water boiler systems for space heating with system input capacities of not less than 1,000,000 Btu/h (293 kW) and not greater than 10,000,000 Btu/h (2931 kW) in new *buildings* shall comply with Sections C403.10.1 and C403.10.2.

Exceptions:

1. Where 25 percent of the annual space heating requirement is provided by *on-site renewable energy*, site-recovered energy or heat recovery chillers.
2. Space heating boilers installed in individual *dwelling units*.
3. Where 50 percent or more of the design heating load is served using perimeter convective heating, radiant ceiling panels or both.
4. Individual gas boilers with input capacity less than 300,000 Btu/h (88 kW) shall not be included in the calculations of the total system input or total system efficiency.

C403.10.1 Boiler efficiency. Gas hot water boilers shall have a thermal efficiency (E_t) of not less than 90 percent where rated in accordance with the test procedures in Table C403.3.2(6). Systems with multiple boilers are allowed to meet this requirement where the space heating input provided by equipment with E_t above or below 90 percent provides an input capacity-weighted average E_t of not less than 90 percent. For boilers rated only for combustion efficiency, the calculation for the input capacity-weighted average E_t shall use the combustion efficiency value.

C403.10.2 Hot water distribution system design. The hot water distribution system shall be designed to meet the following:

1. Coils and other heat exchangers shall be selected so that at design conditions the hot water return temperature entering the boilers is 120°F (49°C) or less.
2. Under all operating conditions, the water temperature entering the boiler is not greater than 120°F (49°C) or the flow rate of supply hot water that recirculates directly into the return system, such as by three-way valves or minimum flow bypass controls, shall be not greater than 20 percent of the design flow of the boilers.

C403.11 Heat rejection equipment. Heat rejection equipment, including air-cooled condensers, dry coolers, open-circuit cooling towers, closed-circuit cooling towers and evaporative condensers, shall comply with this section.

Exception: Heat rejection devices where energy usage is included in the equipment efficiency ratings listed in Tables C403.3.2(6) and C403.3.2(7).

C403.11.1 Fan speed control. Each *fan system* powered by an individual motor or array of motors with connected power, including the motor service factor, totaling 5 hp (3.7 kW) or more shall have controls and devices configured to automatically modulate the fan speed to control the leaving fluid temperature or condensing temperature and pressure of the heat rejection device. Fan motor power input shall be not more than 30 percent of design wattage at 50 percent of the design airflow.

Exceptions:

1. Fans serving multiple refrigerant or fluid cooling circuits.
2. Condenser fans serving flooded condensers.

C403.11.2 Multiple-cell heat rejection equipment. Multiple-cell heat rejection equipment with variable speed fan drives shall be controlled to operate the maximum number of fans allowed that comply with the manufacturer's requirements for all system components and so that all fans operate at the same fan speed required for the instantaneous cooling duty, as opposed to staged on and off operation. The minimum fan speed shall be the minimum allowable speed of the fan drive system in accordance with the manufacturer's recommendations.

C403.11.3 Limitation on centrifugal fan open-circuit cooling towers. Centrifugal fan open-circuit cooling towers with a combined rated capacity of 1,100 gpm (4164 L/m) or greater at 95°F (35°C) condenser water return, 85°F (29°C) condenser water supply, and 75°F (24°C) outdoor air wet-bulb temperature shall meet the energy efficiency requirement for axial fan open-circuit cooling towers listed in Table C403.3.2(7) .

Exception: Centrifugal open-circuit cooling towers that are designed with inlet or discharge *ducts* or require external sound attenuation.

C403.11.4 Tower flow turndown. Open-circuit cooling towers used on water-cooled chiller systems that are configured with multiple- or variable-speed condenser water pumps shall be designed so that all open-circuit cooling tower cells can be run in parallel with the larger of the flow that is produced by the smallest pump at its minimum expected flow rate or at 50 percent of the design flow for the cell.

C403.11.5 Heat recovery for service water heating. Condenser heat recovery shall be installed for heating or reheating of service hot water provided that the facility operates 24 hours a day, the total installed heat capacity of water-cooled systems exceeds 6,000,000 Btu/hr (1758 kW) of heat rejection, and the design service water heating load exceeds 1,000,000 Btu/h (293 kW).

The required heat recovery system shall have the capacity to provide the smaller of the following:

1. Sixty percent of the peak heat rejection load at design conditions.
2. The preheating required to raise the peak service hot water draw to 85°F (29°C).

Exceptions:

1. Facilities that employ condenser heat recovery for space heating or reheat purposes with a heat recovery design exceeding 30 percent of the peak water-cooled condenser load at design conditions.
2. Facilities that provide 60 percent of their service water heating from site solar or site recovered energy or from other sources.

C403.11.6 Heat recovery for space conditioning in health care facilities. Where heated water is used for space heating, a heat pump chiller meeting the requirements of Table C403.3.2(15) for heat recovery and that uses the cooling system return water as the heat source shall be installed where the following are true:

1. The *building* is a Group I-2, Condition 2 occupancy.
2. The total design chilled water capacity for the Group I-2,Condition 2 occupancy, either air cooled or water cooled, required at cooling design conditions exceeds 3,600,000 Btu/h (1100 kw) of cooling.
3. Simultaneous heating, including reheat, and cooling occurs above 60°F (16°C) outdoor air temperature.

The heat recovery system shall have a cooling capacity of not less than 7 percent of the total design chilled water capacity of the Group I-2, Condition 2 occupancy at peak design conditions.

Exceptions:

1. *Buildings* that provide 60 percent or more of their reheat energy from *on-site renewable energy* or other site-recovered energy. *On-site renewable energy* used to meet Section C405.15.1 or C406.3.1 shall not be used to meet this exception.
2. *Buildings* in Climate Zones 5C, 6B, 7 and 8.

C403.12 Refrigeration equipment performance. Refrigeration equipment performance shall be determined in accordance with Sections C403.12.1 and C403.12.2 for commercial refrigerators, freezers, refrigerator-freezers, *walk-in coolers*, *walk-in freezers* and refrigeration equipment. The energy use shall be verified through certification under an *approved* certification program or, where a certification program does not exist, the energy use shall be supported by data furnished by the equipment manufacturer.

Exception: *Walk-in coolers* and *walk-in freezers* regulated under federal law in accordance with Subpart R of DOE 10 CFR 431.

C403.12.1 Commercial refrigerators, refrigerator-freezers and refrigeration. Refrigeration equipment, defined in DOE 10 CFR Part 431.62, shall have an energy use in kWh/day not greater than the values of Table C403.12.1 when tested and rated in accordance with AHRI 1200.

TABLE C403.12.1—MINIMUM EFFICIENCY REQUIREMENTS: COMMERCIAL REFRIGERATORS AND FREEZERS AND REFRIGERATION

EQUIPMENT CATEGORY	CONDENSING UNIT CONFIGURATION	EQUIPMENT FAMILY	RATING TEMP., °F	OPERATING TEMP., °F	EQUIPMENT CLASSIFICATION[a, c]	MAXIMUM DAILY ENERGY CONSUMPTION, kWh/day[d, e]	TEST STANDARD
Remote condensing commercial refrigerators and commercial freezers	Remote (RC)	Vertical open (VOP)	38 (M)	≥ 32	VOP.RC.M	0.64 × TDA + 4.07	AHRI 1200
			0 (L)	< 32	VOP.RC.L	2.20 × TDA + 6.85	
		Semivertical open (SVO)	38 (M)	≥ 32	SVO.RC.M	0.66 × TDA + 3.18	
			0 (L)	< 32	SVO.RC.L	2.20 × TDA + 6.85	
		Horizontal open (HZO)	38 (M)	≥ 32	HZO.RC.M	0.35 × TDA + 2.88	
			0 (L)	< 32	HZO.RC.L	0.55 × TDA + 6.88	
		Vertical closed transparent (VCT)	38 (M)	≥ 32	VCT.RC.M	0.15 × TDA + 1.95	
			0 (L)	< 32	VCT.RC.L	0.49 × TDA + 2.61	
		Horizontal closed transparent (HCT)	38 (M)	≥ 32	HCT.RC.M	0.16 × TDA + 0.13	
			0 (L)	< 32	HCT.RC.L	0.34 × TDA + 0.26	
		Vertical closed solid (VCS)	38 (M)	≥ 32	VCS.RC.M	0.10 × V + 0.26	
			0 (L)	< 32	VCS.RC.L	0.21 × V + 0.54	
		Horizontal closed solid (HCS)	38 (M)	≥ 32	HCS.RC.M	0.10 × V + 0.26	
			0 (L)	< 32	HCS.RC.L	0.21 × V + 0.54	
		Service over counter (SOC)	38 (M)	≥ 32	SOC.RC.M	0.44 × TDA + 0.11	
			0 (L)	< 32	SOC.RC.L	0.93 × TDA + 0.22	
Self-contained commercial refrigerators and commercial freezers with and without doors	Self-contained (SC)	Vertical open (VOP)	38 (M)	≥ 32	VOP.SC.M	1.69 × TDA + 4.71	AHRI 1200
			0 (L)	< 32	VOP.SC.L	4.25 × TDA + 11.82	
		Semivertical open (SVO)	38 (M)	≥ 32	SVO.SC.M	1.70 × TDA + 4.59	
			0 (L)	< 32	SVO.SC.L	4.26 × TDA + 11.51	
		Horizontal open (HZO)	38 (M)	≥ 32	HZO.SC.M	0.72 × TDA + 5.55	
			0 (L)	< 32	HZO.RC.L	1.90 × TDA + 7.08	
		Vertical closed transparent (VCT)	38 (M)	≥ 32	VCT.SC.M	0.10 × V + 0.86	
			0 (L)	< 32	VCT.SC.L	0.29 × V + 2.95	
		Vertical closed solid (VCS)	38 (M)	≥ 32	VCS.SC.M	0.05 × V + 1.36	
			0 (L)	< 32	VCS.SC.L	0.22 × V + 1.38	
		Horizontal closed transparent (HCT)	38 (M)	≥ 32	HCT.SC.M	0.06 × V + 0.37	
			0 (L)	< 32	HCT.SC.L	0.08 × V + 1.23	
		Horizontal closed solid (HCS)	38 (M)	≥ 32	HCS.SC.M	0.05 × V + 0.91	
			0 (L)	< 32	HCS.SC.L	0.06 × V + 1.12	
		Service over counter (SOC)	38 (M)	≥ 32	SOC.SC.M	0.52 × TDA + 1.00	
			0 (L)	< 32	SOC.SC.L	1.10 × TDA + 2.10	
Self-contained commercial refrigerators with transparent doors for pull-down temperature applications	Self-contained (SC)	Pull-down (PD)	38 (M)	≥ 32	PD.SC.M	0.11 × V + 0.81	AHRI 1200

TABLE C403.12.1—MINIMUM EFFICIENCY REQUIREMENTS: COMMERCIAL REFRIGERATORS AND FREEZERS AND REFRIGERATION—continued

EQUIPMENT CATEGORY	CONDENSING UNIT CONFIGURATION	EQUIPMENT FAMILY	RATING TEMP., °F	OPERATING TEMP., °F	EQUIPMENT CLASSIFICATION[a, c]	MAXIMUM DAILY ENERGY CONSUMPTION, kWh/day[d, e]	TEST STANDARD
Commercial ice cream freezers	Remote (RC)	Vertical open (VOP)	-15 (I)	≤ -5[b]	VOP.RC.I	2.79 × TDA + 8.70	AHRI 1200
		Semivertical open (SVO)			SVO.RC.I	2.79 × TDA + 8.70	
		Horizontal open (HZO)			HZO.RC.I	0.70 × TDA + 8.74	
		Vertical closed transparent (VCT)			VCT.RC.I	0.58 × TDA + 3.05	
		Horizontal closed transparent (HCT)			HCT.RC.I	0.40 × TDA + 0.31	
		Vertical closed solid (VCS)			VCS.RC.I	0.25 × V + 0.63	
		Horizontal closed solid (HCS)			HCS.RC.I	0.25 × V + 0.63	
		Service over counter (SOC)			SOC.RC.I	1.09 × TDA + 0.26	
	Self-contained (SC)	Vertical open (VOP)			VOP.SC.I	5.40 × TDA + 15.02	AHRI 1200
		Semivertical open (SVO)			SVO.SC.I	5.41 × TDA + 14.63	
		Horizontal open (HZO)			HZO.SC.I	2.42 × TDA + 9.00	
		Vertical closed transparent (VCT)			VCT.SC.I	0.62 × TDA + 3.29	
		Horizontal closed transparent (HCT)			HCT.SC.I	0.56 × TDA + 0.43	
		Vertical closed solid (VCS)			VCS.SC.I	0.34 × V + 0.88	
		Horizontal closed solid (HCS)			HCS.SC.I	0.34 × V + 0.88	
		Service over counter (SOC)			SOC.SC.I	1.53 × TDA + 0.36	

For SI: 1 square foot = 0.0929 m^2, 1 cubic foot = 0.02832 m^3, °C = (°F – 32)/1.8.

a. The meaning of the letters in this column is indicated in the columns to the left.

b. Ice cream freezer is defined in DOE 10 CFR 431.62 as a commercial freezer that is designed to operate at or below -5 °F and that the manufacturer designs, markets or intends for the storing, displaying or dispensing of ice cream.

c. Equipment class designations consist of a combination [in sequential order separated by periods (AAA).(BB).(C)] of the following:

- (AAA)—An equipment family code (VOP = vertical open, SVO = semivertical open, HZO = horizontal open, VCT = vertical closed transparent doors, VCS = vertical closed solid doors, HCT = horizontal closed transparent doors, HCS = horizontal closed solid doors, and SOC = service over counter);
- (BB)—An operating mode code (RC = remote condensing and SC = self-contained); and
- (C)—A rating temperature code [M = medium temperature (38°F), L = low temperature (0°F), or I = ice cream temperature (-15°F)].
- For example, "VOP.RC.M" refers to the "vertical open, remote condensing, medium temperature" equipment class.

d. V is the volume of the case (ft^3) as measured in AHRI 1200, Appendix C.

e. TDA is the total display area of the case (ft^2) as measured in AHRI 1200, Appendix D.

C403.12.2 Walk-in coolers and walk-in freezers. *Walk-in cooler* and *walk-in freezer* refrigeration systems, except for walk-in process cooling refrigeration systems as defined in DOE 10 CFR 431.302, shall meet the requirements of Tables C403.12.2.1(1), C403.12.2.1(2) and C403.12.2.1(3).

C403.12.2.1 Performance standards. *Walk-in coolers* and *walk-in freezers* shall meet the requirements of Tables C403.12.2.1(1), C403.12.2.1(2) and C403.12.2.1(3).

TABLE C403.12.2.1(1)—WALK-IN COOLER AND FREEZER DISPLAY DOOR EFFICIENCY REQUIREMENTSa

CLASS DESCRIPTOR	CLASS	MAXIMUM ENERGY CONSUMPTION (kWh/day)[a]	TEST PROCEDURE
Display door, medium temperature	DD, M	$0.04 \times A_{dd} + 0.41$	10 CFR 431
Display door, low temperature	DD, L	$0.15 \times A_{dd} + 0.29$	10 CFR 431

a. Add is the surface area of the display door.

TABLE C403.12.2.1(2)—WALK-IN COOLER AND FREEZER NONDISPLAY DOOR EFFICIENCY REQUIREMENTSa

CLASS DESCRIPTOR	CLASS	MAXIMUM ENERGY CONSUMPTION (kWh/day)[a]	TEST PROCEDURE
Passage door, medium temperature	PD, M	$0.05 \times A_{nd} + 1.7$	10 CFR 431
Passage door, low temperature	PD, L	$0.14 \times A_{nd} + 4.8$	10 CFR 431
Freight door, medium temperature	FD, M	$0.04 \times A_{nd} + 1.9$	10 CFR 431
Freight door, low temperature	FD, L	$0.12 \times A_{nd} + 5.6$	10 CFR 431

a. A_{nd} is the surface area of the nondisplay door.

TABLE C403.12.2.1(3)—WALK-IN COOLER AND FREEZER REFRIGERATION SYSTEM EFFICIENCY REQUIREMENTS

CLASS DESCRIPTOR	CLASS	MINIMUM ANNUAL WALK-IN ENERGY FACTOR (AWEF) (Btu/W-h)[a]	TEST PROCEDURE
Dedicated condensing, medium temperature, indoor system	DC.M.I	5.61	AHRI 1250
Dedicated condensing, medium temperature, outdoor system	DC.M.O	7.60	
Dedicated condensing, low temperature, indoor system, net capacity (q_{net}) < 6,500 Btu/h	DC.L.I < 6,500	$9.091 \times 10^{-5} \times q_{net} + 1.81$	
Dedicated condensing, low temperature, indoor system, net capacity (q_{net}) ≥ 6,500 Btu/h	DC.L.I ≥ 6,500	2.40	
Dedicated condensing, low temperature, outdoor system, net capacity (q_{net}) < 6,500 Btu/h	DC.L.O < 6,500	$6.522 \times 10^{-5} \times q_{net} + 2.73$	
Dedicated condensing, low temperature, outdoor system, net capacity (q_{net}) ≥ 6,500 Btu/h	DC.L.O ≥ 6,500	3.15	
Unit cooler, medium	UC.M	9.00	
Unit cooler, low temperature, net capacity (q_{net}) < 15,500 Btu/h	UC.L < 15,500	$1.575 \times 10^{-5} \times q_{net} + 3.91$	
Unit cooler, low temperature, net capacity (q_{net}) ≥ 15,500 Btu/h	UC.L ≥ 15,500	4.15	

For SI: 1 British thermal unit per hour = 0.2931 W.

a. q_{net} is net capacity (Btu/h) as determined in accordance with AHRI 1250.

C403.12.3 Refrigeration systems. Refrigerated display cases, *walk-in coolers* or *walk-in freezers* that are served by remote compressors and remote condensers not located in a condensing unit, shall comply with Sections C403.12.3.1 and C403.12.3.2.

Exception: Systems where the working fluid in the refrigeration cycle goes through both subcritical and super-critical states (transcritical) or that use ammonia refrigerant are exempt.

C403.12.3.1 Condensers serving refrigeration systems. Fan-powered condensers shall comply with the following:

1. The design *saturated condensing temperatures* for air-cooled condensers shall not exceed the design dry-bulb temperature plus 10°F (5.6°C) for low-*temperature refrigeration systems*, and the design dry-bulb temperature plus 15°F (8°C) for *medium temperature refrigeration systems* where the *saturated condensing temperature* for blend refrigerants shall be determined using the average of liquid and vapor temperatures as converted from the condenser drain pressure.

2. Condenser fan motors that are less than 1 hp (0.75 kW) shall use electronically commutated motors, permanent split-capacitor-type motors or 3-phase motors.
3. Condenser fans for air-cooled condensers, evaporatively cooled condensers, air- or water-cooled fluid coolers or cooling towers shall reduce fan motor demand to not more than 30 percent of design wattage at 50 percent of design air volume, and incorporate one of the following continuous variable speed fan control approaches:
   3.1. Refrigeration system condenser control for air-cooled condensers shall use variable setpoint control logic to reset the condensing temperature setpoint in response to ambient dry-bulb temperature.
   3.2. Refrigeration system condenser control for evaporatively cooled condensers shall use variable setpoint control logic to reset the condensing temperature setpoint in response to ambient wet-bulb temperature.
4. Multiple fan condensers shall be controlled in unison.
5. The minimum condensing temperature setpoint shall be not greater than 70°F (21°C).

C403.12.3.2 Compressor systems. Refrigeration compressor systems shall comply with the following:

1. Compressors and multiple-compressor system suction groups shall include control systems that use floating suction pressure control logic to reset the target suction pressure temperature based on the temperature requirements of the attached refrigeration display cases or walk-ins.

   Exception: Controls are not required for the following:
   1. Single-compressor systems that do not have variable capacity capability.
   2. Suction groups that have a design saturated suction temperature of 30°F (-1.1°C) or higher, suction groups that comprise the high stage of a two-stage or cascade system, or suction groups that primarily serve chillers for secondary cooling fluids.
2. Liquid subcooling shall be provided for all low-temperature compressor systems with a design cooling capacity equal to or greater than 100,000 Btu (29.3 kW) with a design-saturated suction temperature of -10°F (-23°C) or lower. The subcooled liquid temperature shall be controlled at a maximum temperature setpoint of 50°F (10°C) at the exit of the subcooler using either compressor economizer (interstage) ports or a separate compressor suction group operating at a saturated suction temperature of 18°F (-7.8°C) or higher.
   2.1. Insulation for liquid lines with a fluid operating temperature less than 60°F (15.6°C) shall comply with Table C403.13.3(1) or C403.13.3(2).
3. Compressors that incorporate internal or external crankcase heaters shall provide a means to cycle the heaters off during compressor operation.

C403.13 Construction of HVAC system elements. *Ducts*, plenums, piping and other elements that are part of an HVAC system shall be constructed and insulated in accordance with Sections C403.13.1 through C403.13.3.1.

C403.13.1 Duct and plenum insulation and sealing. Supply and return air *ducts* and plenums shall be insulated with not less than R-6 insulation where located in unconditioned spaces and where located outside the *building* with not less than R-8 insulation in Climate Zones 0 through 4 and not less than R-12 insulation in Climate Zones 5 through 8. *Ducts* located underground beneath *buildings* shall be insulated as required in this section or have an equivalent *thermal distribution efficiency*. Underground *ducts* utilizing the *thermal distribution efficiency* method shall be *listed* and *labeled* to indicate the *R-value* equivalency. Where located within a *building thermal envelope* assembly, the *duct* or plenum shall be separated from the *building* exterior or unconditioned or exempt spaces by not less than R-8 insulation in Climate Zones 0 through 4 and not less than R-12 insulation in Climate Zones 5 through 8.

Exceptions:

1. Where located within equipment.
2. Where the design temperature difference between the interior and exterior of the *duct* or plenum is not greater than 15°F (8°C).

Ducts, air handlers and filter boxes shall be sealed. Joints and seams shall comply with Section 603.9 of the *International Mechanical Code*.

C403.13.2 Duct construction. Ductwork shall be constructed and erected in accordance with the *International Mechanical Code*.

C403.13.2.1 Low-pressure duct systems. Longitudinal and transverse joints, seams and connections of supply and return *ducts* operating at a static pressure less than or equal to 2 inches water gauge (w.g.) (498 Pa) shall be securely fastened and sealed with welds, gaskets, mastics (adhesives), mastic-plus-embedded-fabric systems or tapes installed in accordance with the manufacturer's instructions. Pressure classifications specific to the *duct system* shall be clearly indicated on the *construction documents* in accordance with the *International Mechanical Code*.

Exception: Locking-type longitudinal joints and seams, other than the snap-lock and button-lock types, need not be sealed as specified in this section.

C403.13.2.2 Medium-pressure duct systems. *Ducts* and plenums designed to operate at a static pressure greater than 2 inches water gauge (w.g.) (498 Pa) but less than 3 inches w.g. (747 Pa) shall be insulated and sealed in accordance with Section C403.13.1. Pressure classifications specific to the *duct system* shall be clearly indicated on the *construction documents* in accordance with the *International Mechanical Code*.

C403.13.2.3 High-pressure duct systems. *Ducts* and plenums designed to operate at static pressures equal to or greater than 3 inches water gauge (747 Pa) shall be insulated and sealed in accordance with Section C403.13.1. In addition, *ducts* and plenums shall be leak tested in accordance with the SMACNA HVAC Air Duct Leakage Test Manual and shown to have a rate of *air leakage* (CL) less than or equal to 4.0 as determined in accordance with Equation 4-7.

Equation 4-7 $CL = F/P^{0.65}$

where:

F = The measured leakage rate in cfm per 100 square feet (9.3 m^2) of duct surface.

P = The static pressure of the test.

Documentation shall be furnished demonstrating that representative sections totaling not less than 25 percent of the *duct* area have been tested and that all tested sections comply with the requirements of this section.

C403.13.3 Piping insulation. Piping serving as part of a heating or cooling system shall be thermally insulated in accordance with Table C403.13.3(1) or C403.13.3(2).

Exceptions:

1. Factory-installed piping within HVAC equipment tested and rated in accordance with a test procedure referenced by this code.
2. Factory-installed piping within room fan coils and unit ventilators tested and rated according to AHRI 440 (except that the sampling and variation provisions of Section 6.5 shall not apply) and AHRI 840, respectively.
3. Piping that conveys fluids that have a design operating temperature range between 60°F (15°C) and 105°F (41°C).
4. Piping that conveys fluids that have not been heated or cooled through the use of fossil fuels or electric power.
5. Strainers, control valves, and balancing valves associated with piping 1 inch (25 mm) or less in diameter.
6. Direct buried piping that conveys fluids at or below 60°F (15°C).
7. In radiant heating systems, sections of piping intended by design to radiate heat.

TABLE C403.13.3(1)—MINIMUM PIPE INSULATION THICKNESS (in inches or *R*-value)[a, c]

FLUID OPERATING TEMPERATURE RANGE AND USAGE (°F)	INSULATION CONDUCTIVITY		INCHES OR *R*-VALUE	NOMINAL PIPE OR TUBE SIZE (inches)				
	Conductivity Btu × in/(h × ft² × °F)[b]	Mean Rating Temperature (°F)		< 1	1 to < 1½	1½ to < 4	4 to < 8	> 8
				Minimum insulation thickness (inches)				
> 350	0.32–0.34	250	Inches	4.5	5.0	5.0	5.0	5.0
			R-value	R-32	R-36	R-34	R-26	R-21
251–350	0.29–0.32	200	Inches	3.0	4.0	4.5	4.5	4.5
			R-value	R-20	R-29	R-32	R-24	R-20
201–250	0.27–0.30	150	Inches	2.5	2.5	2.5	3.0	3.0
			R-value	R-17	R-17	R-17	R-15	R-13
141–200	0.25–0.29	125	Inches	1.5	1.5	2.0	2.0	2.0
			R-value	R-9	R-9	R-11	R-10	R-9
105–140	0.21–0.28	100	Inches	1.0	1.0	1.5	1.5	1.5
			R-value	R-5	R-9	R-8	R-8	R-7
40–60	0.21–0.27	75	Inches	0.5	0.5	1.0	1.0	1.0
			R-value	R-2	R-2	R-5	R-5	R-4
< 40	0.20–0.26	50	Inches	0.5	1.0	1.0	1.0	1.5
			R-value	R-6	R-9	R-9	R-8	R-7

For SI: 1 inch = 25.4 mm, °C = (°F – 32)/1.8.

a. For piping smaller than 1½ inches and located in partitions within conditioned spaces, reduction of these thicknesses by 1 inch shall be permitted (before thickness adjustment required in Note b but not to a thickness less than 1 inch).

b. For insulation outside the stated conductivity range, the minimum thickness (T) shall be determined as follows:

$T = r[(1 + t/r)^{K/k} - 1]$

where:

T = Minimum insulation thickness.

r = Actual outside radius of pipe.

t = Insulation thickness listed in the table for applicable fluid temperature and pipe size.

K = Conductivity of alternate material at mean rating temperature indicated for the applicable fluid temperature (Btu × in/h × ft^2 × °F).

k = The upper value of the conductivity range listed in the table for the applicable fluid temperature.

c. For direct-buried heating and hot water system piping, reduction of these thicknesses by 1½ inches shall be permitted (before thickness adjustment required in Note b but not to thicknesses less than 1 inch).

TABLE C403.13.3(2)—MINIMUM PIPE INSULATION *R*-VALUE[a]

FLUID OPERATING TEMPERATURE RANGE AND USAGE (°F)	NOMINAL PIPE OR TUBE SIZE (inches)				
	< 1	1 to < $1^1/_2$	$1^1/_2$ to < 4	4 to < 8	≥ 8
	Minimum Insulation *R*-Value				
> 350	R-32	R-36	R-34	R-26	R-21
251–350	R-20	R-29	R-32	R-24	R-20
201–250	R-17	R-17	R-17	R-15	R-13
141–200	R-9	R-9	R-11	R-10	R-9
105–140	R-5	R-9	R-8	R-8	R-7
40–60	R-2	R-2	R-5	R-5	R-4
≤ 40	R-6	R-9	R-9	R-8	R-7

For SI: 1 inch = 25.4 mm, R-1 = RSI-0.176228, °C = (°F – 32)/1.8.

a. The *R*-value of cylindrical piping insulation shall be determined as follows:

$R = \{ro[ln(ro/ri)]\}/k$

where:

R = The interior R-value of the cylindrical piping insulation in Btu × ft2 × °F/h.
ro = The outer radius of the piping insulation in inches.
ri = The inner radius of the piping insulation in inches.
k = the thermal conductivity of the insulation material in Btu × in/h × ft2 × °F.

C403.13.3.1 Protection of piping insulation. Piping insulation exposed to the weather shall be protected from damage, including that caused by sunlight, moisture, equipment maintenance and wind. The protection shall provide shielding from solar radiation that can cause degradation of the material. The protection shall be removable and reuseable for not less than 6 inches (152 mm) from the connection to the equipment piping for maintenance. Adhesive tape shall not be permitted as a means of insulation protection.

C403.14 Mechanical systems located outside of the building thermal envelope. Mechanical systems providing heat outside of the *building thermal envelope* of a *building* shall comply with Sections C403.14.1 through C403.14.4.

C403.14.1 Heating outside a building. Systems installed to provide heat outside a *building* shall be radiant systems. Such heating systems shall be controlled by an occupancy sensing device or a timer switch, so that the system is automatically de-energized when occupants are not present.

C403.14.2 Snow- and ice-melt system controls. Snow- and ice-melting systems shall include *automatic* controls configured to shut off the system when the pavement temperature is above 50°F (10°C) and precipitation is not falling, and an *automatic* or *manual* control that is configured to shut off when the outdoor temperature is above 40°F (4°C).

C403.14.3 Roof and gutter deicing controls. Roof and gutter deicing systems, including but not limited to self-regulating cable, shall include *automatic* controls that are configured to shut off the system when the outdoor temperature is above 40°F (4°C) and that include one of the following:

1. A moisture sensor configured to shut off the system in the absence of moisture.
2. A daylight sensor or other means configured to shut off the system between sunset and sunrise.

C403.14.4 Freeze protection system controls. Freeze protection systems, such as heat tracing of outdoor piping and heat exchangers, including self-regulating heat tracing, shall include *automatic* controls configured to shut off the systems when outdoor air temperatures are above 40°F (4°C) or when the conditions of the protected fluid will prevent freezing.

C403.15 Dehumidification in spaces for plant growth and maintenance. Equipment that dehumidifies *indoor grow* and *greenhouse* spaces shall be one or more of the following:

1. *Dehumidifiers* tested in accordance with the test procedure *listed* in DOE 10 CFR 430 and DOE 10 CFR 430, Subpart B, Appendix X or X1.
2. An integrated HVAC system with on-site heat recovery designed to fulfill not less than 75 percent of the annual energy for dehumidification reheat.
3. A chilled water system with on-site heat recovery designed to fulfill not less than 75 percent of the annual energy for dehumidification reheat.
4. A solid or liquid desiccant dehumidification system for system designs that require a dewpoint of not more than 50°F (10°C).

C403.16 Service water pressure-booster systems. Service water pressure-booster systems shall be designed such that the following apply:

1. One or more pressure sensors shall be used to vary pump speed and/or start and stop pumps. The sensors shall either be located near the critical fixtures that determine the pressure required or logic shall be employed that adjusts the setpoint to simulate the operation of remote sensors.
2. No devices shall be installed for the purpose of reducing the pressure of all of the water supplied by any booster system pump or booster system, except for safety devices.
3. No booster system pumps shall operate when there is no service water flow.

C403.17 Clean water pumps. *Clean water pumps* meeting all the following criteria shall achieve a PEI rating not greater than 1.0:

1. Shaft input power is greater than or equal to 1.0 hp (0.75 kW) and less than or equal to 200 hp (149.1 kW) at its best efficiency point (BEP).
2. Designated as either an end-suction close-coupled, end-suction frame-mounted, in-line, radially split vertical or submersible turbine pump.
3. A flow rate of 25 gallons per minute (1.58 L/s) or greater at its BEP at full impeller diameter.
4. Maximum head of 459 feet (139.9 m) at its BEP at full impeller diameter and the number of stages required for testing.
5. Design temperature range from 14°F (-10°C) to 248°F (120°C).
6. Designed to operate with one of the following. Note that for either Item 6.1 or 6.2, the driver and impeller must rotate at the same speed.
   6.1. A 2- or 4-pole induction motor.
   6.2. A noninduction motor with a speed of rotation operating range that includes speeds of rotation between 2,880 and 4,320 rpm and/or 1,440 and 2,160 rpm.
7. For submersible turbine pumps, a 6-inch (152 mm) or smaller bowl diameter.
8. For end-suction close-coupled pumps and end-suction frame-mounted/own bearings pumps, specific speeds less than or equal to 5,000 rpm when calculated using US customary units.

Exceptions: The following pumps are exempt from these requirements:

1. Fire pumps.
2. Self-priming pumps.
3. Prime-assisted pumps.
4. Magnet-driven pumps.
5. Pumps designed to be used in a nuclear facility subject to 10 CFR 50.
6. Pumps meeting the design and construction requirements set forth in US Military Specification MIL-P-17639F (1996), "Pumps, Centrifugal, Miscellaneous Service Naval Shipboard Use" (as amended); MIL-P-17840C (1986), "Pump, Centrifugal, Close Coupled, Navy Standard for Use on Naval Ships" (as amended); MIL-P-17881D (1972), "Pump, Centrifugal, Boiler Feed, (Multi Stage)" (as amended); MIL-P-18472G (1989), "Pumps, Centrifugal, Condensate, Feed Booster, Waste Heat Boiler, and Distilling Plant" (as amended); MIL-P-18682D (1984), "Pump, Centrifugal, Main Condenser Circulating, Naval Shipboard" (as amended).

SECTION C404 —SERVICE WATER HEATING

C404.1 General. This section covers the minimum efficiency of, and controls for, service water-heating equipment and insulation of service hot water piping.

C404.2 Service water-heating equipment performance efficiency. Water-heating equipment and hot water storage tanks shall meet the requirements of Table C404.2. The efficiency shall be verified through data furnished by the manufacturer of the equipment or through certification under an *approved* certification program. Water-heating equipment intended to be used to provide space heating shall meet the applicable provisions of Table C404.2.

TABLE C404.2—MINIMUM PERFORMANCE OF WATER-HEATING EQUIPMENT

EQUIPMENT TYPE	SIZE CATEGORY	SUBCATEGORY OR RATING CONDITION	DRAW PATTERN	PERFORMANCE REQUIRED[a]	TEST PROCEDURE[b]
Electric table-top water heaters[c]	≤ 12 kW	≥ 20 gal ≤ 120 gal[d]	Very small Low Medium High	UEF ≥ 0.6323 – (0.0058 × V_r) UEF ≥ 0.9188 – (0.0031 × V_r) UEF ≥ 0.9577 – (0.0023 × V_r) UEF ≥ 0.9884 – (0.0016 × V_r)	DOE 10 CFR Part 430 App. E
Electric storage water heaters[e, f]: resistance and heat pump	≤ 12 kW	≥ 20 gal ≤ 55 gal[f]	Very small Low Medium High	UEF ≥ 0.8808 – (0.0008 × V_r) UEF ≥ 0.9254 – (0.0003 × V_r) UEF ≥ 0.9307 – (0.0002 × V_r) UEF ≥ 0.9349 – (0.0001 × V_r)	DOE 10 CFR Part 430 App. E
	≤ 12 kW	> 55 gal ≤120 gal[f]	Very small Low Medium High	UEF ≥ 1.9236 – (0.0011 × V_r) UEF ≥ 2.0440 – (0.0011 × V_r) UEF ≥ 2.1171 – (0.0011 × V_r) UEF ≥ 2.2418 – (0.0011 × V_r)	DOE 10 CFR Part 430 App. E
Electric storage water heaters[e, f, m]	> 12 kW	—	—	$(0.3 + 27/V_m)$, %/h	DOE 10 CFR 431.106 App. B
Grid-enabled water heaters[g]	—	> 75 gal[d]	Very small Low Medium High	UEF ≥ 1.0136 – (0.0028 × V_r) UEF ≥ 0.9984 – (0.0014 × V_r) UEF ≥ 0.9853 – (0.0010 × V_r) UEF ≥ 0.9720 – (0.0007 × V_r)	DOE 10 CFR 430 App. E
Electric instantaneous water heaters[h]	≤ 12 kW	< 2 gal[d]	Very small Low Medium High	UEF ≥ 0.91 UEF ≥ 0.91 UEF ≥ 0.91 UEF ≥ 0.92	DOE 10 CFR Part 430
	> 12 kW & ≤ 58.6 kW[i]	≤ 2 gal & ≤180°F	All	UEF ≥ 0.80	DOE 10 CFR Part 430
Gas storage water heaters[e, m]	≤ 75,000 Btu/h	≥20 gal & ≤ 55 gal[d]	Very small Low Medium High	UEF ≥ 0.3456 – (0.0020 × V_r) UEF ≥ 0.5982 – (0.0019 × V_r) UEF ≥ 0.6483 – (0.0017 × V_r) UEF ≥ 0.6920 – (0.0013 × V_r)	DOE 10 CFR Part 430 App. E
	≤ 75,000 Btu/h	> 55 gal & ≤ 100 gal[d]	Very small Low Medium High	UEF ≥ 0.6470 – (0.0006 × V_r) UEF ≥ 0.7689 – (0.0005 × V_r) UEF ≥ 0.7897 – (0.0004 × V_r) UEF ≥ 0.8072 – (0.0003 × V_r)	DOE 10 CFR Part 430 App. E
	> 75,000 Btu/h and ≤ 105,000 Btu/h[j, k]	≤ 120 gal & ≤180°F	Very small Low Medium High	UEF ≥ 0.2674 – (0.0009 × V_r) UEF ≥ 0.5362 – (0.0012 × V_r) UEF ≥ 0.6002 – (0.0011 × V_r) UEF ≥ 0.6597 – (0.0009 × V_r)	DOE 10 CFR Part 430 App. E
	> 105,000 Btu/h[k]	—	—	80% E_t $SL \leq Q/(800 + 110\sqrt{V})$, Btu/h	DOE 10 CFR 431.106

TABLE C404.2—MINIMUM PERFORMANCE OF WATER-HEATING EQUIPMENT—continued

EQUIPMENT TYPE	SIZE CATEGORY	SUBCATEGORY OR RATING CONDITION	DRAW PATTERN	PERFORMANCE REQUIRED[a]	TEST PROCEDURE[b]
Gas instantaneous water heaters[i]	> 50,000 Btu/h and < 200,000 Btu/h[k]	< 2 gal[d]	Very small Low Medium High	UEF ≥ 0.80 UEF ≥ 0.81 UEF ≥ 0.81 UEF ≥ 0.81	DOE 10 CFR Part 430 App. E
	≥ 200,000 Btu/h[k]	< 10 gal	—	80% E_t	DOE 10 CFR 431.106
	≥ 200,000 Btu/h[k]	≥10 gal	—	80% E_t $SL \le Q/(800 + 110\sqrt{V})$, Btu/h	
Oil storage water heaters[e, m]	≤ 105,000 Btu/h	≤ 50 gal[d]	Very small Low Medium High	UEF = 0.2509 – (0.0012 × V_r) UEF = 0.5330 – (0.0016 × V_r) UEF = 0.6078 – (0.0016 × V_r) UEF = 0.6815 – (0.0014 × V_r)	DOE 10 CFR Part 430
	> 105,000 Btu/h and ≤ 140,000 Btu/h[l]	≤ 120 gal & ≤180°F	Very small Low Medium High	UEF ≥ 0.2932 – (0.0015 × V_r) UEF ≥ 0.5596 – (0.0018 × V_r) UEF ≥ 0.6194 – (0.0016 × V_r) UEF ≥ 0.6740 – (0.0013 × V_r)	DOE 10 CFR Part 430 App. E
	> 140,000 Btu/h	All	—	80% E_t $SL \le Q/(800 + 110\sqrt{V})$, Btu/h	DOE 10 CFR 431.106
Oil instantaneous water heaters[h, m]	≤ 210,000 Btu/h	< 2 gal	—	80% E_t EF ≥ 0.59 – (0.0005 × V)	DOE 10 CFR Part 430 App. E
	> 210,000 Btu/h	< 10 gal	—	80% E_t	DOE 10 CFR 431.106
	> 210,000 Btu/h	≥ 10 gal	—	78% E_t $SL \le Q/(800 + 110\sqrt{V})$, Btu/h	DOE 10 CFR 431.106
Hot water supply boilers, gas and oil[h]	≥ 300,000 Btu/h and < 12,500,000 Btu/h	< 10 gal	—	80% E_t	DOE 10 CFR 431.106
Hot water supply boilers, gas[i, m]	≥ 300,000 Btu/h and < 12,500,000 Btu/h	≥ 10 gal	—	80% E_t $SL \le Q/(800 + 110\sqrt{V})$, Btu/h	DOE 10 CFR 431.106
Hot water supply boilers, oil[h, m]	≥ 300,000 Btu/h and < 12,500,000 Btu/h	≥ 10 gal	—	78% E_t $SL \le Q/(800 + 110\sqrt{V})$, Btu/h	DOE 10 CFR 431.106
Pool heaters, gas[d]	All	—[f]	—	82% E_t	DOE 10 CFR Part 430 App. P
Heat pump pool heaters	All	50°F db and 44.2°F wb outdoor air 80.0°F entering water	—	4.0 COP	DOE 10 CFR Part 430 App. P
Unfired storage tanks	All	—	—	Minimum insulation requirement R-12.5 (h × ft² × °F)/Btu	(none)

TABLE C404.2—MINIMUM PERFORMANCE OF WATER-HEATING EQUIPMENT—continued

For SI: 1 foot = 304.8 mm, 1 square foot = 0.0929 m^2, °C = (°F – 32)/1.8, 1 British thermal unit per hour = 0.2931 W, 1 gallon = 3.785 L, 1 British thermal unit per hour per gallon = 0.078 W/L.

a. Thermal efficiency (E_t) is a minimum requirement, while standby loss is a maximum requirement. In the standby loss equation, V is the rated volume in gallons and Q is the nameplate input rate in Btu/h. V_m is the measured volume in the tank in gallons. Standby loss for electric water heaters is in terms of %/h and denoted by the term "S," and standby loss for gas and oil water heaters is in terms of Btu/h and denoted by the term "SL." Draw pattern (DP) refers to the water draw profile in the Uniform Energy Factor (UEF) test. UEF and Energy Factor (EF) are minimum requirements. In the UEF standard equations, V_r refers to the rated volume in gallons.

b. Chapter 6 contains a complete specification, including the year version, of the referenced test procedure.

c. A tabletop water heater is a storage water heater that is enclosed in a rectangular cabinet with a flat top surface not more than 3 feet in height and has a ratio of input capacity (Btu/h) to tank volume (gal) < 4,000.

d. Water heaters or gas pool heaters in this category are regulated as consumer products by the US DOE, as defined in 10 CFR 430.

e. Storage water heaters have a ratio of input capacity (Btu/h) to tank volume (gal) < 4,000.

f. Efficiency requirements for electric storage water heaters ≤ 12 kW apply to both electric-resistance and heat pump water heaters. There are no minimum efficiency requirements for electric heat pump water heaters greater than 12 kW or for gas heat pump water heaters.

g. A grid-enabled water heater is an electric-resistance water heater that meets all of the following:
   1. Has a rated storage tank volume of more than 75 gallons.
   2. Is manufactured on or after April 16, 2015.
   3. Is equipped at the point of manufacture with an activation lock.
   4. Bears a permanent label applied by the manufacturer that complies with all of the following:
      4.1. Is made of material not adversely affected by water.
      4.2. Is attached by means of nonwater soluble adhesive.
      4.3. Advises purchasers and end users of the intended and appropriate use of the product with the following notice printed in 16.5 point Arial Narrow Bold font: "IMPORTANT INFORMATION: This water heater is intended only for use as a part of an electric thermal storage or demand response program. It will not provide adequate hot water unless enrolled in such a program and activated by your utility company or another program operator. Confirm the availability of a program in your local area before purchasing or installing this product."

h. Instantaneous water heaters and hot water supply boilers have an input capacity (Btu/h) divided by storage volume (gal) ≥ 4,000 (Btu/h)/gal.

i. Electric instantaneous water heaters with input capacity >12 kW and ≤ 58.6 kW that (1) have a storage volume > 2 gallons, (2) are designed to provide outlet hot water at temperatures greater than 180°F, or (3) use three-phase power have no efficiency standard.

j. Gas storage water heaters with input capacity > 75,000 Btu/h and ≤ 105,000 Btu/h must comply with the requirements for the > 105,000 Btu/h if the water heater (1) has a storage volume > 120 gallons, (2) is designed to provide outlet hot water at temperatures greater than 180°F, or (3) uses three-phase power.

k. Refer to Section C404.2.1 for additional requirements for gas storage and instantaneous water heaters and gas hot water supply boilers.

l. Oil storage water heaters with input capacity > 105,000 Btu/h and ≤ 140,000 Btu/h must comply with the requirements for the > 140,000 Btu/h if the water heater either (1) has a storage volume > 120 gallons, (2) is designed to provide outlet hot water at temperatures greater than 180ºF, or (3) uses three-phase power.

m. Water heaters and hot water supply boilers with more than 140 gallons of storage capacity need not meet the standby loss requirement where: (1) the tank surface area is thermally insulated to R-12.5 or more, (2) there is no standing pilot light, and (3) for gas- or oil-fired storage water heaters, the heater is equipped with a fire damper or fan-assisted combustion.

C404.2.1 High-input service water-heating systems. Gas-fired *water heaters* installed in new *buildings* where the total input capacity provided by high-capacity gas-fired *water heaters* is 1,000,000 Btu/h (293 kW) or greater shall comply with either or both of the following requirements:

1. Where a singular piece of a high-capacity gas-fired *water heater* is installed, the water heater shall have a thermal efficiency, E_t, of not less than 92 percent.
2. Where multiple pieces of high-capacity gas-fired *water heaters* are connected to the same service water-heating system, the combined input-capacity-weighted average thermal efficiency, E_t, shall be not less than 90 percent, and a minimum of 30 percent of the input to the high-capacity gas-fired *water heaters* in the service water-heating system shall have an E_t of not less than 92 percent.

Exceptions:

1. The input rating of *water heaters* installed in individual *dwelling units* shall not be required to be included in the total input rating of service water-heating equipment for a *building*.
2. The input rating of water heaters with an input rating of not greater than 105,000 Btu/h (30.8 kW) shall not be required to be included in the total input rating of service water-heating equipment for a *building*.
3. Where not less than 25 percent of the annual *service water-heating* requirement is provided by *on-site renewable energy* or site-recovered energy, the minimum E_t requirements of this section shall not apply. On-site renewable energy used to meet Section C405.15.1 or C406.3.1 shall not be used to meet this exception.

C404.3 Heat traps for hot water storage tanks. Storage tank-type *water heaters* and hot water storage tanks that have vertical water pipes connecting to the inlet and outlet of the tank shall be provided with integral *heat traps* at those inlets and outlets or shall have pipe-configured *heat traps* in the piping connected to those inlets and outlets. Tank inlets and outlets associated with solar water-heating system circulation loops shall not be required to have *heat traps*.

C404.4 Service water heating system piping insulation. *Service water heating* system piping shall be surrounded by uncompressed insulation. The wall thickness of the insulation shall be not less than the thickness shown in Table C404.4.1. Where the insulation thermal conductivity is not within the range in the table, Equation 4-8 shall be used to calculate the minimum insulation thickness:

Equation 4-8 $t_{alt} = r \times [(1 + t_{table}/r)k_{alt}/k_{upper} - 1$

where:

t_{alt} = Minimum insulation thickness of the alternate material (in) (mm).

r = Actual outside radius of the pipe (in) (mm).

t_{table} = Insulation thickness listed in this table for applicable fluid temperature and pipe size.

k_{alt} = Thermal conductivity of the alternate material at mean rating temperature indicated for the applicable fluid temperature [Btu × in/h × ft^2 × °F] [W(m × °C)].

k_{upper} = The upper value of the thermal conductivity range listed in this table for the applicable fluid temperature [Btu × in/h × ft^2 × °F] [W(m × °C)].

For nonmetallic piping thicker than Schedule 80 and having thermal resistance greater than that of steel pipe, reduced insulation thicknesses are permitted if documentation is provided showing that the pipe with the proposed insulation has no more heat transfer per foot (meter) than a steel pipe of the same size with the insulation thickness shown in the table.

Exception: Tubular pipe insulation shall not be required on the following:

1. Factory-installed piping within *water heaters* and hot water storage tanks.
2. Valves, pumps, strainers and threaded unions in piping that is 1 inch (25 mm) or less in nominal diameter.
3. Piping that conveys hot water that has not been heated through the use of fossil fuels or electricity.
4. Piping from user-controlled shower and bath mixing valves to the water outlets.
5. Cold-water piping of a *demand recirculation water system*.
6. Piping in existing buildings where *alterations* are made to existing *service water heating* systems where there is insufficient space or access to meet the requirements.
7. Piping at locations where a vertical support of the piping is installed.
8. Where piping passes through a framing member if it requires increasing the size of the framing member.

C404.4.1 Installation requirements. The following piping shall be insulated per the requirements of this section:

1. Recirculating system piping, including the supply and return piping.
2. The first 8 feet (2.4 m) of outlet piping from:
   - 2.1. Storage *water heaters*.
   - 2.2. Hot water storage tanks.
   - 2.3. Any *water heater* and hot water supply boiler containing not less than 10 gallons (37.9 L) of water heated by a direct heat source, an indirect heat source, or both a direct heat source and an indirect heat source.
3. The first 8 feet (2.4 m) of branch piping connecting to recirculated, heat traced or impedance-heated piping.

4. The makeup water inlet piping between *heat traps* and the storage *water heaters* and the storage tanks they are serving, in a nonrecirculating *service water heating* storage system.
5. Hot water piping between multiple *water heaters*, between multiple hot water storage tanks, and between *water heaters* and hot water storage tanks.
6. Piping that is externally heated (such as heat trace or impedance heating).
7. For direct-buried *service water heating* system piping, reduction of these thicknesses by $1^1/_2$ inches (38.1 mm) shall be permitted (before thickness adjustment required in Section C404.4) but not to thicknesses less than 1 inch (25.4 mm).

TABLE C404.4.1—MINIMUM PIPING INSULATION THICKNESS FOR SERVICE WATER HEATING SYSTEMS[a]

SERVICE HOT-WATER TEMPERATURE RANGE	INSULATION THERMAL CONDUCTIVITY		NOMINAL PIPE OR TUBE SIZE (inches)				
	Conductivity (Btu × in/h × ft² × °F)	Mean Rating Temperature (°F)	< 1	1 to < $1^1/_2$	$1^1/_2$ to < 4	4 to < 8	≥ 8
			Insulation Thickness (inches)				
105°F to 140°F	0.22 to 0.28	100	1.0	1.0	1.5	1.5	1.5
> 140°F to 200°F	0.25 to 0.29	125	1.0	1.0	2.0	2.0	2.0
> 200°F	0.27 to 0.30	150	1.5	1.5	2.5	3.0	3.0

For SI: 1 inch = 25.4 mm, 1 Btu/h × ft × °F = 1.73 W/mK, °C = [(°F) – 32]/1.8.

a. These thicknesses are based on energy efficiency considerations only. Additional insulation may be necessary for safety.

C404.5 Heated water supply piping. Heated water supply piping shall be in accordance with Section C404.5.1 or C404.5.2. The flow rate through $^1/_4$-inch (6.4 mm) piping shall be not greater than 0.5 gpm (1.9 L/m). The flow rate through $^5/_{16}$-inch (7.9 mm) piping shall be not greater than 1 gpm (3.8 L/m). The flow rate through $^3/_8$-inch (9.5 mm) piping shall be not greater than 1.5 gpm (5.7 L/m).

C404.5.1 Maximum allowable pipe length method. The maximum allowable piping length from the nearest source of heated water to the termination of the fixture supply pipe shall be in accordance with the following. Where the piping contains more than one size of pipe, the largest size of pipe within the piping shall be used for determining the maximum allowable length of the piping in Table C404.5.1.

1. For a public lavatory faucet, use the "Public lavatory faucets" column in Table C404.5.1.
2. For all other plumbing fixtures and plumbing appliances, use the "Other fixtures and appliances" column in Table C404.5.1.

TABLE C404.5.1—PIPING VOLUME AND MAXIMUM PIPING LENGTHS

NOMINAL PIPE SIZE (inches)	VOLUME (liquid ounces per foot length)	MAXIMUM PIPING LENGTH (feet)	
		Public Lavatory Faucets	Other Fixtures and Appliances
$^1/_4$	0.33	6	50
$^5/_{16}$	0.5	4	50
$^3/_8$	0.75	3	50
$^1/_2$	1.5	2	43
$^5/_8$	2	1	32
$^3/_4$	3	0.5	21
$^7/_8$	4	0.5	16
1	5	0.5	13
$1^1/_4$	8	0.5	8
$1^1/_2$	11	0.5	6
2 or larger	18	0.5	4

For SI: 1 inch = 25.4 mm, 1 foot = 304.8 mm, 1 liquid ounce = 0.030 L, 1 gallon = 128 ounces.

C404.5.2 Maximum allowable pipe volume method. The water volume in the piping shall be calculated in accordance with Section C404.5.2.1. *Water heaters*, circulating water systems and heat trace temperature maintenance systems shall be considered to be sources of heated water.

The volume from the nearest source of heated water to the termination of the fixture supply pipe shall be as follows:

1. For a public lavatory faucet: not more than 2 ounces (0.06 L).
2. For other plumbing fixtures or plumbing appliances; not more than 0.5 gallon (1.89 L).

C404.5.2.1 Water volume determination. The volume shall be the sum of the internal volumes of pipe, fittings, valves, meters and manifolds between the nearest source of heated water and the termination of the fixture supply pipe. The volume in the piping shall be determined from the "Volume" column in Table C404.5.1 or from Table C404.5.2.1. The volume contained within fixture shutoff valves, within flexible water supply connectors to a fixture fitting and within a fixture fitting shall not be included in the water volume determination. Where heated water is supplied by a recirculating system or heat-traced piping, the volume shall include the portion of the fitting on the branch pipe that supplies water to the fixture.

TABLE C404.5.2.1—INTERNAL VOLUME OF VARIOUS WATER DISTRIBUTION TUBING

Nominal Size (inches)	OUNCES OF WATER PER FOOT OF TUBE								
	Copper Type M	Copper Type L	Copper Type K	CPVC CTS SDR 11	CPVC SCH 40	CPVC SCH 80	PE-RT SDR 9	Composite ASTM F1281	PEX CTS SDR 9
$^3/_8$	1.06	0.97	0.84	N/A	1.17	—	0.64	0.63	0.64
$^1/_2$	1.69	1.55	1.45	1.25	1.89	1.46	1.18	1.31	1.18
$^3/_4$	3.43	3.22	2.90	2.67	3.38	2.74	2.35	3.39	2.35
1	5.81	5.49	5.17	4.43	5.53	4.57	3.91	5.56	3.91
$1^1/_4$	8.70	8.36	8.09	6.61	9.66	8.24	5.81	8.49	5.81
$1^1/_2$	12.18	11.83	11.45	9.22	13.20	11.38	8.09	13.88	8.09
2	21.08	20.58	20.04	15.79	21.88	19.11	13.86	21.48	13.86

For SI: 1 foot = 304.8 mm, 1 inch = 25.4 mm, 1 liquid ounce = 0.030 L, 1 oz/ft^2= 305.15 g/m^2.
N/A = Not Available.

C404.6 Heated-water circulating and temperature maintenance systems. Heated-water circulation systems shall be in accordance with Section C404.6.1. Heat trace temperature maintenance systems shall be in accordance with Section C404.6.2. Controls for hot water storage shall be in accordance with Section C404.6.3. *Automatic* controls, temperature sensors and pumps shall be in a location with *access*. *Manual* controls shall be in a location with *ready access*.

C404.6.1 Circulation systems. Heated-water circulation systems shall be provided with a circulation pump. Gravity and thermosyphon circulation systems shall be prohibited. The system return pipe shall be a dedicated return pipe. Controls shall be configured to automatically turn off the pump when the water in the circulation loop is at the desired temperature and when there is not a demand for hot water. Where a circulation pump serves multiple risers or piping zones, controls shall include self-actuating thermostatic balancing valves or another means of flow control to automatically balance the flow rate through each riser or piping zone.

C404.6.1.1 Demand recirculation controls. Demand recirculation water systems shall have controls that start the pump upon receiving a signal from the action of a user of a fixture or appliance, sensing the presence of a user of a fixture, or sensing the flow of hot or tempered water to a fixture fitting or appliance.

C404.6.2 Heat trace systems. Electric heat trace systems shall comply with IEEE 515.1. Controls for such systems shall be able to automatically adjust the energy input to the heat tracing to maintain the desired water temperature in the piping in accordance with the times when heated water is used in the occupancy. Heat trace shall be arranged to be turned off automatically when there is not a demand for hot water.

C404.6.3 Controls for hot water storage. The controls on pumps that circulate water between a *water heater* and a heated-water storage tank shall limit operation of the pump from heating cycle startup to not greater than 5 minutes after the end of the cycle.

C404.7 Drain water heat recovery units. Drain water heat recovery units shall comply with CSA B55.2. Potable water-side pressure loss shall be less than 10 psi (69 kPa) at maximum design flow. For *Group R* occupancies, the efficiency of drain water heat recovery unit efficiency shall be in accordance with CSA B55.1.

C404.8 Energy consumption of pools and permanent spas. The energy consumption of pools and permanent spas shall be controlled by the requirements in Sections C404.8.1 through C404.8.3.

C404.8.1 Heaters. The electric power to all heaters shall be controlled by an on-off switch that is an integral part of the heater, mounted on the exterior of the heater, or external to and within 3 feet (914 mm) of the heater in a location with *ready access*. Operation of such switch shall not change the setting of the heater *thermostat*. Such switches shall be in addition to a circuit breaker for the power to the heater. Gas-fired heaters shall not be equipped with continuously burning ignition pilots.

C404.8.2 Time switches. Time switches or other control methods that can automatically turn off and on heaters and pump motors according to a preset schedule shall be installed for heaters and pump motors. Heaters and pump motors that have built-in time switches shall be in compliance with this section.

Exceptions:

1. Where public health standards require 24-hour pump operation.
2. Pumps that operate solar- and waste-heat-recovery pool heating systems.

C404.8.3 Covers. Outdoor heated pools and outdoor permanent spas shall be provided with a vapor-retardant cover or other *approved* vapor-retardant means.

Exception: Where more than 75 percent of the energy for heating, computed over an operating season of not fewer than 3 calendar months, is from a heat pump or an on-site renewable energy system, covers or other vapor-retardant means shall not be required. *On-site renewable energy* used to meet Section C405.15.1 or C406.3.1 shall not be used to meet this exception.

C404.9 Portable spas. The energy consumption of electric-powered portable spas shall be controlled by the requirements of APSP 14.

SECTION C405 —ELECTRICAL POWER AND LIGHTING SYSTEMS

C405.1 General. Electrical power and lighting systems and generation shall comply with this section. General lighting shall consist of all lighting included when calculating the total connected interior lighting power in accordance with Section C405.3.1 and which does not require specific application controls in accordance with Section C405.2.5.

Exception: *Dwelling units* and *sleeping units* that comply with Sections C405.2.10, C405.3.3 and C405.6.

C405.2 Lighting controls. Lighting systems in *interior parking areas* shall be provided with controls that comply with Section C405.2.9. All other lighting systems powered through the energy service for the *building* and building site lighting for which the *building owner* is responsible shall be provided with controls that comply with Sections C405.2.1 through C405.2.8.

Exceptions: Lighting controls are not required for the following:

1. Spaces where an *automatic* shutoff could endanger occupant safety or security.
2. Interior exit stairways, interior exit ramps and exit passageways.
3. Emergency lighting that is automatically off during normal operations.
4. Emergency lighting required by the *International Building Code* in exit access components that are not provided with fire alarm systems.
5. Up to 0.02 watts per square foot (0.22 W/m^2) of lighting in exit access components that are provided with fire alarm systems.

C405.2.1 Occupant sensor controls. Occupant *sensor controls* shall be installed to control lights in the following space types:

1. Classrooms/lecture/training rooms.
2. *Computer room, data center.*
3. Conference/meeting/multipurpose rooms.
4. Copy/print rooms.
5. Lounges/breakrooms.
6. Medical supply room in a health care facility.
7. Enclosed offices.
8. Laundry/washing area.
9. Open plan office areas.
10. Restrooms.
11. Storage rooms.
12. Telemedicine room in a health care facility.
13. Locker rooms.
14. Corridors.
15. Warehouse storage areas.
16. Other spaces 300 square feet (28 m^2) or less that are enclosed by floor-to-ceiling height partitions.

Exception: Luminaires that are required to have specific application controls in accordance with Section C405.2.5.

C405.2.1.1 Occupant sensor control function. Occupant sensor controls in warehouse storage areas shall comply with Section C405.2.1.2. Occupant sensor controls in open plan office areas shall comply with Section C405.2.1.3. Occupant sensor controls in corridors shall comply with Section C405.2.1.4. Occupant sensor controls for all other spaces specified in Section C405.2.1 shall comply with the following:

1. They shall automatically turn off lights within 20 minutes after all occupants have left the space.
2. They shall be manual on or controlled to automatically turn on the lighting to not more than 50 percent power.
3. They shall incorporate a manual control to allow occupants to turn off lights.

Exception: Full automatic-on controls with no manual control shall be permitted in *interior parking areas*, stairways, restrooms, locker rooms, lobbies, library stacks and areas where *manual* operation would endanger occupant safety or security.

C405.2.1.2 Occupant sensor control function in warehouse storage areas. Lighting in warehouse storage areas shall be controlled as follows:

1. Lighting in each aisleway shall be controlled independently of lighting in all other aisleways and open areas.

2. Occupant sensors shall automatically reduce lighting power within each controlled area to an unoccupied setpoint of not more than 50 percent of full power within 20 minutes after all occupants have left the controlled area.
3. Lights that are not turned off by occupant sensors shall be turned off by *time-switch control* complying with Section C405.2.2.1.
4. A *manual* control shall be provided to allow occupants to turn off lights in the space.

C405.2.1.3 Occupant sensor control function in open plan office areas. Occupant sensor controls in open plan office spaces less than 300 square feet (28 m^2) in area shall comply with Section C405.2.1.1. Occupant sensor controls in all other open plan office spaces shall comply with all of the following:

1. The controls shall be configured so that *general lighting* can be controlled separately in control zones with floor areas not greater than 600 square feet (55 m^2) within the open plan office space.
2. *General lighting* in each control zone shall be permitted to automatically turn on upon occupancy within the control zone. *General lighting* in other unoccupied zones within the open plan office space shall be permitted to turn on to not more than 20 percent of full power or remain unaffected.
3. The controls shall automatically turn off *general lighting* in all control zones within 20 minutes after all occupants have left the open plan office space.

   Exception: Where general lighting is turned off by *time-switch control* complying with Section C405.2.2.1.
4. General lighting in each control zone shall turn off or uniformly reduce lighting power to an unoccupied setpoint of not more than 20 percent of full power within 20 minutes after all occupants have left the control zone.

C405.2.1.4 Occupant sensor control function in corridors. Occupant sensor controls in corridors shall uniformly reduce lighting power to an unoccupied setpoint not more than 50 percent of full power within 20 minutes after all occupants have left the space.

Exception: Corridors provided with less than two footcandles of illumination on the floor at the darkest point with all lights on.

C405.2.2 Time-switch controls. Each area of the *building* that is not provided with *occupant sensor controls* complying with Section C405.2.1.1 shall be provided with *time-switch controls* complying with Section C405.2.2.1.

Exceptions:

1. Luminaires that are required to have specific application controls in accordance with Section C405.2.4.
2. Spaces where patient care is directly provided.

C405.2.2.1 Time-switch control function. *Time-switch controls* shall comply with all of the following:

1. Programmed to automatically turn off lights when the space is scheduled to be unoccupied.
2. Have a minimum 7-day clock.
3. Be capable of being set for seven different day types per week.
4. Incorporate an automatic holiday "shutoff" feature, which turns off all controlled lighting loads for not fewer than 24 hours and then resumes normally scheduled operations.
5. Have program backup capabilities, which prevent the loss of program and time settings for not fewer than 10 hours, if power is interrupted.
6. Include an override switch that complies with the following:
   - 6.1. The override switch shall be a *manual* control.
   - 6.2. The override switch, when initiated, shall permit the controlled lighting to remain on for not more than 2 hours.
   - 6.3. Any individual override switch shall control the lighting for an area not larger than 5,000 square feet (465 m^2).
7. For spaces where schedules are not available, *time switch controls* are programmed to a schedule that turns off lights not less than 12 hours per day.

Exception: Within mall concourses, auditoriums, sales areas, manufacturing facilities and sports arenas:

1. The time limit shall be permitted to be greater than 2 hours, provided that the switch is a captive key device.
2. The area controlled by the override switch shall not be limited to 5,000 square feet (465 m^2) provided that such area is less than 20,000 square feet (1860 m^2).

C405.2.3 Dimming controls. Dimming controls complying with Section C405.2.3.1 are required for general lighting in the following space types:

1. Classroom/lecture hall/training room.
2. Conference/multipurpose/meeting room.
3. In a dining area for bar/lounge or leisure, family dining.
4. Laboratory.
5. Lobby.
6. Lounge/break room.

7. Offices.
8. Gymnasium/fitness center.
9. Library reading room.
10. In a health care facility for imaging rooms, exam rooms, nursery and nurses' station.
11. Spaces not provided with occupant sensor controls complying with Section C405.2.1.1.

Exception: Luminaires controlled by special application controls complying with Section C405.2.5.

C405.2.3.1 Dimming control function. Spaces required to have dimming control shall be provided with *manual* controls that allow lights to be dimmed from full output to 10 percent of full power or lower with continuous dimming, as well as turning off lights. *Manual* control shall be provided within each room to dim lights.

Exceptions: *Manual* dimming control is not required in spaces where *high-end trim* lighting controls are provided that comply with the following:

1. The calibration adjustment equipment is located for ready *access* only by authorized personnel.
2. Lighting controls with ready *access* for users cannot increase the lighting power above the maximum level established by the *high-end trim* controls.

C405.2.4 Daylight responsive controls. *Daylight responsive controls* complying with Section C405.2.4.1 shall be provided to control the *general lighting* within *daylight zones* in the following spaces:

1. Spaces with a total of more than 75 watts of *general lighting* within primary sidelit *daylight zones* complying with Section C405.2.4.2.
2. Spaces with a total of more than 150 watts of *general lighting* within sidelit *daylight zones* complying with Section C405.2.4.2.
3. Spaces with a total of more than 75 watts of *general lighting* within toplit *daylight zones* complying with Section C405.2.4.3.

Exceptions: *Daylight responsive controls* are not required for the following:

1. Spaces in health care facilities where patient care is directly provided.
2. Sidelit *daylight zones* on the first floor above grade in Group A-2 and Group M occupancies.
3. Enclosed office spaces less than 250 square feet (23.2 m^2).

C405.2.4.1 Daylight responsive control function. Where required, *daylight responsive controls* shall be provided within each space for control of lights in that space and shall comply with all of the following:

1. Lights in *toplit daylight zones* in accordance with Section C405.2.4.3 shall be controlled independently of lights in sidelit *daylight zones* in accordance with Section C405.2.4.2.
2. Lights in the primary sidelit *daylight zone* shall be controlled independently of lights in the secondary sidelit *daylight zone*.
3. *Daylight responsive controls* within each space shall be configured so that they can be calibrated from within that space by authorized personnel.
4. Calibration mechanisms shall be in a location with *ready access*.
5. *Daylight responsive controls* shall dim lights continuously from full light output to 15 percent of full light output or lower.
6. *Daylight responsive controls* shall be configured to completely shut off all controlled lights.
7. When occupant sensor controls have reduced the lighting power to an unoccupied setpoint in accordance with Sections C405.2.1.2 through C405.2.1.4, *daylight responsive controls* shall continue to adjust electric light levels in response to available daylight, but shall be configured to not increase the lighting power above the specified unoccupied setpoint.
8. Lights in *sidelit daylight zones* in accordance with Section C405.2.4.2 facing different cardinal orientations [within 45 degrees (0.79 rad) of due north, east, south, west] shall be controlled independently of each other.

Exceptions:

1. Within each space, up to 150 watts of lighting within the primary sidelit *daylight zone* is permitted to be controlled together with lighting in a primary sidelit *daylight zone* facing a different cardinal orientation.
2. Within each space, up to 150 watts of lighting within the secondary sidelit *daylight zone* is permitted to be controlled together with lighting in a secondary sidelit *daylight zone* facing a different cardinal orientation.

C405.2.4.2 Sidelit daylight zone. The sidelit *daylight zone* is the floor area adjacent to vertical *fenestration* that complies with all of the following:

1. Where the *fenestration* is located in a wall, the primary sidelit *daylight zone* shall extend laterally to the nearest full-height wall, or up to 1.0 times the height from the floor to the top of the *fenestration*, and longitudinally from the edge of the *fenestration* to the nearest full-height wall, or up to 0.5 times the height from the floor to the top of the *fenestration*, whichever is less, as indicated in Figure C405.2.4.2(1).
2. Where the *fenestration* is located in a *rooftop monitor*, the primary sidelit *daylight zone* shall extend laterally to the nearest obstruction that is taller than 0.7 times the ceiling height, or up to 1.0 times the height from the floor to the

bottom of the *fenestration*, whichever is less, and longitudinally from the edge of the *fenestration* to the nearest obstruction that is taller than 0.7 times the ceiling height, or up to 0.25 times the height from the floor to the bottom of the *fenestration*, whichever is less, as indicated in Figures C405.2.4.2(2) and C405.2.4.2(3).

3. Where the *fenestration* is located in a wall, the secondary sidelit *daylight zone* is directly adjacent to the primary sidelit *daylight zone* and shall extend laterally to 2.0 times the height from the floor to the top of the *fenestration* or to the nearest full height wall, whichever is less, and longitudinally from the edge of the *fenestration* to the nearest full height wall, or up to 0.5 times the height from the floor to the top of the *fenestration*, whichever is less, as indicated in Figure C405.2.4.2(1).
4. The area of the *fenestration* is not less than 24 square feet (2.23 m^2).
5. The distance from the *fenestration* to any *building* or geological formation that would block *access to* daylight is greater than one-half of the height from the bottom of the *fenestration* to the top of the *building* or geologic formation.
6. The *visible transmittance* of the *fenestration* is not less than 0.20.
7. The projection factor (determined in accordance with Equation 4-4) for any overhanging projection that is shading the *fenestration* is not greater than 1.0 for *fenestration* oriented 45 degrees or less from true north and not greater than 1.5 for all other orientations.

FIGURE C405.2.4.2(1)—PRIMARY AND SECONDARY SIDELIT DAYLIGHT ZONES

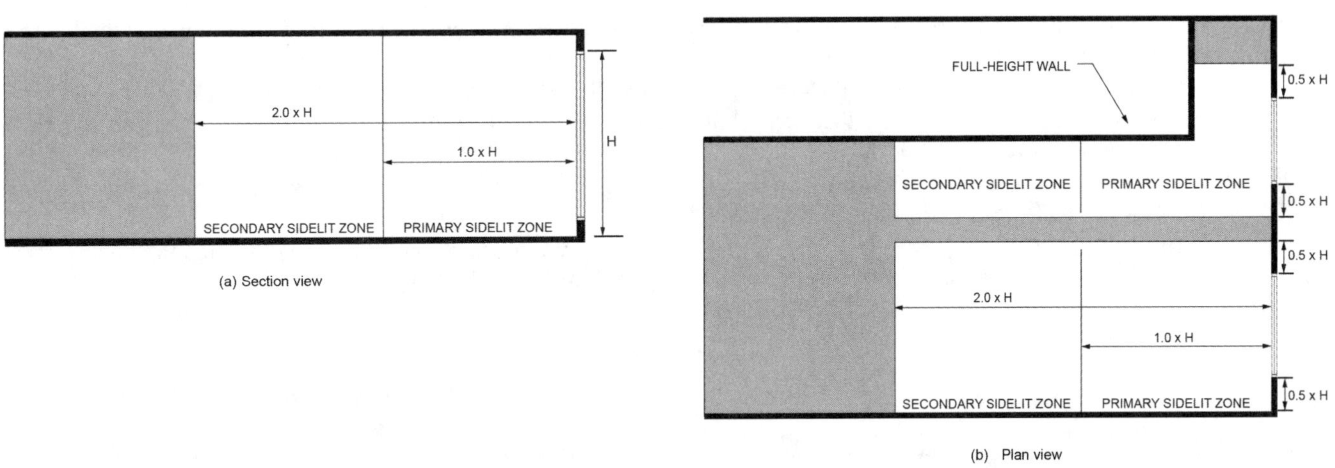

FIGURE C405.2.4.2(2)—DAYLIGHT ZONE UNDER A ROOFTOP MONITOR

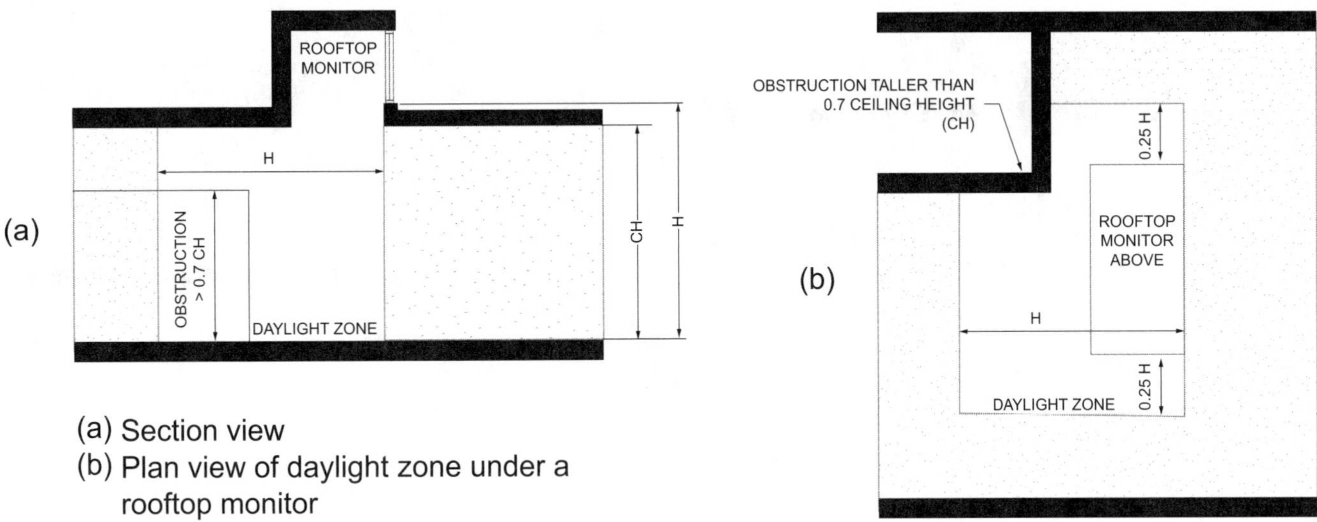

FIGURE C405.2.4.2(3)—DAYLIGHT ZONE UNDER A SLOPED ROOFTOP MONITOR

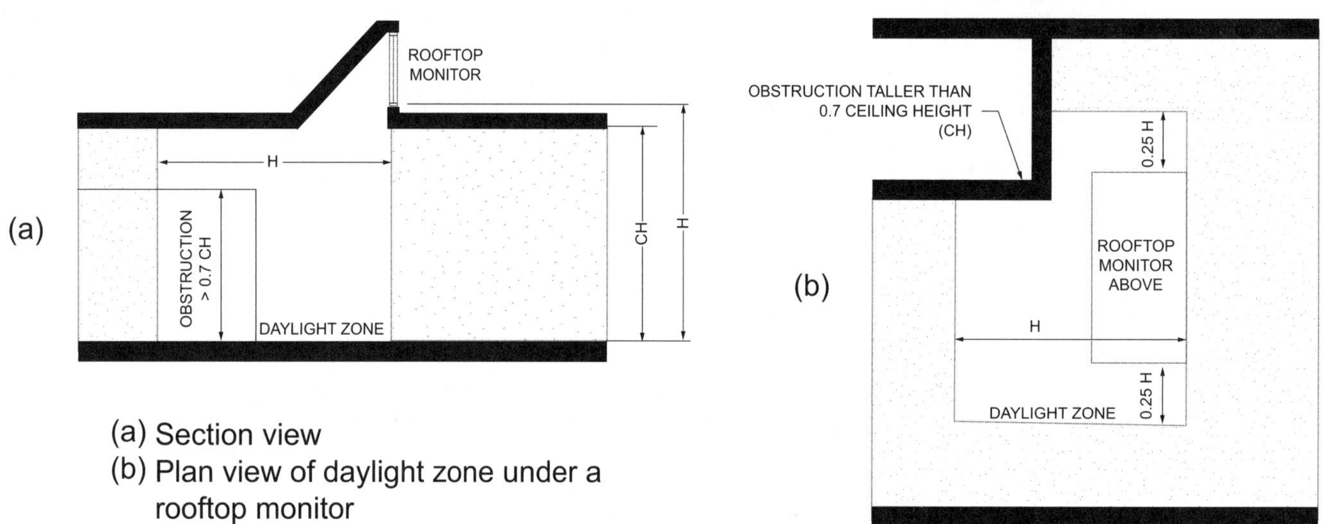

(a) Section view
(b) Plan view of daylight zone under a rooftop monitor

C405.2.4.3 Toplit daylight zone. The *toplit daylight zone* is the floor area underneath a roof *fenestration* assembly that complies with all of the following:

1. The toplit *daylight zone* shall extend laterally and longitudinally beyond the edge of the roof *fenestration* assembly to the nearest obstruction that is taller than 0.7 times the ceiling height, or up to 0.7 times the ceiling height, whichever is less, as indicated in Figure C405.2.4.3.
2. Direct sunlight is not blocked from hitting the roof *fenestration* assembly at the peak solar angle on the summer solstice by *buildings* or geological formations.
3. The product of the *visible transmittance* of the roof *fenestration* assembly and the area of the rough opening of the roof *fenestration* assembly divided by the area of the *toplit* zone is not less than 0.008.

FIGURE C405.2.4.3—TOPLIT DAYLIGHT ZONE

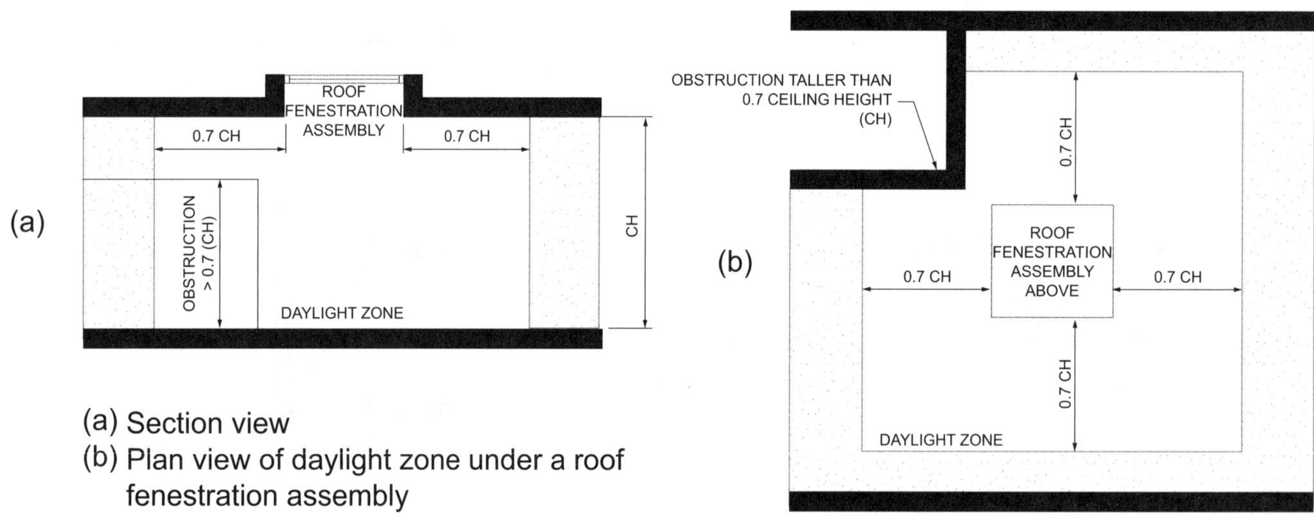

(a) Section view
(b) Plan view of daylight zone under a roof fenestration assembly

C405.2.4.4 Atriums. *Daylight zones* at atrium spaces shall be established at the top floor surrounding the atrium and at the floor of the atrium space, and not on intermediate floors, as indicated in Figure C405.2.4.4.

FIGURE C405.2.4.4—DAYLIGHT ZONES AT A MULTISTORY ATRIUM

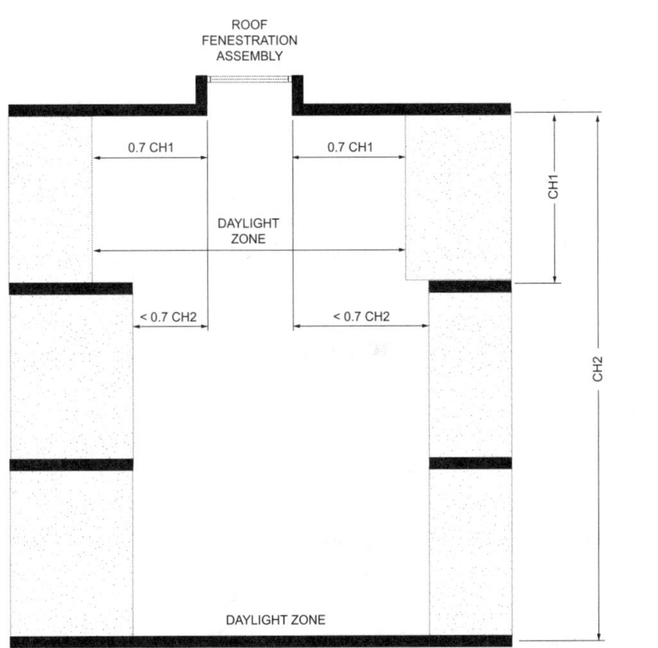

(a) Section view of roof fenestration assembly at atrium

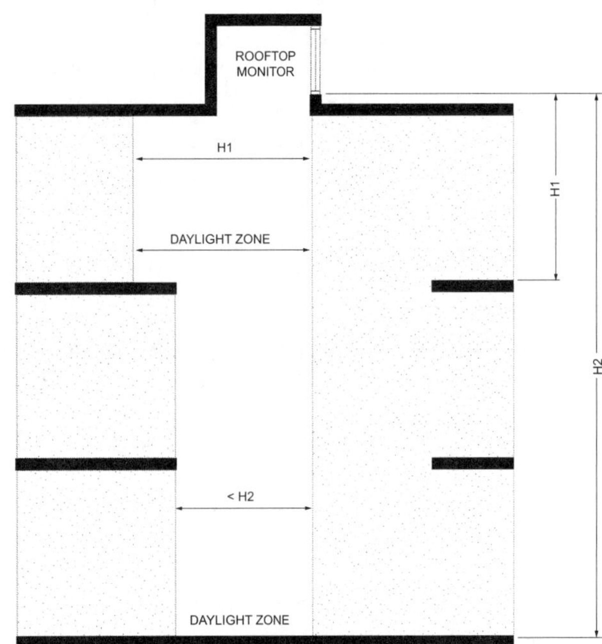

(b) Section view of roof monitor at atrium

C405.2.5 Specific application controls. Specific application controls shall be provided for the following:

1. The following lighting shall be controlled by an occupant sensor complying with Section C405.2.1.1 or a *time-switch control* complying with Section C405.2.2.1. In addition, a *manual* control shall be provided to control such lighting separately from the *general lighting* in the space:
   1.1. Luminaires for which additional lighting power is claimed in accordance with Section C405.3.2.2.1.
   1.2. Display and accent, including lighting in display cases.
   1.3. Supplemental task lighting, including permanently installed under-shelf or under-cabinet lighting.
   1.4. Lighting equipment that is for sale or demonstration in lighting education.
2. Lighting for nonvisual applications, such as plant growth and food warming, shall be controlled by a *time switch control* complying with Section C405.2.2.1 that is independent of the controls for other lighting within the room or space.
3. Task lighting for medical and dental purposes that is in addition to *general lighting* shall be provided with a *manual control*.
4. Lighting integrated into range hoods and exhaust fans shall be controlled independently of fans.

C405.2.6 Manual controls. Where required by this code, *manual* controls for lights shall comply with the following:

1. They shall be in a location with *ready access* to occupants.
2. They shall be located where the controlled lights are visible, or shall identify the area served by the lights and indicate their status.

C405.2.7 Exterior lighting controls. Exterior lighting systems shall be provided with controls that comply with Sections C405.2.7.1 through C405.2.7.4.

Exceptions:

1. Lighting for covered vehicle entrances to *buildings* where required for eye adaptation.
2. Lighting controlled from within *dwelling units*.

C405.2.7.1 Daylight shutoff. Lights shall be automatically turned off when daylight is present and satisfies the lighting needs.

C405.2.7.2 Building facade and landscape lighting. *Building* facade and landscape lighting shall automatically shut off from not later than 1 hour after *building* or business closing to not earlier than 1 hour before *building* or business opening.

C405.2.7.3 Lighting setback. Lighting that is not controlled in accordance with Section C405.2.7.2 shall comply with the following:

1. Be controlled so that the total wattage of such lighting is automatically reduced by not less than 50 percent by selectively switching off or dimming luminaires at one of the following times:
   1.1. From not later than midnight to not earlier than 6 a.m.
   1.2. From not later than 1 hour after *building* or business closing to not earlier than 1 hour before *building* or business opening.
   1.3. During any time where activity has not been detected for 15 minutes or more.

2. Luminaires serving exterior parking areas and having a rated input wattage of greater than 40 watts and a mounting height of 24 feet (7315 mm) or less above the ground shall be controlled so that the total wattage of such lighting is automatically reduced by not less than 50 percent during any time where activity has not been detected for 15 minutes or more. Not more than 1,500 watts of lighting power shall be controlled together.

C405.2.7.4 Exterior time-switch control function. *Time-switch controls* for exterior lighting shall comply with the following:

1. They shall have a clock capable of being programmed for not fewer than 7 days.
2. They shall be capable of being set for seven different day types per week.
3. They shall incorporate an automatic holiday setback feature.
4. They shall have program backup capabilities that prevent the loss of program and time settings for a period of not less than 10 hours in the event that power is interrupted.

C405.2.8 Reserved.

C405.2.8.1 Demand responsive lighting controls function. Demand responsive controls for lighting shall be capable of the following:

1. Automatically reducing the output of controlled lighting to 80 percent or less of full power or light output upon receipt of a *demand response signal*.
2. Where *high-end trim* has been set, automatically reducing the output of controlled lighting to 80 percent or less of the *high-end trim* setpoint upon receipt of a *demand response signal*.
3. Dimming controlled lights gradually and continuously over a period of not longer than 15 minutes to achieve their demand response setpoint.
4. Returning controlled lighting to its normal operational settings at the end of the demand response period.

Exception: Storage rooms and warehouse storage areas shall be permitted to switch off 25 percent or more of general lighting power rather than dimming.

C405.2.9 Interior parking area lighting control. Interior parking area lighting shall be controlled by an *occupant sensor* complying with Section C405.2.1.1 or a *time-switch control* complying with Section C405.2.2.1. Additional lighting controls shall be provided as follows:

1. Lighting power of each luminaire shall be automatically reduced by not less than 30 percent when there is no activity detected within a lighting zone for 20 minutes. Lighting zones for this requirement shall be not larger than 3,600 square feet (334.5 m^2).

   Exception: Lighting zones provided with less than 1.5 footcandles of illumination on the floor at the darkest point with all lights on are not required to have *automatic* light-reduction controls.
2. Where lighting for eye adaptation is provided at vehicle entrances to *buildings*, such lighting shall be separately controlled by a device that automatically reduces lighting power by at least 50 percent from sunset to sunrise.
3. The power to luminaires within 20 feet (6096 mm) of perimeter wall openings shall automatically reduce in response to daylight by at least 50 percent.

   Exceptions:

   1. Where the opening-to-wall ratio is less than 40 percent as viewed from the interior and encompassing the vertical distance from the driving surface to the lowest structural element.
   2. Where the distance from the opening to any exterior daylight blocking obstruction is less than one-half the height from the bottom of the opening or *fenestration* to the top of the obstruction.
   3. Where openings are obstructed by permanent screens or architectural elements restricting daylight entering the interior space.

C405.2.10 Sleeping unit and dwelling unit lighting and switched receptacle controls. *Sleeping units* and *dwelling units* shall be provided with lighting controls and switched receptacles as specified in Sections C405.2.10.1 and C405.2.10.2.

C405.2.10.1 Sleeping units and dwelling units in hotels, motels and vacation timeshare properties. *Sleeping units* and *dwelling units* in hotels, motels and vacation timeshare properties shall be provided with the following:

1. Not less than two 125V, 15- and 20-amp switched receptacles in each room, except for bathrooms, kitchens, foyers, hallways and closets.
2. Lighting controls that automatically turn off all lighting and switched receptacles within 20 minutes after all occupants have left the unit.

Exception: Automatic shutoff is not required where captive key override controls all lighting and switched receptacles in units with five or fewer permanently installed lights and switched receptacles.

C405.2.10.2 Sleeping units in congregate living facilities. *Sleeping units* in *congregate living facilities* shall be provided with the following controls:

1. Lighting in bathrooms shall be controlled by an *occupant sensor control* that automatically turns off lights within 20 minutes after all occupants have left the space.
2. Each unit shall have a *manual* control by the entrance that turns off all lighting and switched receptacles in the unit, except for lighting in bathrooms and kitchens. The *manual* control shall be marked to indicate its function.

C405.3 Interior lighting power requirements. A *building* complies with this section where its total connected interior lighting power calculated under Section C405.3.1 is not greater than the interior lighting power allowance calculated under Section C405.3.2. *Sleeping units* and *dwelling units* shall comply with Section C405.3.3.

C405.3.1 Total connected interior lighting power. The total connected interior lighting power shall be determined in accordance with Equation 4-9.

Equation 4-9 TCLP = [LVL + BLL + LED + TRK + Other]

where:

TCLP = Total connected lighting power (watts).

LVL = For luminaires with lamps connected directly to building power, such as line voltage lamps, the rated wattage of the lamp.

BLL = For luminaires incorporating a ballast or transformer, the rated input wattage of the ballast or transformer when operating that lamp.

LED = For light-emitting diode luminaires with either integral or remote drivers, the rated wattage of the luminaire.

TRK = For lighting track, cable conductor, rail conductor, and plug-in busway systems that allow the addition and relocation of luminaires without rewiring, the wattage shall be one of the following:

1. The specified wattage of the luminaires, but not less than 8 W per linear foot (25 W/lin m).
2. The wattage limit of the permanent current-limiting devices protecting the system.
3. The wattage limit of the transformer supplying the system.

Other = The wattage of all other luminaires and lighting sources not covered previously and associated with interior lighting verified by data supplied by the manufacturer or other approved sources.

The connected power associated with the following lighting equipment and applications is not included in calculating total connected lighting power.

1. Emergency lighting automatically off during normal building operation.
2. Lighting in spaces specifically designed for use by occupants with special lighting needs, including those with visual impairment and other medical and age-related issues.
3. Mirror lighting in makeup or dressing areas used for video broadcasting, video or film recording, or live theatrical and music performance.
4. Task lighting for medical and dental purposes that is in addition to *general lighting*.
5. Display lighting for exhibits in galleries, museums and monuments that is in addition to *general lighting*.
6. Lighting in any location that is specifically used for video broadcasting, video or film recording, or live theatrical and music performance.
7. Lighting for photographic processes.
8. Lighting integral to equipment or instrumentation and installed by the manufacturer.
9. Task lighting for plant growth or maintenance.
10. Advertising signage or directional signage.
11. Lighting for food warming.
12. Lighting equipment that is for sale.
13. Lighting demonstration equipment in lighting education facilities.
14. Lighting approved because of safety considerations.
15. Lighting in retail display windows, provided that the display area is enclosed by ceiling-height partitions.
16. Furniture-mounted supplemental task lighting that is controlled by automatic shutoff.
17. Exit signs.
18. Antimicrobial lighting used for the sole purpose of disinfecting a space.
19. Lighting in *sleeping units* and *dwelling units*.
20. For exit access and exit stairways, including landings, where the applicable code requires an illuminance of 10 footcandles or more on the walking surface, the power in excess of the allowed power calculated according to Section C405.3.2.2 is not included.

C405.3.2 Interior lighting power allowance. The total interior lighting power allowance (watts) for an entire *building* shall be determined according to Table C405.3.2(1) using the Building Area Method or Table C405.3.2(2) using the Space-by-Space Method. The interior lighting power allowance for projects that involve only portions of a *building* shall be determined according to Table C405.3.2(2) using the Space-by-Space Method. *Buildings* with unfinished spaces shall use the Space-by-Space Method.

TABLE C405.3.2(1)—INTERIOR LIGHTING POWER ALLOWANCES: BUILDING AREA METHOD

BUILDING AREA TYPE	LPD (watts/ft^2)
Automotive facility	0.73
Convention center	0.64
Courthouse	0.75
Dining: bar lounge/leisure	0.74
Dining: cafeteria/fast food	0.70
Dining: family	0.65
Dormitory	0.52
Exercise center	0.72
Fire station	0.56
Gymnasium	0.75
Health care clinic	0.77
Hospital	0.92
Hotel/Motel	0.53
Library	0.83
Manufacturing facility	0.82
Motion picture theater	0.43
Multiple-family	0.46
Museum	0.56
Office	0.62
Parking garage	0.17
Penitentiary	0.65
Performing arts theater	0.82
Police station	0.62
Post office	0.64
Religious building	0.66
Retail	0.78
School/university	0.70
Sports arena	0.73
Town hall	0.67
Transportation	0.56
Warehouse	0.45
Workshop	0.86

For SI: 1 watt per square foot = 10.76 w/m^2.

TABLE C405.3.2(2)—INTERIOR LIGHTING POWER ALLOWANCES: SPACE-BY-SPACE METHOD

COMMON SPACE TYPES[a]	LPD (watts/ft²)
Atrium	
Less than or equal to 40 feet in height	0.41
Greater than 40 feet in height	0.51
Audience seating area	
In an auditorium	0.57
In a gymnasium	0.23
In a motion picture theater	0.27
In a penitentiary	0.56
In a performing arts theater	1.09
In a religious building	0.72
In a sports arena	0.27
Otherwise	0.33
Banking activity area	0.56
Breakroom (See Lounge/breakroom)	
Classroom/lecture hall/training room	
In a penitentiary	0.74
Otherwise	0.72
Computer room, data center	0.75
Conference/meeting/multipurpose room	0.88
Copy/print room	0.56
Corridor	
In a facility for the visually impaired (and not used primarily by the staff)[b]	0.71
In a hospital	0.61
Otherwise	0.44
Courtroom	1.08
Dining area	
In bar/lounge or leisure dining	0.76
In cafeteria or fast food dining	0.36
In a facility for the visually impaired (and not used primarily by the staff)[b]	1.22
In family dining	0.52
In a penitentiary	0.35
Otherwise	0.42
Electrical/mechanical room	0.71
Emergency vehicle garage	0.51
Food preparation area	1.19
Laboratory	
In or as a classroom	1.05
Otherwise	1.21
Laundry/washing area	0.51
Loading dock, interior	0.88
Lobby	
For an elevator	0.64
In a facility for the visually impaired (and not used primarily by the staff)[b]	1.44
In a motion picture theater	0.20
In a performing arts theater	1.21
Otherwise	0.80

TABLE C405.3.2(2)—INTERIOR LIGHTING POWER ALLOWANCES: SPACE-BY-SPACE METHOD—continued	
COMMON SPACE TYPES[a]	**LPD (watts/ft²)**
Locker room	0.43
Lounge/breakroom	
In a health care facility	0.77
Mother's wellness room	0.68
Otherwise	0.55
Office	
Enclosed	0.73
Open plan	0.56
Parking area daylight transition zone	1.06
Parking area, interior	0.11
Pharmacy area	1.59
Restroom	
In a facility for the visually impaired (and not used primarily by the staff)[b]	0.96
Otherwise	0.74
Sales area	0.85
Seating area, general	0.21
Security screening general areas	0.64
Security screening in transportation facilities	0.93
Security screening transportation waiting area	0.56
Stairwell	0.47
Storage room	0.35
Vehicular maintenance area	0.59
Workshop	1.17
BUILDING TYPE SPECIFIC SPACE TYPES[a]	**LPD (watts/ft²)**
Automotive (see Vehicular maintenance area)	
Convention Center—exhibit space	0.50
Dormitory—living quarters	0.48
Facility for the visually impaired[b]	
In a chapel (and not used primarily by the staff)	0.58
In a recreation room (and not used primarily by the staff)	1.20
Fire station—sleeping quarters	0.48
Gaming establishments	
High limits game	1.68
Slots	0.54
Sportsbook	0.82
Table games	1.09
Gymnasium/fitness center	
In an exercise area	0.82
In a playing area	0.82
Health care facility	
In an exam/treatment room	1.33
In an imaging room	0.94
In a medical supply room	0.56
In a nursery	0.87
In a nurse's station	1.07
In an operating room	2.26
In a patient room	0.78

TABLE C405.3.2(2)—INTERIOR LIGHTING POWER ALLOWANCES: SPACE-BY-SPACE METHOD—continued	
COMMON SPACE TYPES[a]	**LPD (watts/ft^2)**
Health care facility—continued	
In a physical therapy room	0.82
In a recovery room	1.18
In a telemedicine room	1.44
Library	
In a reading area	0.86
In the stacks	1.18
Manufacturing facility	
In a detailed manufacturing area	0.75
In an equipment room	0.73
In an extra-high-bay area (greater than 50 feet floor-to-ceiling height)	1.36
In a high-bay area (25–50 feet floor-to-ceiling height)	1.24
In a low-bay area (less than 25 feet floor-to-ceiling height)	0.86
Museum	
In a general exhibition area	0.31
In a restoration room	1.24
Performing arts theater—dressing room	0.39
Post office—sorting area	0.71
Religious buildings	
In a fellowship hall	0.50
In a worship/pulpit/choir area	0.75
Retail facilities	
In a dressing/fitting room	0.45
Hair salon	0.65
Nail salon	0.75
In a mall concourse	0.57
Massage space	0.81
Sports arena—playing area	
For a Class I facility[c]	2.86
For a Class II facility[d]	1.98
For a Class III facility[e]	1.29
For a Class IV facility[f]	0.86
Sports arena—pools	
For a Class I facility	2.20
For a Class II facility	1.47
For a Class III facility	0.99
For a Class IV facility	0.59
Transportation facility	
Airport hanger	1.36
At a terminal ticket counter	0.40
In a baggage/carousel area	0.28
In an airport concourse	0.49
Passenger loading area	0.71
Warehouse—storage area	
For medium to bulky, palletized items	0.33
For smaller, hand-carried items	0.69

TABLE C405.3.2(2)—INTERIOR LIGHTING POWER ALLOWANCES: SPACE-BY-SPACE METHOD—continued
For SI: 1 foot = 304.8 mm, 1 watt per square foot = 10.76 w/m^2.
a. In cases where both a common space type and a building area specific space type are listed, the building area specific space type shall apply.
b. A 'Facility for the Visually Impaired' is a facility that is licensed or will be licensed by local or state authorities for senior long-term care, adult day care, senior support or people with special visual needs.
c. Class I facilities consist of professional facilities; and semiprofessional, collegiate or club facilities with seating for 5,000 or more spectators.
d. Class II facilities consist of collegiate and semiprofessional facilities with seating for fewer than 5,000 spectators, club facilities with seating for between 2,000 and 5,000 spectators, and amateur league and high school facilities with seating for more than 2,000 spectators.
e. Class III facilities consist of club, amateur league and high school facilities with seating for 2,000 or fewer spectators.
f. Class IV facilities consist of elementary school and recreational facilities; and amateur league and high school facilities without provision for spectators.

C405.3.2.1 Building Area Method. For the Building Area Method, the interior lighting power allowance is calculated as follows:

1. For each building area type inside the *building*, determine the applicable building area type and the allowed lighting power density for that type from Table C405.3.2(1). For building area types not listed, select the building area type that most closely represents the use of that area. For the purposes of this method, an "area" shall be defined as all contiguous spaces that accommodate or are associated with a single building area type.
2. Determine the floor area for each building area type listed in Table C405.3.2(1) and multiply this area by the applicable value from Table C405.3.2(1) to determine the lighting power (watts) for each building area type. *Sleeping units* and *dwelling units* are excluded from lighting power allowance calculations by application of Section C405.3.3. The area of *sleeping units* and *dwelling units* is not included in the calculation.
3. The total interior lighting power allowance (watts) for the entire *building* is the sum of the lighting power from each building area type.

C405.3.2.2 Space-by-Space Method. Where a *building* has unfinished spaces, the lighting power allowance for the unfinished spaces shall be the total connected lighting power for those spaces, or 0.1 watts per square foot (1.08 w/m^2), whichever is less. For the Space-by-Space Method, the interior lighting power allowance is calculated as follows:

1. For each space enclosed by partitions that are not less than 80 percent of the ceiling height, determine the applicable space type from Table C405.3.2(2). For space types not listed, select the space type that most closely represents the proposed use of the space. Where a space has multiple functions, that space may be divided into separate spaces.
2. Determine the total floor area of all the spaces of each space type and multiply by the value for the space type in Table C405.3.2(2) to determine the allowed lighting power (watts) for each space type. *Sleeping units* and *dwelling units* are excluded from lighting power allowance calculations by application of Section C405.3.3. The area of *sleeping units* and *dwelling units* is not included in the calculation.
3. The total interior lighting power allowance (watts) shall be the sum of the lighting power allowances for all space types.

C405.3.2.2.1 Additional interior lighting power. Where using the Space-by-Space Method, an increase in the interior lighting power allowance is permitted for specific lighting functions. Additional power shall be permitted only where the specified lighting is installed and controlled in accordance with Section C405.2.5. These additional power allowances shall be used only for the luminaires serving the specific lighting function and shall not be used for any other purpose. An increase in the interior lighting power allowance is permitted in the following cases:

1. For lighting equipment to be installed in sales areas specifically to highlight merchandise, the additional lighting power allowance shall be the connected lighting power of the luminaires specifically highlighting merchandise, calculated in accordance with Equation 4-9, or the additional power allowance calculated in accordance with Equation 4-10, whichever is less.

Equation 4-10 Additional lighting power allowance = 750 W + (Retail Area 1 × 0.40 W/ft^2) + (Retail Area 2 × 0.40W/ft^2) + (Retail Area 3 × 0.70 W/ft^2) + (Retail Area 4 × 1.00 W/ft^2)

For SI units:

Additional lighting power allowance = 750 W + (Retail Area 1 × 4.3 W/m^2) + (Retail Area 2 × 4.3 W/m^2) + (Retail Area 3 × 7.5 W/m^2) + (Retail Area 4 × 10.8 W/m^2)

where:

Retail Area 1 = The floor area for all products not listed in Retail Area 2, 3 or 4.

Retail Area 2 = The floor area used for the sale of vehicles, sporting goods and small electronics.

Retail Area 3 = The floor area used for the sale of furniture, clothing, cosmetics and artwork.

Retail Area 4 = The floor area used for the sale of jewelry, crystal and china.

Exception: Other merchandise categories are permitted to be included in Retail Areas 2 through 4, provided that justification documenting the need for additional lighting power based on visual inspection, contrast or other critical display is approved by the code official.

2. For spaces in which lighting is specified to be installed in addition to the *general lighting* for the purpose of decorative appearance or for highlighting art or exhibits, the additional lighting power allowance for that space shall be the smallest of the following:
   2.1. 0.66 W/ft^2 (7.1W/m^2) in lobbies,
   2.2. 0.55 W/ft^2 (5.9 W/m ^{2}) in other spaces, or
   2.3. The connected lighting power of the luminaires specifically for decorative appearance or for highlighting art or exhibits, calculated according to Equation 4-9.

C405.3.3 Lighting power for sleeping units and dwelling units. *Sleeping units* in Group I-2 occupancies that are patient rooms shall comply with Sections C405.3.1 and C405.3.2. For all other *sleeping units* and *dwelling units*, permanently installed lighting, including lighting integrated into range hoods and exhaust fans, shall be provided by lamps capable of operating with an efficacy of not less than 65 lumens per watt or luminaires capable of operating with an efficacy of not less than 45 lumens per watt.

Exceptions:

1. Lighting integral to other appliances.
2. Antimicrobial lighting used for the sole purpose of disinfecting.
3. Luminaires with an input rating of less than 3 watts.

C405.4 Horticultural lighting. Permanently installed luminaires shall have a *photosynthetic photon efficacy* of not less than 1.7 micromoles per joule (μmol/J) for *horticultural lighting* in greenhouses and not less than 1.9 μmol/J for all other *horticultural lighting*. Luminaires for *horticultural lighting* in greenhouses shall be controlled by a device that automatically turns off the luminaire when sufficient daylight is available. Luminaires for *horticultural lighting* shall be controlled by a device that automatically turns off the luminaire at specific programmed times.

C405.5 Exterior lighting power requirements. The total connected exterior lighting power calculated in accordance with Section C405.5.1 shall be not greater than the exterior lighting power allowance calculated in accordance with Section C405.5.2.

C405.5.1 Total connected exterior lighting power. The total exterior connected lighting power shall be the total maximum rated wattage of all exterior lighting that is powered through the energy service for the *building* and building site lighting for which the *building owner* is responsible.

Exception: Lighting used for the following applications shall not be included.

1. Lighting *approved* because of safety considerations.
2. Emergency lighting automatically off during normal business operation.
3. Exit signs.
4. Specialized signal, directional and marker lighting associated with transportation.
5. Advertising signage or directional signage.
6. Integral to equipment or instrumentation and installed by its manufacturer.
7. Lighting in any location that is specifically used for video broadcasting, video or film recording, or live theatrical and music performances.
8. Athletic playing areas.
9. Temporary lighting.
10. Industrial production, material handling, transportation sites and associated storage areas.
11. Theme elements in theme/amusement parks.
12. Used to highlight features of art, public monuments and the national flag.
13. Lighting for water features and swimming pools.
14. Lighting controlled from within *sleeping units* and *dwelling units*.
15. Lighting of the exterior means of egress as required by the *International Building Code*.

C405.5.2 Exterior lighting power allowance. The exterior lighting power allowance (watts) is calculated as follows:

1. Determine the Lighting Zone (LZ) for the *building* according to Table C405.5.2(1), unless otherwise specified by the *code official*.
2. For each exterior area that is to be illuminated by lighting that is powered through the energy service for the *building* and building site lighting for which the building *owner* is responsible, determine the applicable area type from Table C405.5.2(2). For area types not listed, select the area type that most closely represents the proposed use of the area.
3. Determine the total area or length of each area type and multiply by the value for the area type in Table C405.5.2(2) to determine the lighting power (watts) allowed for each area type.
4. The total exterior lighting power allowance (watts) is the sum of the base site allowance determined according to Table C405.5.2(2), plus the watts from each area type.

TABLE C405.5.2(1)—EXTERIOR LIGHTING ZONES	
LIGHTING ZONE	**DESCRIPTION**
1	Developed areas of national parks, state parks, forest land and rural areas
2	Areas predominantly consisting of residential zoning, neighborhood business districts, light industrial with limited night-time use and residential mixed-use areas
3	All other areas not classified as lighting zone 1, 2 or 4
4	High-activity commercial districts in major metropolitan areas as designated by the local land use planning authority

TABLE C405.5.2(2)—LIGHTING POWER ALLOWANCES FOR BUILDING EXTERIORS				
	LIGHTING ZONES			
	Zone 1	**Zone 2**	**Zone 3**	**Zone 4**
Base Site Allowance	160 W	280 W	400 W	560 W
Uncovered Parking Areas				
Parking area, exterior	0.015 W/ft^2	0.026 W/ft^2	0.037 W/ft^2	0.052 W/ft^2
Building Grounds				
Walkways and ramps less	0.50 W/linear foot	0.50 W/linear foot	0.55 W/linear foot	0.60 W/linear foot
Plaza areas	0.028 W/ft^2	0.049 W/ft^2	0.070 W/ft^2	0.098 W/ft^2
Dining areas	0.156 W/ft^2	0.273 W/ft^2	0.390 W/ft^2	0.546 W/ft^2
Stairways	Exempt	Exempt	Exempt	Exempt
Pedestrian tunnels	0.063 W/ft^2	0.110 W/ft^2	0.157 W/ft^2	0.220 W/ft^2
Landscaping	0.014 W/ft^2	0.025 W/ft^2	0.036 W/ft^2	0.050 W/ft^2
Building Entrances and Exits				
Pedestrian and vehicular entrances and exits	5.6 W/linear foot of opening	9.8 W/linear foot of opening	14 W/linear foot of opening	19.6 W/linear foot of opening
Entry canopies	0.072 W/ft^2	0.126 W/ft^2	0.180 W/ft^2	0.252W/ft^2
Loading docks	0.104 W/ft^2	0.182 W/ft^2	0.260 W/ft^2	0.364 W/ft^2
Sales Canopies				
Free-standing and attached	0.20 W/ft^2	0.35 W/ft^2	0.50 W/ft^2	0.70 W/ft^2
Outdoor Sales				
Open areas (including vehicle sales lots)	0.072 W/ft^2	0.126 W/ft^2	0.180 W/ft^2	0.252 W/ft^2
Street frontage for vehicle sales lots in addition to "open area" allowance	No allowance	7.2 W/linear foot	10.3 W/linear foot	14.4 W/linear foot

For SI: 1 foot = 304.8 mm, 1 watt per square foot = 10.76 W/m^2.
W = Watt.

TABLE C405.5.2(3)—INDIVIDUAL LIGHTING POWER ALLOWANCES FOR BUILDING EXTERIORS				
	LIGHTING ZONES			
	Zone 1	**Zone 2**	**Zone 3**	**Zone 4**
Building facades	No allowance	0.075 W/ft^2 of gross above-grade wall area	0.113 W/ft^2 of gross above-grade wall area	0.15 W/ft^2 of gross above-grade wall area
Automated teller machines (ATM) and night depositories	90 W per location plus 35W per additional ATM per location			
Uncovered entrances and gatehouse inspection stations at guarded facilities	0.144 W/ft^2 of area	0.252 W/ft^2 of area	0.360 W/ft^2 of area	0.504 W/ft^2 of area
Uncovered loading areas for law enforcement, fire, ambulance and other emergency service vehicles	0.104 W/ft^2 of area	0.182 W/ft^2 of area	0.260 W/ft^2 of area	0.364 W/ft^2 of area
Drive-up windows and doors	53 W per drive-through	92 W per drive-through	132 W per drive-through	185 W per drive-through

TABLE C405.5.2(3)—INDIVIDUAL LIGHTING POWER ALLOWANCES FOR BUILDING EXTERIORS—continued				
	LIGHTING ZONES			
	Zone 1	Zone 2	Zone 3	Zone 4
Parking area near 24-hour retail entrances	80 W per main entry	140 W per main entry	200 W per main entry	280 W per main entry

For SI: 1 watt per square foot = 10.76 W/m^2.
W = Watts.

C405.5.2.1 Additional exterior lighting power. Additional exterior lighting power allowances are available for the specific lighting applications listed in Table C405.5.2(3). These additional power allowances shall be used only for the luminaires serving these specific applications and shall not be used to increase any other lighting power allowance.

C405.5.3 Gas lighting. Gas-fired lighting appliances shall not be equipped with continuously burning pilot ignition systems.

C405.6 Dwelling electrical meter. Each *dwelling unit* located in a Group R-2 *building* shall have a separate electrical meter.

C405.7 Electrical transformers. Low-voltage dry-type distribution electric transformers shall meet the minimum efficiency requirements of Table C405.7 as tested and rated in accordance with the test procedure *listed* in DOE 10 CFR 431. The efficiency shall be verified through certification under an *approved* certification program or, where a certification program does not exist, the equipment efficiency ratings shall be supported by data furnished by the transformer manufacturer.

Exceptions: The following transformers are exempt in accordance with the DOE definition of Distribution Transformers found in 10 CFR 431.192:

1. Transformers with a tap range of 20 percent or more.
2. Drive (isolation) transformers.
3. Rectifier transformers.
4. Auto-transformers.
5. Uninterruptible power system transformers.
6. Special impedance transformers.
7. Regulating transformers.
8. Sealed transformers.
9. Machine tool (control) transformers.
10. Welding transformers.
11. Grounding transformers.
12. Testing transformers.
13. Nonventilated transformers.

TABLE C405.7—MINIMUM NOMINAL EFFICIENCY LEVELS FOR DOE 10 CFR 431 LOW-VOLTAGE DRY-TYPE DISTRIBUTION TRANSFORMERS

SINGLE-PHASE TRANSFORMERS[a]		THREE-PHASE TRANSFORMERS[a]	
kVA[b]	Efficiency (%)[c]	kVA[b]	Efficiency (%)[c]
15	97.70	15	97.89
25	98.00	30	98.23
37.5	98.20	45	98.40
50	98.30	75	98.60
75	98.50	112.5	98.74
100	98.60	150	98.83
167	98.70	225	98.94
250	98.80	300	99.02
333	98.90	500	99.14
—	—	750	99.23
—	—	1000	99.28

a. A low-voltage dry-type distribution transformer with a kVA rating not listed in the table shall have its minimum efficiency level determined by linear interpolation of the kVA and efficiency values listed in the table immediately above and below its kVA rating. Extrapolation shall not be used below the minimum values or above the maximum values shown for single-phase transformers and three-phase transformers.
b. kiloVolt-Amp rating.
c. Nominal efficiencies shall be established in accordance with the DOE 10 CFR 431 test procedure for low-voltage dry-type transformers.

C405.8 Electric motors. Electric motors shall meet the minimum efficiency requirements of Tables C405.8(1) through C405.8(4) when tested and rated in accordance with the DOE 10 CFR 431. The efficiency shall be verified through certification under an *approved* certification program or, where a certification program does not exist, the equipment efficiency ratings shall be supported by data furnished by the motor manufacturer.

Exception: The standards in this section shall not apply to the following exempt electric motors:

1. Air-over electric motors.
2. Component sets of an electric motor.
3. Liquid-cooled electric motors.
4. Submersible electric motors.
5. Inverter-only electric motors.
6. Definite-purpose machines within the scope of ANSI/NEMA MG 1, Part 18.

TABLE C405.8(1)—MINIMUM NOMINAL FULL-LOAD EFFICIENCY FOR NEMA DESIGN A, NEMA DESIGN B, AND IEC DESIGN N MOTORS (EXCLUDING FIRE PUMP) ELECTRIC MOTORS AT 60 HZ[a, b]

MOTOR HORSEPOWER (STANDARD KILOWATT EQUIVALENT)	NOMINAL FULL-LOAD EFFICIENCY (%) AS OF JUNE 1, 2016							
	2 Pole		4 Pole		6 Pole		8 Pole	
	Enclosed	Open	Enclosed	Open	Enclosed	Open	Enclosed	Open
1 (0.75)	77.0	77.0	85.5	85.5	82.5	82.5	75.5	75.5
1.5 (1.1)	84.0	84.0	86.5	86.5	87.5	86.5	78.5	77.0
2 (1.5)	85.5	85.5	86.5	86.5	88.5	87.5	84.0	86.5
3 (2.2)	86.5	85.5	89.5	89.5	89.5	88.5	85.5	87.5
5 (3.7)	88.5	86.5	89.5	89.5	89.5	89.5	86.5	88.5
7.5 (5.5)	89.5	88.5	91.7	91.0	91.0	90.2	86.5	89.5
10 (7.5)	90.2	89.5	91.7	91.7	91.0	91.7	89.5	90.2
15 (11)	91.0	90.2	92.4	93.0	91.7	91.7	89.5	90.2
20 (15)	91.0	91.0	93.0	93.0	91.7	92.4	90.2	91.0
25 (18.5)	91.7	91.7	93.6	93.6	93.0	93.0	90.2	91.0
30 (22)	91.7	91.7	93.6	94.1	93.0	93.6	91.7	91.7
40 (30)	92.4	92.4	94.1	94.1	94.1	94.1	91.7	91.7
50 (37)	93.0	93.0	94.5	94.5	94.1	94.1	92.4	92.4
60 (45)	93.6	93.6	95.0	95.0	94.5	94.5	92.4	93.0
75 (55)	93.6	93.6	95.4	95.0	94.5	94.5	93.6	94.1
100 (75)	94.1	93.6	95.4	95.4	95.0	95.0	93.6	94.1
125 (90)	95.0	94.1	95.4	95.4	95.0	95.0	94.1	94.1
150 (110)	95.0	94.1	95.8	95.8	95.8	95.4	94.1	94.1
200 (150)	95.4	95.0	96.2	95.8	95.8	95.4	94.5	94.1
250 (186)	95.8	95.0	96.2	95.8	95.8	95.8	95.0	95.0
300 (224)	95.8	95.4	96.2	95.8	95.8	95.8	—	—
350 (261)	95.8	95.4	96.2	95.8	95.8	95.8	—	—
400 (298)	95.8	95.8	96.2	95.8	—	—	—	—
450 (336)	95.8	96.2	96.2	96.2	—	—	—	—
500 (373)	95.8	96.2	96.2	96.2	—	—	—	—

a. Nominal efficiencies shall be established in accordance with DOE 10 CFR 431.

b. For purposes of determining the required minimum nominal full-load efficiency of an electric motor that has a horsepower or kilowatt rating between two horsepower or two kilowatt ratings listed in this table, each such motor shall be deemed to have a listed horsepower or kilowatt rating, determined as follows:
1. A horsepower at or above the midpoint between the two consecutive horsepowers shall be rounded up to the higher of the two horsepowers.
2. A horsepower below the midpoint between the two consecutive horsepowers shall be rounded down to the lower of the two horsepowers.
3. A kilowatt rating shall be directly converted from kilowatts to horsepower using the formula: 1 kilowatt = (1/0.746) horsepower. The conversion should be calculated to three significant decimal places, and the resulting horsepower shall be rounded in accordance with No. 1 or No. 2 above, as applicable.

TABLE C405.8(2)—MINIMUM NOMINAL FULL-LOAD EFFICIENCY FOR NEMA DESIGN C AND IEC DESIGN H MOTORS AT 60 HZ[a, b]

MOTOR HORSEPOWER (STANDARD KILOWATT EQUIVALENT)	NOMINAL FULL-LOAD EFFICIENCY (%) AS OF JUNE 1, 2016					
	4 Pole		6 Pole		8 Pole	
	Enclosed	Open	Enclosed	Open	Enclosed	Open
1 (0.75)	85.5	85.5	82.5	82.5	75.5	75.5
1.5 (1.1)	86.5	86.5	87.5	86.5	78.5	77.0
2 (1.5)	86.5	86.5	88.5	87.5	84.0	86.5
3 (2.2)	89.5	89.5	89.5	88.5	85.5	87.5
5 (3.7)	89.5	89.5	89.5	89.5	86.5	88.5
7.5 (5.5)	91.7	91.0	91.0	90.2	86.5	89.5
10 (7.5)	91.7	91.7	91.0	91.7	89.5	90.2
15 (11)	92.4	93.0	91.7	91.7	89.5	90.2
20 (15)	93.0	93.0	91.7	92.4	90.2	91.0
25 (18.5)	93.6	93.6	93.0	93.0	90.2	91.0
30 (22)	93.6	94.1	93.0	93.6	91.7	91.7
40 (30)	94.1	94.1	94.1	94.1	91.7	91.7
50 (37)	94.5	94.5	94.1	94.1	92.4	92.4
60 (45)	95.0	95.0	94.5	94.5	92.4	93.0
75 (55)	95.4	95.0	94.5	94.5	93.6	94.1
100 (75)	95.4	95.4	95.0	95.0	93.6	94.1
125 (90)	95.4	95.4	95.0	95.0	94.1	94.1
150 (110)	95.8	95.8	95.8	95.4	94.1	94.1
200 (150)	96.2	95.8	95.8	95.4	94.5	94.1

a. Nominal efficiencies shall be established in accordance with DOE 10 CFR 431.

b. For purposes of determining the required minimum nominal full-load efficiency of an electric motor that has a horsepower or kilowatt rating between two horsepower or two kilowatt ratings listed in this table, each such motor shall be deemed to have a listed horsepower or kilowatt rating, determined as follows:
1. A horsepower at or above the midpoint between the two consecutive horsepowers shall be rounded up to the higher of the two horsepowers.
2. A horsepower below the midpoint between the two consecutive horsepowers shall be rounded down to the lower of the two horsepowers.
3. A kilowatt rating shall be directly converted from kilowatts to horsepower using the formula: 1 kilowatt = (1/0.746) horsepower. The conversion should be calculated to three significant decimal places, and the resulting horsepower shall be rounded in accordance with No. 1 or No. 2 above, as applicable.

TABLE C405.8(3)—MINIMUM AVERAGE FULL-LOAD EFFICIENCY POLYPHASE SMALL ELECTRIC MOTORS[a]

MOTOR HORSEPOWER	OPEN MOTORS			
	Number of Poles	2	4	6
	Synchronous Speed (RPM)	3600	1800	1200
0.25	—	65.6	69.5	67.5
0.33	—	69.5	73.4	71.4
0.50	—	73.4	78.2	75.3
0.75	—	76.8	81.1	81.7
1	—	77.0	83.5	82.5
1.5	—	84.0	86.5	83.8
2	—	85.5	86.5	N/A
3	—	85.5	86.9	N/A

N/A = Not Applicable.

a. Average full-load efficiencies shall be established in accordance with DOE 10 CFR 431.

TABLE C405.8(4)—MINIMUM AVERAGE FULL-LOAD EFFICIENCY FOR CAPACITOR-START CAPACITOR-RUN AND CAPACITOR-START INDUCTION-RUN SMALL ELECTRIC MOTORS[a]

MOTOR HORSEPOWER	OPEN MOTORS			
	Number of Poles	2	4	6
	Synchronous Speed (RPM)	3600	1800	1200
0.25	—	66.6	68.5	62.2
0.33	—	70.5	72.4	66.6
0.50	—	72.4	76.2	76.2
0.75	—	76.2	81.8	80.2
1	—	80.4	82.6	81.1
1.5	—	81.5	83.8	N/A
2	—	82.9	84.5	N/A
3	—	84.1	N/A	N/A

N/A = Not Applicable.
a. Average full-load efficiencies shall be established in accordance with DOE 10 CFR 431.

C405.9 Data centers and computer rooms. Electrical equipment in *data centers* and *computer rooms* shall comply with this section.

C405.9.1 Data centers. Transformers, uninterruptable power supplies, motors and electrical power processing equipment in *data centers* shall comply with Section 8 of ASHRAE 90.4 in addition to this code.

C405.9.2 Computer rooms. Uninterruptable power supplies in *computer rooms* shall comply with the requirements in Tables 8.5 and 8.6 of ASHRAE 90.4 in addition to this code.

Exception: AC-output UPS that utilizes standardized NEMA 1-15P or NEMA 5-15P input plug, as specified in ANSI/NEMA WD-6.

C405.10 Vertical and horizontal transportation systems and equipment. Vertical and horizontal transportation systems and equipment shall comply with this section.

C405.10.1 Elevator cabs. For the luminaires in each elevator cab, not including signals and displays, the sum of the lumens divided by the sum of the watts shall be not less than 35 lumens per watt. Ventilation fans in elevators that do not have their own air-conditioning system shall not consume more than 0.33 watts/cfm at the maximum rated speed of the fan. Controls shall be provided that will de-energize ventilation fans and lighting systems when the elevator is stopped, unoccupied and with its doors closed for over 15 minutes.

C405.10.2 Escalators and moving walks. Escalators and moving walks shall comply with ASME A17.1/CSA B44 and shall have *automatic* controls that reduce speed as permitted in accordance with ASME A17.1/CSA B44 and applicable local code.

Exception: A variable voltage drive system that reduces operating voltage in response to light loading conditions is an alternative to the reduced speed function.

C405.10.2.1 Energy recovery. Escalators shall be designed to recover electrical energy when resisting overspeed in the down direction.

C405.11 Voltage drop. The total *voltage drop* across the combination of customer-owned service conductors, feeder conductors and branch circuit conductors shall not exceed 5 percent.

C405.12 Automatic receptacle control. The following shall have *automatic* receptacle control complying with Section C405.12.1:

1. At least 50 percent of all 125V, 15- and 20-amp receptacles installed in enclosed offices, conference rooms, rooms used primarily for copy or print functions, breakrooms, classrooms and individual workstations, including those installed in modular partitions and module office workstation systems.
2. At least 25 percent of branch circuit feeders installed for modular furniture not shown on the *construction documents*.

C405.12.1 Automatic receptacle control function. *Automatic* receptacle controls shall comply with the following:

1. Either split controlled receptacles shall be provided with the top receptacle controlled, or a controlled receptacle shall be located within 12 inches (304.8 mm) of each uncontrolled receptacle.
2. One of the following methods shall be used to provide control:
   2.1. A scheduled basis using a time-of-day operated control device that turns receptacle power off at specific programmed times and can be programmed separately for each day of the week. The control device shall be configured to provide an independent schedule for each portion of the *building* of not more than 5,000 square feet (464.5 m^2) and not more than one floor. The occupant shall be able to manually override an area for not more than 2 hours. Any individual override switch shall control the receptacles of not more than 5,000 feet (1524 m).
   2.2. An *occupant sensor control* that shall turn off receptacles within 20 minutes of all occupants leaving a space.

2.3. An automated signal from another control or alarm system that shall turn off receptacles within 20 minutes after determining that the area is unoccupied.

3. All controlled receptacles shall be permanently marked in accordance with NFPA 70 and be uniformly distributed throughout the space.
4. Plug-in devices shall not comply.

Exceptions: *Automatic* receptacle controls are not required for the following:

1. Receptacles specifically designated for equipment requiring continuous operation (24 hours per day, 365 days per year).
2. Spaces where an *automatic* control would endanger the safety or security of the room or building occupants.
3. Within a single modular office workstation, noncontrolled receptacles are permitted to be located more than 12 inches (304.8 mm), but not more than 72 inches (1828 mm) from the controlled receptacles serving that workstation.

C405.13 Energy monitoring. New *buildings* with a gross *conditioned floor area* of not less than 10,000 square feet (929 m^2) shall be equipped to measure, monitor, record and report energy consumption in accordance with Sections C405.13.1 through C405.13.6 for load categories indicated in Table C405.13.2 and Sections C405.13.7 through C405.13.11 for end-use categories indicated in Table C405.13.8.

Exceptions:

1. *Dwelling units* in R-2 occupancies.
2. Individual tenant spaces are not required to comply with this section provided that the space has its own utility services and meters and has less than 5,000 square feet (464.5 m^2) of *conditioned floor area*.

C405.13.1 Electrical energy metering. For electrical energy supplied to the *building* and its associated site, including but not limited to site lighting, parking, recreational facilities and other areas that serve the *building* and its occupants, meters or other measurement devices shall be provided to collect energy consumption data for each end-use category required by Section C405.13.2.

C405.13.2 End-use electric metering categories. Meters or other *approved* measurement devices shall be provided to collect energy use data for each end-use category indicated in Table C405.13.2. Where multiple meters are used to measure any end-use category, the data acquisition system shall total all of the energy used by that category. Not more than 5 percent of the design load for each of the end-use categories indicated in Table C405.13.2 shall be permitted to be from a load that is not within that category.

Exceptions:

1. HVAC and water heating equipment serving only an individual *dwelling unit* shall not require end-use metering.
2. End-use metering shall not be required for fire pumps, stairwell pressurization fans or any system that operates only during testing or emergency.
3. End-use metering shall not be required for an individual tenant space having a floor area not greater than 2,500 square feet (232 m^2) where a dedicated source meter complying with Section C405.13.3 is provided.

TABLE C405.13.2—ELECTRICAL ENERGY USE CATEGORIES

LOAD CATEGORY	DESCRIPTION OF ENERGY USE
Total HVAC system	Heating, cooling and ventilation, including but not limited to fans, pumps, boilers, chillers and water heating. Energy used by 120-volt equipment, or by 208/120-volt equipment that is located in a building where the main service is 480/277-volt power, is permitted to be excluded from total HVAC system energy use.
Interior lighting	Lighting systems located within the building.
Exterior lighting	Lighting systems located on the building site but not within the building.
Plug loads	Devices, appliances and equipment connected to convenience receptacle outlets.
Process load	Any single load that is not included in an HVAC, lighting or plug load category and that exceeds 5 percent of the peak connected load of the whole building, including but not limited to data centers, manufacturing equipment and commercial kitchens.
Building operations and other miscellaneous loads	The remaining loads not included elsewhere in this table, including but not limited to vertical transportation systems, automatic doors, motorized shading systems, ornamental fountains, fireplaces, swimming pools, spas and snow-melt systems.
Electric hot water heating for uses other than space conditioning	Electricity used to generate hot water. **Exception:** Electric water heating with design capacity that is less than 10 percent of the building service rating.

C405.13.3 Electrical meters. Meters or other measurement devices required by this section shall be configured to automatically communicate energy consumption data to the data acquisition system required by Section C405.13.4. Source meters shall be allowed to be any digital-type meter. Lighting, HVAC or other building systems that can self-monitor their energy consumption shall be permitted instead of meters. Current sensors shall be permitted, provided that they have a tested accuracy of ±2 percent. Required metering systems and equipment shall have the capability to provide at least hourly data that is fully integrated into the data acquisition system and graphical energy report in accordance with Sections C405.13.4 and C405.13.5. Nonintrusive load monitoring (NILM) packages that extract energy consumption data from detailed electric waveform analysis shall be permitted to substitute for individual meters if the equivalent data is available for collection in Section C405.13.4 and reporting in Section C405.13.5.

C405.13.4 Electrical energy data acquisition system. A data acquisition system shall have the capability to store the data from the required meters and other sensing devices for a minimum of 36 months. The data acquisition system shall have the capability to store real-time energy consumption data and provide hourly, daily, monthly and yearly logged data for each end-use category required by Section C405.13.2. The data acquisition system shall have the capability of providing *building* total peak electric demand and the time(s) of day and time(s) per month at which the peak occurs. Peak demand shall be integrated over the same time period as the underlying whole-building meter reading rate.

C405.13.5 Graphical energy report. A permanent and readily available reporting mechanism shall be provided in the *building* for access by *building* operation and management personnel. The reporting mechanism shall have the capability to graphically provide the energy consumption for each end-use category required by Section C405.13.2 not less than every hour, day, month and year for the previous 36 months.

C405.13.6 Renewable energy. On-site renewable energy sources shall be metered with no less frequency than nonrenewable energy systems in accordance with Section C405.13.3.

C405.13.7 Nonelectrical energy submetering. For all nonelectrical energy supplied to the *building* and its associated site that serves the *building* and its occupants, submeters or other measurement devices shall be provided to collect energy consumption data for each end-use category required by Section C405.13.8.

Exceptions:

1. HVAC and water heating equipment serving only an individual *dwelling unit* shall not require end-use submetering.
2. End-use submetering shall not be required for fire pumps, stairwell pressurization fans or any system that operates only during testing or emergency.
3. End-use submetering shall not be required for an individual tenant space having a floor area not greater than 2,500 square feet (232 m^2) where a dedicated source meter complying with Section C405.13.9 is provided.
4. Equipment powered primarily by solid fuels serving loads other than *building* heating and service water heating loads.

C405.13.8 End-use nonelectrical submetering categories. Submeters or other *approved* measurement devices shall be provided to collect energy use data for each end-use category indicated in Table C405.13.8. Where multiple submeters are used to measure any end-use category, the data acquisition system shall total all of the energy used by that category. Not more than 5 percent of the design load for each of the end-use categories indicated in Table C405.13.8 shall be permitted to be from a load that is not within that category.

TABLE C405.13.8—NONELECTRICAL ENERGY USE CATEGORIES

END USE CATEGORY	DESCRIPTION OF END USE
Total HVAC system	Heating and cooling systems, including but not limited to boilers, chillers and furnaces. District heating and cooling energy entering the building's distribution system shall be monitored at the point of entry to the building distribution system.
Process loads	Any single load that is not included in the HVAC or service water heating categories where the rated fuel gas or fuel oil input of the load and that is not less than 5 percent of the sum of the rated fuel gas or fuel oil input of all monitored equipment, including but not limited to manufacturing equipment, process equipment, commercial kitchens, and commercial laundry equipment.
Other miscellaneous loads	The remaining loads not included elsewhere in this table, including but not limited to fireplaces, swimming pools, spas, gas lighting, and snow-melt systems.
Service water heating	Fuel used to heat potable water. **Exception:** Water heating with design capacity that is less than 10 percent of the sum of the rated fuel gas or fuel oil input of all monitored equipment.

C405.13.9 Nonelectrical submeters. Submeters or other measurement devices required by this section shall be configured to automatically communicate energy consumption data to the data acquisition system required by Section C405.13.10. Source submeters shall be allowed to be any digital-type meter that can provide a digital output to the data acquisition system. Required submetering systems and equipment shall be fully integrated into the data acquisition system and graphical energy report that updates at least hourly in accordance with Sections C405.13.10 and C405.13.11.

C405.13.10 Nonelectrical energy data acquisition system. A data acquisition system shall have the capability to store the data from the required submeters and other sensing devices for not less than 36 months. The data acquisition system shall have the capability to store real-time energy consumption data and provide hourly, daily, monthly and yearly logged data for each end-use category required by Section C405.13.8. The data acquisition system shall have the capability of providing building total nonelectrical peak demand and the time(s) of day and time(s) per month at which the peak occurs. Where applicable as determined by the authority having jurisdiction (AHJ), peak demand shall be integrated over the same time period as the underlying whole-building meter reading rate.

C405.13.11 Graphical energy report. A permanent and readily accessible reporting mechanism shall be provided in the *building* that is accessible by building operation and management personnel. The reporting mechanism shall have the capability to graphically provide the nonelectrical energy consumption for each end-use category required by Section C405.13.8 not less than every hour, day, month and year for the previous 36 months. The graphical report shall incorporate natural gas interval data from the submeter or the ability to enter gas utility bills into the report.

C405.14 Reserved.

C405.15 Renewable energy systems. *Buildings* in Climate Zones 0 through 7 shall comply with Sections C405.15.1 through C405.15.4.

C405.15.1 On-site renewable energy systems. *Buildings* shall be provided with on-site renewable electricity generation systems with a direct current (DC) nameplate power rating of not less than 0.75 watts per square foot (8.1 W/m^2) multiplied by the sum of the gross *conditioned floor area* of all floors, not to exceed the combined gross *conditioned floor area* of the three largest floors.

Exceptions: The following *buildings* or building sites shall comply with Section C405.15.2:

1. A *building site* located where an unshaded flat plate collector oriented toward the equator and tilted at an angle from horizontal equal to the latitude receives an annual daily average incident solar radiation less than 1.1 kBtu/ft^2 per day (3.5 kWh/m^2/day).
2. A *building* where more than 80 percent of the roof area is covered by any combination of permanent obstructions such as, but not limited to, mechanical equipment, vegetated space, access pathways or occupied roof terrace.
3. Any *building* where more than 50 percent of the roof area is shaded from direct-beam sunlight by natural objects or by structures that are not part of the *building* for more than 2,500 annual hours between 8:00 a.m. and 4:00 p.m.
4. A *building* with gross *conditioned floor area* less than 5,000 square feet (465 m^2).

C405.15.2 Off-site renewable energy. *Buildings* that qualify for one or more of the exceptions to Section C405.15.1 or do not meet the requirements of Section C405.15.1 with an on-site renewable energy system shall procure off-site renewable electrical energy, in accordance with Sections C405.15.2.1 and C405.15.2.2, that shall be not less than the total off-site renewable electrical energy determined in accordance with Equation 4-11.

Equation 4-11 $TRE_{off} = (REN_{off} \times 0.75\ \text{W/ft}^2 \times FLRA - IRE_{on}) \times 15$

where:

TRE_{off} = Total off-site renewable electrical energy in kilowatt-hours (kWh) to be procured in accordance with Table C405.15.2.

REN_{off} = Annual off-site renewable electrical energy from Table C405.15.2, in units of kilowatt-hours per watt of array capacity.

FLRA = The sum of the gross conditioned floor area of all floors not to exceed the combined floor area of the three largest floors.

IRE_{on} = Annual on-site renewable electrical energy generation of a new on-site renewable energy system, to be installed as part of the building project, whose rated capacity is less than the rated capacity required in Section C405.15.1.

TABLE C405.15.2—ANNUAL OFF-SITE RENEWABLE ENERGY REQUIREMENTS

CLIMATE ZONE	ANNUAL OFF-SITE RENEWABLE ELECTRICAL ENERGY (kWh/W)
1A, 2B, 3B, 3C, 4B and 5B	1.75
0A, 0B, 1B, 2A, 3A and 6B	1.55
4A, 4C, 5A, 5C, 6A and 7	1.35

C405.15.2.1 Off-site procurement. The building *owner,* as defined in the *International Building Code* , shall procure and be credited for the total amount of off-site renewable electrical energy, not less than required in accordance with Equation 4-11, with one or more of the following:

1. *Physical renewable energy power purchase agreement.*
2. *Financial renewable energy power purchase agreement.*
3. *Community renewable energy facility.*
4. Off-site renewable energy system owned by the building property owner.
5. *Renewable energy investment fund.*
6. *Green retail tariff.*

The generation source shall be located where the energy can be delivered to the *building site* by any of the following:

1. Direct connection to the off-site renewable energy facility.
2. The local utility or distribution entity.
3. An interconnected electrical network where energy delivery capacity between the generator and the building site is available.

C405.15.2.2 Off-site contract. The renewable energy shall be delivered or credited to the *building site* under an energy contract with a duration of not less than 10 years. The contract shall be structured to survive a partial or full transfer of ownership of the building property.

C405.15.3 Renewable energy certificate (REC) documentation. The property *owner* or owner's authorized agent shall demonstrate that where renewable energy certificates (RECs) or energy attribute certificates (EACs) are associated with on-site and off-site renewable energy production required by Sections C405.15.1 and C405.15.2, all of the following criteria for RECs and EACs shall be met:

1. The RECs and EACs are retained and retired by or on behalf of the property owner or tenant for a period of not less than 15 years or the duration of the contract in Section C405.15.2.2, whichever is less.
2. The RECs and EACs are created within a 12-month period of the use of the REC.
3. The RECs and EACs are from a generating asset placed in service not more than 5 years before the issuance of the certificate of occupancy.

C405.15.4 Renewable energy certificate purchase. A *building* that qualifies for one or more of the exceptions to Section C405.15.1, and where it can be demonstrated to the *code official* that the requirements of Section C405.15.2 cannot be met, the building owner shall contract the purchase of renewable electricity products before the certificate of occupancy is issued. The purchase of renewable electricity products shall comply with the Green-e Energy National Standard for renewable electricity products equivalent to five times the amount of total off-site renewable energy calculated in accordance with Equation 4-11.

C405.16 Inverters. Direct-current-to-alternating-current inverters serving on-site renewable energy systems or on-site electrical energy storage systems (ESS) shall be compliant with IEEE 1547 and UL 1741.

SECTION C406—ADDITIONAL EFFICIENCY, RENEWABLE AND LOAD MANAGEMENT REQUIREMENTS

C406.1 Compliance. *Buildings* shall comply as follows:

1. *Buildings* with greater than 2,000 square feet (186 m^2) of *conditioned floor area* shall comply with Section C406.1.1.
2. *Buildings* with greater than 5,000 square feet (465 m^2) of *conditioned floor area* shall comply with Sections C406.1.1 and C406.1.2.
3. Build-out construction greater than 1,000 square feet (93 m^2) of *conditioned floor area* that does not have final lighting or final HVAC systems installed under a prior building permit shall comply with Section C406.1.1.2.

Exceptions: Core and shell *buildings* where not less than 20 percent of the net floor area is without final lighting or final HVAC that comply with all of the following:

1. *Buildings* with greater than 5,000 square feet (465 m^2) of *conditioned floor area* shall comply with Section C406.1.2.
2. Portions of the *building* where the net floor area is without final lighting or final HVAC shall comply with Section C406.1.1.2.
3. Portions of the *building* where the net floor area has final lighting and final HVAC systems shall comply with Section C406.1.1.

C406.1.1 Additional energy efficiency credit requirements. *Buildings* shall comply with measures from Section C406.2 to achieve not less than the number of required efficiency credits from Table C406.1.1(1) based on building occupancy group and *climate zone.* Where a project contains multiple occupancies, the total required energy credits from each building occupancy shall be weighted by the gross *conditioned floor area* to determine the weighted-average project energy credits required. Accessory occupancies shall be included with the primary occupancy group for the purposes of Section C406.

Exceptions:

1. Portions of *buildings* devoted to manufacturing or industrial use.
2. Where a *building* achieves more renewable and load management credits in Section C406.3 than are required in Section C406.1.2, surplus credits shall be permitted to reduce the required energy efficiency credits as follows:

Equation 4-12 $EEC_{red} = EEC_{tbl} - \{\text{the lesser of: } [SLRM_{lim}, SRLM_{adj} \times (RLM_{ach} - RLM_{req})]\}$

where:

EEC_{red} = Reduced required energy efficiency credits.

EEC_{tbl} = Required energy efficiency credits from Table C406.1.1(1).

$SRLM_{lim}$ = Surplus renewable and load management credit limit from Table C406.1.1(2).

$SRLM_{adj}$ = 1.0 for all-electric or all-renewable buildings (excluding emergency generation); 0.7 for buildings with fossil fuel equipment (excluding emergency generation).

RLM_{ach} = Achieved renewable and load management credits from Section C406.3.

RLM_{req} = Required renewable and load management credits from Section C406.1.2.

TABLE C406.1.1(1)—ENERGY CREDIT REQUIREMENTS BY BUILDING OCCUPANCY GROUP

BUILDING OCCUPANCY GROUP	CLIMATE ZONE																		
	0A	0B	1A	1B	2A	2B	3A	3B	3C	4A	4B	4C	5A	5B	5C	6A	6B	7	8
R-2, R-4 and I-1	65	66	67	77	80	86	80	81	90	86	90	90	86	90	90	70	89	80	78
I-2	43	42	38	37	36	38	32	32	30	36	36	35	43	43	44	46	47	50	53
R-1	63	62	66	65	70	71	77	80	84	81	83	88	85	86	90	83	87	87	85
B	62	62	64	66	66	65	64	64	68	70	72	74	71	73	77	71	74	74	71
A-2	70	70	72	72	75	75	70	73	82	69	74	78	67	72	78	60	67	57	51
M	80	79	83	79	81	84	67	74	87	80	66	65	79	62	50	75	67	75	58
E	56	57	55	58	58	57	59	62	59	61	66	62	64	67	67	65	67	63	58
S-1 and S-2	61	60	61	60	58	57	44	54	62	85	68	75	90	82	72	90	89	90	90
All other	31	31	31	32	32	33	30	32	36	35	35	35	37	36	36	36	37	36	34

TABLE C406.1.1(2)—LIMIT TO ENERGY EFFICIENCY CREDIT CARRYOVER FROM RENEWABLE AND LOAD MANAGEMENT CREDITS

BUILDING OCCUPANCY GROUP	CLIMATE ZONE																		
	0A	0B	1A	1B	2A	2B	3A	3B	3C	4A	4B	4C	5A	5B	5C	6A	6B	7	8
R-2, R-4 and I-1	5	5	5	5	5	5	5	24	19	5	22	18	5	5	19	5	5	5	5
I-2	16	14	11	8	6	5	5	10	6	8	14	10	17	26	29	21	21	22	39
R-1	7	5	8	5	19	5	32	40	41	24	41	42	17	37	41	5	24	15	22
B	7	5	5	8	6	6	14	26	31	23	39	34	19	35	45	5	19	17	27
A-2	18	16	14	15	13	9	11	23	32	5	23	23	5	5	26	5	5	5	5
M	5	5	5	5	5	5	5	5	20	5	5	5	5	5	5	5	5	5	5
E	13	13	18	16	17	14	21	35	40	25	43	29	23	32	27	11	17	25	5
S-1 and S-2	5	5	5	5	5	5	5	5	13	5	17	20	5	35	23	5	5	11	40
All other	5	5	5	5	5	5	5	7	17	5	10	7	5	6	11	5	5	5	5

C406.1.1.1 Reserved.

C406.1.1.2 Building core/shell and build-out construction. Where separate permits are issued for core and shell *buildings* and build-out construction, compliance shall be in accordance with the following requirements.

1. Core and shell *buildings* or portions of *buildings* shall comply with one of the following:
   - 1.1. Where the permit includes a central HVAC system or *service water heating* system with chillers, heat pumps, boilers, *service water heating* equipment or loop pumping systems with heat rejection, the project shall achieve not less than 50 percent of the energy credits required by Section C406.1.1 in accordance with Section C406.2.
   - 1.2. Alternatively, the project shall achieve not less than 33 percent of the energy credits required by Section C406.1.1.
2. For core and shell buildings or portions of *buildings*, the energy credits achieved shall be subject to the following adjustments:
   - 2.1. Lighting measure credits shall be determined only for areas with final lighting installed.
   - 2.2. Where HVAC or *service water heating* systems are designed to serve the entire *building*, full HVAC or *service water heating* measure credits shall be achieved.
   - 2.3. Where HVAC or *service water heating* systems are designed to serve individual areas, HVAC or *service water heating* measure credits achieved shall be reduced in proportion to the floor area with final HVAC systems or final *service water heating* systems installed.

3. Build-out construction shall be deemed to comply with Section C406.1 where one of the following applies:
   3.1. Where heating and cooling generation is provided by a previously installed central system, the energy credits achieved in accordance with Section C406.2 under the build-out project are not less than 33 percent of the credits required by Section C406.1.1.
   3.2. Where heating and cooling generation is provided by an HVAC system installed in the build-out, the energy credits achieved in accordance with Section C406.2 under the build-out project are not less than 50 percent of the credits required by Section C406.1.1.
   3.3. Where the core and shell *building* is *approved* in accordance with Section C407 under the 2021 IECC or later.

C406.1.2 Additional renewable and load management credit requirements. *Buildings* shall comply with measures from Section C406.3 to achieve not less than the number of required renewable and load management credits from Table C406.1.2 based on building occupancy group and *climate zone*. Where a project contains multiple occupancies, credits in Table C406.1.2 from each building occupancy shall be weighted by the gross floor area to determine the weighted-average project energy credits required. Accessory occupancies shall be included with the primary occupancy group for the purposes of Section C406.

Exception: Where a *building* achieves more energy efficiency credits in Section C406.2 than are required in Section C406.1.1, the renewable and load management credits required in Table C406.1.2 shall be permitted to be reduced by the amount of surplus energy efficiency credits.

TABLE C406.1.2—RENEWABLE AND LOAD MANAGEMENT CREDIT REQUIREMENTS BY BUILDING OCCUPANCY GROUP

BUILDING OCCUPANCY GROUP	CLIMATE ZONE																		
	0A	0B	1A	1B	2A	2B	3A	3B	3C	4A	4B	4C	5A	5B	5C	6A	6B	7	8
R-2, R-4 and I-1	34	37	31	46	48	56	49	56	38	31	42	32	26	33	34	23	27	25	25
I-2	23	24	25	25	25	28	26	30	22	25	32	24	25	28	29	26	28	22	20
R-1	30	28	35	30	34	36	34	37	41	32	37	27	28	33	32	25	29	22	18
B	38	39	45	42	45	49	47	56	57	44	55	42	38	47	46	38	45	38	31
A-2	8	8	9	9	8	9	9	11	13	8	11	9	8	10	9	8	9	8	3
M	32	32	42	37	39	47	44	58	57	42	54	46	38	48	5	42	45	38	34
E	27	34	38	37	39	47	44	58	57	42	54	46	38	48	50	42	45	38	34
S-1 and S-2	89	90	90	90	90	90	90	90	90	90	90	90	70	90	90	84	86	71	54
All other	35	39	46	42	46	52	49	56	56	40	52	42	37	44	44	36	39	32	28

C406.2 Additional energy efficiency credits achieved. Each energy efficiency credit measure used to meet credit requirements for the project shall have efficiency that is greater than the requirements in Sections C402 through C405. Measures installed in the project that meet the requirements in Sections C406.2.1 through C406.2.6 shall achieve the base credits listed for the measure and occupancy type in Tables C406.2(1) through C406.2(9) or, where calculations required by Sections C406.2.1 through C406.2.6 create or modify the table credits, the credits achieved shall be based on the calculations. Energy credits achieved for measures shall be determined by one of the following, as applicable:

1. The measure's energy credit shall be the base energy credit from Tables C406.2(1) through C406.2(9) for the measure where no adjustment factor or calculation is included in the description of the measure in Section C406.2.
2. The measure's energy credit shall be the base energy credit for the measure adjusted by a factor or equation as stated in the description of the measure in Section C406.2. Where adjustments are applied, each measure's energy credit shall be rounded to the nearest whole number.
3. The measure's energy credit shall be calculated as stated in the measure's description in Section C406.2, where each individual measure credit shall be rounded to the nearest whole number.

Energy credits achieved for the project shall be the sum of the individual measure's energy credits. Credits are available for the measures listed in this section. Where a project contains multiple building occupancy groups:

1. Credits achieved for each occupancy group shall be summed and then weighted by the *conditioned floor area* of each occupancy group to determine the weighted average project energy credits achieved.
2. Improved envelope efficiency (E01 through E06), HVAC performance (H01) and lighting reduction (L06) measure credits shall be determined for the *building* or permitted *conditioned floor area* as a whole. Credits for other measures shall be determined for each occupancy separately. Credits shall be taken from applicable tables or calculations for each occupancy and weighted by the building occupancy group floor area.

TABLE C406.2(1)—BASE ENERGY CREDITS FOR GROUP R-2, R-4 AND I-1 OCCUPANCIES[a]

ID	ENERGY CREDIT MEASURE	SECTION	CLIMATE ZONE																		
			0A	0B	1A	1B	2A	2B	3A	3B	3C	4A	4B	4C	5A	5B	5C	6A	6B	7	8
E01	Envelope performance	C406.2.1.1	Determined in accordance with Section C406.2.1.1																		
E02	UA reduction (15%)	C406.2.1.2	7	6	2	4	1	1	4	1	1	22	1	3	29	10	1	32	27	30	39
E03	Reduced air leakage	C406.2.1.3	15	10	12	8	6	16	13	5	1	7	7	9	65	16	11	73	43	52	26
E04	Add roof insulation	C406.2.1.4	1	1	1	1	1	1	4	3	1	5	3	4	6	5	4	7	7	6	8
E05	Add wall insulation	C406.2.1.5	10	10	6	8	5	6	8	4	1	8	3	4	11	7	3	14	12	13	13
E06	Improve fenestration	C406.2.1.6	7	7	4	6	9	11	13	3	1	22	5	10	27	18	7	41	33	22	21
H01	HVAC performance	C406.2.2.1	20	19	16	17	14	13	11	11	5	13	10	8	15	12	7	18	14	17	19
H02	Heating efficiency	C406.2.2.2	x	x	x	x	x	x	3	1	1	6	2	3	10	5	2	14	10	13	16
H03	Cooling efficiency	C406.2.2.3	7	6	4	4	3	3	1	1	1	1	1	1	1	1	x	x	x	x	x
H04	Residential HVAC control	C406.2.2.4	9	10	8	22	20	25	16	17	32	21	24	17	23	27	16	21	24	18	18
H05	DOAS/fan control	C406.2.2.5	32	31	27	28	23	23	28	21	12	42	24	24	56	36	19	73	54	70	79
W01	SHW preheat recovery	C406.2.3.1 a	61	63	74	74	85	88	101	100	121	103	109	122	102	111	130	93	106	99	96
W02	Heat pump water heater	C406.2.3.1 b	50	52	62	61	72	74	86	85	104	88	94	106	88	96	112	81	92	87	84
W03	Efficient gas water heater	C406.2.3.1 c	38	39	46	46	53	55	63	62	76	64	68	76	64	69	81	58	66	62	60
W04	SHW pipe insulation	C406.2.3.2	7	7	8	7	8	8	8	9	10	8	9	9	7	8	9	6	7	6	6
W05	Point of use water heaters	C406.2.3.3 a	x	x	x	x	x	x	x	x	x	x	x	x	x	x	x	x	x	x	x
W06	Thermostatic bal. valves	C406.2.3.3 b	3	3	3	3	3	3	3	3	4	3	3	4	3	3	4	3	3	3	2
W07	SHW heat trace system	C406.2.3.3 c	12	12	13	13	14	15	15	15	18	14	15	16	13	14	16	11	13	11	10
W08	SHW submeters	C406.2.3.4	11	11	13	13	15	16	18	18	22	19	20	22	19	20	24	17	20	18	18
W09	SHW flow reduction	C406.2.3.5	22	22	27	26	31	32	37	37	45	38	40	45	38	41	48	35	39	37	36
W10	Shower heat recovery	C406.2.3.6	15	16	19	19	22	23	26	26	32	27	29	32	27	29	34	25	28	27	26
P01	Energy monitoring	C406.2.4	3	3	2	3	2	2	2	2	2	2	2	2	2	2	2	3	2	2	3
L01	Lighting performance	C406.2.5.1	x	x	x	x	x	x	x	x	x	x	x	x	x	x	x	x	x	x	x
L02	Lighting dimming & tuning	C406.2.5.2	1	1	1	1	1	1	1	1	1	1	1	1	1	1	1	1	1	1	1
L03	Increase occp. sensor	C406.2.5.3	3	3	4	4	4	4	3	4	3	2	3	2	1	1	2	1	1	1	1
L04	Increase daylight area	C406.2.5.4	x	x	x	x	x	x	x	x	x	x	x	x	x	x	x	x	x	x	x
L05	Residential light control	C406.2.5.5	8	8	9	9	9	9	8	8	10	6	8	7	4	6	8	3	5	4	3
L06	Light power reduction	C406.2.5.6	2	2	2	2	2	2	2	2	2	1	2	1	1	1	1	1	1	1	1
Q01	Efficient elevator	C406.2.6.1	4	4	4	4	5	5	5	5	5	4	5	5	4	4	5	4	4	4	3
Q02	Commercial kitchen equip.	C406.2.6.2	x	x	x	x	x	x	x	x	x	x	x	x	x	x	x	x	x	x	x
Q03	Residential kitchen equip.	C406.2.6.3	15	15	17	16	17	18	17	18	20	16	17	18	15	16	18	13	15	13	12
Q04	Fault detection	C406.2.6.4	3	3	2	3	2	2	2	2	1	2	2	1	1	2	1	3	2	3	3

DOAS = Dedicated Outside Air System; HVAC = Heating, Ventilation and Air Conditioning; SHW = Service Hot Water; UA = *U*-factor × Area.

a. "x" indicates credit is not available in that climate zone for that measure.

TABLE C406.2(2)—BASE ENERGY CREDITS FOR GROUP I-2 OCCUPANCIES[a]

ID	ENERGY CREDIT MEASURE	SECTION	CLIMATE ZONE																		
			0A	0B	1A	1B	2A	2B	3A	3B	3C	4A	4B	4C	5A	5B	5C	6A	6B	7	8
E01	Envelope performance	C406.2.1.1	Determined in accordance with Section C406.2.1.1																		
E02	UA reduction (15%)	C406.2.1.2	1	1	1	1	2	1	1	1	3	1	3	11	27	7	10	3	3	2	10
E03	Reduced air leakage	C406.2.1.3	5	3	4	3	5	8	8	3	2	6	2	2	7	3	1	9	7	19	5
E04	Add roof insulation	C406.2.1.4	1	1	1	1	1	1	1	1	1	1	1	1	2	1	1	2	1	2	3
E05	Add wall insulation	C406.2.1.5	1	3	1	3	2	2	9	4	1	4	1	1	3	1	1	3	3	3	3
E06	Improve fenestration	C406.2.1.6	1	1	1	1	1	1	1	1	1	4	3	5	5	1	1	5	5	2	2
H01	HVAC performance	C406.2.2.1	x	x	x	x	x	x	x	x	x	x	x	x	x	x	x	x	x	x	x
H02	Heating efficiency	C406.2.2.2	x	x	x	x	2	3	4	3	7	6	4	6	8	6	10	11	12	15	19
H03	Cooling efficiency	C406.2.2.3	6	6	4	4	3	3	2	2	1	1	1	1	1	1	1	x	x	x	x
H04	Residential HVAC control	C406.2.2.4	x	x	x	x	x	x	x	x	x	x	x	x	x	x	x	x	x	x	x
H05	DOAS/fan control	C406.2.2.5	41	41	40	40	42	36	42	37	39	49	40	46	56	46	61	65	68	82	93
W01	SHW preheat recovery	C406.2.3.1 a	4	4	4	4	5	5	5	5	6	6	6	6	6	6	6	6	5	5	5
W02	Heat pump water heater	C406.2.3.1 b	2	2	2	2	2	2	3	3	3	3	3	3	3	3	3	3	3	3	3
W03	Efficient gas water heater	C406.2.3.1 c	2	2	2	2	2	3	3	3	3	3	3	3	3	3	3	3	3	3	3
W04	SHW pipe insulation	C406.2.3.2	1	1	1	1	1	1	1	1	1	1	1	1	1	1	1	1	1	1	1
W05	Point of use water heaters	C406.2.3.3 a	x	x	x	x	x	x	x	x	x	x	x	x	x	x	x	x	x	x	x
W06	Thermostatic bal. valves	C406.2.3.3 b	1	1	1	1	1	1	1	1	1	1	1	1	1	1	1	1	1	1	1
W07	SHW heat trace system	C406.2.3.3 c	1	1	2	2	2	2	2	2	2	2	2	2	2	2	2	1	1	1	1
W08	SHW submeters	C406.2.3.4	x	x	x	x	x	x	x	x	x	x	x	x	x	x	x	x	x	x	x
W09	SHW flow reduction	C406.2.3.5	x	x	x	x	x	x	x	x	x	x	x	x	x	x	x	x	x	x	x
W10	Shower heat recovery	C406.2.3.6	1	1	1	1	1	1	1	1	1	1	1	1	1	1	1	1	1	1	1
P01	Energy monitoring	C406.2.4	3	3	3	3	3	3	3	3	3	3	3	3	3	3	3	3	3	3	3
L01	Lighting performance	C406.2.5.1	x	x	x	x	x	x	x	x	x	x	x	x	x	x	x	x	x	x	x
L02	Lighting dimming & tuning	C406.2.5.2	5	5	5	5	5	6	5	6	6	5	6	6	5	5	5	4	4	3	2
L03	Increase occp. sensor	C406.2.5.3	5	5	5	5	5	5	5	5	6	5	5	6	5	5	5	4	4	3	2
L04	Increase daylight area	C406.2.5.4	x	x	x	x	x	x	x	x	x	x	x	x	x	x	x	x	x	x	x
L05	Residential light control	C406.2.5.5	x	x	x	x	x	x	x	x	x	x	x	x	x	x	x	x	x	x	x
L06	Light power reduction	C406.2.5.6	7	7	7	7	7	7	7	7	9	7	7	8	6	7	7	5	5	4	3
Q01	Efficient elevator	C406.2.6.1	1	2	2	2	2	2	2	2	2	2	2	2	2	2	2	2	2	1	1
Q02	Commercial kitchen equip.	C406.2.6.2	x	x	x	x	x	x	x	x	x	x	x	x	x	x	x	x	x	x	x
Q03	Residential kitchen equip.	C406.2.6.3	x	x	x	x	x	x	x	x	x	x	x	x	x	x	x	x	x	x	x
Q04	Fault detection	C406.2.6.4	3	3	3	3	3	3	3	3	2	3	3	2	3	3	3	3	3	4	4

DOAS = Dedicated Outside Air System; HVAC = Heating, Ventilation and Air Conditioning; SHW = Service Hot Water; UA = *U*-Factor × Area.

a. "x" indicates credit is not available in that climate zone for that measure.

TABLE C406.2(3)—BASE ENERGY CREDITS FOR GROUP R-1 OCCUPANCIES[a]

ID	ENERGY CREDIT MEASURE	SECTION	CLIMATE ZONE																		
			0A	0B	1A	1B	2A	2B	3A	3B	3C	4A	4B	4C	5A	5B	5C	6A	6B	7	8
E01	Envelope performance	C406.2.1.1	Determined in accordance with Section C406.2.1.1																		
E02	UA reduction (15%)	C406.2.1.2	2	3	1	2	1	3	3	2	1	5	2	2	7	4	2	9	7	9	11
E03	Reduced air leakage	C406.2.1.3	15	9	12	8	6	16	7	5	10	14	3	1	19	5	1	28	16	28	18
E04	Add roof insulation	C406.2.1.4	1	1	1	2	2	1	2	1	1	2	1	2	2	1	2	3	2	2	3
E05	Add wall insulation	C406.2.1.5	18	26	11	25	3	4	5	3	1	6	2	4	7	4	4	8	6	8	5
E06	Improve fenestration	C406.2.1.6	2	2	1	2	2	3	5	3	1	6	3	4	9	7	6	13	8	6	6
H01	HVAC performance	C406.2.2.1	21	20	17	18	16	13	12	12	11	11	11	8	11	11	8	13	11	14	16
H02	Heating efficiency	C406.2.2.2	x	x	x	x	x	x	1	1	6	2	1	1	3	2	2	6	4	8	11
H03	Cooling efficiency	C406.2.2.3	7	6	4	4	3	2	1	2	1	1	2	1	1	1	1	x	x	x	x
H04	Residential HVAC control	C406.2.2.4	x	x	x	x	x	x	x	x	x	x	x	x	x	x	x	x	x	x	x
H05	DOAS/fan control	C406.2.2.5	32	30	26	28	25	23	24	22	28	26	22	20	30	26	19	41	34	48	62
W01	SHW preheat recovery	C406.2.3.1 a	18	19	22	22	25	27	31	21	32	34	34	38	37	36	40	36	37	36	35
W02	Heat pump water heater	C406.2.3.1 b	14	15	18	17	20	22	25	25	27	29	29	32	31	31	34	30	32	31	30
W03	Efficient gas water heater	C406.2.3.1 c	11	12	14	14	16	17	19	19	20	21	21	24	23	23	25	22	23	23	22
W04	SHW pipe insulation	C406.2.3.2	3	3	4	3	4	4	4	4	4	4	4	4	4	4	4	4	4	4	3
W05	Point of use water heaters	C406.2.3.3 a	x	x	x	x	x	x	x	x	x	x	x	x	x	x	x	x	x	x	x
W06	Thermostatic bal. valves	C406.2.3.3 b	1	1	1	1	1	2	2	2	2	2	2	2	2	2	2	2	2	1	1
W07	SHW heat trace system	C406.2.3.3 c	5	6	6	6	6	7	7	7	7	7	7	8	7	7	8	7	7	6	6
W08	SHW submeters	C406.2.3.4	x	x	x	x	x	x	x	x	x	x	x	x	x	x	x	x	x	x	x
W09	SHW flow reduction	C406.2.3.5	6	7	8	8	9	10	11	11	12	13	13	14	14	13	15	13	14	14	13
W10	Shower heat recovery	C406.2.3.6	4	5	5	5	6	7	8	8	8	9	9	10	10	9	10	9	10	10	9
P01	Energy monitoring	C406.2.4	3	3	2	2	2	2	2	2	2	2	2	2	2	2	2	2	2	2	2
L01	Lighting performance	C406.2.5.1	x	x	x	x	x	x	x	x	x	x	x	x	x	x	x	x	x	x	x
L02	Lighting dimming & tuning	C406.2.5.2	1	1	1	1	1	1	1	1	1	1	1	1	1	1	1	1	1	1	1
L03	Increase occp. sensor	C406.2.5.3	3	3	3	3	3	3	3	3	3	4	2	3	2	2	3	2	2	1	1
L04	Increase daylight area	C406.2.5.4	x	x	x	x	x	x	x	x	x	x	x	x	x	x	x	x	x	x	x
L05	Residential light control	C406.2.5.5	x	x	x	x	x	x	x	x	x	x	x	x	x	x	x	x	x	x	x
L06	Light power reduction	C406.2.5.6	1	1	2	2	2	2	2	2	2	2	1	2	1	1	2	1	1	1	1
Q01	Efficient elevator	C406.2.6.1	2	2	2	2	2	2	2	2	3	3	3	3	3	3	3	2	2	2	2
Q02	Commercial kitchen equip.	C406.2.6.2	x	x	x	x	x	x	x	x	x	x	x	x	x	x	x	x	x	x	x
Q03	Residential kitchen equip.	C406.2.6.3	9	9	10	10	10	11	11	11	11	11	11	12	11	11	12	10	11	10	9
Q04	Fault detection	C406.2.6.4	3	3	3	3	2	2	2	2	2	2	2	2	2	2	1	2	2	2	2

DOAS = Dedicated Outside Air System; HVAC = Heating, Ventilation and Air Conditioning; SHW = Service Hot Water; UA = *U*-Factor × Area.

a. “x” indicates credit is not available in that climate zone for that measure.

TABLE C406.2(4)—BASE ENERGY CREDITS FOR GROUP B OCCUPANCIES[a]

ID	ENERGY CREDIT MEASURE	SECTION	CLIMATE ZONE																		
			0A	0B	1A	1B	2A	2B	3A	3B	3C	4A	4B	4C	5A	5B	5C	6A	6B	7	8
E01	Envelope performance	C406.2.1.1	Determined in accordance with Section C406.2.1.1																		
E02	UA reduction (15%)	C406.2.1.2	7	8	3	6	5	3	7	3	1	13	4	8	21	15	11	13	24	37	43
E03	Reduced air leakage	C406.2.1.3	5	3	4	2	2	2	5	1	x	8	x	2	13	4	x	18	9	18	7
E04	Add roof insulation	C406.2.1.4	2	2	2	2	2	2	3	2	1	3	1	2	3	2	2	3	3	2	3
E05	Add wall insulation	C406.2.1.5	13	14	8	11	4	4	7	4	1	5	2	4	6	4	3	9	7	10	8
E06	Improve fenestration	C406.2.1.6	5	5	4	5	7	7	8	2	1	8	2	4	10	5	1	21	17	10	9
H01	HVAC performance	C406.2.2.1	22	22	19	20	17	17	15	15	11	15	15	11	16	15	11	19	17	18	20
H02	Heating efficiency	C406.2.2.2	X	x	x	x	x	x	1	1	1	3	2	2	5	4	3	9	7	8	12
H03	Cooling efficiency	C406.2.2.3	7	6	4	5	3	3	1	2	1	1	2	1	1	1	1	x	x	x	x
H04	Residential HVAC control	C406.2.2.4	X	x	x	x	x	x	x	x	x	x	x	x	x	x	x	x	x	x	x
H05	DOAS/fan control	C406.2.2.5	31	31	27	29	25	25	28	26	18	35	28	28	47	38	29	64	53	58	74
W01	SHW preheat recovery	C406.2.3.1 a	8	9	10	9	11	11	12	12	14	13	13	14	13	13	15	12	13	14	14
W02	Heat pump water heater	C406.2.3.1 b	3	3	3	3	4	4	5	4	5	5	5	6	5	5	6	5	5	6	6
W03	Efficient gas water heater	C406.2.3.1 c	5	5	6	6	7	7	8	7	8	8	8	9	8	8	9	8	8	9	8
W04	SHW pipe insulation	C406.2.3.2	3	3	4	4	4	4	4	4	5	4	4	5	4	4	5	4	4	4	4
W05	Point of use water heaters	C406.2.3.3 a	12	15	17	16	18	18	19	19	22	20	20	22	20	20	22	18	19	20	19
W06	Thermostatic bal. valves	C406.2.3.3 b	1	1	1	1	1	1	1	1	1	1	1	1	1	1	1	1	1	1	1
W07	SHW heat trace system	C406.2.3.3 c	4	4	4	4	5	5	5	5	6	5	5	6	5	5	6	5	5	5	5
W08	SHW submeters	C406.2.3.4	x	x	x	x	x	x	x	x	x	x	x	x	x	x	x	x	x	x	x
W09	SHW flow reduction	C406.2.3.5	x	x	x	x	x	x	x	x	x	x	x	x	x	x	x	x	x	x	x
W10	Shower heat recovery	C406.2.3.6	x	x	x	x	x	x	x	x	x	x	x	x	x	x	x	x	x	x	x
P01	Energy monitoring	C406.2.4	3	3	3	3	3	3	3	3	3	3	3	3	3	3	3	3	3	3	3
L01	Lighting performance	C406.2.5.1	x	x	x	x	x	x	x	x	x	x	x	x	x	x	x	x	x	x	x
L02	Lighting dimming & tuning	C406.2.5.2	5	5	6	6	6	6	6	6	7	6	6	6	5	5	6	4	5	3	2
L03	Increase occp. sensor	C406.2.5.3	5	6	6	6	6	6	6	6	8	6	6	6	5	5	6	4	5	4	3
L04	Increase daylight area	C406.2.5.4	7	7	8	8	8	8	8	8	9	6	7	7	6	6	6	6	6	7	5
L05	Residential light control	C406.2.5.5	x	x	x	x	x	x	x	x	x	x	x	x	x	x	x	x	x	x	x
L06	Light power reduction	C406.2.5.6	7	7	8	8	8	8	8	8	9	7	8	8	6	7	8	5	6	5	3
Q01	Efficient elevator	C406.2.6.1	4	4	4	4	5	5	5	5	5	5	5	5	5	5	5	4	5	4	4
Q02	Commercial kitchen equip.	C406.2.6.2	x	x	x	x	x	x	x	x	x	x	x	x	x	x	x	x	x	x	x
Q03	Residential kitchen equip.	C406.2.6.3	x	x	x	x	x	x	x	x	x	x	x	x	x	x	x	x	x	x	x
Q04	Fault detection	C406.2.6.4	3	3	3	3	3	2	2	2	2	2	2	2	2	2	2	3	3	3	3

DOAS = Dedicated Outside Air System; HVAC = Heating, Ventilation and Air Conditioning; SHW = Service Hot Water; UA = *U*-Factor × Area.

a. "x" indicates credit is not available in that climate zone for that measure.

TABLE C406.2(5)—BASE ENERGY CREDITS FOR GROUP A-2 OCCUPANCIES[a]

ID	ENERGY CREDIT MEASURE	SECTION	CLIMATE ZONE																		
			0A	0B	1A	1B	2A	2B	3A	3B	3C	4A	4B	4C	5A	5B	5C	6A	6B	7	8
E01	Envelope performance	C406.2.1.1	Determined in accordance with Section C406.2.1.1																		
E02	UA reduction (15%)	C406.2.1.2	1	1	1	1	13	1	3	2	1	4	4	5	5	5	6	6	6	6	6
E03	Reduced air leakage	C406.2.1.3	2	1	1	1	2	3	11	2	1	24	4	6	33	9	3	42	29	36	16
E04	Add roof insulation	C406.2.1.4	1	1	x	1	1	1	2	1	1	1	1	1	2	2	1	2	2	1	2
E05	Add wall insulation	C406.2.1.5	1	1	x	1	1	2	3	3	1	2	1	1	2	2	2	2	2	2	2
E06	Improve fenestration	C406.2.1.6	1	1	1	1	1	1	2	2	1	1	2	2	3	2	1	4	4	1	1
H01	HVAC performance	C406.2.2.1	x	x	x	x	x	x	x	x	x	x	x	x	x	x	x	x	x	x	x
H02	Heating efficiency	C406.2.2.2	x	x	x	x	1	1	6	3	3	10	6	8	15	11	10	19	15	23	28
H03	Cooling efficiency	C406.2.2.3	6	5	3	4	3	2	1	1	1	1	1	1	1	1	1	x	x	x	x
H04	Residential HVAC control	C406.2.2.4	x	x	x	x	x	x	x	x	x	x	x	x	x	x	x	x	x	x	x
H05	DOAS/fan control	C406.2.2.5	29	27	20	25	24	21	36	27	15	51	35	38	67	53	45	84	70	97	115
W01	SHW preheat recovery	C406.2.3.1 a	24	26	31	29	33	35	37	38	45	38	41	44	37	40	44	34	38	33	30
W02	Heat pump water heater	C406.2.3.1 b	15	16	19	18	21	23	25	25	29	26	28	30	26	28	31	25	27	24	22
W03	Efficient gas water heater	C406.2.3.1 c	15	16	19	18	21	22	23	24	28	24	25	27	23	25	27	21	24	21	18
W04	SHW pipe insulation	C406.2.3.2	2	3	3	3	3	3	3	3	3	3	3	3	2	3	3	2	2	2	2
W05	Point of use water heaters	C406.2.3.3 a	x	x	x	x	x	x	x	x	x	x	x	x	x	x	x	x	x	x	x
W06	Thermostatic bal. valves	C406.2.3.3 b	1	1	1	1	1	1	1	1	1	1	1	1	1	1	1	1	1	1	1
W07	SHW heat trace system	C406.2.3.3 c	3	4	4	4	4	4	4	4	4	4	4	4	3	4	4	3	3	3	3
W08	SHW submeters	C406.2.3.4	x	x	x	x	x	x	x	x	x	x	x	x	x	x	x	x	x	x	x
W09	SHW flow reduction	C406.2.3.5	x	x	x	x	x	x	x	x	x	x	x	x	x	x	x	x	x	x	x
W10	Shower heat recovery	C406.2.3.6	x	x	x	x	x	x	x	x	x	x	x	x	x	x	x	x	x	x	x
P01	Energy monitoring	C406.2.4	2	2	2	2	2	1	2	1	1	2	1	1	2	2	1	2	2	2	3
L01	Lighting performance	C406.2.5.1	x	x	x	x	x	x	x	x	x	x	x	x	x	x	x	x	x	x	x
L02	Lighting dimming & tuning	C406.2.5.2	2	2	2	2	2	2	2	2	2	2	2	2	1	2	1	1	1	1	x
L03	Increase occp. sensor	C406.2.5.3	2	2	2	2	2	2	2	2	2	1	1	1	1	1	1	1	1	1	x
L04	Increase daylight area	C406.2.5.4	x	x	x	x	x	x	x	x	x	x	x	x	x	x	x	x	x	x	x
L05	Residential light control	C406.2.5.5	x	x	x	x	x	x	x	x	x	x	x	x	x	x	x	x	x	x	x
L06	Light power reduction	C406.2.5.6	3	3	3	3	3	3	3	3	3	2	2	2	2	2	2	1	2	1	1
Q01	Efficient elevator	C406.2.6.1	1	1	1	1	1	1	1	1	1	1	1	1	1	1	1	1	1	1	1
Q02	Commercial kitchen equip.	C406.2.6.2	24	26	28	27	28	29	27	29	32	26	28	29	24	26	28	21	23	19	17
Q03	Residential kitchen equip.	C406.2.6.3	x	x	x	x	x	x	x	x	x	x	x	x	x	x	x	x	x	x	x
Q04	Fault detection	C406.2.6.4	3	2	2	2	2	2	2	2	1	2	2	1	2	2	2	3	2	3	4

DOAS = Dedicated Outside Air System; HVAC = Heating, Ventilation and Air Conditioning; SHW = Service Hot Water; UA = *U*-Factor × Area.

a. "x" indicates credit is not available in that climate zone for that measure.

TABLE C406.2(6)—BASE ENERGY CREDITS FOR GROUP M OCCUPANCIES[a]

ID	ENERGY CREDIT MEASURE	SECTION	CLIMATE ZONE																		
			0A	0B	1A	1B	2A	2B	3A	3B	3C	4A	4B	4C	5A	5B	5C	6A	6B	7	8
E01	Envelope performance	C406.2.1.1	Determined in accordance with Section C406.2.1.1																		
E02	UA reduction (15%)	C406.2.1.2	14	14	8	13	7	9	20	15	1	35	18	28	41	37	40	43	44	46	31
E03	Reduced air leakage	C406.2.1.3	3	3	2	2	3	3	19	3	1	44	6	11	56	13	6	64	44	43	19
E04	Add roof insulation	C406.2.1.4	8	6	5	7	7	7	18	16	4	19	18	20	21	22	23	24	26	24	30
E05	Add wall insulation	C406.2.1.5	64	65	48	62	13	15	23	18	4	27	21	27	25	24	25	23	24	24	16
E06	Improve fenestration	C406.2.1.6	4	3	3	3	4	4	6	5	2	7	5	7	7	5	7	10	10	3	3
H01	HVAC performance	C406.2.2.1	31	30	26	28	23	21	23	20	14	27	21	22	29	25	23	32	28	30	33
H02	Heating efficiency	C406.2.2.2	x	x	x	x	x	x	10	3	1	19	8	15	26	17	18	29	24	27	31
H03	Cooling efficiency	C406.2.2.3	10	9	7	7	5	4	2	2	1	1	2	1	1	1	1	x	x	x	x
H04	Residential HVAC control	C406.2.2.4	x	x	x	x	x	x	x	x	x	x	x	x	x	x	x	x	x	x	x
H05	DOAS/fan control	C406.2.2.5	48	48	42	47	40	38	66	46	31	98	61	82	120	91	90	134	115	125	141
W01	SHW preheat recovery	C406.2.3.1 a	12	13	16	15	18	20	19	21	26	17	21	21	16	19	21	13	16	15	13
W02	Heat pump water heater	C406.2.3.1 b	3	3	4	3	4	5	5	5	7	5	6	6	4	5	6	4	4	4	4
W03	Efficient gas water heater	C406.2.3.1 c	6	7	8	8	10	10	10	11	14	9	11	11	8	10	11	7	8	8	7
W04	SHW pipe insulation	C406.2.3.2	3	3	4	4	4	4	4	4	5	4	4	5	4	4	5	4	4	4	4
W05	Point of use water heaters	C406.2.3.3 a	x	x	x	x	x	x	x	x	x	x	x	x	x	x	x	x	x	x	x
W06	Thermostatic bal. valves	C406.2.3.3 b	1	1	1	1	1	1	1	1	1	1	1	1	1	1	1	1	1	1	1
W07	SHW heat trace system	C406.2.3.3 c	4	4	4	4	5	5	5	5	6	5	5	6	5	5	6	5	5	5	5
W08	SHW submeters	C406.2.3.4	x	x	x	x	x	x	x	x	x	x	x	x	x	x	x	x	x	x	x
W09	SHW flow reduction	C406.2.3.5	x	x	x	x	x	x	x	x	x	x	x	x	x	x	x	x	x	x	x
W10	Shower heat recovery	C406.2.3.6	x	x	x	x	x	x	x	x	x	x	x	x	x	x	x	x	x	x	x
P01	Energy monitoring	C406.2.4	5	5	5	5	5	5	5	5	5	5	5	5	5	5	5	5	5	5	5
L01	Lighting performance	C406.2.5.1	x	x	x	x	x	x	x	x	x	x	x	x	x	x	x	x	x	x	x
L02	Lighting dimming & tuning	C406.2.5.2	9	9	11	10	12	13	11	13	15	9	12	11	7	9	10	5	7	5	3
L03	Increase occp. sensor	C406.2.5.3	9	9	11	10	12	13	12	13	15	10	12	11	7	10	11	6	8	5	4
L04	Increase daylight area	C406.2.5.4	12	13	15	14	16	17	15	16	20	11	14	13	9	12	11	8	10	10	8
L05	Residential light control	C406.2.5.5	x	x	x	x	x	x	x	x	x	x	x	x	x	x	x	x	x	x	x
L06	Light power reduction	C406.2.5.6	12	12	14	14	15	16	12	15	19	8	12	9	6	10	7	6	7	6	5
Q01	Efficient elevator	C406.2.6.1	3	3	4	3	4	4	4	4	5	3	4	4	3	4	4	3	3	3	2
Q02	Commercial kitchen equip.	C406.2.6.2	x	x	x	x	x	x	x	x	x	x	x	x	x	x	x	x	x	x	x
Q03	Residential kitchen equip.	C406.2.6.3	x	x	x	x	x	x	x	x	x	x	x	x	x	x	x	x	x	x	x
Q04	Fault detection	C406.2.6.4	3	2	2	2	2	2	2	2	1	2	2	1	2	2	2	3	2	3	4

DOAS = Dedicated Outside Air System; HVAC = Heating, Ventilation and Air Conditioning; SHW = Service Hot Water; UA = *U*-Factor × Area.

a. "x" indicates credit is not available in that climate zone for that measure.

TABLE C406.2(7)—BASE ENERGY CREDITS FOR GROUP E OCCUPANCIES[a]

ID	ENERGY CREDIT MEASURE	SECTION	CLIMATE ZONE																		
			0A	0B	1A	1B	2A	2B	3A	3B	3C	4A	4B	4C	5A	5B	5C	6A	6B	7	8
E01	Envelope performance	C406.2.1.1	Determined in accordance with Section C406.2.1.1																		
E02	UA reduction (15%)	C406.2.1.2	8	18	7	19	12	13	20	17	11	24	20	17	33	32	29	40	38	46	44
E03	Reduced air leakage	C406.2.1.3	4	3	3	3	2	5	2	1	1	1	1	1	1	1	1	2	1	1	1
E04	Add roof insulation	C406.2.1.4	8	8	4	9	5	7	16	7	1	14	7	10	18	13	13	23	25	22	28
E05	Add wall insulation	C406.2.1.5	5	7	4	8	3	6	8	6	2	6	3	6	5	5	6	7	6	7	8
E06	Improve fenestration	C406.2.1.6	8	10	6	9	11	11	15	9	1	16	8	15	22	18	19	33	29	19	18
H01	HVAC performance	C406.2.2.1	30	28	25	26	23	21	20	18	15	19	18	17	19	20	15	23	20	25	29
H02	Heating efficiency	C406.2.2.2	x	x	x	x	x	x	4	3	3	5	5	10	9	11	6	15	11	18	26
H03	Cooling efficiency	C406.2.2.3	9	8	6	7	5	4	2	2	1	1	1	1	1	1	1	x	x	x	x
H04	Residential HVAC control	C406.2.2.4	x	x	x	x	x	x	x	x	x	x	x	x	x	x	x	x	x	x	x
H05	DOAS/fan control	C406.2.2.5	45	42	37	41	36	34	41	39	30	43	46	58	57	65	40	79	63	88	117
W01	SHW preheat recovery	C406.2.3.1 a	7	7	9	8	10	11	13	13	15	14	15	15	15	14	17	13	15	14	12
W02	Heat pump water heater	C406.2.3.1 b	4	4	6	5	7	7	9	9	10	10	10	11	11	10	12	10	11	10	9
W03	Efficient gas water heater	C406.2.3.1 c	4	4	6	5	6	7	8	8	9	9	9	10	9	9	11	8	10	9	7
W04	SHW pipe insulation	C406.2.3.2	3	3	4	4	4	4	4	5	6	5	5	6	5	5	7	4	5	4	4
W05	Point of use water heaters	C406.2.3.3 a	3	4	4	4	4	5	5	5	6	5	5	5	5	5	6	4	5	4	3
W06	Thermostatic bal. valves	C406.2.3.3 b	1	1	1	1	1	1	1	2	2	2	2	2	2	2	2	1	2	1	1
W07	SHW heat trace system	C406.2.3.3 c	4	4	4	4	5	5	5	6	7	6	6	7	6	6	8	5	7	5	5
W08	SHW submeters	C406.2.3.4	x	x	x	x	x	x	x	x	x	x	x	x	x	x	x	x	x	x	x
W09	SHW flow reduction	C406.2.3.5	x	x	x	x	x	x	x	x	x	x	x	x	x	x	x	x	x	x	x
W10	Shower heat recovery	C406.2.3.6	2	2	2	2	3	3	3	3	4	3	3	4	3	3	4	3	3	3	3
P01	Energy monitoring	C406.2.4	4	4	3	3	3	3	3	3	3	3	3	3	3	3	3	3	3	3	4
L01	Lighting performance	C406.2.5.1	x	x	x	x	x	x	x	x	x	x	x	x	x	x	x	x	x	x	x
L02	Lighting dimming & tuning	C406.2.5.2	5	5	5	6	6	6	5	6	7	6	6	6	5	5	6	4	4	3	2
L03	Increase occp. sensor	C406.2.5.3	4	4	5	5	5	6	6	6	7	6	6	5	4	4	5	3	4	3	2
L04	Increase daylight area	C406.2.5.4	6	6	7	7	7	7	7	7	8	6	6	6	5	5	6	5	5	5	4
L05	Residential light control	C406.2.5.5	x	x	x	x	x	x	x	x	x	x	x	x	x	x	x	x	x	x	x
L06	Light power reduction	C406.2.5.6	6	7	7	7	8	8	8	8	10	7	8	7	6	7	8	5	6	4	2
Q01	Efficient elevator	C406.2.6.1	3	4	4	4	4	5	5	5	5	5	5	5	5	5	5	4	5	4	3
Q02	Commercial kitchen equip.	C406.2.6.2	x	x	x	x	x	x	x	x	x	x	x	x	x	x	x	x	x	x	x
Q03	Residential kitchen equip.	C406.2.6.3	x	x	x	x	x	x	x	x	x	x	x	x	x	x	x	x	x	x	x
Q04	Fault detection	C406.2.6.4	4	4	4	4	3	3	3	3	2	3	3	3	3	3	2	4	3	4	4

DOAS = Dedicated Outside Air System; HVAC = Heating, Ventilation and Air Conditioning; SHW = Service Hot Water; UA = *U*-Factor × Area.

a. "x" indicates measure is not available in that climate zone for that measure.

TABLE C406.2(8)—BASE ENERGY CREDITS FOR GROUP S-1 AND S-2 OCCUPANCIES[a]

ID	ENERGY CREDIT MEASURE	SECTION	CLIMATE ZONE																		
			0A	0B	1A	1B	2A	2B	3A	3B	3C	4A	4B	4C	5A	5B	5C	6A	6B	7	8
E01	Envelope performance	C406.2.1.1	Determined in accordance with Section C406.2.1.1																		
E02	UA reduction (15%)	C406.2.1.2	14	14	1	12	1	9	27	16	2	37	29	39	44	47	50	43	52	55	74
E03	Reduced air leakage	C406.2.1.3	2	2	1	2	1	3	31	3	1	77	14	17	92	25	8	95	71	69	26
E04	Add roof insulation	C406.2.1.4	13	12	10	11	10	11	18	17	7	14	19	18	14	20	22	10	14	12	19
E05	Add wall insulation	C406.2.1.5	19	23	13	21	7	10	15	12	3	10	12	13	9	12	12	7	9	9	8
E06	Improve fenestration	C406.2.1.6	7	5	8	7	6	6	2	4	2	4	1	6	5	1	7	3	4	4	7
H01	HVAC performance	C406.2.2.1	x	x	x	x	x	x	x	x	x	x	x	x	x	x	x	x	x	x	x
H02	Heating efficiency	C406.2.2.2	x	x	x	x	x	x	16	3	1	33	17	22	41	31	21	44	38	43	43
H03	Cooling efficiency	C406.2.2.3	7	7	4	5	3	3	1	1	1	1	1	1	1	1	1	x	x	x	x
H04	Residential HVAC control	C406.2.2.4	x	x	x	x	x	x	x	x	x	x	x	x	x	x	x	x	x	x	x
H05	DOAS/fan control	C406.2.2.5	35	37	26	33	24	27	77	35	14	141	83	96	168	132	90	180	157	177	178
W01	SHW preheat recovery	C406.2.3.1 a	8	7	9	8	10	10	8	10	12	5	8	8	4	6	9	3	4	3	3
W02	Heat pump water heater	C406.2.3.1 b	2	2	2	2	2	2	2	2	3	1	2	2	1	2	2	1	1	1	1
W03	Efficient gas water heater	C406.2.3.1 c	4	4	5	4	5	5	4	5	6	3	4	4	2	3	5	2	2	2	2
W04	SHW pipe insulation	C406.2.3.2	3	3	4	3	3	3	2	3	4	2	2	3	1	2	3	1	1	1	1
W05	Point of use water heaters	C406.2.3.3 a	x	x	x	x	x	x	x	x	x	x	x	x	x	x	x	x	x	x	x
W06	Thermostatic bal. valves	C406.2.3.3 b	1	1	1	1	1	1	1	1	1	1	1	1	1	1	1	1	1	1	1
W07	SHW heat trace system	C406.2.3.3 c	4	4	4	3	4	4	3	4	5	2	3	3	2	2	4	2	2	2	2
W08	SHW submeters	C406.2.3.4	x	x	x	x	x	x	x	x	x	x	x	x	x	x	x	x	x	x	x
W09	SHW flow reduction	C406.2.3.5	x	x	x	x	x	x	x	x	x	x	x	x	x	x	x	x	x	x	x
W10	Shower heat recovery	C406.2.3.6	x	x	x	x	x	x	x	x	x	x	x	x	x	x	x	x	x	x	x
P01	Energy monitoring	C406.2.4	5	5	6	6	6	6	5	6	6	5	5	5	5	5	6	5	5	5	5
L01	Lighting performance	C406.2.5.1	x	x	x	x	x	x	x	x	x	x	x	x	x	x	x	x	x	x	x
L02	Lighting dimming & tuning	C406.2.5.2	10	10	12	11	12	14	9	12	14	6	9	9	3	6	9	3	5	3	2
L03	Increase occp. sensor	C406.2.5.3	12	12	14	13	15	14	12	14	17	7	11	11	5	7	11	4	6	3	3
L04	Increase daylight area	C406.2.5.4	15	14	18	16	18	17	13	16	21	7	12	11	5	8	10	4	6	6	5
L05	Residential light control	C406.2.5.5	x	x	x	x	x	x	x	x	x	x	x	x	x	x	x	x	x	x	x
L06	Light power reduction	C406.2.5.6	14	14	17	16	17	17	13	17	19	8	13	12	5	8	12	4	6	4	2
Q01	Efficient elevator	C406.2.6.1	15	14	18	16	18	18	15	18	21	9	14	14	7	10	14	5	7	5	5
Q02	Commercial kitchen equip.	C406.2.6.2	x	x	x	x	x	x	x	x	x	x	x	x	x	x	x	x	x	x	x
Q03	Residential kitchen equip.	C406.2.6.3	x	x	x	x	x	x	x	x	x	x	x	x	x	x	x	x	x	x	x
Q04	Fault detection	C406.2.6.4	3	3	2	3	2	2	3	2	1	5	3	3	5	4	3	6	5	6	6

DOAS = Dedicated Outside Air System; HVAC = Heating, Ventilation and Air Conditioning; SHW = Service Hot Water; UA = *U*-Factor × Area.

a. "x" indicates creditis not available in that climate zone for that measure.

TABLE C406.2(9)—BASE ENERGY CREDITS FOR OTHER OCCUPANCIES[a, b]

ID	ENERGY CREDIT MEASURE	SECTION	CLIMATE ZONE																		
			0A	0B	1A	1B	2A	2B	3A	3B	3C	4A	4B	4C	5A	5B	5C	6A	6B	7	8
E01	Envelope performance	C406.2.1.1	Determined in accordance with Section C406.2.1.1																		
E02	UA reduction (15%)	C406.2.1.2	7	8	3	7	5	5	11	7	2	18	10	14	26	20	19	24	25	29	32
E03	Reduced air leakage	C406.2.1.3	6	4	5	4	3	7	12	3	2	28	5	6	36	9	3	41	27	33	15
E04	Add roof insulation	C406.2.1.4	4	4	3	4	4	4	8	6	2	7	6	7	9	8	9	9	10	9	12
E05	Add wall insulation	C406.2.1.5	16	19	11	17	5	6	10	7	2	9	6	8	9	7	7	9	9	10	8
E06	Improve fenestration	C406.2.1.6	4	4	3	4	5	6	6	4	1	9	4	7	11	7	6	16	14	8	8
H01	HVAC performance	C406.2.2.1	x	x	x	x	x	x	x	x	x	x	x	x	x	x	x	x	x	x	x
H02	Heating efficiency	C406.2.2.2	x	x	x	x	x	x	6	2	3	11	6	8	15	11	9	18	15	19	23
H03	Cooling efficiency	C406.2.2.3	7	7	5	5	4	3	1	2	1	x	x	x	x	x	x	x	x	x	x
H04	Residential HVAC control	C406.2.2.4	x	x	x	x	x	x	x	x	x	x	x	x	x	x	x	x	x	x	x
H05	DOAS/fan control	C406.2.2.5	37	36	31	34	30	28	43	32	23	61	42	49	75	61	49	90	77	93	90
W01	SHW preheat recovery	C406.2.3.1 a	18	19	22	21	25	26	28	29	34	29	31	34	29	31	35	26	29	27	26
W02	Heat pump water heater	C406.2.3.1 b	12	12	15	14	17	17	20	20	24	21	22	25	21	23	26	20	22	21	20
W03	Efficient gas water heater	C406.2.3.1 c	11	11	13	13	15	16	17	17	21	18	19	21	18	19	22	16	18	17	16
W04	SHW pipe insulation	C406.2.3.2	3	3	4	4	4	4	4	4	5	4	4	5	4	4	5	3	4	3	3
W05	Point of use water heaters	C406.2.3.3 a	8	10	11	10	11	12	12	12	14	13	13	14	13	13	14	11	12	12	11
W06	Thermostatic bal. valves	C406.2.3.3 b	1	1	1	1	1	1	1	1	2	1	1	2	1	1	2	1	1	1	1
W07	SHW heat trace system	C406.2.3.3 c	5	5	5	5	6	6	6	6	7	6	6	7	5	6	7	5	5	5	5
W08	SHW submeters	C406.2.3.4	x	x	x	x	x	x	x	x	x	x	x	x	x	x	x	x	x	x	x
W09	SHW flow reduction	C406.2.3.5	x	x	x	x	x	x	x	x	x	x	x	x	x	x	x	x	x	x	x
W10	Shower heat recovery	C406.2.3.6	6	6	7	7	8	9	10	10	11	10	11	12	10	11	12	10	11	10	10
P01	Energy monitoring	C406.2.4	4	4	3	3	3	3	3	3	3	3	3	3	3	3	3	3	3	3	4
L01	Lighting performance	C406.2.5.1	x	x	x	x	x	x	x	x	x	x	x	x	x	x	x	x	x	x	x
L02	Lighting dimming & tuning	C406.2.5.2	5	5	5	5	6	6	5	6	7	5	5	5	4	4	5	3	4	3	2
L03	Increase occp. sensor	C406.2.5.3	5	6	6	6	7	7	6	7	8	5	6	6	4	5	6	3	4	3	2
L04	Increase daylight area	C406.2.5.4	x	x	x	x	x	x	x	x	x	x	x	x	x	x	x	x	x	x	x
L05	Residential light control	C406.2.5.5	x	x	x	x	x	x	x	x	x	x	x	x	x	x	x	x	x	x	x
L06	Light power reduction	C406.2.5.6	7	7	8	7	8	8	7	8	9	5	7	6	4	5	6	4	4	3	2
Q01	Efficient elevator	C406.2.6.1	4	4	5	4	5	5	5	5	6	4	5	5	4	4	5	3	4	3	3
Q02	Commercial kitchen equip.	C406.2.6.2	x	x	x	x	x	x	x	x	x	x	x	x	x	x	x	x	x	x	x
Q03	Residential kitchen equip.	C406.2.6.3	x	x	x	x	x	x	x	x	x	x	x	x	x	x	x	x	x	x	x
Q04	Fault detection	C406.2.6.4	3	3	3	3	3	2	3	2	2	3	3	2	3	3	2	4	3	4	4

DOAS = Dedicated Outside Air System; HVAC = Heating, Ventilation and Air Conditioning; SHW = Service Hot Water; UA = *U*-Factor × Area.

a. "x" indicates credit is not available in that climate zone for that measure.

b. Other occupancy groups include all groups except Groups A-2, B, E, I, M, S and R.

C406.2.1 More efficient building thermal envelope. A project shall achieve credits for improved envelope performance by complying with one of the following measures:

1. Section C406.2.1.1: E01.
2. Section C406.2.1.2: E02.
3. Section C406.2.1.3: E03.
4. Both E02 and E03.
5. Any combination of:
    5.1. Section C406.2.1.3: E03.
    5.2. Section C406.2.1.4: E04.
    5.3. Section C406.2.1.5: E05.
    5.4. Section C406.2.1.6: E06.

C406.2.1.1 E01 Improved envelope performance ASHRAE 90.1 Appendix C. *Building thermal envelope* measures shall be installed to improve the energy performance of the project. The achieved energy credits shall be determined using Equation 4-13.

Equation 4-13 $EC_{ENV} = 1{,}000 \times (EPF_B - EPF_P)/EPF_B$

where:

EC_{ENV} = E01 measure energy credits.

EPF_B = base envelope performance factor calculated in accordance with ASHRAE 90.1 Appendix C.

EPF_P = proposed envelope performance factor calculated in accordance with ASHRAE 90.1 Appendix C.

C406.2.1.2 E02 Component performance envelope reduction. Energy credits shall be achieved where the component performance of the *building thermal envelope* as designed is not less than 15 percent below the component performance of the *building thermal envelope* in accordance with Section C402.1.4.

C406.2.1.3 E03 Reduced air leakage. Energy credits shall be achieved where tested building *air leakage* is not less than 10 percent of the maximum leakage permitted by Section C402.6.2, provided that the *building* is tested in accordance with the applicable method in Section C402.6.2. Energy credits achieved for measure E03 shall be determined as follows:

Equation 4-14 $EC_{E03} = EC_B \times EC_{adj}$

where:

EC_{E03} = Energy efficiency credits achieved for envelope leakage reduction.

EC_B = Section C406.2.1.3 credits from Tables C406.2(1) through C406.2(9).

$EC_{adj} = L_s / EC_a$

L_s = Leakage savings fraction: the lesser of $[(L_r - L_m)/L_r]$ or 0.8.

L_r = Maximum leakage permitted for tested buildings, by occupancy group, in accordance with Section C402.6.2.

L_m = Measured leakage in accordance with Section C402.6.2.1 or C402.6.2.2.

EC_a = Energy credit alignment factor: 0.37 for whole-building tests in accordance with Section C402.6.2.1 or 0.25 for dwelling and sleeping unit enclosure tests in accordance with Section C402.6.2.2.

C406.2.1.4 E04 Added roof insulation. Energy credits shall be achieved for insulation that is in addition to the required insulation in Table C402.1.3. All roof areas in the project shall have additional R-10 *continuous insulation* included in the *roof assembly*. For attics, this is permitted to be achieved with fill or batt insulation rated at R-10 that is continuous and not interrupted by ceiling or roof joists. Where interrupted by joists, the added insulation shall be not less than R-13. Alternatively, one-half of the base credits shall be achieved where the added *R-value* is one-half of the additional *R-value* required by this section.

C406.2.1.5 E05 Added wall insulation. Energy credits shall be achieved for insulation applied to not less than 90 percent of all opaque wall area in the project that is in addition to the required insulation in Table C402.1.3. Opaque walls shall have additional R-5 *continuous insulation* included in the wall assembly. Alternatively, one-half of the base credits shall be achieved where the added *R-value* is R-2.5.

C406.2.1.6 E06 Improve fenestration. Energy credits shall be achieved for improved energy characteristics of all vertical *fenestration* in the project meeting the requirements in Table C406.2.1.6. The area-weighted average *U-factor* and SHGC of all vertical *fenestration* shall be equal to or less than the value shown in the table. Where vertical *fenestration* is located under a permanently attached shading projection with a projection factor (PF) not less than 0.2 as determined in accordance with Section C402.5.3, the SHGC for that *fenestration* shall be permitted to be divided by 1.2. The area-weighted average *visible transmittance* (VT) of all vertical *fenestration* shall be equal to or greater than the value shown in the table.

TABLE C406.2.1.6—VERTICAL FENESTRATION REQUIREMENTS FOR ENERGY CREDIT E06

APPLICABLE CLIMATE ZONE	MAXIMUM *U*-FACTOR		MAXIMUM SHGC	MINIMUM VT
	Fixed	Operable		
0–2	0.45	0.52	0.21	0.28
3	0.33	0.44	0.23	0.30
4–5	0.31	0.38	0.34	0.41
6–7	0.26	0.32	0.38	0.44
8	0.24	0.28	0.38	0.44

C406.2.2 More efficient HVAC equipment performance. All heating and cooling systems shall meet the minimum requirements of Section C403 and efficiency improvements shall be referenced to minimum efficiencies listed in tables referenced by Section C403.3.2. Where multiple efficiency requirements are listed, equipment shall meet the seasonal or part-load efficiencies including SEER, integrated energy efficiency ratio (IEER), *integrated part load value* (IPLV) or AFUE. Equipment that is larger than the maximum capacity range indicated in tables referenced by Section C403.3.2 shall utilize the values listed for the largest capacity equipment for the associated equipment type shown in the table. Where multiple individual heating or cooling systems serve the project, the improvement shall be the weighted-average improvement based on individual system capacity. Systems are permitted to achieve HVAC energy credits by meeting the requirements of one of the following:

1. C406.2.2.1 H01.
2. C406.2.2.2 H02.
3. C406.2.2.3 H03.
4. C406.2.2.4 H04.
5. C406.2.2.5 H05.
6. Any combination of H02, H03, H04 and H05.
7. The combination of H01 and H04.

C406.2.2.1 H01 HVAC Total System Performance Ratio (TSPR). H01 energy credits shall be earned where systems are permitted to use Section C409 and where the savings ($TSPR_s$) based on the proposed TSPR ($TSPR_p$) compared to the target TSPR is 5 percent or more. If savings are greater than 5 percent, determine H01 earned credits using Equation 4-15. Energy credits for H01 shall not be combined with energy credits from HVAC measures H02, H03 or H05.

Equation 4-15 $EC_{TSPR} = EC_{BASE} \times AREA_{TSPR} \times TSPR_s/0.05$

EC_{TSPR} = Energy credits achieved for H01.

EC_{BASE} = H01 base energy credits from Tables C406.2(1) through C406.2(9).

$TSPR_s$ = The lesser of 0.20 and [1 – ($TSPR_t$ /$TSPR_p$)].

where:

$AREA_{TSPR}$ = (floor area served by systems included in TSPR)/(total building conditioned floor area)

$TSPR_p$ = HVAC TSPR of the proposed design calculated in accordance with Sections C409.4, C409.5 and C409.6.

$TSPR_t = TSPR_r/MPF$.

$TSPR_r$ = HVAC TSPR of the reference building design calculated in accordance with Sections C409.4, C409.5 and C409.6.

MPF = Mechanical performance factor from Table C409.4 based on climate zone and building use type. Where a building has multiple building use types, *MPF* shall be area weighted in accordance with Section C409.4.

C406.2.2.2 H02 More efficient HVAC equipment heating performance. In accordance with Section C406.1.1, not less than 90 percent of the total HVAC heating capacity serving the total *conditioned floor area* of the entire *building* or tenant space shall comply with the requirements of this section.

1. Equipment installed shall be types that have their efficiency listed in tables referenced by Section C403.3.2. Electric resistance heating capacity shall be limited to 20 percent of system capacity, with the exception of heat pump supplemental heating.
2. Equipment shall exceed the minimum heating efficiency requirements listed in tables referenced by Section C403.3.2 by not less than 5 percent. Where equipment exceeds the minimum annual heating efficiency requirements by more

than 5 percent, energy efficiency credits for heating shall be determined using Equation 4-16, rounded to the nearest whole number.

Equation 4-16 $EEC_{HEH} = EEC_{H5} \times (HEI/0.05)$

where:

EEC_{HEH} = Energy efficiency credits for heating efficiency improvement.

EEC_{H5} = Section C406.2.2.2 credits from Tables C406.2(1) through C406.2(9).

HEI = The lesser of the improvement above minimum heating efficiency requirements, expressed as a fraction, or 20 percent (0.20). Where heating equipment with different minimum efficiencies are included in the building, a heating capacity weighted-average improvement shall be used. Where electric resistance primary heating or reheat is included in the building, it shall be included in the weighted-average improvement with an HEI of 0. Supplemental gas and electric heat for heat pump systems shall be excluded from the weighted HEI. For heat pumps rated at multiple ambient temperatures, the efficiency at 47°F (8.3°C) shall be used. For metrics that increase as efficiency increases, HEI shall be calculated as follows:

$HEI = (HM_{DES} / HM_{MIN}) - 1$

where:

HM_{DES} = Design heating efficiency metric, part-load or annualized where available.

HM_{MIN} = Minimum required heating efficiency metric, part-load or annualized where available from Section C403.3.2.

Exception: In low-energy spaces complying with Section C402.1.1, not less than 90 percent of the installed heating capacity is provided by electric infrared or gas-fired radiant heating equipment for localized heating applications. Such spaces shall achieve base energy credits only for EEC_{H5}.

C406.2.2.3 H03 More efficient HVAC cooling equipment and fan performance. In accordance with Section C406.1.1, not less than 90 percent of the total HVAC cooling capacity serving the total *conditioned floor area* of the entire *building* or tenant space shall comply with all of the requirements of this section.

1. Equipment installed shall be types that are listed in tables referenced by Section C403.3.2.
2. Equipment shall exceed the minimum cooling efficiency requirements listed in tables referenced by Section C403.3.2 by not less than 5 percent. For water-cooled chiller plants, heat-rejection equipment performance in Table C403.3.2(7) shall also be increased by at least the chiller efficiency improvement. Where equipment exceeds both the minimum annual cooling efficiency and heat-rejection efficiency requirements by more than 5 percent, energy efficiency credits for cooling shall be determined using Equation 4-17, rounded to the nearest whole number.

Where fan energy is not included in the packaged equipment rating or it is and the fan size has been increased from the as-rated equipment condition, fan power or horsepower shall be less than 95 percent of the allowed fan power in Section C403.8.1.

Equation 4-17 $EEC_{HEH} = EEC_{H5} \times (CEI/0.05)$

EEC_{HEC} = Energy efficiency credits for cooling efficiency improvement.

EEC_5 = Section C406.2.2.3 base energy credits from Tables C406.2(1) through C406.2(9).

CEI = The lesser of the improvement above minimum cooling efficiency and heat-rejection performance requirements, expressed as a fraction, or 20 percent (0.20). Where cooling equipment with different minimum efficiencies is included in the building, a cooling capacity weighted-average improvement shall be used. Where multiple cooling efficiency or performance requirements are provided, the equipment shall exceed the annualized energy or part-load requirement. Meeting both part-load and full-load efficiencies is not required.

For metrics that increase as efficiency increases, *CEI* shall be calculated as follows:

$CEI = (CM_{DES} / CM_{MIN}) - 1$. For metrics that decrease as efficiency increases, *CEI* shall be calculated as follows:

$CEI = (CM_{MIN} / CM_{DES}) - 1$.

where:

CM_{DES} = Design cooling efficiency metric, part-load or annualized where available.

CM_{MIN} = Minimum required cooling efficiency metric, part-load or annualized where available from Section C403.3.2.

For data centers using ASHRAE Standard 90.4, *CEI* shall be calculated as follows:

$CEI = (AMLC_{MAX} / AMLC_{DES}) - 1$

where:

$AMLC_{DES}$ = As-designed annualized mechanical load component calculated in accordance with ASHRAE Standard 90.4, Section 6.5.

$AMLC_{MAX}$ = Maximum annualized mechanical load component from ASHRAE Standard 90.4, Table 6.5.

C406.2.2.4 H04 Residential HVAC control. HVAC systems serving *dwelling units* or *sleeping units* shall be controlled to automatically activate a setback at least 5°F (3°C) for both heating and cooling. The temperature controller shall be configured to provide setback during occupied sleep periods. The unoccupied setback mode shall be configured to operate in conjunction with one of the following:

1. A *manual* main control device by each *dwelling unit* main entrance that initiates setback and nonventilation mode for all HVAC units in the *dwelling unit* and is clearly identified as "Heating/Cooling Master Setback."
2. Occupancy sensors in each room of the *dwelling unit* combined with a door switch to initiate setback and nonventilation mode for all HVAC units in the dwelling within 20 minutes of all spaces being vacant immediately after a door switch operation. Where separate room HVAC units are used, an individual occupancy sensor on each unit that is configured to provide setback shall meet this requirement.
3. An advanced learning *thermostat* or controller that recognizes occupant presence and automatically creates a schedule for occupancy and provides a dynamic setback schedule based on when the spaces are generally unoccupied.
4. An automated control and sensing system that uses geographic fencing connected to the *dwelling unit* occupants' cell phones and initiates the setback condition when all occupants are away from the *building*.

C406.2.2.5 H05 Dedicated outdoor air system. Credits for this measure are allowed only where single-*zone* HVAC units are not required to have multispeed or variable-speed fan control in accordance with Section C403.8.6.1. HVAC controls and ventilation systems shall include all of the following:

1. *Zone* controls shall cycle the heating/cooling unit fans off when not providing required heating and cooling or shall limit fan power to 0.12 watts/cfm (0.056 w/l/s) of *zone* supply air.
2. Outdoor air shall be supplied by an independent ventilation system designed to provide not more than 130 percent of the minimum outdoor air to each individual occupied *zone*, as specified by the *International Mechanical Code*.

   Exception: Outdoor airflow is permitted to increase during emergency or economizer operation, implemented as described in Item 4.
3. The ventilation system shall have energy recovery with an *enthalpy recovery ratio* of 65 percent or more at heating design conditions in Climate Zones 3 through 8 and an *enthalpy recovery ratio* of 65 percent or more at cooling design conditions in Climate Zones 0, 1, 2, 3A, 3B, 4A, 4B, 5A and 6A. In "A" *climate zones*, energy recovery shall include latent recovery. Where no humidification is provided, heating energy recovery effectiveness is permitted to be based on *sensible energy recovery ratio*. Where energy recovery effectiveness is less than the 65 percent required for full credit, adjust the credits from Section C406.2 by the factors in Table C406.2.2.5.
4. Where the ventilation system serves multiple *zones* and the system is not in a latent recovery outside air dehumidification mode, partial economizer cooling through an outdoor air bypass or wheel speed control shall automatically do one of the following:
   4.1. Set the energy recovery leaving-air temperature 55°F (13°C) or 100 percent outdoor air bypass when a majority of *zones* require cooling and outdoor air temperature is below 70°F (21°C).
   4.2. The HVAC ventilation system shall include supply-air temperature controls that automatically reset the supply-air temperature in response to representative *building* loads, or to outdoor air temperatures. The controls shall reset the supply-air temperature not less than 25 percent of the difference between the design supply-air temperature and the design room-air temperature.
5. Ventilation systems providing mechanical dehumidification shall use recovered energy for reheat within the limits of Item 4. This shall not limit the use of latent energy recovery for dehumidification.

Where only a portion of the *building* is permitted to be served by constant air volume units or the *enthalpy recovery ratio* or *sensible energy recovery ratio* is less than 65 percent, the base energy credits shown in Section C406.2 shall be prorated as follows:

Equation 4-18 $EC_{DOAS} = EC_{BASE} \times FLOOR_{CAV} \times ERE_{ADJ}$

where:

EC_{DOAS} = Energy credits achieved for H05.

EC_{BASE} = H05 base energy credits in Section C406.2.

$FLOOR_{CAV}$ = Fraction of whole-project gross conditioned floor area not required to have variable-speed or multi-speed fan airflow control in accordance with Section C403.8.6.

ERE_{adj} = The energy recovery adjustment from Table C406.2.2.5 based on the lower of actual cooling or heating enthalpy recovery ratio or sensible energy recovery ratio where required for the climate zone. Where recovery ratios vary, use a weighted average by supply airflow.

TABLE C406.2.2.5—DOAS ENERGY RECOVERY ADJUSTMENTS		
ERE_{adj} BASED ON LOWER OF ACTUAL HEATING OR COOLING ENERGY RECOVERY EFFECTIVENESS WHERE REQUIRED		
Cooling *Err* Is at Least	**Heating Enthalpy Recovery Ratio or Sensible Energy Recovery Ratio Is at Least**	**Energy Recovery Effectiveness Adjustment (ERE_{adj})**
65%	65%	1.00
60%	60%	0.67
55%	55%[a]	0.33
50%	50%[a]	0.25

a. In climate zones where heating recovery is required in Section C403, a heating recovery effectiveness below 60 percent is not allowed for dwelling units.

C406.2.3 Reduced energy use in service water heating. For projects with service water heating equipment that serves the whole *building*, a *building addition* or a tenant space shall achieve credits through compliance with the requirements of this section. Systems are permitted to achieve energy credits by meeting the requirements of one of the following:

1. Section C406.2.3.1 by selecting one allowed measure W01, W02, W03 or a combination in accordance with Section C406.2.3.1.4.
2. Section C406.2.3.2 W04.
3. Section C406.2.3.3 by selecting one allowed measure: W05, W06 or W07.
4. Section C406.2.3.4 W08.
5. Section C406.2.3.5 W09.
6. Section C406.2.3.6 W10.
7. Any combination of measures in Sections C406.2.3.1 through C406.2.3.6 as long as not more than one allowed measure from Sections C406.2.3.1 and C406.2.3.3 are selected.

C406.2.3.1 Service water heating system efficiency. A project is allowed to achieve energy credits from only one of Sections C406.2.3.1.1 through C406.2.3.1.4.

C406.2.3.1.1 W01 Recovered or renewable water heating. The *building* service water-heating system shall have one or more of the following that are sized to provide not less than 30 percent of the *building's* annual hot water requirements, or sized to provide not less than 70 percent of the *building's* annual hot water requirements if the *building* is required to comply with Section C403.11.5:

1. Waste heat recovery from service hot water, heat recovery chillers, *building* equipment or process equipment.
2. A water-to-water heat pump that precools chilled water return for *building* cooling while heating SHW.
3. On-site renewable energy water-heating systems.

C406.2.3.1.2 W02 Heat pump water heater. Air-source heat pump *water heaters* shall be installed according to the manufacturer's instructions and at least 30 percent of design end-use *service water heating* requirements shall be met using only heat pump heating at an ambient condition of 67.5°F (19.7°C), db without supplemental electric resistance or fossil fuel heating. For a heat pump *water heater* with supplemental electric resistance heating, the heat pump-only capacity shall be deemed at 40 percent of first-hour draw. Where the heat pump-only capacity exceeds 50 percent of the design end-use load, excluding recirculating system losses, the credits from the Section C406.2 tables shall be prorated as follows:

Equation 4-19 $EC_{HPWH} = (EC_{BASE}/0.5) \times \{(CAP_{HPWH})/(\text{Endload})[\text{not greater than } 2]\}$

where:

ECH_{PWH} = Energy credits achieved for W02.

EC_{BASE} = W02 base energy credits from Tables C406.2(1) through C406.2(9).

Endload = End-use peak hot water load, excluding load for heat trace or recirculation, Btu/h or kW.

CAP_{HPWH} = The heat pump-only capacity at 50°F (10°C) entering air and 70°F (21°C) entering potable water without supplemental electric resistance or fossil fuel heat, Btu/h or kW.

The heat pump service water heating system shall comply with the following requirements:

1. For systems with an installed total output capacity of more than 100,000 Btu/h (29 kW) at an ambient condition of 67.5°F (19.7°C) db, a preheat storage tank with greater than or equal to 0.75 gallons per 1,000 Btu/h (≥ 9.7 L/kW) of design end-use service water-heating requirements shall be heated only with a heat pump heating when the ambient temperature is greater than 45°F (7.2°C).
2. For systems with piping temperature maintenance, either a heat trace system or a separate water heater in series for recirculating system and final heating shall be installed.
3. Heat pump water heater efficiency shall meet or exceed one of the following:
   3.1. Output-capacity-weighted-average UEF of 3.0 in accordance with 10 CFR 430 Appendix E.

3.2. Output-capacity-weighted-average COP of not less than 4.0 tested at 50°F (10°C) entering air and 70°F (21°C) entering potable water in accordance with ANSI/AHRI 1300.

Where the heat pump capacity at 50°F (10°C) entering air and 70°F (21°C) entering water exceeds 50 percent of the design end-use load, excluding recirculating system losses, the base credits from Section C406.2 shall be prorated based on Equation 4-20.

Equation 4-20 W02 credit = base W02 table credit × (HP_{lf}/50%)

where:

HP_{lf} = Heat pump capacity as a fraction of the design end-use SHW requirements, excluding recirculating system losses, not to exceed 80 percent.

C406.2.3.1.3 W03 Efficient fossil fuel water heater. The combined input-capacity-weighted-average equipment rating of all gas water heating equipment in the *building* shall be not less than 95 percent E_t or 0.93 UEF. Adjustments shall apply as follows:

1. Where the *service water heating* system is required to comply with Section C404.2.1, this measure shall achieve 30 percent of the listed base W03 energy credits in Tables C406.2(1) through C406.2(9).
2. Where the installed *building service water heating* capacity is less than 200,000 Btu/h (59 kW) and weighted UEF is less than 0.93 UEF and not less than 0.82, this measure shall achieve 25 percent of the base W03 credit in Tables C406.2(1) through C406.2(9).

C406.2.3.1.4 Combination service water heating systems. Combination *service water heating* systems shall achieve credits using one of the measure combinations as follows:

1. (W01 + W02) Where *service water heating* employs both energy recovery and heat pump water heating, W01 may be combined with W02 and receive the sum of both credits.
2. (W01 + W03) Where *service water heating* employs both energy recovery and efficient gas water heating, W01 may be combined with W03 and receive the sum of the W01 credit and the portion of the W03 credit based on Item 4.
3. (W02 + W03) Where *service water heating* employs both heat pump water heating and efficient gas water heating, W02 may be combined with W03 and receive the sum of the W02 credit and the portion of the W03 credit based on Item 4.
4. For Items 2 and 3, the achieved W03 credit shall be the Section C406.2.3.1.3 W03 credit multiplied by the fractional share of total water-heating installed capacity served by gas water heating that is not less than 95 percent E_t or 0.93 UEF. In no case shall the achieved W03 credit exceed 60 percent of the W03 credit in the Section C406.2 tables. In *buildings* that have a service water heating design generating capacity greater than 900,000 Btu/h (264 kW), that proportioned W03 credit shall be further multiplied by 30 percent.

C406.2.3.2 W04 Service hot water piping insulation increase. Where service hot water is provided by a central water heating system, the hot water pipe insulation thickness shall be at least 1.5 times the thickness required in Section C404.4. All service hot water piping shall be insulated from the hot water source to the fixture shutoff. Where 50 percent or more of hot water piping does not have increased insulation due to installation in partitions, the credit shall be prorated as a percentage of lineal feet of piping with increased insulation.

C406.2.3.3 Service water-heating distribution temperature maintenance. A project is allowed to claim energy credits from only one of the following SHW distribution temperature maintenance measures.

1. **W05 Point of use *water heaters*.** Credits are available for Group B or E *buildings* larger than 5,000 square feet (465 m^2) where *service water heating* systems meet the following requirements:
   1.1. Fixtures requiring hot water shall be supplied from a local *water heater* with no recirculating system or heat trace piping.

   Exception: Commercial kitchens or showers in locker rooms shall be permitted to have a local recirculating system or heat trace piping where water heaters are located not more than 50 lineal feet (15 m) from the farthest fixture served.

   1.2. Supply piping from the *water heater* to the termination of the fixture supply pipe shall be insulated to the levels shown in Table C404.4.1.

   Exceptions:

   1. Piping at locations where a vertical support of the piping is installed.
   2. Where piping passes through a framing member and insulation requires increasing the size of the framing member.

   1.3. The water volume in the piping from the *water heater* to the termination of any individual fixture shall be limited as follows:

   1.3.1. Nonresidential public lavatory faucets that are available for use by members of the general public: not more than 2 ounces (59 mL).

   1.3.2. Commercial kitchens or showers in locker rooms with recirculating systems or heat trace piping: not more than 24 ounces (710 mL) from the recirculating system or heat trace piping.

   1.3.3. All other plumbing fixtures or appliances: not more than 16 ounces (473 mL).

2. **W06 Thermostatic balancing valves.** Credits are available where *service water heating* is provided centrally and distributed throughout the *building* with a recirculating system. Each recirculating system branch return connection to the main SHW supply piping shall have an *automatic* thermostatic balancing valve set to a minimal return water flow when the branch return temperature is greater than 120°F (49°C).
3. **W07 Heat trace system.** Credits are available for projects with gross floor area greater than 10,000 square feet (929 m^2) and a central water heating system. The energy credits achieved shall be from Tables C406.2(1) through C406.2(9). This system shall include self-regulating electric heat cables, connection kits and electronic controls. The cable shall be installed directly on the hot water supply pipes underneath the insulation to replace standby losses.

C406.2.3.4 W08 Water-heating system submeters. Each individual *dwelling unit* in a Group R-2 occupancy served by a central service water heating system shall be provided with a service hot water meter connected to a reporting system that provides individual *dwelling unit* reporting of actual domestic hot water use. Preheated water serving the cold water inlet to showers need not be metered.

C406.2.3.5 W09 Service hot water flow reduction. *Dwelling unit*, *sleeping unit* and guestroom plumbing fixtures that are connected to the service water-heating system shall have a flow or consumption rating less than or equal to the values shown in Table C406.2.3.5.

TABLE C406.2.3.5—MAXIMUM FLOW RATING FOR RESIDENTIAL PLUMBING FIXTURES WITH HEATED WATER

PLUMBING FIXTURE	MAXIMUM FLOW RATE
Faucet for private lavatory,[a] hand sinks, or bar sinks	1.2 gpm at 60 psi
Faucet for residential kitchen sink[a, b, c]	1.8 gpm at 60 psi
Shower head (including hand-held shower spray)[a, b, d]	1.8 gpm at 80 psi

For SI: 1 gallon per minute = 3.785 L/min, 1 pound per square inch = 6.89 kPa.

a. Showerheads, lavatory faucets and kitchen faucets are subject to US federal requirements listed in 10 CFR 430.32(o)–(p).

b. Maximum flow allowed is less than required by flow rates listed in 10 CFR 430.32(o)–(p) for showerheads and kitchen faucets.

c. Residential kitchen faucets may temporarily increase the flow above the maximum rate, but not above 2.2 gallons per minute at 60 psi (8.3 L/min at 414 kPa), and must default to the maximum flow rate listed.

d. Where a shower is served by multiple shower heads, the combined flow rate of all shower heads controlled by a single valve shall not exceed the maximum flow rate listed or the shower shall be designed to allow only one shower head to operate at a time.

C406.2.3.6 W10 Shower drain heat recovery. Cold water serving *building* showers shall be preheated by shower drain heat recovery units that comply with Section C404.7. The efficiency of drain heat recovery units shall be 54 percent or greater measured in accordance with CSA B55.1. Full credits are applicable to the following building uses: I-2, I-4, R-1, R-2 and also Group E where there are more than eight showers. Partial credits are applicable to *buildings* where all but ground floor showers are served where the base energy credit from Section C406.2 is adjusted by Equation 4-21.

Equation 4-21 W10 credit = W10 base energy credit × (showers with drain heat recovery/total showers in building)

C406.2.4 P01 Energy monitoring. A project not required to comply with Section C405.13 can achieve energy credits for installing an energy monitoring system that complies with all the requirements of Sections C405.13.1 through C405.13.5.

C406.2.5 Energy savings in lighting systems. Projects are permitted to achieve energy credits for increased lighting system performance by meeting the requirements of one of the following:

1. Section C406.2.5.2 L02.
2. Section C406.2.5.3 L03.
3. Section C406.2.5.4 L04.
4. Section C406.2.5.5 L05.
5. Section C406.2.5.6 L06.
6. Any combination of L03, L04, L05 and L06.
7. Any combination of L02, L03 and L04.

C406.2.5.1 L01 Lighting system performance (reserved). Reserved for future use.

C406.2.5.2 L02 High-end trim lighting controls. Measure credits shall be achieved where qualifying spaces are not less than 50 percent of the project interior floor area exclusive of *dwelling* and *sleeping units*. Qualifying spaces are those where *general lighting* is controlled by *high-end trim* lighting controls complying with the following:

1. The calibration adjustment equipment is located for ready access only by authorized personnel.
2. Lighting controls with ready access for users cannot increase the lighting power above the maximum level established by the *high-end trim* controls.
3. Construction documents shall state that maximum light output or power of general lighting in spaces contributing to the qualifying floor area shall be not greater than 85 percent of full power or light output.
4. *High-end trim* lighting controls shall be tested in accordance with Section C408.3.1.5.

The base credits from Tables C406.2(1) through C406.2(9) shall be prorated as follows:

HET × [Base energy credits for C406.2.5.2]/50%

where:

HET = Floor area of qualifying spaces where *general lighting* is provided with high-end trim lighting controls complying with this section, expressed as a percentage of total interior floor area, excluding *dwelling* and *sleeping units*.

C406.2.5.3 L03 Increase occupancy sensor. Lighting controls shall comply with Sections C406.2.5.3.1, C406.2.5.3.2 and C406.2.5.3.3.

C406.2.5.3.1 Occupant sensor controls. Occupant sensor controls shall be installed to control lights in the following space types:

1. Food preparation area.
2. Laboratory.
3. Elevator lobby.
4. Pharmacy area.
5. Vehicular maintenance area.
6. Workshop.
7. Recreation room in a facility for the visually impaired.
8. Exercise area in a fitness center.
9. Playing area in a fitness center.
10. Exam/treatment room in a health care facility.
11. Imaging room in a health care facility.
12. Physical therapy room in a health care facility.
13. Library reading area.
14. Library stacks.
15. Detailed manufacturing area.
16. *Equipment room* in a manufacturing facility.
17. Low-bay area in a manufacturing facility.
18. Post office sorting area.
19. Religious fellowship hall.
20. Hair salon.
21. Nail salon.
22. Banking activity area.
23. Museum restoration room.

C406.2.5.3.2 Occupant sensor control function. Occupant sensors in library stacks and laboratories shall comply with Section C405.2.1.2. Occupant sensors in elevator lobbies shall comply with Section C405.2.1.4. All other occupant sensors required by Section C406.2.5.3.1 shall comply with Section C405.2.1.1.

Exception: In spaces where an *automatic* shutoff could endanger occupant safety or security, occupant sensor controls shall uniformly reduce lighting power to not more than 20 percent of full power within 10 minutes after all occupants have left the space. Time-switch controls complying with Section C405.2.2.1 shall automatically turn off lights.

C406.2.5.3.3 Occupant sensor time delay and setpoint. Occupant sensor controls installed in accordance with Sections C405.2.1.1, C405.2.1.2, C405.2.1.3 and C405.2.1.4 shall automatically turn off lights or reduce lighting power within 10 minutes after all occupants have left the space. Occupant sensor controls installed in accordance with Section C405.2.1.2 shall have an unoccupied setpoint of not greater than 20 percent of full power.

C406.2.5.4 L04 Increased daylight area. The total daylight area of the *building* (DLA_{BLDG}) determined by Equation 4-22 shall be at least 5 percent greater than the typical daylight area (DLA_{TYP}) from Table C406.2.5.4. Credits for measure L04 shall be determined by Equation 4-23 or Equation 4-24, whichever is less:

Equation 4-22 $DLA_{BLDG} = DLZ/LFA$

where:

DLZ = The total building floor area located within sidelit and toplit daylight zones complying with Section C405.2.4.2 or C405.2.4.3 and provided with daylight responsive controls complying with Section C405.2.4.1, ft^2 or m^2.

LFA = The total building floor area used to determine the lighting power allowance in Section C405.3.2, ft^2 or m^2.

Equation 4-23 $EC_{DL} = EC_{DL5} \times 20 \times (DLA_{BLDG} - DLA_{TYP})$

where:

EC_{DL} = The lesser of actual area of daylight zones in the *building* with continuous daylight dimming, ft^2 or m^2 and (*GLFA* × *DLA*); see Table C406.2.5.4. Daylight zones shall meet the criteria in Sections C405.2.4.2 and C405.2.4.3 for primary sidelit daylight zones, secondary sidelit daylight zones and toplit daylight zones.

DLA_{TYP} = Typical percent of building area with daylight control (as a fraction) from Table C406.2.5.4.

EC_{DL5} = Section C406.2.5.4 L04 base energy credits from Section C406.2.

Equation 4-24 $EC_{DL} = EC_{DL5} \times 20 \times (DLA_{MAX} - DLA_{TYP})$

where:

EC_{DL} = The number of credits achieved by this measure.

EC_{DL5} = Section C406.2.5.4 L04 base energy credits from Section C406.2 and Tables C406.2(4), C406.2(6), C406.2(7) and C406.2(8).

DLA_{TYP} = Typical percent of building floor area with daylight control (as a fraction) from Table C406.2.5.4.

DLA_{MAX} = Maximum percent of building floor area with daylight control that can be counted for compliance with this measure, from Table C406.2.5.4.

TABLE C406.2.5.4—ADDED DAYLIGHTING PARAMETERS

BUILDING-USE TYPE	DLA_{TYP}	DLA_{MAX}
Group B; ≤ 5,000 ft² (460 m²)	10%	20%
Group B; > 5,000 ft² (460 m²)	21%	31%
Group M; with ≤ 1,000 ft² (900 m²) roof area	0%	20%
Group M; with > 1,000 ft² (900 m²) roof area	60%	80%
Group E; education	42%	52%
Groups S-1 and S-2; warehouse	50%	70%
Groups S-1 and S-2; other than warehouse	NA	NA

NA = Not Available.

C406.2.5.5 L05 Residential light control. In *buildings* with Group R-2 occupancy spaces, interior lighting systems shall comply with the following:

1. In *common areas*, the following space types shall have occupant sensor controls that comply with the requirements of Section C405.2.1.1:
   1.1. Laundry/washing areas.
   1.2. Dining areas.
   1.3. Food preparation areas.
   1.4. Seating areas.
   1.5. Exercise areas.
   1.6. Massage spaces.
2. In *dwelling units*, not less than one receptacle in each living room and each sleeping room shall be controlled by a switch in that room.
3. Lights and switched receptacles in bathrooms and kitchens shall be controlled by an occupant sensor complying with Section C405.2.1.1. All other lights and switched receptacles in each *dwelling unit* shall be controlled by a switch at the main entrance. The switch shall be marked to indicate its function.

   Exception: Lighting and switched receptacles controlled by an occupant sensor complying with Section C405.2.1.1 are not required to be controlled by the switch at the main entrance.

C406.2.5.6 L06 Reduced lighting power. Interior lighting within all *building* areas shall comply with this section.

1. The connected interior lighting power (LP) determined in accordance with Section C405.3.1 shall be 95 percent or less than the interior lighting power allowance (LPA) determined in accordance with Section C405.3.2 using the same method used to comply with Section C405.3. Energy credits shall not be greater than four times the L06 base credit from Section C406.2 and shall be determined using Equation 4-25.
2. All permanently installed lighting serving *dwelling units* and *sleeping units*, including ceiling fan light kits and lighting integrated into range hoods and exhaust fans shall be provided by lamps with an efficacy of not less than 90 lumens per watt or by luminaires that have an efficacy of not less than 65 lumens per watt.

   Exceptions:
   1. Lighting integral to other appliances.
   2. Antimicrobial lighting used for the sole purpose of disinfecting.

Equation 4-25 $EC_{LPA} = EC_5 \times 20 \times (LPA - LP)/LPA$

where:

EC_{LPA} = Additional energy credit for lighting power reduction.

LP = Connected interior lighting power calculated in accordance with Section C405.3.1, watts.

LPA = Interior lighting power allowance calculated in accordance with the requirements of Section C405.3.2, watts.

EC_5 = L06 base credit from Section C406.2.

C406.2.6 Efficient equipment credits. Projects are permitted to achieve energy credits using any combination of Efficient Equipment Credits Q01 through Q04.

C406.2.6.1 Q01 Efficient elevator equipment. Qualifying elevators in the *building* shall be energy efficiency class A per ISO 25745-2, Table 7. Only *buildings* three or more floors above grade may use this credit. Credits shall be prorated based on Equation 4-26, rounded to the nearest whole credit. Projects with a compliance ratio below 0.5 do not qualify for this credit.

Equation 4-26 $EC_e = EC_t \times CR_e$

where:

EC_e = Elevator energy credit achieved for the building.

EC_t = Q01 base energy credit from applicable Table C406.2(1) through Table C406.2(9).

CR_e = Compliance ratio = FA/FB.

FA = Sum of floors served by class A elevators.

FB = Sum of floors served by all building elevators and escalators.

C406.2.6.2 Q02 Efficient commercial kitchen equipment. For *buildings* and spaces designated as Group A-2, or facilities whose primary business type involves the use of a commercial kitchen where at least one gas or electric fryer is installed before the issuance of the certificate of occupancy, all fryers, dishwashers, steam cookers and ovens installed before the issuance of the certificate of occupancy shall comply with all of the following:

1. Achieve performance levels in accordance with the equipment specifications listed in Tables C406.2.6.2(1) through C406.2.6.2(4) where rated in accordance with the applicable test procedure.
2. Have associated performance levels listed on the *construction documents* submitted for permitting.

TABLE C406.2.6.2(1)—MINIMUM EFFICIENCY REQUIREMENTS: COMMERCIAL FRYERS

	HEAVY-LOAD COOKING ENERGY EFFICIENCY	IDLE ENERGY RATE	TEST PROCEDURE
Standard open deep-fat gas fryers	≥ 50%	≤ 9,000 Btu/h	ASTM F1361
Standard open deep-fat electric fryers	≥ 83%	≤ 800 watts	
Large vat open deep-fat gas fryers	≥ 50%	≤ 12,000 Btu/h	ASTM F2144
Large vat open deep-fat electric fryers	≥ 80%	≤ 1,100 watts	

For SI: 1 British thermal unit per hour = 0.293 watts.

TABLE C406.2.6.2(2)—MINIMUM EFFICIENCY REQUIREMENTS: COMMERCIAL STEAM COOKERS

FUEL TYPE	PAN CAPACITY	COOKING ENERGY EFFICIENCY[a]	IDLE ENERGY RATE	TEST PROCEDURE
Electric steam	3-pan	50%	400 watts	ASTM F1484
	4-pan	50%	530 watts	
	5-pan	50%	670 watts	
	6-pan and larger	50%	800 watts	
Gas steam	3-pan	38%	6,250 Btu/h	
	4-pan	38%	8,350 Btu/h	
	5-pan	38%	10,400 Btu/h	
	6-pan and larger	38%	12,500 Btu/h	

For SI: 1 British thermal unit per hour = 0.293 watts.

a. Cooking energy efficiency is based on heavy-load (potato) cooking capacity.

TABLE C406.2.6.2(3)—MINIMUM EFFICIENCY REQUIREMENTS: COMMERCIAL DISHWASHERS							
MACHINE TYPE	HIGH-TEMPERATURE EFFICIENCY REQUIREMENTS			LOW-TEMPERATURE EFFICIENCY REQUIREMENTS			TEST PROCEDURE
	Idle Energy Rate[a]	Washing Energy	Water Consumption[b]	Idle Energy Rate[a]	Washing Energy	Water Consumption[b]	
Under counter	≤ 0.30 kW	≤ 0.35 kWh/rack	≤ 0.86 GPR (≤ 3.3 LPR)	≤ 0.25 kW	≤ 0.15 kWh/rack	≤ 1.19 GPR ≤ 4.5 LPR	ASTM F1696 ASTM F1920
Stationary single-tank door	≤ 0.55 kW	≤ 0.35 kWh/rack	≤ 0.89 GPR (≤ 3.4 LPR)	≤ 0.30 kW	≤ 0.15 kWh/rack	≤ 1.18 GPR ≤ 4.47 LPR	
Pot, pan and utensil	≤ 0.90 kW	kWh/rack ≤ 0.55 + 0.05 × SF_{rack}[c] (≤ 0.55 + 0.0046 × SM_{rack}[c])	≤ 0.58 GPSF (≤ 2.2 LPSM)	NA	NA	NA	
Single-tank conveyor	≤ 1.20 kW	≤ 0.36 kWh/rack	≤ 0.70 GPR (≤ 2.6 LPR)	≤ 0.85 kW	≤ 0.16 kWh/rack	≤ 0.79 GPR ≤ 3.0 LPR	
Multiple-tank conveyor	≤ 1.85 kW	≤ 0.36 kWh/rack	≤ 0.54 GPR (≤ 2.0 LPR)	≤ 1.00 kW	≤ 0.22 kWh/rack	≤ 0.54 GPR ≤ 2.0 LPR	
Single-tank flight type	Reported	Reported	GPH ≤ 2.975c + 55.0 (LPH ≤ 0.276d + 208)	NA	NA	NA	
Multiple-tank flight type	Reported	Reported	GPH ≤ 4.96c+ 17.00 (LPH ≤ 0.461d + 787)	NA	NA	NA	

a. Idle results should be measured with the door closed and represent the total idle energy consumed by the machine, including all tank heaters and controls. The most energy-consumptive configuration in the product family shall be selected to test the idle energy rate. Booster heater (internal or external) energy consumption shall be measured and reported separately, if possible, per ASTM F1696 and ASTM F1920, Sections 10.8 and 10.9, respectively. However, if booster energy cannot be measured separately, it will be included in the idle energy rate measurements.

b. GPR = gallons per rack, LPR = liters per rack, GPSF = gallons per square foot of rack, LPSM = liters per square meter of rack, GPH = gallons per hour, c = [maximum conveyor belt speed (feet/minute)] × [conveyor belt width (feet)], LPH = liters per hour, d = [maximum conveyor belt speed (m/minute)] × [conveyor belt width (m)].

c. Pot, pan and utensil (PPU) washing energy is still in the format kWh/rack when evaluated; SF_{rack} (SM_{rack}) is square feet of rack area (square meters of rack area), the same as in the PPU water consumption metric.

TABLE C406.2.6.2(4)—MINIMUM EFFICIENCY REQUIREMENTS: COMMERCIAL OVENS				
FUEL TYPE	CLASSIFICATION	IDLE RATE	COOKING ENERGY EFFICIENCY, %	TEST PROCEDURE
Convection Ovens				
Gas	Full-size	≤ 12,000 Btu/h	≥ 46	ASTM F1496
Electric	Half-size	≤ 1.0 kW	≥ 71	
Electric	Full-size	≤ 1.60 kW		
Combination Ovens				
Gas	Steam mode	≤ 200 P[a] + 6,511 Btu/h	≥ 41	ASTM F2861
	Convection mode	≤ 150 P[a] + 5,425 Btu/h	≥ 56	
Electric	Steam mode	≤ 0.133 P[a] + 0.6400 kW	≥ 55	
	Convection mode	≤ 0.080 P[a] + 0.4989 kW	≥ 76	
Rack Ovens				
Gas	Single	≤ 25,000 Btu/h	≥ 48	ASTM F2093
	Double	≤ 30,000 Btu/h	≥ 52	

For SI: 1 British thermal unit per hour = 0.293 watts.

a. P = Pan capacity: the number of steam table pans the combination oven is able to accommodate in accordance with ASTM F1495.

C406.2.6.3 Q03 Efficient residential kitchen equipment. For projects with Group R-1 and R-2 occupancies, energy credits shall be achieved where all dishwashers, refrigerators and freezers comply with all of the following:

1. Achieve the Energy Star Most Efficient 2021 label in accordance with the specifications current as of:
   1.1. Refrigerators and freezers 5.0, 9/15/2014.
   1.2. Dishwashers 6.0, 1/29/2016.

2. Be installed before the issuance of the certificate of occupancy.

For Group R-1 where only some guestrooms are equipped with both refrigerators and dishwashers, the table credits shall be prorated as follows:

Equation 4-27 [Section C406.2 base credits] × [floor area of guestrooms with kitchens]/[total guestroom floor area]

C406.2.6.4 Q04 Fault detection and diagnostics system. A project not required to comply with Section C403.2.3 can achieve energy credits for installing a fault detection and diagnostics system to monitor the HVAC system's performance and automatically identify faults. The installed system shall comply with Items 1 through 6 in Section C403.2.3.

C406.3 Renewable and load management credits achieved. Renewable energy and load management measures shall achieve credits as follows:

1. General measure requirements. Credits are achieved for measures installed in the *building* that comply with Sections C406.3.1 through C406.3.8.
2. Achieved credits are determined as follows:
    2.1. Measure credits achieved shall be determined in one of two ways, depending on the measure:
        2.1.1. The measure credit shall be the base credit listed by occupancy group and *climate zone* for the measure in Tables C406.3(1) through C406.3(9) where no adjustment factor or formula is shown in the description of the measure in Section C406.3.
        2.1.2. The measure credit shall be the base energy credit for the measure adjusted by a factor or formula as stated in the description of the measure in Section C406.3. Where adjustments are applied, each energy credit shall be rounded to the nearest whole number.
    2.2. Load management and renewable credits achieved for the project shall be the sum of credits for individual measures included in the project. Credits are available for the measures listed in this section.
    2.3. Where a project contains multiple building use groups, credits achieved for each building use group shall be summed and then weighted by the gross floor area of each building use group to determine the weighted-average project energy credits achieved.
3. Load management control requirements. The load management measures in Sections C406.3.2 (G01) through C406.3.7 (G06) require load management control sequences that are capable of and configured to automatically provide the load management operation specified based on indication of a peak period related to high short-term electric prices, grid condition or peak building load. Such a peak period shall, where possible, be initiated by a *demand response signal* from the controlling entity, such as a utility or service operator. Where communications are disabled or unavailable, all demand-responsive controls shall continue backup demand response based on a local schedule or building-demand monitoring. The local building schedule shall be adjustable without programming and reflect the electric rate peak period dates and times. The load management control sequences shall be activated for peak period control by one of the following:
    3.1. A certified OpenADR 2.0a or OpenADR 2.0b Virtual End Node (VEN), as specified under Clause 11, Conformance, in the applicable OpenADR 2.0 Specification.
    3.2. A device certified by the manufacturer as being capable of responding to a *demand response signal* from a certified OpenADR 2.0b VEN by automatically implementing the control functions requested by the VEN for the equipment it controls.
    3.3. The physical configuration and communication protocol of ANSI/CTA-2045-A or ANSI/CTA-2045-B.
    3.4. For air conditioners and heat pumps with two or more stages of control and cooling capacity of less than 65,000 Btu/h (19 kW), *thermostats* with a *demand responsive control* that complies with the communication and performance requirements of AHRI 1380.
    3.5. A device that complies with IEC 62746-10-1, an international standard for the open automated demand response system interface between the appliance, system, or energy management system and the controlling entity.
    3.6. An interface that complies with the communication protocol required by a controlling entity to participate in an automated demand response program.
    3.7. Where the controlling entity does not have a *demand response signal* available for the *building* type and size, local load management control shall be provided based on either:
        3.7.1. *Building* demand management controls that monitor *building* electrical demand and initiate controls to minimize monthly or peak time period demand charges.
        3.7.2. A local *building* schedule that reflects the electric rate peak period dates and times where *buildings* are less than 25,000 gross square feet (2322 m^2).

        In this case, a binary input to the control system shall be provided that activates the demand response sequence.

TABLE C406.3(1)—RENEWABLE AND LOAD MANAGEMENT CREDITS FOR GROUP R-2, R-4 AND I-1 OCCUPANCIES

ID	ENERGY CREDIT ABBREVIATED TITLE	SECTION	CLIMATE ZONE																		
			0A	0B	1A	1B	2A	2B	3A	3B	3C	4A	4B	4C	5A	5B	5C	6A	6B	7	8
R01	Renewable energy	C406.3.1	9	15	11	17	18	20	19	21	13	10	13	9	9	11	10	9	10	9	7
G01	Lighting load management	C406.3.2	16	7	9	12	12	16	11	14	12	11	16	14	8	11	14	5	7	7	11
G02	HVAC load management	C406.3.3	42	41	21	35	23	37	30	28	28	17	33	24	20	22	23	10	13	15	17
G03	Automated shading	C406.3.4	11	x	7	18	10	13	5	13	12	2	14	7	10	13	11	1	8	8	16
G04	Electric energy storage	C406.3.5	10	10	10	11	10	13	13	14	17	16	13	17	14	13	17	14	14	14	15
G05	Cooling energy storage	C406.3.6	28	6	31	13	22	21	21	37	11	12	22	11	9	17	9	7	17	2	3
G06	SHW energy storage	C406.3.7	17	17	19	18	19	19	20	20	22	19	19	21	19	19	20	18	19	18	17
G07	Building thermal mass	C406.3.8	7	2	11	5	16	28	22	27	60	19	43	46	32	58	37	27	45	40	19

HVAC = Heating, Ventilation and Air Conditioning; SHW = Service Hot Water.
x = Credits excluded from this building use type and climate zone.

TABLE C406.3(2)—RENEWABLE AND LOAD MANAGEMENT CREDITS FOR GROUP I-2 OCCUPANCIES

ID	ENERGY CREDIT ABBREVIATED TITLE	SECTION	CLIMATE ZONE																		
			0A	0B	1A	1B	2A	2B	3A	3B	3C	4A	4B	4C	5A	5B	5C	6A	6B	7	8
R01	Renewable energy	C406.3.1	6	6	6	6	6	8	7	9	8	6	8	6	6	7	7	6	7	5	4
G01	Lighting load management	C406.3.2	11	12	13	13	13	12	12	12	6	13	16	12	13	14	15	14	14	12	12
G02	HVAC load management	C406.3.3	10	11	10	10	8	21	10	10	13	11	18	11	12	14	13	12	11	9	7
G03	Automated shading	C406.3.4	1	1	1	1	x	x	x	1	x	x	2	x	x	2	x	x	1	1	x
G04	Electric energy storage	C406.3.5	13	13	13	13	14	15	14	15	15	14	15	15	14	15	15	13	14	13	12
G05	Cooling energy storage	C406.3.6	25	6	33	14	25	19	27	37	27	16	22	19	14	18	11	11	20	2	3
G06	SHW energy storage	C406.3.7	4	4	4	4	4	4	4	4	4	4	4	5	4	4	4	4	4	4	4
G07	Building thermal mass	C406.3.8	6	2	10	4	15	25	20	24	57	18	39	44	31	53	33	25	40	34	14

HVAC = Heating, Ventilation and Air Conditioning; SHW = Service Hot Water.
x = Credits excluded from this building use type and climate zone.

TABLE C406.3(3)—RENEWABLE AND LOAD MANAGEMENT CREDITS FOR GROUP R-1 OCCUPANCIES

ID	ENERGY CREDIT ABBREVIATED TITLE	SECTION	CLIMATE ZONE																		
			0A	0B	1A	1B	2A	2B	3A	3B	3C	4A	4B	4C	5A	5B	5C	6A	6B	7	8
R01	Renewable energy	C406.3.1	9	8	12	9	11	11	10	12	13	9	12	8	9	11	9	8	9	7	5
G01	Lighting load management	C406.3.2	12	12	11	12	12	14	14	13	15	14	13	11	10	11	14	9	11	8	8
G02	HVAC load management	C406.3.3	x	x	x	x	x	x	x	x	x	x	x	x	x	x	x	x	x	x	x
G03	Automated shading	C406.3.4	2	2	2	3	1	2	3	2	4	3	2	1	1	1	3	1	2	1	1
G04	Electric energy storage	C406.3.5	9	9	10	10	9	13	13	15	13	14	13	14	14	12	16	13	12	12	13
G05	Cooling energy storage	C406.3.6	31	7	38	17	29	24	31	44	26	18	26	16	15	21	11	12	24	2	4
G06	SHW energy storage	C406.3.7	25	25	28	26	28	29	29	30	31	29	30	31	28	29	31	26	28	25	24
G07	Building thermal mass	C406.3.8	6	1	10	4	14	24	19	23	53	17	38	41	30	52	33	26	42	37	17

HVAC = Heating, Ventilation and Air Conditioning; SHW = Service Hot Water.
x = Credits excluded from this building use type and climate zone.

TABLE C406.3(4)—RENEWABLE AND LOAD MANAGEMENT CREDITS FOR GROUP B OCCUPANCIES

ID	ENERGY CREDIT ABBREVIATED TITLE	SECTION	CLIMATE ZONE																		
			0A	0B	1A	1B	2A	2B	3A	3B	3C	4A	4B	4C	5A	5B	5C	6A	6B	7	8
R01	Renewable energy	C406.3.1	14	14	17	15	17	19	18	22	24	17	22	16	14	18	18	14	17	14	11
G01	Lighting load management	C406.3.2	10	11	11	12	11	11	11	12	9	10	11	10	10	11	10	10	11	10	9
G02	HVAC load management	C406.3.3	x	10	10	9	9	3	8	12	7	12	8	11	9	10	12	8	9	10	2
G03	Automated shading	C406.3.4	4	7	7	8	7	8	5	6	6	4	6	5	4	5	5	5	5	4	7
G04	Electric energy storage	C406.3.5	14	15	14	14	16	16	17	16	18	17	16	18	17	17	18	16	15	17	18
G05	Cooling energy storage	C406.3.6	28	7	36	16	27	24	28	45	27	17	27	15	15	20	9	12	25	2	4
G06	SHW energy storage	C406.3.7	5	5	6	6	6	6	7	7	8	7	7	7	7	7	8	6	7	6	6
G07	Building thermal mass	C406.3.8	3	1	5	2	6	9	6	7	14	4	11	8	9	15	5	8	12	15	7

HVAC = Heating, Ventilation and Air Conditioning; SHW = Service Hot Water.
x = Credits excluded from this building use type and climate zone.

TABLE C406.3(5)—RENEWABLE AND LOAD MANAGEMENT CREDITS FOR A-2 OCCUPANCIES

ID	ENERGY CREDIT ABBREVIATED TITLE	SECTION	CLIMATE ZONE																		
			0A	0B	1A	1B	2A	2B	3A	3B	3C	4A	4B	4C	5A	5B	5C	6A	6B	7	8
R01	Renewable energy	C406.3.1	2	2	2	2	2	2	2	3	4	2	3	2	2	3	2	2	2	2	1
G01	Lighting load management	C406.3.2	4	4	5	5	4	5	5	5	5	4	5	5	4	4	5	4	5	4	1
G02	HVAC load management	C406.3.3	32	26	37	28	31	26	27	22	23	20	17	14	19	14	10	16	14	14	1
G03	Automated shading	C406.3.4	x	x	x	x	x	x	x	x	x	x	x	x	x	x	x	x	x	x	x
G04	Electric energy storage	C406.3.5	4	4	4	4	5	5	5	5	4	4	4	4	3	4	4	4	3	3	2
G05	Cooling energy storage	C406.3.6	15	4	17	8	12	10	10	16	6	5	7	3	3	4	1	2	4	x	x
G06	SHW energy storage	C406.3.7	13	13	15	14	15	16	16	17	19	16	17	19	16	17	18	15	16	14	13
G07	Building thermal mass	C406.3.8	3	1	5	2	7	12	8	10	21	6	15	14	8	18	10	6	12	8	3

HVAC = Heating, Ventilation and Air Conditioning; SHW = Service Hot Water.
x = Credits excluded from this building use type and climate zone.

TABLE C406.3(6)—RENEWABLE AND LOAD MANAGEMENT CREDITS FOR GROUP M OCCUPANCIES

ID	ENERGY CREDIT ABBREVIATED TITLE	SECTION	CLIMATE ZONE																		
			0A	0B	1A	1B	2A	2B	3A	3B	3C	4A	4B	4C	5A	5B	5C	6A	6B	7	8
R01	Renewable energy	C406.3.1	8	8	12	9	11	12	12	17	17	11	13	9	10	11	10	9	10	9	6
G01	Lighting load management	C406.3.2	16	16	18	19	17	19	19	21	17	18	21	21	18	21	22	18	22	18	16
G02	HVAC load management	C406.3.3	x	15	16	15	15	6	15	21	13	23	15	23	17	19	26	14	17	18	3
G03	Automated shading	C406.3.4	7	11	11	12	11	13	10	11	11	7	11	11	8	10	11	8	9	8	12
G04	Electric energy storage	C406.3.5	6	10	8	10	11	12	11	10	14	11	10	12	10	11	12	11	9	10	8
G05	Cooling energy storage	C406.3.6	40	9	51	22	35	31	34	53	21	17	28	10	11	19	4	9	18	2	2
G06	SHW energy storage	C406.3.7	3	3	4	3	4	4	4	4	5	4	4	5	4	4	5	4	4	4	3
G07	Building thermal mass	C406.3.8	5	1	6	3	8	12	10	10	20	7	17	15	14	24	10	13	20	24	12

HVAC = Heating, Ventilation and Air Conditioning; SHW = Service Hot Water.
x = Credits excluded from this building use type and climate zone.

TABLE C406.3(7)—RENEWABLE AND LOAD MANAGEMENT CREDITS FOR GROUP E OCCUPANCIES

ID	ENERGY CREDIT ABBREVIATED TITLE	SECTION	CLIMATE ZONE																		
			0A	0B	1A	1B	2A	2B	3A	3B	3C	4A	4B	4C	5A	5B	5C	6A	6B	7	8
R01	Renewable energy	C406.3.1	10	11	13	12	13	16	15	21	22	15	19	15	14	17	16	13	16	12	10
G01	Lighting load management	C406.3.2	7	12	12	13	13	15	14	16	13	12	16	16	10	14	18	16	13	14	14
G02	HVAC load management	C406.3.3	18	22	32	23	25	31	26	26	20	23	31	24	20	31	12	18	27	16	9
G03	Automated shading	C406.3.4	7	13	16	12	18	17	17	18	13	12	17	17	10	15	13	14	10	16	17
G04	Electric energy storage	C406.3.5	16	16	18	17	19	21	21	23	26	22	24	24	23	24	24	20	22	19	19
G05	Cooling energy storage	C406.3.6	36	9	46	21	36	32	39	62	39	24	37	22	20	28	13	16	31	3	4
G06	SHW energy storage	C406.3.7	5	5	6	5	6	6	7	7	8	7	7	8	7	7	8	7	7	7	6
G07	Building thermal mass	C406.3.8	7	2	11	5	17	28	23	27	63	21	44	48	37	60	38	31	50	47	21

HVAC = Heating, Ventilation and Air Conditioning; SHW = Service Hot Water.
x = Credits excluded from this building use type and climate zone.

TABLE C406.3(8)—RENEWABLE AND LOAD MANAGMENT CREDITS FOR GROUP S-1 AND S-2 OCCUPANCIES

ID	ENERGY CREDIT ABBREVIATED TITLE	SECTION	CLIMATE ZONE																		
			0A	0B	1A	1B	2A	2B	3A	3B	3C	4A	4B	4C	5A	5B	5C	6A	6B	7	8
R01	Renewable energy	C406.3.1	38	37	55	45	53	53	49	58	66	36	56	38	29	41	36	24	32	23	16
G01	Lighting load management	C406.3.2	13	26	32	28	32	35	36	33	36	31	27	37	32	23	28	36	22	25	22
G02	HVAC load management	C406.3.3	18	46	37	37	28	36	29	26	22	23	17	12	16	13	5	14	8	10	3
G03	Automated shading	C406.3.4	x	x	x	x	x	x	x	x	x	x	x	x	x	x	x	x	x	x	x
G04	Electric energy storage	C406.3.5	40	40	47	41	47	44	40	44	42	30	38	31	21	31	26	24	29	23	21
G05	Cooling energy storage	C406.3.6	20	5	21	11	14	14	11	21	5	5	9	2	2	5	1	1	3	x	x
G06	SHW energy storage	C406.3.7	3	3	3	3	4	3	4	4	4	3	4	4	3	3	4	2	2	2	2
G07	Building thermal mass	C406.3.8	7	2	12	5	17	29	23	28	66	18	44	47	28	56	37	20	39	29	13

HVAC = Heating, Ventilation and Air Conditioning; SHW = Service Hot Water.
x indicates measure is not available for building occupancy in that climate zone.

TABLE C406.3(9)—RENEWABLE AND LOAD MANAGEMENT CREDITS FOR OTHER[a] OCCUPANCIES

ID	ENERGY CREDIT ABBREVIATED TITLE	SECTION	CLIMATE ZONE																		
			0A	0B	1A	1B	2A	2B	3A	3B	3C	4A	4B	4C	5A	5B	5C	6A	6B	7	8
R01	Renewable energy	C406.3.1	12	13	16	14	16	18	17	20	21	13	18	13	12	15	14	11	13	10	8
G01	Lighting load management	C406.3.2	11	13	14	14	14	16	15	16	14	14	16	16	13	14	16	14	13	12	12
G02	HVAC load management	C406.3.3	24	24	23	22	20	23	21	21	18	18	20	17	16	18	14	13	14	13	6
G03	Automated shading	C406.3.4	5	6	7	9	8	9	7	9	8	5	9	7	5	8	7	5	6	6	9
G04	Electric energy storage	C406.3.5	14	15	16	15	16	17	17	18	19	16	17	17	15	16	17	14	15	14	14
G05	Cooling energy storage	C406.3.6	28	7	34	15	25	22	25	39	20	14	22	12	11	17	7	9	18	2	3
G06	SHW energy storage	C406.3.7	9	9	11	10	11	11	11	12	13	11	12	13	11	11	12	10	11	10	9
G07	Building thermal mass	C406.3.8	6	2	9	4	13	21	16	20	44	14	31	33	24	42	25	20	33	29	13

HVAC = Heating, Ventilation and Air Conditioning; SHW = Service Hot Water.

a. Other occupancy groups include all Groups except for Groups A-2, B, E, I, M and R.

C406.3.1 R01 Renewable energy. Projects installing on-site renewable energy systems with a capacity of at least 0.1 watts per gross square foot (1.08 W/m^2) of *building* area or securing off-site renewable energy shall achieve energy credits for this measure calculated as follows:

Equation 4-28 $EC_R = EC_{0.1} \times (R_t + R_{off} - R_{ex})/(0.1 \times PGFA)$

where:

EC_R = Section C406.3.1 R01 energy credits achieved for this project.

$EC_{0.1}$ = Section C406.3.1 R01 base credits from Tables C406.3(1) through C406.3(9).

R_t = Actual total rating of on-site renewable energy systems (W).

R_{off} = Actual total equivalent rating of off-site renewable energy contracts (W), calculated as follows:

$R_{off} = TRE/(REN \times 20)$

where:

TRE = Total off-site renewable electrical energy in kilowatt-hours (kWh) that is procured in accordance with Sections C405.15.2.1 through C405.15.4.

REN = Annual off-site renewable electrical energy from Table C405.15.2, in units of kilowatt-hours per watt of array capacity.

R_{ex} = Rating (W) of renewable energy resources capacity excluded from credit calculated as follows:

$R_{ex} = RR_r + RR_x + RR_c$

where:

RR_r = Rating of on-site renewable energy systems required by Section C405.15.1, without exception (W).

RR_x = Rating of renewable energy resources used to meet any exceptions of this code (W).

RR_c = Rating of renewable energy resources used to achieve other energy credits in Section C406 (W).

PGFA = Project gross floor area, ft^2. Where renewable requirements, exceptions or credits are expressed in annual kWh or Btu rather than watts of output capacity, they shall be converted as 3413 Btu = 1 kWh and converted to W equivalent capacity as follows:

RR_w = Actual total equivalent rating of renewable energy capacity (W), calculated as follows:

$RR_w = TRE_x/(REN \times PGFA)$

where:

TRE_x = Total renewable energy in kilowatt-hours (kWh) that is excluded from R01 energy credits.

C406.3.2 G01 Lighting load management. A project not required to comply with Section C405.2.8 can achieve energy credits for installing demand-responsive lighting controls for interior *general lighting* that comply with Section C405.2.8.1. The demand-responsive lighting controls shall automatically reduce the light output or power of controlled lighting to not more than 80 percent of full output, or 80 percent of the *high-end trim* setpoint, whichever is less. Energy credits can be earned where demand-responsive lighting controls are installed for the following:

1. Not less than 10 percent of the interior floor area in Group R or I occupancies.
2. Not less than 50 percent of the interior floor area in all other occupancies.

G01 credits shall be prorated using Equation 4-29 with not more than 75 percent of the interior floor area being counted.

Equation 4-29 [interior floor area with lighting load management, %] × [table credits for Section C406.3.2]/75%

C406.3.3 G02 HVAC load management. *Automatic* load management controls shall be configured as follows:

1. Cooling temperature shift: Where electric cooling is in use, controls shall gradually increase the cooling setpoint by at least 3°F (1.7°C) over a minimum of 3 hours or reduce effective cooling capacity to 60 percent of installed capacity during the peak period or adjust the cooling temperature setpoint as described in Section C403.6.1.
2. Heating temperature shift: Where electric heating is in use, controls shall gradually decrease the heating setpoint by at least 3°F (1.7°C) over a minimum of 3 hours or reduce effective heating capacity to 60 percent of installed capacity during the peak period or adjust the heating temperature setpoint as described in Section C403.6.1.
3. Ventilation shift: Where HVAC systems serve multiple zones and have less than 70 percent outdoor air required, include controls that provide excess outdoor air preceding the peak period and reduce outdoor air by at least 30 percent during the peak period, in accordance with ASHRAE Standard 62.1 Section 6.2.5.2 or provisions for approved engineering analysis in Section 403.3.1.1 of the International Mechanical Code.

Credits achieved for measure G02 shall be calculated as follows:

Equation 4-30 $EC_{G02\_ach} = EC_{G02\_base} \times EC_{G02\_adj}$

where:

EC_{G02_ach} = Demand responsive control credit achieved for project.

EC_{G02_base} = G02 Base energy credit from Section C406.3.

EC_{G02_adj} = Energy credit adjustment factor from Table C406.3.3.

TABLE C406.3.3—ENERGY CREDIT ADJUSTMENT BASED ON USE OF VENTILATION SHIFT OR DEMAND RESPONSE

DEMAND RESPONSE SIGNAL AVAILABLE[a]	DEMAND RESPONSE REQUIRED BY SECTION C403.4.6.1[b]	INCLUDES VENTILATION SHIFT[c]	EC_{G02_adj}
No	No	Yes	100%
No	Yes	Yes	80%
Yes	No	Yes	80%
Yes	Yes	Yes	40%
No	No	No	70%
No	Yes	No	50%
Yes	No	No	50%
Yes	Yes	No	0%

a. "Demand Response Signal Available" is "Yes" where a controlling entity other than the owner makes a demand response signal available to the building.
b. Where the exception is invoked in Section C403.4.6.1 for buildings that comply with Load Management measure G02, then "Demand Response Required" is "Yes."
c. Ventilation shift controls in accordance with Section C406.3.3, Item 3.

C406.3.4 G03 Automated shading load management. Where *fenestration* on east, south and west exposures is greater than 20 percent of the wall area, load management credits shall be achieved as follows:

1. *Automatic* exterior shading devices or *dynamic glazing* that is capable of reducing solar gain through sunlit *fenestration* by not less than 50 percent when fully closed shall receive the full credits in Tables C406.3(1) through C406.3(9). The exterior shades shall have fully open and fully closed *solar heat gain coefficient* (SHGC) determined in accordance with AERC 1.
2. *Automatic* interior shading devices with a solar reflectance of not less than 0.50 for the surface facing the *fenestration* shall receive 40 percent of the credits in Tables C406.3(1) through C406.3(9).
3. All shading devices, *dynamic glazing* or shading attachments shall:
    3.1. Provide not less than 90 percent coverage of the total *fenestration* on east, south and west exposures in the *building* to achieve the credits determined in Item 1 or 2. Alternatively, provide not less than 70 percent coverage of the total *fenestration* on the south and west exposures in the *building* to achieve 50 percent of the credits determined in Item 1 or 2.
    3.2. Be automatically controlled and shall modulate in multiple steps or continuously the amount of solar gain and light transmitted into the space in response to peak periods and either daylight levels or solar intensity.

3.3. Include a *manual* override located in the same *enclosed space* as the shaded vertical *fenestration* that shall override operation of *automatic* controls for not longer than 4 hours. Such override shall be locked out during peak periods.

For this section, directional exposures shall exclude *fenestration* that has an orientation deviating by more than 45 degrees of facing the cardinal direction. In the southern hemisphere, where the south exposure is referred to, it shall be replaced by the north exposure.

C406.3.5 G04 Electric energy storage. Electric storage devices shall be charged and discharged by *automatic* load management controls to store energy during nonpeak periods and use stored energy during peak periods to reduce *building* demand. Electric storage devices shall have a minimum capacity of 1.5 watt-hours per square foot (Wh/ft^2) (16 Wh/m^2) of gross *building* area. Base credits in Tables C406.3(1) through C406.3(9) are based on installed electric storage of 5 Wh/ft^2 (54 Wh/m^2) and shall be prorated for actual installed storage capacity between 1.5 and 15 Wh/ft^2 (16 to 161 Wh/m^2), as follows:

Equation 4-31 [installed electric storage capacity, Wh/ft^2 (Wh/m^2)]/5(54) × [table credits for Section C406.3.5]

Larger energy storage shall be permitted; however, credits are limited to the range of 1.5 to 15 Wh/ft^2 (16 to 161 Wh/m^2).

C406.3.6 G05 Cooling energy storage. *Automatic* load management controls shall be capable of activating ice or chilled water storage equipment to reduce demand during summer peak periods. Storage tank standby loss shall be demonstrated through analysis to be not more than 2 percent of storage capacity over a 24-hour period for the cooling design day. Base credits in Section C406.3 are based on storage capacity of the design peak hour cooling load with a 1.15 sizing factor. Credits shall be prorated for installed storage systems sized between 0.5 and 4.0 times the design day peak hour cooling load, rounded to the nearest whole credit. Larger storage shall be permitted but the associated credits are limited to the range provided in this section. Energy credits shall be determined as follows:

Equation 4-32 $EC_S = EC_{1.0} \times (1.44 \times SR + 0.71)/2.15$

where:

EC_s = Cooling storage credit achieved for project.

$EC_{1.0}$ = G05 base energy credit for building use type and climate zone based on 1.0 ton-hours storage per design day ton (kWh/kW) of cooling load.

SR = Storage ratio in ton-hours storage per design day ton (kWh/kW) of cooling load where $0.5 \leq SR \leq 4.0$.

C406.3.7 G06 Service hot water energy storage. Where service hot water (SHW) is heated by electricity, *automatic* load management controls complying with ANSI/CTA-2045-B shall preheat stored SHW before the peak period and suspend electric water heating during the peak period. Storage capacity shall be provided by either:

1. Preheating water above 140°F (60°C) delivery temperature with at least 1.34 kWh of energy storage per kW of water-heating capacity. Tempering valves shall be provided at the water heater delivery location.
2. Providing additional heated water tank storage capacity above peak SHW demand with equivalent peak storage capacity to Item 1.

Credits earned for measure G06 shall be calculated using Equation 4-33:

Equation 4-33 $EC_{G06\_ach} = EC_{G06\_base} \times EC_{G06\_adj}$

where:

EC_{G06_ach} = SWH energy storage credit achieved for project.

EC_{G06_base} = G06 Base energy credit from Section C406.3.

EC_{G06_adj} = Energy credit adjustment factor from Table C406.3.7.

TABLE C406.3.7—ENERGY CREDIT ADJUSTMENT BASED ON USE OF HEAT PUMP WATER HEATER OR DEMAND RESPONSE

DEMAND RESPONSE READY PER SECTION C404.10	DEMAND RESPONSE SIGNAL AVAILABLE[a]	HAS HPWH	EC_{G06_adj}[b]
No	NA	No	100%
No	NA	Yes	33%
Yes	No	No	50%
Yes	No	Yes	17%
Yes	Yes	NA	0%

HPWH = Heat Pump Water Heater, NA = Not available.

a. "Demand Response Signal Available" is "Yes" where a controlling entity currently makes a demand response signal available to the building.

b. The lower values of EC_{G06_adj} in this column apply where not less than 67 percent of the whole-building design end use service water heating requirements are met using only heat pump heating at the conditions described in Section C406.2.3.1.2.

C406.3.8 G07 Building thermal mass. The project shall have additional passive interior mass and a night flush control of the HVAC system. The credit is available to projects that have at least 80 percent of gross floor area unoccupied between midnight and 6:00 a.m. The project shall meet the following requirements:

1. Interior to the *building thermal envelope* insulation, provide 10 pounds per foot (15 kg/m) of project *conditioned floor area* of passive thermal mass in the *building* interior wall, the inside of the *exterior wall* or the interior floor construction. Mass construction shall have mass surfaces directly contacting the air in *conditioned spaces* with directly attached gypsum panels allowed. Mass with carpet or furred gypsum panels or exterior wall mass that is on the exterior of the insulation layer [e.g., the portion of concrete masonry unit (CMU) block on the exterior of insulation-filled cell cavities] shall not be included toward the building mass required.
2. HVAC units for 80 percent or more of the supply airflow in the project shall be equipped with outdoor air economizers and fans that have variable or low speed capable of operating at 66 percent or lower airflow and be included in the night flush control sequence.
3. Night flush controls shall be configured with the following sequence or another night flush strategy shall be permitted where demonstrated to be effective, avoids added morning heating and is *approved* by the authority having jurisdiction.
    3.1. Summer mode shall be activated when outdoor air temperature exceeds 70°F (21°C) and shall continue uninterrupted until deactivated when outdoor air temperature falls below 45°F (7°C). During summer mode, the occupied cooling setpoint shall be set 1°F (0.6°C) higher than normal and the occupied heating setpoint shall be reset 2°F (1.1°C) lower than normal.
    3.2. Where all the following conditions exist, night flush shall be activated:
        3.2.1. Summer mode is active in accordance with Item 3.1.
        3.2.2. Outdoor air temperature is 5°F (2.8°C) or more below indoor average *zone* temperature.
        3.2.3. Indoor average *zone* temperature is greater than morning occupied heating setpoint.
        3.2.4. In Climate Zones 0A, 1A, 2A and 3A, outdoor dewpoint is below 50°F (10°C) or outdoor air enthalpy is less than indoor air enthalpy.
        3.2.5. Local time is between 10:00 p.m. and 6:00 a.m.
    3.3. When night flush is active, *automatic* night flush controls shall operate outdoor air economizers at low fan speed not exceeding 66 percent during the unoccupied period with mechanical cooling and heating locked out.

SECTION C407—SIMULATED BUILDING PERFORMANCE

C407.1 Scope. This section establishes criteria for compliance using *simulated building performance*. The following systems and loads shall be included in determining the *simulated building performance*: heating systems, cooling systems, *service water heating*, *fan systems*, lighting power, receptacle loads and process loads.

Exception: Energy used to recharge or refuel vehicles that are used for on-road and off-site transportation purposes.

C407.2 Mandatory requirements. Compliance based on *simulated building performance* requires that a *proposed design* meet all of the following:

1. The requirements of the sections indicated within Table C407.2(1).
2. An annual *energy cost* that is less than or equal to the percentage of the annual *energy cost* (PAEC) of the *standard reference design* calculated in Equation 4-34. Energy prices shall be taken from a source *approved* by the *code official*, such as the Department of Energy, Energy Information Administration's *State Energy Data System Prices and Expenditures* reports. *Code officials* shall be permitted to require time-of-use pricing in *energy cost* calculations. The reduction in *energy cost* of the *proposed design* associated with *on-site renewable energy* shall be not more than 5 percent of the total *energy cost*. The amount of renewable energy purchased from off-site sources shall be the same in the *standard reference design* and the *proposed design*.

Equation 4-34 $PAEC = 100 \times (0.80 + 0.025 - EC_r/1{,}000)$

where:

PAEC = The percentage of the annual energy cost of the standard reference design.

EC_r = Energy efficiency credits required for the building in accordance with Section C406.1 (do not include load management and renewable credits).

Exceptions:

1. Jurisdictions that require site energy (1 kWh = 3413 Btu) rather than *energy cost* as the metric of comparison.
2. Where energy use based on source energy expressed in Btu or Btu per square foot of *conditioned floor area* is substituted for the *energy cost*, the energy use shall be calculated using source energy factors from Table C407.2(2). For electricity, US locations shall use values from eGRID subregions. Locations outside the United States shall use the value for "All other electricity" or locally derived values.

TABLE C407.2(1)—REQUIREMENTS FOR SIMULATED BUILDING PERFORMANCE	
SECTION[a]	**TITLE**
Envelope	
C401.3	Building thermal envelope certificate
C402.2.1.1	Joints staggered
C402.2.1.2	Skylight curbs
C402.2.6	Insulation of radiant heating system panels
C402.6	Air leakage—building thermal envelope
Mechanical	
C403.1.1	Calculation of heating and cooling loads
C403.1.2	Data centers
C403.2	System design
C403.3	Heating and cooling equipment efficiencies
C403.4.1	Thermostatic controls
C403.4.2	Off-hour controls
C403.4.7	Heating and cooling system controls for operable openings to the outdoors
C403.5.5	Economizer fault detection and diagnostics
C403.7, except C403.7.4.1	Ventilation and exhaust systems
C403.8, except C403.8.6	Fan and fan controls
C403.9	Large-diameter ceiling fans
C403.12, except C403.12.3	Refrigeration equipment performance
C403.13	Construction of HVAC system elements
C403.14	Mechanical systems located outside of the building thermal envelope
C404	Service water heating
C405, except C405.3	Electrical power and lighting systems
C406.1.2	Additional renewable and load management credit requirements
C408	Maintenance information and system commissioning

a. Reference to a code section includes all the relative subsections except as indicated in the table.

TABLE C407.2(2)—SOURCE ENERGY CONVERSION FACTORS FOR ELECTRICITY	
SOURCE ENERGY	**CONVERSION FACTOR**
Fossil Fuels Delivered to Buildings	
Natural gas	1.092
LPG or propone	1.151
Fuel oil (residual)	1.191
Fuel oil (distillate)	1.158
Coal	1.048
Gasoline	1.187
Other fuels not specified in this table	1.048
Electricity	
AKGD-ASCC Alaska Grid	2.47
AKMS-ASCC Miscellaneous	1.35
AZNM-WECC Southwest	2.57
CAMX-WECC California	1.66
ERCT-ERCOT All	2.32
FRCC-FRCC All	2.78

TABLE C407.2(2)—SOURCE ENERGY CONVERSION FACTORS FOR ELECTRICITY—continued	
SOURCE ENERGY	**CONVERSION FACTOR**
Electricity—continued	
HIMS-HICC Miscellaneous	3.15
HIOA-HICC Oahu	3.87
MROE-MRO East	2.92
MROW-MRO West	2.21
NEWE-NPCC New England	2.66
NWPP-WECC Northwest	1.48
NYCW-NPCC NYC/Westchester	2.89
NYLI-NPCC Long Island	2.84
NYUP-NPCC Upstate NY	1.81
PRMS-Puerto Rico Miscellaneous	3.27
RFCE-RFC East	2.90
RFCM-RFC Michigan	2.93
RFCW-RFC West	2.97
RMPA-WECC Rockies	2.16
SPNO-SPP North	2.21
SPSO-SPP South	2.05
SRMV-SERC Mississippi Valley	2.84
SRMW-SERC Midwest	3.09
SRSO-SERC South	2.89
SRTV-SERC Tennessee Valley	2.82
SRVC-SERC Virginia/Carolina	2.91
All other electricity	2.51
Thermal Energy	
Chilled water	0.60
Steam	1.84
Hot water	1.73

C407.3 Documentation. Documentation verifying that the methods and accuracy of compliance software tools conform to the provisions of this section shall be provided to the *code official.*

C407.3.1 Compliance report. Permit submittals shall include a report documenting that the *proposed design* has annual energy costs less than or equal to the annual energy costs of the *standard reference design*. The compliance documentation shall include the following information:

1. Address of the *building*.
2. An inspection checklist documenting the building component characteristics of the *proposed design* as specified in Table C407.4.1(1). The inspection checklist shall show the estimated annual *energy cost* for both the *standard reference design* and the *proposed design*.
3. Name of individual completing the compliance report.
4. Name and version of the compliance software tool.

C407.3.2 Additional documentation. The *code official* shall be permitted to require the following documents:

1. Documentation of the building component characteristics of the *standard reference design*.
2. Thermal zoning diagrams consisting of floor plans showing the thermal zoning scheme for *standard reference design* and *proposed design*.
3. Input and output reports from the *energy analysis* simulation program containing the complete input and output files, as applicable. The output file shall include energy use totals and energy use by energy source and end-use served, total hours that space conditioning loads are not met and any errors or warning messages generated by the simulation tool as applicable.
4. An explanation of any error or warning messages appearing in the simulation tool output.

5. A certification signed by the builder providing the building component characteristics of the *proposed design* as given in Table C407.4.1(1).
6. Documentation of the reduction in energy use associated with *on-site renewable energy*.

C407.4 Calculation procedure. Except as specified by this section, the *standard reference design* and *proposed design* shall be configured and analyzed using identical methods and techniques.

C407.4.1 Building specifications. The *standard reference design* and *proposed design* shall be configured and analyzed as specified by Table C407.4.1(1). Table C407.4.1(1) shall include by reference all notes contained in Table C402.1.2.

TABLE C407.4.1(1)—SPECIFICATIONS FOR THE STANDARD REFERENCE AND PROPOSED DESIGNS

BUILDING COMPONENT CHARACTERISTICS	STANDARD REFERENCE DESIGN	PROPOSED DESIGN
Space use classification	Same as proposed	The space use classification shall be chosen in accordance with Table C405.3.2(1) or C405.3.2(2) for all areas of the building covered by this permit. Where the space use classification for a building is not known, the building shall be categorized as an office building.
Roofs	Type: insulation entirely above deck	As proposed
	Gross area: same as proposed	As proposed
	U-factor: as specified in Table C402.1.2	As proposed
	Solar reflectance: 0.25, except as specified in Section C402.4 and Table C402.4 for Climate Zones 0, 1, 2 and 3	As proposed
	Emittance: 0.90, except as specified in Section C402.4 and Table C402.4 for Climate Zones 0, 1, 2 and 3	As proposed
Walls, above-grade	Type: same as proposed	As proposed
	Gross area: same as proposed	As proposed
	U-factor: as specified in Table C402.1.2	As proposed
	Thermal bridges: account for heat transfer consistent with compliant psi- and chi-factors from Table C402.1.4 for thermal bridges as identified in Section C402.7 that are present in the proposed design	As proposed; psi- and chi-factors for proposed thermal bridges shall be determined in accordance with requirements in Section C402.1.4.
	Solar reflectance: 0.25	As proposed
	Emittance: 0.90	As proposed
Walls, below-grade	Type: mass wall	As proposed
	Gross area: same as proposed	As proposed
	U-factor: as specified in Table C402.1.2 with insulation layer on interior side of walls	As proposed
Floors, above-grade	Type: joist/framed floor	As proposed
	Gross area: same as proposed	As proposed
	U-factor: as specified in Table C402.1.2	As proposed
Floors, slab-on-grade	Type: unheated	As proposed
	F-factor: as specified in Table C402.1.2	As proposed
Opaque doors	Type: swinging	As proposed
	Area: Same as proposed	As proposed
	U-factor: as specified in Table C402.1.2	As proposed

TABLE C407.4.1(1)—SPECIFICATIONS FOR THE STANDARD REFERENCE AND PROPOSED DESIGNS—continued		
BUILDING COMPONENT CHARACTERISTICS	**STANDARD REFERENCE DESIGN**	**PROPOSED DESIGN**
Vertical fenestration other than opaque doors	Area 1. The proposed vertical fenestration area; where the proposed vertical fenestration area is less than 40 percent of above-grade wall area. 2. Forty percent of above-grade wall area; where the proposed vertical fenestration area is 40 percent or more of the above-grade wall area.	As proposed
	U-factor: as specified in Table C402.5	As proposed
	SHGC: as specified in Table C402.5 except that for climates with no requirement (NR), SHGC = 0.40 shall be used	As proposed
	External shading and PF: none	As proposed
Skylights	Area 1. The proposed skylight area; where the proposed skylight area is less than that permitted by Section C402.1. 2. The area permitted by Section C402.1; where the proposed skylight area exceeds that permitted by Section C402.1.	As proposed
	U-factor: as specified in Table C402.5	As proposed
	SHGC: as specified in Table C402.5 except that for climates with no requirement (NR), SHGC = 0.40 shall be used	As proposed
Lighting, interior	The interior lighting power shall be determined in accordance with Section C405.3.2. Where the occupancy of the building is not known, the lighting power density shall be 1.0 watt per square foot based on the categorization of buildings with unknown space classification as offices.	As proposed
Lighting, exterior	The lighting power shall be determined in accordance with Tables C405.5.2(1), C405.5.2(2) and C405.5.2(3). Areas and dimensions of surfaces shall be the same as proposed.	As proposed
Internal gains	Same as proposed	Receptacle, motor and process loads shall be modeled and estimated based on the space use classification. End-use load components within and associated with the building shall be modeled to include, but not be limited to, the following: exhaust fans, parking garage ventilation fans, exterior building lighting, swimming pool heaters and pumps, elevators, escalators, refrigeration equipment and cooking equipment.
Schedules	Same as proposed **Exception:** Thermostat settings and schedules for HVAC systems that utilize radiant heating, radiant cooling and elevated air speed, provided that equivalent levels of occupant thermal comfort are demonstrated by means of equal Standard Effective Temperature as calculated in Normative Appendix B of ASHRAE Standard 55.	Operating schedules shall include hourly profiles for daily operation and shall account for variations between weekdays, weekends, holidays and any seasonal operation. Schedules shall model the time-dependent variations in occupancy, illumination, receptacle loads, thermostat settings, mechanical ventilation, HVAC equipment availability, service hot water usage and any process loads. The schedules shall be typical of the proposed building type as determined by the designer and approved by the jurisdiction.
Outdoor airflow	Where the proposed design specifies mechanical ventilation: 1. For systems 1–4 as specified in Tables C407.4.1(2) and C407.4.1(3), the outdoor airflow rate shall be determined in accordance with Section C403.7 and Section 403.3.1.1.2.3.4, Equation 4-8, of the *International Mechanical Code* using a system ventilation efficiency (E_v) of 0.75. 2. For systems 5–11 as specified in Tables C407.4.1(2) and C407.4.1(3), the outdoor airflow rate shall be determined in accordance with Section C403.7 and Section 403.3 of the *International Mechanical Code*. Where the proposed design specifies natural ventilation, as proposed.	As proposed, in accordance with Section C403.2.2.

TABLE C407.4.1(1)—SPECIFICATIONS FOR THE STANDARD REFERENCE AND PROPOSED DESIGNS—continued		
BUILDING COMPONENT CHARACTERISTICS	**STANDARD REFERENCE DESIGN**	**PROPOSED DESIGN**
Heating systems	Fuel type: same as proposed design	As proposed
	Equipment type[a]: as specified in Tables C407.4.1(2) and C407.4.1(3)	As proposed
	Efficiency: as specified in the tables in Section C403.3.2.	As proposed
	Capacity[b]: sized proportionally to the capacities in the proposed design based on sizing runs, and shall be established such that no smaller number of unmet heating load hours and no larger heating capacity safety factors are provided than in the proposed design.	As proposed
Cooling systems	Fuel type: same as proposed design	As proposed
	Equipment type[c]: as specified in Tables C407.4.1(2) and C407.4.1(3)	As proposed
	Efficiency: as specified in Tables C403.3.2(1), C403.3.2(2) and C403.3.2(3)	As proposed
	Capacity[b]: sized proportionally to the capacities in the proposed design based on sizing runs, and shall be established such that no smaller number of unmet cooling load hours and no larger cooling capacity safety factors are provided than in the proposed design.	As proposed
	Economizer[d]: same as proposed, in accordance with Section C403.5.	As proposed
Service water heating[e]	Fuel type: same as proposed	As proposed
	Efficiency: as specified in Table C404.2	For Group R, as proposed multiplied by SWHF. For other than Group R, as proposed multiplied by efficiency as provided by the manufacturer of the DWHR unit.
	Capacity: same as proposed	As proposed
	Where no service water hot water system exists or is specified in the proposed design, no service hot water heating shall be modeled.	
Energy recovery	Where the proposed design specifies mechanical ventilation, as specified in Section C403.7.4 based on the standard reference design airflows.	As proposed
	Where the proposed design specifies natural ventilation, as proposed.	
Fan power	As specified in Section C403.8 for the proposed design. **Exceptions:** 1. Where the fan power of the proposed design is exempted from the requirements of Section C403.8, as proposed. 2. Fan systems addressed by Section C403.8.1: fan system BHP shall be as proposed or to the limits specified in Section C403.8.1, whichever is smaller. If the limit is reached, the power of each fan shall be reduced proportionally until the limit is met. 3. Fan systems serving areas where the mechanical ventilation is provided in accordance with an engineered ventilation system design of Section 403.2 of the *International Mechanical Code* shall not use the particulate filtration or air cleaner pressure drop adjustment available in Table C403.8.1(1) when calculating the fan system BHP limit for the portion of the airflow being treated to comply with the engineered ventilation system design.	As proposed

TABLE C407.4.1(1)—SPECIFICATIONS FOR THE STANDARD REFERENCE AND PROPOSED DESIGNS—continued

BUILDING COMPONENT CHARACTERISTICS	STANDARD REFERENCE DESIGN	PROPOSED DESIGN
On-site renewable energy	Where a system providing on-site renewable energy has been modeled in the proposed design, the same system shall be modeled identically in the standard reference design except the rated capacity shall meet the requirements of Section C405.15.1 Where no system is designed or included in the proposed design, model an unshaded photovoltaic system with the following characteristics: Size: rated capacity per Section C405.15.1. Module type: crystalline silicone panel with glass cover, 19.1% nominal efficiency and temperature coefficient of -0.35%/°C. Performance shall be based on a reference temperature of 77°F (25°C), air mass of 1.5 atmosphere and irradiance of 317 Btu/h × ft^2 (1000 W/m^2). Array type: rack-mounted array with installed nominal operating cell temperature (INOCT) of 103°F (45°C). Total system losses (DC output to AC output): 11.3%. Tilt: 0 degrees (mounted horizontally). Azimuth: 180 degrees.	As proposed

For SI: 1 watt per square foot = 10.7 w/m^2.

SWHF = Service Water Heat Recovery factor, DWHR = Drain Water Heat Recovery.

a. Where no heating system exists or has been specified, the heating system shall be modeled as fossil fuel. The system characteristics shall be identical in both the standard reference design and proposed design.

b. The ratio between the capacities used in the annual simulations and the capacities determined by sizing runs shall be the same for both the standard reference design and proposed design.

c. Where no cooling system exists or no cooling system has been specified, the cooling system shall be modeled as an air-cooled single-zone system, one unit per thermal zone. The system characteristics shall be identical in both the standard reference design and proposed design.

d. If an economizer is required in accordance with Table C403.5(1) and where no economizer exists or is specified in the proposed design, then a supply-air economizer shall be provided in the standard reference design in accordance with Section C403.5.

e. The SWHF shall be applied as follows:
   1. Where potable water from the DWHR unit supplies not less than one shower and not greater than two showers, of which the drain water from the same showers flows through the DWHR unit then SWHF = [1 - (DWHR unit efficiency × 0.36)].
   2. Where potable water from the DWHR unit supplies not less than three showers and not greater than four showers, of which the drain water from the same showers flows through the DWHR unit then SWHF = [1 - (DWHR unit efficiency × 0.33)].
   3. Where potable water from the DWHR unit supplies not less than five showers and not greater than six showers, of which the drain water from the same showers flows through the DWHR unit, then SWHF = [1 - (DWHR unit efficiency × 0.26)].
   4. Where Items 1 through 3 are not met, SWHF = 1.0.

TABLE C407.4.1(2)—HVAC SYSTEMS MAP

CONDENSER COOLING SOURCE[a]	HEATING SYSTEM CLASSIFICATION[b]	STANDARD REFERENCE DESIGN HVC SYSTEM TYPE[c]		
		Single-zone Residential System	Single-zone Nonresidential System	All Other
Water/ground	Electric resistance	System 5	System 5	System 1
	Heat pump	System 6	System 6	System 6
	Fossil fuel	System 7	System 7	System 2
Air/none	Electric resistance	System 8	System 9	System 3
	Heat pump	System 8	System 9	System 3
	Fossil fuel	System 10	System 11	System 4

a. Select "water/ground" where the proposed design system condenser is water or evaporatively cooled; select "air/none" where the condenser is air cooled. Closed-circuit dry coolers shall be considered to be air cooled. Systems utilizing district cooling shall be treated as if the condenser water type were "water." Where mechanical cooling is not specified or the mechanical cooling system in the proposed design does not require heat rejection, the system shall be treated as if the condenser water type were "Air." For proposed designs with ground-source or groundwater-source heat pumps, the standard reference design HVAC system shall be water-source heat pump (System 6).

b. Select the path that corresponds to the proposed design heat source: electric resistance, heat pump (including air source and water source), or fuel fired. Systems utilizing district heating (steam or hot water) and systems without heating capability shall be treated as if the heating system type were "fossil fuel." For systems with mixed fuel heating sources, the system or systems that use the secondary heating source type (the one with the smallest total installed output capacity for the spaces served by the system) shall be modeled identically in the standard reference design and the primary heating source type shall be used to determine standard reference design HVAC system type.

c. Select the standard reference design HVAC system category: The system under "single-zone residential system" shall be selected where the HVAC system in the proposed design is a single-zone system and serves a Group R occupancy. The system under "single-zone nonresidential system" shall be selected where the HVAC system in the proposed design is a single-zone system and serves other than Group R occupancy. The system under "all other" shall be selected for all other cases.

TABLE C407.4.1(3)—SPECIFICATIONS FOR THE STANDARD REFERENCE DESIGN HVAC SYSTEM DESCRIPTIONS

SYSTEM NO.	SYSTEM TYPE	FAN CONTROL	COOLING TYPE	HEATING TYPE
1	Variable air volume with parallel fan-powered boxes[a]	VAV[d]	Chilled water[e]	Electric resistance
2	Variable air volume with reheat[b]	VAV[d]	Chilled water[e]	Hot water fossil fuel boiler[f]
3	Packaged variable air volume with parallel fan-powered boxes[a]	VAV[d]	Direct expansion[c]	Electric resistance
4	Packaged variable air volume with reheat[b]	VAV[d]	Direct expansion[c]	Hot water fossil fuel boiler[f]
5	Two-pipe fan coil	Constant volume[i]	Chilled water[e]	Electric resistance
6	Water-source heat pump	Constant volume[i]	Direct expansion[c]	Electric heat pump and boiler[g]
7	Four-pipe fan coil	Constant volume[i]	Chilled water[e]	Hot water fossil fuel boiler[f]
8	Packaged terminal heat pump	Constant volume[i]	Direct expansion[c]	Electric heat pump[h]
9	Packaged rooftop heat pump	Constant volume[i]	Direct expansion[c]	Electric heat pump[h]
10	Packaged terminal air conditioner	Constant volume[i]	Direct expansion	Hot water fossil fuel boiler[f]
11	Packaged rooftop air conditioner	Constant volume[i]	Direct expansion	Fossil fuel furnace

For SI: 1 foot = 304.8 mm, 1 cfm = 0.4719 L/s, 1 Btu/h = 0.293/W, °C = [(°F) – 32]/1.8.

a. **VAV with parallel boxes:** Fans in parallel VAV fan-powered boxes shall be sized for 50 percent of the peak design flow rate and shall be modeled with 0.35 W/cfm fan power. Minimum volume setpoints for fan-powered boxes shall be equal to the minimum rate for the space required for ventilation consistent with Section C403.6.1, Item 3. Supply air temperature setpoint shall be constant at the design condition.

b. **VAV with reheat:** Minimum volume setpoints for VAV reheat boxes shall be 0.4 cfm/ft^2 of floor area. Supply air temperature shall be reset based on zone demand from the design temperature difference to a 10°F temperature difference under minimum load conditions. Design airflow rates shall be sized for the reset supply air temperature (i.e., a 10°F temperature difference).

c. **Direct expansion:** The fuel type for the cooling system shall match that of the cooling system in the proposed design.

d. **VAV:** Where the proposed design system has a supply, return or relief fan motor 25 hp or larger, the corresponding fan in the VAV system of the standard reference design shall be modeled assuming a variable-speed drive. For smaller fans, a forward-curved centrifugal fan with inlet vanes shall be modeled. Where the proposed design's system has a direct digital control system at the zone level, static pressure setpoint reset based on zone requirements in accordance with Section C403.8.6 shall be modeled.

e. **Chilled water:** For systems using purchased chilled water, the chillers are not explicitly modeled and chilled water costs shall be based as determined in Sections C407.2 and C407.4.2. Otherwise, the standard reference design's chiller plant shall be modeled with chillers having the number as indicated in Table C407.4.1(4) as a function of standard reference building chiller plant load and type as indicated in Table C407.4.1(5) as a function of individual chiller load. Where chiller fuel source is mixed, the system in the standard reference design shall have chillers with the same fuel types and with capacities having the same proportional capacity as the proposed design's chillers for each fuel type. Chilled water supply temperature shall be modeled at 44°F design supply temperature and 56°F return temperature. Piping losses shall not be modeled in either building model. Chilled water supply water temperature shall be reset in accordance with Section C403.4.4. Pump system power for each pumping system shall be the same as the proposed design; where the proposed design has no chilled water pumps, the standard reference design pump power shall be 22 W/gpm (equal to a pump operating against a 75-foot head, 65 percent combined impeller and motor efficiency). The chilled water system shall be modeled as primary-only variable flow with flow maintained at the design rate through each chiller using a bypass. Chilled water pumps shall be modeled as riding the pump curve or with variable-speed drives where required in Section C403.4.4. The heat rejection device shall be an axial fan cooling tower with two-speed fans where required in Section C403.11. Condenser water design supply temperature shall be 85°F or 10°F approach to design wet-bulb temperature, whichever is lower, with a design temperature rise of 10°F. The tower shall be controlled to maintain a 70°F leaving water temperature where weather permits, floating up to leaving water temperature at design conditions. Pump system power for each pumping system shall be the same as the proposed design; where the proposed design has no condenser water pumps, the standard reference design pump power shall be 19 W/gpm (equal to a pump operating against a 60-foot head, 60 percent combined impeller and motor efficiency). Each chiller shall be modeled with separate condenser water and chilled water pumps interlocked to operate with the associated chiller.

f. **Fossil fuel boiler:** For systems using purchased hot water or steam, the boilers are not explicitly modeled and hot water or steam costs shall be based on actual utility rates. Otherwise, the boiler plant shall use the same fuel as the proposed design and shall be natural draft. The standard reference design boiler plant shall be modeled with a single boiler where the standard reference design plant load is 600,000 Btu/h and less and with two equally sized boilers for plant capacities exceeding 600,000 Btu/h. Boilers shall be staged as required by the load. Hot water supply temperature shall be modeled at 180°F design supply temperature and 130°F return temperature. Piping losses shall not be modeled in either building model. Hot water supply water temperature shall be reset in accordance with Section C403.4.4. Pump system power for each pumping system shall be the same as the proposed design; where the proposed design has no hot water pumps, the standard reference design pump power shall be 19 W/gpm (equal to a pump operating against a 60-foot head, 60 percent combined impeller and motor efficiency). The hot water system shall be modeled as primary only with continuous variable flow. Hot water pumps shall be modeled as riding the pump curve or with variable speed drives where required by Section C403.4.4.

g. **Electric heat pump and boiler:** Water-source heat pumps shall be connected to a common heat pump water loop controlled to maintain temperatures between 60°F and 90°F. Heat rejection from the loop shall be provided by an axial fan closed-circuit evaporative fluid cooler with two-speed fans where required in Section C403.8.6. Heat addition to the loop shall be provided by a boiler that uses the same fuel as the proposed design and shall be natural draft. Where no boilers exist in the proposed design, the standard reference building boilers shall be fossil fuel. The standard reference design boiler plant shall be modeled with a single boiler where the standard reference design plant load is 600,000 Btu/h or less and with two equally sized boilers for plant capacities exceeding 600,000 Btu/h. Boilers shall be staged as required by the load. Piping losses shall not be modeled in either building model. Pump system power shall be the same as the proposed design; where the proposed design has no pumps, the standard reference design pump power shall be 22 W/gpm, which is equal to a pump operating against a 75-foot head, with a 65 percent combined impeller and motor efficiency. Loop flow shall be variable with flow shutoff at each heat pump when its compressor cycles off as required by Section C403.4.4. Loop pumps shall be modeled as riding the pump curve or with variable speed drives where required by Section C403.11.

h. **Electric heat pump:** Electric air-source heat pumps shall be modeled with electric auxiliary heat. The system shall be controlled with a multistage space thermostat and an outdoor air thermostat wired to energize auxiliary heat only on the last thermostat stage and when outdoor air temperature is less than 40°F.

i. **Constant volume:** Fans shall be controlled in the same manner as in the proposed design (i.e., fan operation whenever the space is occupied or fan operation cycled on calls for heating and cooling). Where the fan is modeled as cycling and the fan energy is included in the energy efficiency rating of the equipment, fan energy shall not be modeled explicitly.

TABLE C407.4.1(4)—NUMBER OF CHILLERS	
TOTAL CHILLER PLANT CAPACITY	**NUMBER OF CHILLERS**
≤ 300 tons	1
> 300 tons, < 600 tons	2, sized equally
≥ 600 tons	2 minimum, with chillers added so that all are sized equally and none is larger than 800 tons
For SI: 1 ton = 3517 W.	

TABLE C407.4.1(5)—WATER CHILLER TYPES		
INDIVIDUAL CHILLER PLANT CAPACITY	**ELECTRIC CHILLER TYPE**	**FOSSIL FUEL CHILLER TYPE**
≤ 100 tons	Reciprocating	Single-effect absorption, direct fired
> 100 tons, < 300 tons	Screw	Double-effect absorption, direct fired
≥ 300 tons	Centrifugal	Double-effect absorption, direct fired
For SI: 1 ton = 3517 W.		

C407.4.2 Thermal blocks. The *standard reference design* and *proposed design* shall be analyzed using identical *thermal blocks* as specified in Section C407.4.2.1, C407.4.2.2 or C407.4.2.3.

C407.4.2.1 HVAC zones designed. Where HVAC *zones* are defined on HVAC design drawings, each HVAC *zone* shall be modeled as a separate *thermal block*.

Exception: Different HVAC *zones* shall be allowed to be combined to create a single *thermal block* or identical *thermal blocks* to which multipliers are applied, provided that:

1. The space use classification is the same throughout the *thermal block*.
2. All HVAC *zones* in the *thermal block* that are adjacent to glazed exterior walls face the same orientation or their orientations are within 45 degrees (0.79 rad) of each other.
3. All of the *zones* are served by the same HVAC system or by the same kind of HVAC system.

C407.4.2.2 HVAC zones not designed. Where HVAC *zones* have not yet been designed, *thermal blocks* shall be defined based on similar internal load densities, occupancy, lighting, thermal and temperature schedules, and in combination with the following guidelines:

1. Separate *thermal blocks* shall be assumed for interior and perimeter spaces. Interior spaces shall be those located more than 15 feet (4572 mm) from an *exterior wall*. Perimeter spaces shall be those located closer than 15 feet (4572 mm) from an *exterior wall*.
2. Separate *thermal blocks* shall be assumed for spaces adjacent to glazed *exterior walls*: a separate *zone* shall be provided for each orientation, except orientations that differ by not more than 45 degrees (0.79 rad) shall be permitted to be considered to be the same orientation. Each *zone* shall include floor area that is 15 feet (4572 mm) or less from a glazed perimeter wall, except that floor area within 15 feet (4572 mm) of glazed perimeter walls having more than one orientation shall be divided proportionately between *zones*.
3. Separate *thermal blocks* shall be assumed for spaces having floors that are in contact with the ground or exposed to ambient conditions from *zones* that do not share these features.
4. Separate *thermal blocks* shall be assumed for spaces having exterior ceiling or *roof assemblies* from *zones* that do not share these features.

C407.4.2.3 Group R-2 occupancy buildings. Group R-2 occupancy spaces shall be modeled using one *thermal block* per space except that those facing the same orientations are permitted to be combined into one *thermal block*. Corner units and units with roof or floor loads shall only be combined with units sharing these features.

C407.5 Calculation software tools. Calculation procedures used to comply with Section C407 shall apply an *approved* version of a performance analysis software tool capable of calculating the annual energy consumption of all building elements that differ between the *standard reference design* and the *proposed design*. The same *approved* version of the performance analysis tool shall be used to calculate the *proposed design* and *standard reference design*.

C407.5.1 Software tool approval. Any version of a performance analysis tool meeting the requirements of Sections C407.5.1.1 and C407.5.1.2 shall be permitted to be *approved*. Tools are permitted to be *approved* based on meeting a specified threshold for a jurisdiction. The *code official* shall be permitted to approve tools for a specified application or limited scope.

C407.5.1.1 Software tool capabilities. *Approved* software tools shall include the following capabilities:

1. *Building* operation for a full calendar year (8,760 hours).
2. Climate data for a full calendar year (8,760 hours) and shall reflect *approved* coincident hourly data for temperature, solar radiation, humidity and wind speed for the *building* location.
3. Ten or more thermal zones.

4. Thermal mass effects.
5. Hourly variations in occupancy, illumination, receptacle loads, *thermostat* settings, mechanical ventilation, HVAC equipment availability, service hot water usage and any process loads.
6. Part-load performance curves for mechanical equipment.
7. Capacity and efficiency correction curves for mechanical heating and cooling equipment.
8. Printed *code official* inspection checklist listing each of the *proposed design* component characteristics from Table C407.4.1(1) determined by the analysis to provide compliance, along with their respective performance ratings, including but not limited to *R*-value, *U-factor*, SHGC, HSPF, AFUE, SEER and EF.

C407.5.1.2 Testing required by software vendors. Prior to approval, software tools shall be tested by the software vendor in accordance with ASHRAE Standard 140, except Sections 7 and 8. During testing, hidden inputs that are not normally available to the user shall be permitted to avoid introducing source code changes strictly used for testing. Software vendors shall publish, on a publicly available website, the following ASHRAE Standard 140 test results, input files and modeler reports for each tested version of a software tool:

1. Test results that demonstrate the software tool was tested in accordance with ASHRAE Standard 140 and that meet or exceed the values for "The Minimum Number of Range Cases within the Test Group to Pass" for all test groups in ASHRAE Standard 140, Table A3-14.
2. Test results of the performance analysis tool and input files used for generating the ASHRAE Standard 140 test cases along with the results of the other performance analysis tools included in ASHRAE Standard 140, Annexes B8 and B16.
3. The modeler report in ASHRAE Standard 140, Annex A2, Attachment A2.7, Report Blocks A and G shall be completed for results exceeding the maximum or falling below the minimum of the reference values shown in ASHRAE Standard 140, Tables A3-1 through A3-13, and Report Blocks A and E shall be completed for any omitted results.

C407.5.2 Algorithms not tested. Algorithms not tested in accordance with Section C407.5.1.2, including algorithms that are alternatives to those that were tested, and numerical settings not tested, such as time steps and tolerances, shall be permitted to be used where modeling the *proposed design* and *standard reference design*.

C407.5.3 Input values. Where calculations require input values not specified by Sections C402, C403, C404 and C405, those input values shall be taken from an *approved* source.

C407.5.4 Exceptional calculation methods. Where the simulation program does not model a design, material or device of the *proposed design*, an exceptional calculation method shall be used where *approved* by the *code official*. Where there are multiple designs, materials or devices that the simulation program does not model, each shall be calculated separately and exceptional savings determined for each. The total exceptional savings shall not constitute more than half of the difference between the baseline *simulated building performance* and the proposed *simulated building performance*. Applications for approval of an exceptional method shall include all of the following:

1. Step-by-step documentation of the exceptional calculation method performed, detailed enough to reproduce the results.
2. Copies of all spreadsheets used to perform the calculations.
3. A sensitivity analysis of energy consumption where each of the input parameters is varied from half to double the value assumed.
4. The calculations shall be performed on a time step basis consistent with the simulation program used.
5. The performance rating calculated with and without the exceptional calculation method.

SECTION C408 —MAINTENANCE INFORMATION AND SYSTEM COMMISSIONING

C408.1 General. This section covers the provision of maintenance information and the commissioning of, and the functional testing requirements for, *building* systems.

C408.1.1 Building operations and maintenance information. The *building* operations and maintenance documents shall be provided to the *owner* and shall consist of manufacturers' information, specifications and recommendations; programming procedures and data points; narratives; and other means of illustrating to the *owner* how the *building*, equipment and systems are intended to be installed, maintained and operated. Required regular maintenance actions for equipment and systems shall be clearly stated on a readily visible label. The label shall include the title or publication number for the operation and maintenance manual for that particular model and type of product.

C408.2 Mechanical systems and service water-heating systems commissioning and completion requirements. Prior to the final mechanical and plumbing inspections, the *registered design professional or approved agency* shall provide evidence of mechanical systems *commissioning* and completion in accordance with the provisions of this section.

Construction document notes shall clearly indicate provisions for *commissioning* and completion requirements in accordance with this section and are permitted to refer to specifications for further requirements. Copies of all documentation shall be given to the *owner* or owner's authorized agent and made available to the *code official* upon request in accordance with Sections C408.2.4 and C408.2.5.

Exceptions: The following systems are exempt:

1. *Buildings* with less than 10,000 square feet (929 m^2) gross *conditioned floor area* and combined heating, cooling and service water heating capacity of less than 960,000 Btu/h (281 kW).

2. Components within *dwelling units* and *sleeping units* served by one of the following systems:
    2.1. Simple unitary or packaged HVAC equipment listed in Table C403.3.2(1), C403.3.2(2), C403.3.2(4) or C403.3.2(5), each serving one *zone* and controlled by a single *thermostat* in the *zone* served.
    2.2. Two-pipe heating systems installed in the dwelling, serving one or more *zones*.

C408.2.1 Commissioning plan. A *commissioning plan* shall be developed by a *registered design professional* or *approved agency* and shall include the following items:

1. A narrative description of the activities that will be accomplished during each phase of *commissioning,* including the personnel intended to accomplish each of the activities.
2. A listing of the specific equipment, appliances or systems to be tested and a description of the tests to be performed.
3. Functions to be tested including, but not limited to, calibrations and economizer controls.
4. Conditions under which the test will be performed. Testing shall affirm winter and summer design conditions and full outside air conditions.
5. Measurable criteria for performance.

C408.2.2 Systems adjusting and balancing. HVAC systems shall be balanced in accordance with generally accepted engineering standards. Air and water flow rates shall be measured and adjusted to deliver final flow rates within the tolerances provided in the product specifications. Test and balance activities shall include air system and hydronic system balancing.

C408.2.2.1 Air systems balancing. Each supply air outlet and *zone* terminal device shall be equipped with means for air balancing in accordance with the requirements of Chapter 6 of the *International Mechanical Code*. Discharge dampers used for air-system balancing are prohibited on constant-volume fans and variable volume fans with motors 10 hp (18.6 kW) and larger. Air systems shall be balanced in a manner to first minimize throttling losses then, for fans with system power of greater than 1 hp (0.746 kW), fan speed shall be adjusted to meet design flow conditions.

Exception: Fans with fan motors of 1 hp (0.74 kW) or less are not required to be provided with a means for air balancing.

C408.2.2.2 Hydronic systems balancing. Individual hydronic heating and cooling coils shall be equipped with means for balancing and measuring flow. Hydronic systems shall be proportionately balanced in a manner to first minimize throttling losses, then the pump impeller shall be trimmed or pump speed shall be adjusted to meet design flow conditions. Each hydronic system shall have either the capability to measure pressure across the pump, or test ports at each side of each pump.

Exception: The following equipment is not required to be equipped with a means for balancing or measuring flow:

1. Pumps with pump motors of 5 hp (3.7 kW) or less.
2. Where throttling results in not greater than 5 percent of the *nameplate horsepower* draw above that required if the impeller were trimmed.

C408.2.3 Functional performance testing. Functional performance testing specified in Sections C408.2.3.1 through C408.2.3.3 shall be conducted.

C408.2.3.1 Equipment. Equipment functional performance testing shall demonstrate the installation and operation of components, systems and system-to-system interfacing relationships in accordance with *approved* plans and specifications such that operation, function and maintenance serviceability for each of the commissioned systems are confirmed. Testing shall include all modes and *sequence of operation*, including under full-load, part-load and the following emergency conditions:

1. All modes as described in the *sequence* of *operation*.
2. Redundant or *automatic* back-up mode.
3. Performance of alarms.
4. Mode of operation upon a loss of power and restoration of power.

Exception: Unitary or packaged HVAC equipment listed in the tables in Section C403.3.2 that do not require supply air economizers.

C408.2.3.2 Controls. HVAC and service water-heating control systems shall be tested to document that control devices, components, equipment and systems are calibrated and adjusted and operate in accordance with *approved* plans and specifications. Sequences of operation shall be functionally tested to document they operate in accordance with *approved* plans and specifications.

C408.2.3.3 Economizers. Air economizers shall undergo a functional test to determine that they operate in accordance with manufacturer's specifications.

C408.2.4 Preliminary commissioning report. A preliminary report of *commissioning* test procedures and results shall be completed and certified by the *registered design professional* or *approved agency* and provided to the *building owner* or owner's authorized agent. The report shall be organized with mechanical and service hot water findings in separate sections to allow independent review. The report shall be identified as "Preliminary Commissioning Report," shall include the completed Commissioning Compliance Checklist, Figure C408.2.4, and shall identify:

1. Itemization of deficiencies found during testing required by this section that have not been corrected at the time of report preparation.
2. Deferred tests that cannot be performed at the time of report preparation because of climatic conditions.

3. Climatic conditions required for performance of the deferred tests.
4. Results of functional performance tests.
5. Functional performance test procedures used during the commissioning process, including measurable criteria for test acceptance.

FIGURE C408.2.4—COMMISSIONING COMPLIANCE CHECKLIST

Project Information: ______________________________ Project Name: ____________________________________

Project Address: ____

Commissioning Authority: ____

Commissioning Plan (Section C408.2.1)

☐ Commissioning Plan was used during construction and includes all items required by Section C408.2.1

☐ Systems Adjusting and Balancing has been completed.

☐ HVAC Equipment Functional Testing has been executed. If applicable, deferred and follow-up testing is scheduled to be provided on: ____________________________

☐ HVAC Controls Functional Testing has been executed. If applicable, deferred and follow-up testing is scheduled to be provided on: ____________________________

☐ Economizer Functional Testing has been executed. If applicable, deferred and follow-up testing is scheduled to be provided on: ____________________________

☐ Lighting Controls Functional Testing has been executed. If applicable, deferred and follow-up testing is scheduled to be provided on: ____________________________

☐ Service Water Heating System Functional Testing has been executed. If applicable, deferred and follow-up testing is scheduled to be provided on: ____________________________

☐ Manual, record documents and training have been completed or scheduled

☐ Preliminary Commissioning Report submitted to owner and includes all items required by Section C408.2.4

I hereby certify that the commissioning provider has provided me with evidence of mechanical, service water heating and lighting systems commissioning in accordance with the 2021 IECC.

Signature of Building Owner or Owner's Representative ____________________________ Date ______________

C408.2.4.1 Acceptance of report. *Buildings*, or portions thereof, shall not be considered as acceptable for a final inspection pursuant to Section C107.2.6 until the *code official* has received the Preliminary Commissioning Report from the *building owner* or owner's authorized agent.

C408.2.4.2 Copy of report. The *code official* shall be permitted to require that a copy of the Preliminary Commissioning Report be made available for review by the *code official*.

C408.2.5 Documentation requirements. The *construction documents* shall specify that the documents described in this section be provided to the *building owner* or owner's authorized agent within 90 days of the date of receipt of the *certificate of occupancy*.

C408.2.5.1 System balancing report. A written report describing the activities and measurements completed in accordance with Section C408.2.2.

C408.2.5.2 Final commissioning report. A report of test procedures and results identified as "Final Commissioning Report" shall be delivered to the *building owner* or owner's authorized agent. The report shall be organized with mechanical system and service hot water system findings in separate sections to allow independent review. The report shall include the following:

1. Results of functional performance tests.
2. Disposition of deficiencies found during testing, including details of corrective measures used or proposed.

3. Functional performance test procedures used during the commissioning process including measurable criteria for test acceptance, provided herein for repeatability.

Exception: Deferred tests that cannot be performed at the time of report preparation due to climatic conditions.

C408.3 Functional testing of lighting and receptacle controls. *Automatic* lighting and receptacle controls required by this code shall comply with this section.

C408.3.1 Functional testing. Prior to passing final inspection, the *registered design professional* or *approved agency* shall provide evidence that the lighting and receptacle control systems have been tested to ensure that control hardware and software are calibrated, adjusted, programmed and in proper working condition in accordance with the *construction documents* and manufacturer's instructions. Functional testing shall be in accordance with Sections C408.3.1.1 through C408.3.1.3 for the applicable control type.

C408.3.1.1 Occupant sensor controls. Where *occupant sensor controls* are provided, the following procedures shall be performed:

1. Certify that the *occupant sensor* has been located and aimed in accordance with manufacturer recommendations.
2. For projects with seven or fewer *occupant sensors*, each sensor shall be tested.
3. For projects with more than seven *occupant sensors,* testing shall be done for each unique combination of sensor type and space geometry. Where multiples of each unique combination of sensor type and space geometry are provided, not less than 10 percent and in no case fewer than one, of each combination shall be tested unless the *code official* or design professional requires a higher percentage to be tested. Where 30 percent or more of the tested controls fail, all remaining identical combinations shall be tested.

   For *occupant sensor controls* to be tested, verify the following:

   3.1. Where *occupant sensor controls* include status indicators, verify correct operation.

   3.2. The lights and receptacles controlled by *occupant sensor controls* turn off or down to the permitted level within the required time upon vacancy of the space.

   3.3. For auto-on *occupant sensor controls*, the lights and receptacles controlled by *occupant sensor controls* turn on to the permitted level when an occupant enters the space.

   3.4. For manual-on *occupant sensor controls*, the lights and receptacles controlled by *occupant sensor controls* turn on only when manually activated.

   3.5. The lights are not incorrectly turned on by movement in adjacent areas or by HVAC operation.

C408.3.1.2 Time-switch controls. Where *time-switch controls* are provided, Items 1 through 5 shall be performed for all *time-switch controls*. For projects with more than seven spaces where lighting or receptacles are controlled by *time-switch controls*, not less than 10 percent of spaces and in no case fewer than one space shall be tested according to Items 6 and 7 unless the *code official* or *registered design professional* requires a higher percentage to be tested. Where 30 percent or more of the tested spaces fail any of the requirements in Items 6 and 7, all remaining spaces shall be tested.

1. Confirm that the *time-switch control* is programmed with accurate weekday, weekend and holiday schedules.
2. Provide documentation to the *owner* of *time-switch controls* programming including weekday, weekend, holiday schedules, and set-up and preference program settings.
3. Verify the correct time and date in the time switch.
4. Verify that any battery back-up is installed and energized.
5. Verify that the override time limit is set to not more than 2 hours.
6. Simulate occupied condition. Verify and document the following:

   6.1. All lights can be turned on and off by their respective area control switch.

   6.2. The switch only operates lighting in the *enclosed space* in which the switch is located.

   6.3. Receptacles in the space controlled by the *time-switch controls* turn on.
7. Simulate unoccupied condition. Verify and document the following:

   7.1. Nonexempt lighting turns off.

   7.2. *Manual* override switch allows only the lights and receptacles controlled by the time-switch controls in the *enclosed space* where the override switch is located to turn on controlled lighting and receptacles for more than 2 hours.

   7.3. Receptacles controlled by the *time-switch controls* turn off.
8. Additional testing as specified by the *registered design professional.*

C408.3.1.3 Daylight responsive controls. Where *daylight responsive controls* are provided, the following shall be verified:

1. Control devices have been properly located, field calibrated and set for accurate setpoints and threshold light levels.
2. Daylight controlled lighting loads adjust to light level setpoints in response to available daylight.
3. The calibration adjustment equipment is located for *ready access* only by authorized personnel.

C408.3.1.4 High-end trim controls. Where lighting controls are configured for *high-end trim*, verify the following:

1. *High-end trim* maximum level has been set.

2. The calibration adjustment equipment is located for *ready access* only by authorized personnel.
3. Lighting controls with *ready access* for users cannot increase the lighting power above the maximum level established by the *high-end trim* controls.

C408.3.1.5 High-end trim lighting control verification for L02 Additional Efficiency Credit. For the qualifying spaces associated with the project receiving the additional efficiency credits in Section C406.2.5.2, the following shall be documented while *daylight responsive controls* are not reducing lighting power:

1. The maximum setting for power or light output for each control group of general lighting luminaires.
2. The *high-end trim* setting for power or light output for each control group of general lighting luminaires.
3. For projects with seven or fewer claimed qualifying spaces, the reduction in light output or reduction in power due to *high-end trim* shall be tested in all spaces and shown to reduce the general lighting power or light output to not greater than 85 percent of full power or light output. For projects with more than seven claimed qualifying spaces, the reduction in light output or reduction in power due to *high-end trim* shall be tested in not less than 10 percent of spaces, and not less than seven spaces, and be shown to reduce general lighting power or light output to not greater than 85 percent of full power or light output. Where more than 30 percent of the tested spaces fail, the remaining qualifying spaces shall be tested.
4. Summarize the reduction in general lighting power or light output resulting from the *high-end trim* setting for each qualifying space and the floor area of each qualifying space.
5. Summarize the fraction of total floor area for spaces where *high-end trim* reduces general lighting power or light output to not greater than 85 percent of full power or light output.

C408.3.1.6 Demand responsive lighting controls G01. For spaces associated with the project receiving renewable and load management credits in Section C406.3.2, the following procedures shall be performed:

1. Confirm the maximum setpoint upon receipt of the *demand response signal* has been established for each space.
2. For projects with seven or fewer spaces with controls, each space shall be tested.
3. For projects with more than seven spaces with controls, testing shall be done for each unique space type. Where multiple spaces of each space type exist, not less than 10 percent of each space type, and in no case fewer than one space, shall be tested unless the *code official* requires a higher percentage to be tested. Where 30 percent or more of the tested controls fail in a space type, all remaining identical space types shall be tested.
4. For demand responsive controls to be tested, verify the following:
    - 4.1. Where *high-end trim* controls are used, the *high-end trim* shall be set before testing.
    - 4.2. Turn off all nongeneral lighting in the space.
    - 4.3. Set *general lighting* to its maximum illumination level. Where *high-end trim* is set, this will be the maximum illumination level at the *high-end trim* setpoint.
    - 4.4. An illumination measurement shall be taken in an area of the space not controlled by daylight responsive controlled lighting. If there is not an area without *daylight responsive controls*, the *daylight responsive controls* shall be overridden from reducing the lighting level during the test.
    - 4.5. Measure and document the maximum illumination level of the space.
5. Simulate a *demand response signal* and measure the illumination level at the same location as for the measurement in Section C408.3.1.6, Item 4.5. Verify the illumination level has been reduced to not greater than 80 percent of the maximum illumination level documented in Section C408.3.1.6, Item 4.5.
6. Simulate the end of a demand event by turning off the *demand response signal*; confirm controls automatically return to their normal operational settings at the end of the demand response event.

C408.3.2 Documentation requirements. The *construction documents* shall specify that the documents described in this section be provided to the *building owner* or owner's authorized agent within 90 days of the date of receipt of the *certificate of occupancy*.

C408.3.2.1 Drawings. *Construction documents* shall include the location and catalogue number of each piece of equipment.

C408.3.2.2 Manuals. An operating and maintenance manual shall be provided and include the following:

1. Name and address of not less than one service agency for installed equipment.
2. A narrative of how each system is intended to operate, including recommended setpoints.
3. Submittal data indicating all selected options for each piece of lighting equipment and lighting controls.
4. Operation and maintenance manuals for each piece of lighting equipment. Required routine maintenance actions, cleaning and recommended relamping shall be clearly identified.
5. A schedule for inspecting and recalibrating all lighting controls.

C408.3.2.3 Report. A report of test results shall be provided and include the following:

1. Results of functional performance tests.
2. Disposition of deficiencies found during testing, including details of corrective measures used or proposed.

SECTION C409—CALCULATION OF THE HVAC TOTAL SYSTEM PERFORMANCE RATIO

C409.1 Applicability. Use of the *HVAC total system performance ratio* (TSPR) method shall comply with this section.

C409.2 Permitted uses. Only HVAC systems that serve building occupancies and uses in Table C409.4 and are not excluded by Section C409.2.1 shall be permitted to use the TSPR method.

C409.2.1 Systems not permitted. The following HVAC systems are not permitted to use Section C403.1, Item 3:

1. HVAC systems using:
   1.1. District heating water, chilled water or steam.
   1.2. Small-duct high-velocity air-cooled, space-constrained air-cooled, or single-package vertical air conditioner; single-package vertical heat pump; or double-duct air conditioner or double-duct heat pump, as defined in subpart F to 10 CFR Part 431.
   1.3. Packaged terminal air conditioners and packaged terminal heat pumps that have a cooling capacity greater than 12,000 Btu/h (3.5 kW).
   1.4. A common heating source serving both HVAC and *service water heating* equipment.
2. HVAC systems that provide recovered heat for *service water heating.*
3. HVAC systems not specified in Table C409.6.1.10.1.
4. HVAC systems specified in Table C409.6.1.10.1 with characteristics or parameters in Table C409.6.1.10.2(1), not identified as applicable to that HVAC system type.
5. HVAC systems with chilled water supplied by absorption chillers, heat recovery chillers, water-to-water heat pumps, air-to-water heat pumps, or a combination of air- and water-cooled chillers on the same chilled water loop.
6. HVAC systems served by heating water systems that include air-to-water or water-to-water heat pumps.
7. Underfloor air distribution and displacement ventilation HVAC systems.
8. Space-conditioning systems that do not include mechanical cooling.
9. HVAC systems serving laundry rooms, elevator rooms, mechanical rooms, electrical rooms, *data centers* and *computer rooms.*
10. *Buildings* or areas of medical office buildings required to use ASHRAE Standard 170.
11. *Buildings* or areas that are required by regulation to have continuous air-handling unit operation.
12. HVAC systems serving laboratories with fume hoods.
13. Locker rooms with more than two showers.
14. Natatoriums and rooms with saunas.
15. Restaurants and commercial kitchens with a total cooking capacity greater than 100,000 Btu/h (29 kW).
16. Areas of *buildings* with commercial refrigeration equipment exceeding 100 kW of power input.
17. Cafeterias and dining rooms

C409.3 HVAC TSPR compliance. HVAC systems permitted to use TSPR shall comply with Section C409.4 and the following:

1. HVAC systems shall comply with applicable requirements of Section C403 as follows:
   1.1. Air economizers shall meet the requirements of Sections C403.5.3.4 and C403.5.5.
   1.2. Variable-air-volume systems shall meet the requirements of Sections C403.6.5, C403.6.6 and C403.6.9.
   1.3. Hydronic systems shall meet the requirements of Section C403.4.4.
   1.4. Plants with multiple chillers or boilers shall meet the requirements of Section C403.4.5.
   1.5. Hydronic (water loop) heat pumps and water-cooled unitary air conditioners shall meet the requirements of Section C403.4.3.3.
   1.6. Cooling tower turndown shall meet the requirements of Section C403.11.4.
   1.7. Heating of unenclosed spaces shall meet the requirements of Section C403.14.1.
   1.8. Hot-gas bypass shall meet the requirements of Section C403.3.3.
   1.9. Systems shall meet the operable openings interlock requirements of Section C403.4.7. Refrigeration systems shall meet the requirements of Section C403.12.
2. Systems shall comply with the applicable provisions of Section C403 required by Table C407.2.

C409.4 Performance target. For HVAC systems serving uses or portions of uses listed in Section C409.2 that are not served by systems listed in Section C409.2.1, the *HVAC TSPR* of the *proposed design* shall be greater than or equal to the *HVAC TSPR* of the *standard reference design* divided by the mechanical performance factor (MPF) using Equation 4-35.

Equation 4-35 $TSPR_p > TSPR_r/MPF$

where:

$TSPR_p$ = HVAC TSPR of the proposed design calculated in accordance with Sections C409.4, C409.5 and C409.6.

$TSPR_r$ = HVAC TSPR of the reference building design calculated in accordance with Sections C409.4, C409.5 and C409.6.

MPF = Mechanical performance factor from Table C409.4 based on climate zone and building use type.

Equation 4-36 $MPF = (A_1 \times MPF_1 + A_2 \times MPF_2 + \ldots + A_n \times MPF_n)/(A_1 + A_2 + \ldots + A_n)$

where:

MPF_1, MPF_2 through MPF_n = Mechanical performance factors from Table C409.4 based on climate zone and building use types 1, 2 through *n*.

A_1, A_2 through A_n = Conditioned floor areas for building use types 1, 2 through *n*.

TABLE C409.4—MECHANICAL PERFORMANCE FACTORS

BUILDING USE	OCCUPANCY GROUP	CLIMATE ZONE																		
		0A	0B	1A	1B	2A	2B	3A	3B	3C	4A	4B	4C	5A	5B	5C	6A	6B	7	8
Office (all others)[a]	B	0.72	0.715	0.70	0.705	0.685	0.65	0.71	0.68	0.645	0.805	0.70	0.78	0.845	0.765	0.805	0.865	0.835	0.875	0.895
Office (large)[a]	B	0.83	0.83	0.84	0.84	0.79	0.82	0.72	0.81	0.77	0.67	0.76	0.63	0.71	0.72	0.63	0.73	0.71	0.71	0.71
Retail	M	0.60	0.57	0.50	0.55	0.46	0.46	0.43	0.51	0.40	0.45	0.57	0.68	0.46	0.68	0.67	0.50	0.45	0.44	0.38
Hotel/motel	R-1	0.62	0.62	0.63	0.63	0.62	0.68	0.61	0.71	0.73	0.45	0.59	0.52	0.38	0.47	0.51	0.35	0.38	0.31	0.26
Multi-family/ dormitory	R-2	0.64	0.63	0.67	0.63	0.65	0.64	0.59	0.72	0.55	0.53	0.50	0.44	0.54	0.47	0.38	0.55	0.50	0.51	0.47
School/ education and libraries	E (A-3)	0.82	0.81	0.80	0.79	0.75	0.72	0.71	0.72	0.67	0.73	0.72	0.68	0.82	0.73	0.61	0.89	0.80	0.83	0.77

a. Large-office conditioned floor area greater than 150,000 square feet or more than five stories.

C409.4.1 HVAC TSPR. *HVAC TSPR* is calculated according to Equation 4-37.

Equation 4-37 HVAC TSPR = heating and cooling load/building HVAC system energy

where:

Building HVAC system energy = Sum of the annual site energy consumption for heating, cooling, fans, energy recovery, pumps and heat rejection in thousands of Btu (kWh).

Heating and cooling load = Sum of the annual heating and cooling loads met by the building HVAC system in thousands of Btu (kWh).

C409.5 General. Projects shall use the procedures of this section when calculating compliance using HVAC total system performance ratio.

C409.5.1 Simulation program. Simulation tools used to calculate the *HVAC TSPR* of the *standard reference design* shall comply with the following:

1. The simulation program shall calculate the *HVAC TSPR* based only on the input for the *proposed design* and the requirements of Section C409. The calculation procedure shall not allow the user to directly modify the building component characteristics of the *standard reference design*.
2. Performance analysis tools shall meet the applicable subsections of Section C409 and be tested in accordance with ASHRAE Standard 140, except for Sections 7 and 8. The required tests shall include the *building thermal envelope* and fabric load test (Sections 5.2.1, 5.2.2 and 5.2.3), ground-coupled slab-on-grade analytical verification tests (Section 5.2.4), space-cooling equipment performance tests (Section 5.3), space-heating equipment performance tests (Section 5.4), and air-side HVAC equipment analytical verification test (Section 5.5), along with the associated reporting (Section 6).
3. The test results and modeler reports shall be publicly available and shall include the test results of the simulation programs and input files used for generating the results along with the results of the other simulation programs included in ASHRAE Standard 140, Annexes B8 and B16. The modeler report in ASHRAE Standard 140 Annex A2 Attachment A2.7 shall be completed for results exceeding the maximum or falling below the minimum of the reference values and for omitted results.
4. The simulation program shall have the ability to model part-load performance curves or other part-load adjustment methods based on manufacturer's part-load performance data for mechanical equipment.
5. The *code official* shall be permitted to approve specific software deemed to meet these requirements in accordance with Section C101.4.1.

C409.5.2 Climatic data. The simulation program shall perform the simulation using hourly values of climatic data for a full calendar year (8,760 hours) and shall reflect *approved* coincident hourly data for temperature, solar radiation, humidity and wind speed for the building location.

C409.5.3 Documentation. Documentation or web links to documentation conforming to the provisions of this section shall be provided to the *code official*.

C409.5.3.1 Compliance report. Building permit submittals shall include:

1. A report produced by the simulation software that includes the following:
   1.1. Address of the *building*.
   1.2. Name of the individual completing the compliance report.
   1.3. Name and version of the compliance software tool.
   1.4. The dimensions, floor heights and number of floors for each *thermal block*.
   1.5. By thermal block, the *U-factor*, *C-factor* or *F-factor* for each simulated opaque envelope component and the *U-factor* and SHGC for each *fenestration* component.
   1.6. By *thermal block* or by surface for each *thermal block*, the *fenestration* area.
   1.7. By *thermal block*, a list of the HVAC equipment simulated in the *proposed design*, including the equipment type, fuel type, equipment efficiencies and system controls.
   1.8. Annual site HVAC energy use by end use for the proposed and baseline building.
   1.9. Annual sum of heating and cooling loads for the baseline building.
   1.10. The *HVAC TSPR* for both the *standard reference design* and the *proposed design*.
2. A mapping of the actual *building* HVAC component characteristics and those simulated in the *proposed design* showing how individual pieces of HVAC equipment identified in Item 1 have been combined into average inputs as required by Section C409.6.1.10, including:
   2.1. Fans.
   2.2. Hydronic pumps.
   2.3. Air handlers.
   2.4. Packaged cooling equipment.
   2.5. Furnaces.
   2.6. Heat pumps.
   2.7. Boilers.
   2.8. Chillers.
   2.9. Heat rejection equipment (open- and closed-circuit cooling towers, dry coolers).
   2.10. Electric resistance coils.
   2.11. Condensing units.
   2.12. Motors for fans and pumps.
   2.13. Energy recovery devices.
3. For each piece of equipment identified in Item 2, include the following, as applicable:
   3.1. Equipment name or tag consistent with that found on the design documents.
   3.2. Rated efficiency level.
   3.3. Rated capacity.
   3.4. Where not provided by the simulation program report in Item 1, documentation of the calculation of any weighted equipment efficiencies input into the program.
   3.5. Electrical input power for fans and pumps (before any speed or frequency control device) at design condition and calculation of input value (W/cfm or W/gpm) or W/gpm (W/Lps).
4. Floor plan of the *building*, identifying:
   4.1. How portions of the *buildings* are assigned to the simulated thermal blocks.
   4.2. Areas of the *building* that are not covered under the requirements of Section C403.1.1.

C409.6 Calculation procedures. Except as specified by this section, the *standard reference design* and *proposed design* shall be configured and analyzed using identical methods and techniques.

C409.6.1 Simulation of the proposed building design. The *proposed design* shall be configured and analyzed as specified in this section.

C409.6.1.1 Thermal block geometry. The geometry of *buildings* shall be configured using one or more thermal blocks. Each *thermal block* shall define attributes, including *thermal block* dimensions, number of floors, floor-to-floor height and floor-to-ceiling height. Simulation software may allow the use of simplified shapes (such as rectangle, L-shaped, H-shaped, U-shaped or T-shaped) to represent thermal blocks. Where actual building shape does not match these predefined shapes, simplifications are permitted, provided that the following requirements are met:

1. The *conditioned floor area* and volume of each *thermal block* shall match the *proposed design* within 10 percent.
2. The area of each exterior envelope component from Table C402.1.4 is accounted for within 10 percent of the actual design.

3. The area of vertical *fenestration* and skylights is accounted for within 10 percent of the actual design.
4. The orientation of each component in Items 2 and 3 is accounted for within 45 degrees of the actual design.

The creation of additional *thermal blocks* may be necessary to meet these requirements. A more complex zoning of the *building* shall be allowed where all *thermal zones* in the reference and proposed models are the same, and rules related to thermal block geometry and HVAC system assignment to *thermal blocks* are met with appropriate assignment to thermal zones.

Exception: Portions of the *building* that are unconditioned or served by systems not covered by the requirements of Section C403.1.1 shall be omitted.

C409.6.1.1.1 Number of thermal blocks. One or more *thermal blocks* may be required per *building* based on the following restrictions:

1. Each *thermal block* shall have not more than one building use.
2. Each *thermal block* shall be served by not more than one type of HVAC system. A single block shall be created for each unique HVAC system and building use combination, and multiple HVAC units or components of the same type shall be combined in accordance with Section C409.6.1.10.2.
3. Each *thermal block* shall have not more than a single defined floor-to-floor or floor-to-ceiling height. Where floor heights differ by more than 2 feet, separate thermal blocks shall be created.
4. Each block shall include either above-grade or below-grade stories. For *buildings* with both above-grade and below-grade stories, separate blocks shall be created for each. Where blocks have *exterior walls* partially below grade, if greater than 50 percent of the exterior wall surface is below grade, then simulate the block as below grade; otherwise, simulate as above grade.
5. Where a block includes multiple stories, separate blocks shall be created, if needed, to comply with both the following *fenestration* modeling requirements:
   5.1. The product of the *proposed design U-factor* times the area of windows ($U \times A$) on a given story of each facade shall not differ by more than 15 percent of the average $U \times A$ for that modeled facade in each block.
   5.2. The product of the *proposed design* SHGC times the area of windows (SHGC × A) on a given story of each facade shall not differ by more than 15 percent of the average SHGC × A for that modeled facade in each block.
6. For a building model with multiple blocks, the blocks shall be configured together to have the same adjacencies as the actual building design.

C409.6.1.2 Thermal zoning. Each story in a *thermal block* shall be modeled as follows:

1. Below-grade stories shall be modeled as a single thermal zone.
2. Where any facade in the block is less than 45 feet (13.7 m) in length, it shall be modeled as a single thermal zone per story.
3. For stories not covered by Item 1 or Item 2, each story shall be modeled with five thermal zones. A perimeter zone shall be created, extending from each facade to a depth of 15 feet (4572 mm). Where facades intersect, the zone boundary shall be formed by a 45-degree angle with the two facades. The remaining area of each story shall be modeled as a core zone with no *exterior walls*.

C409.6.1.2.1 Core and shell, build-out and future system construction analysis. Where the building permit applies to only a portion of the HVAC system in a *building* and the remaining components will be designed under a future building permit or were previously installed, such components shall be modeled as follows:

1. Blocks including existing or future HVAC *zone* served by independent systems and not part of the construction project shall not be modeled.
2. Where the HVAC *zones* that do not include complete HVAC systems in the permit are intended to receive HVAC services from systems that are part of the construction project, their proposed zonal systems shall be modeled with equipment that meets, but does not exceed, the requirements of Section C403.
3. Where existing HVAC systems serve permitted *zone* equipment, the existing systems shall be modeled with equipment matching the manufacturer's stated efficiency for the installed equipment or equipment that meets, but does not exceed, the requirements of Section C403.
4. Where the central plant heating and cooling equipment is completely replaced and HVAC *zones* with existing systems receive HVAC services from systems in the permit, their proposed zonal systems shall be modeled with equipment that meets, but does not exceed, the requirements of Section C403.

C409.6.1.3 Occupancy. Building occupancies modeled in the *standard reference design* and the *proposed design* shall comply with the following requirements.

C409.6.1.3.1 Occupancy type. The occupancy type for each *thermal block* shall be consistent with the building occupancy and uses specified in Table C409.4. Portions of the building occupancy and uses other than those specified in Table C409.4 shall not be included in the simulation. Surfaces adjacent to such excluded *building* portions shall be modeled as adiabatic in the simulation program.

C409.6.1.3.2 Occupancy schedule, density and heat gain. The occupant density, heat gain and schedule shall be for multifamily, offices, retail spaces, libraries, hotels/motels or schools as specified by ANSI/ASHRAE/IES 90.1, Normative Appendix C.

C409.6.1.4 Building thermal envelope components. *Building thermal envelope* components modeled in the *standard reference design* and the *proposed design* shall comply with the requirements of this section.

C409.6.1.4.1 Roofs. The roof *U-factor* and area shall be modeled as in the *proposed design*. If different roof thermal properties are present in a single *thermal block*, an area-weighted *U-factor* shall be used. Roofs shall be modeled with insulation above a steel roof deck, with a solar reflectance of 0.25 and an *emittance* of 0.90.

Exception: For Climate Zones 0, 1, 2 and 3, solar reflectance and *emittance* shall be as specified in Section C402.4 and Table C402.4.

C409.6.1.4.2 Above-grade walls. The *U-factor* and area of *above-grade walls* shall be modeled as in the *proposed design*. If different wall constructions exist on the facade of a thermal block, an area-weighted *U-factor* shall be used. Walls will be modeled as steel-frame construction.

C409.6.1.4.3 Below-grade walls. The *C-factor* and area of *below-grade walls* shall be modeled as in the *proposed design*. If different *below-grade wall* constructions exist in a *thermal block*, an area-weighted *C-factor* shall be used.

C409.6.1.4.4 Above-grade exterior floors. The *U-factor* and area of floors shall be modeled as in the *proposed design*. If different floor constructions exist in the *thermal block*, an area-weighted *U-factor* shall be used. Exterior floors shall be modeled as steel frame.

C409.6.1.4.5 Slab-on-grade floors. The *F-factor* and perimeter of slab-on-grade floors shall be modeled as in the *proposed design*. If different slab-on-grade floor constructions exist in a *thermal block*, a perimeter-weighted *F-factor* shall be used.

C409.6.1.4.6 Vertical fenestration. The window area and area-weighted *U-factor* and SHGC shall be modeled for each facade based on the *proposed design*. Each exterior surface in a *thermal block* must comply with Section C409.6.1.1.1, Item 5. Windows shall be combined into a single window centered on each facade based on the area and sill height input by the user. Where different *U-values*, SHGC or sill heights exist on a single facade in a block, the area-weighted average for each shall be input by the user.

C409.6.1.4.7 Skylights. The skylight area and area-weighted *U-factor* and SHGC shall be modeled for each roof based on the *proposed design*. Skylights shall be combined into a single skylight centered on the roof of each zone based on the area input by the user.

C409.6.1.4.8 Exterior shading. Permanent window overhangs shall be modeled. Where windows with and without overhangs or windows with different overhang projection factors exist on a facade, window width-weighted projection factors shall be input by the user as follows:

Equation 4-38 $P_{avg} = (A_1 \times L_{o1} + A_2 \times L_{o2} \ldots + A_n \times L_{on})/(L_{w1} + L_{w2} \ldots + L_{wn})$

where:

P_{avg} = Average overhang projection modeled in the simulation tool.

A = Distance measured horizontally from the farthest continuous extremity of any overhang, eave or permanently attached shading device to the vertical surface of the glazing.

L_o = Length off the overhang.

L_w = Length of the window.

C409.6.1.5 Lighting. Interior lighting power density shall be equal to the allowance in Table C405.3.2(1) for multifamily buildings, offices, retail spaces, libraries or schools. The lighting schedule shall be for multifamily buildings, offices, retail spaces, libraries or schools as specified by ANSI/ASHRAE/IES 90.1, Normative Appendix C. The impact of lighting controls is assumed to be captured by the lighting schedule and no explicit controls shall be modeled. Exterior lighting shall not be modeled.

C409.6.1.6 Miscellaneous equipment. The miscellaneous equipment schedule and power shall be for multifamily buildings, offices, retail spaces, libraries or schools as specified by ANSI/ASHRAE/IES 90.1, Normative Appendix C. The impact of miscellaneous equipment controls is assumed to be captured by the equipment schedule and no explicit controls shall be modeled.

Exceptions:

1. Multiple-family dwelling units shall have a miscellaneous load density of 0.42 watts per square foot.
2. Multiple-family *common areas* shall have a miscellaneous load density of 0 watts per square foot.

C409.6.1.7 Elevators. Elevators shall not be modeled.

C409.6.1.8 Service water heating equipment. *Service water heating* shall not be modeled.

C409.6.1.9 On-site renewable energy systems. On-site renewable energy systems shall not be modeled.

C409.6.1.10 HVAC equipment. Where proposed or where reference system parameters are not specified in Section C409, HVAC systems shall be modeled to meet the minimum requirements of Section C403.

C409.6.1.10.1 Supported HVAC systems. At a minimum, the HVAC systems shown in Table C409.6.1.10.1 shall be supported by the simulation program.

TABLE C409.6.1.10.1—PROPOSED BUILDING HVAC SYSTEMS SUPPORTED BY HVAC TSPR SIMULATION SOFTWARE	
SYSTEM NO.	**SYSTEM NAME**
1	Packaged terminal air conditioner (with electric or hydronic heat)
2	Packaged terminal air heat pump
3	Packaged single-zone gas furnace[a] and/or air-cooled air conditioner (includes split systems)[b]
4	Packaged single-zone heat pump (air to air only)(includes split systems[b] and electric or gas supplemental heat)
5	Variable refrigerant flow (air cooled only)
6	Four pipe fan coil
7	Water-source heat pump (water loop), water-source *variable refrigerant flow system* or water-source air conditioner
8	Ground source heat pump
9	Packaged variable air volume (DX cooling)[a]
10	Variable air volume (hydronic cooling)[a]
11	Variable air volume with fan-powered terminal units
12	Dedicated outdoor air system (in conjunction with systems 1–8)

a. Reheat or primary heat may be electric, hydronic or gas furnace.
b. Condensing units with DX air handlers are modeled as package furnaces with air conditioners or heat pumps.

C409.6.1.10.2 Proposed building HVAC system simulation. The HVAC systems shall be modeled as in the *proposed design* at design conditions unless otherwise stated, with clarifications and simplifications as described in Tables C409.6.1.10.2(1) and C409.6.1.10.2(2). System parameters not described in the following sections shall be simulated to meet the minimum requirements of Section C403. All *zones* within a thermal block shall be served by the same HVAC system type as described in Section C409.6.1.1.1, Item 2. Heat loss from *ducts* and pipes shall not be modeled. The proposed building system parameters in Table C409.6.1.10.2(1) are based on input of full-load equipment efficiencies with adjustments using part-load curves integrated into the simulation program. Where other approaches to part-load adjustments are used, it is permitted for specific input parameters to vary. The simulation program shall model part-load HVAC equipment performance using one of the following:

1. Full-load efficiency adjusted for fan power input that is modeled separately and typical part-load performance adjustments for the proposed equipment.
2. Part-load adjustments based on input of both full-load and part-load metrics.
3. Equipment-specific adjustments based on performance data provided by the equipment manufacturer for the proposed equipment.

Where multiple system components serve a thermal block, average values weighted by the appropriate metric as described in this section shall be used.

1. Where multiple fan systems serve a single thermal block, fan power shall be based on a weighted average using the design supply air (cfm).
2. Where multiple cooling systems serve a single thermal block, the coefficient of performance (COP) shall be based on a weighted average using cooling capacity. Direct expansion (DX) coils shall be entered as multistage if more than 50 percent of coil capacity serving the thermal block is multistage with staged controls.
3. Where multiple heating systems serve a single thermal block, thermal efficiency or heating COP shall be based on a weighted average using heating capacity.
4. Where multiple boilers or chillers serve a heating water or chilled water loop, efficiency shall be based on a weighted average for using heating or cooling capacity.
5. Where multiple cooling towers serving a condenser water loop are combined, the cooling tower efficiency, cooling tower design approach and design range are based on a weighted average of the design water flow rate through each cooling tower.
6. Where multiple pumps serve a heating water, chilled water or condenser water loop, pump power shall be based on a weighted average for using design water flow rate.
7. Where multiple system types with and without economizers are combined, the economizer maximum outside air fraction of the combined system shall be based on the weighted average of 100 percent supply air for systems with economizers and design outdoor air for systems without economizers.
8. Multiple systems with and without ERVs cannot be combined.
9. Systems with and without supply-air temperature reset controls cannot be combined.
10. Systems with different fan controls (constant volume, multispeed or VAV) for supply fans cannot be combined.

TABLE C409.6.1.10.2(1)—PROPOSED BUILDING SYSTEM PARAMETERS

CATEGORY	PARAMETER	FIXED OR USER DEFINED	REQUIRED	APPLICABLE SYSTEMS
HVAC system type	System type	User defined	Selected from Table C409.6.1.10.1	All
System sizing	Design day information	Fixed	99.6% heating design and 1% dry-bulb and 1% wet-bulb cooling design	All
	Zone coil capacity	Fixed	Sizing factors used are 1.25 for heating equipment and 1.15 for cooling equipment	All
	Supply airflow	Fixed	Based on a supply-air-to-room-air temperature setpoint difference of 20°F (11.2°C)	1–11
		Fixed	Equal to required outdoor air ventilation	12
Outdoor ventilation air	Portion of supply air with proposed filter ≥ MERV 13	User defined	Percentage of supply airflow subject to higher filtration (adjusts baseline fan power higher; prorated)	All
	Outdoor ventilation airflow rate	Fixed	As specified in ANSI/ASHRAE/IES 90.1, Normative Appendix C; adjusted for proposed DCV control	All
	Outdoor ventilation supply airflow rate adjustments	Fixed	Based on ASHRAE 62.1 Section 6.2.4.3, system ventilation efficiency (E_v) is 0.75	9–11
		Fixed	System ventilation efficiency (E_v) is 1.0	1–8, 12
		Fixed	Basis is 1.0 zone air distribution effectiveness	All
System operation	Space temperature setpoints	Fixed	As specified in ANSI/ASHRAE/IES 90.1, Normative Appendix C, except: • Multiple-family, which shall use 68°F heating and 76°F cooling setpoints. • Hotel/motel setpoints, which shall be 70°F heating and 72°F cooling.	1–11
	Fan operation—occupied	User defined	Runs continuously during occupied hours or cycles to meet load. Multispeed fans reduce airflow related to thermal loads.	1–11
	Fan operation—occupied	Fixed	Fan runs continuously during occupied hours	12
	Fan operation—night cycle	Fixed	Fan cycles on to meet setback temperatures	1–11
Packaged equipment efficiency	DX cooling efficiency	User defined	Cooling COP without fan energy calculated in accordance with Section C409.6.1.10.2	1, 2, 3, 4, 5,7, 8, 9, 11,12
	DX coil number of stages	User defined	Single stage or multistage	3, 4, 9, 10, 11, 12
	Heat pump efficiency	User defined	Heating COP without fan energy calculated in accordance with Section C409.6.1.10.2	2, 4, 5, 7, 8, 12
	Furnace efficiency	User defined	Furnace thermal efficiency	3, 9, 11, 12
Heat pump supplemental heat	Heat source	User defined	Electric resistance or gas furnace	2, 4, 7, 8, 12
	Control	Fixed	Supplemental electric heat locked out above 40°F OAT. Runs as needed in conjunction with compressor between 40°F and 0°F. Gas heat operates in place of the heat pump when the heat pump cannot meet load.	2, 4, 7, 8, 12
System fan power and controls	Part-load fan controls[a]: • Constant volume. • Two speed or three speed. • VAV.	User defined	Static pressure reset included for VAV	1–8 (CAV, two or three speed), 9, 10, 11 (VAV), 12 (CAV and VAV)
	Design fan power (W/cfm)	User defined	Input electric power for all fans required to operate at fan system design conditions divided by the supply airflow rate. This is a wire-to-air value, including all drive, motor efficiency and other losses.	All

TABLE C409.6.1.10.2(1)—PROPOSED BUILDING SYSTEM PARAMETERS—continued

CATEGORY	PARAMETER	FIXED OR USER DEFINED	REQUIRED	APPLICABLE SYSTEMS
System fan power and controls—continued	Low-speed and medium-speed fan power	User defined	Low-speed input electric power for all fans required to operate at low-speed conditions divided by the low-speed supply airflow rate. This is a wire-to-air value, including all drive, motor efficiency and other losses. Also provide medium-speed values for three-speed fans.	1-8
Variable air volume systems	Supply air temperature (SAT) controls	User defined	If not SAT reset, then constant at 55°F. Options for reset based on OAT or warmest zone. If warmest zone, then the user can specify the minimum and maximum temperatures. If OAT reset, SAT is reset higher to 60°F at an outdoor low of 50°F. SAT is 55°F at an outdoor high of 70°F.	9, 10, 11
	Minimum terminal unit airflow percentage	User defined	Average minimum terminal unit airflow percentage for thermal block weighted by cfm or minimum required for outdoor air ventilation, whichever is higher.	9, 10, 11
	Terminal unit heating source	User defined	Electric or hydronic	9, 10, 11
	Dual setpoint minimum VAV damper position	User defined	Heating maximum airflow fraction	9, 10
	Fan-powered terminal unit (FPTU) type	User defined	Series or parallel FPTU	11
	Parallel FPTU fan	Fixed	Sized for 50% peak primary air at 0.35 W/cfm	11
	Series FPTU fan	Fixed	Sized for 50% peak primary air at 0.35 W/cfm	11
Economizer	Economizer presence	User defined	Yes or no	3, 4, 5, 6, 9, 10, 11
	Economizer control type	Fixed	Lockout on differential db temperature (OAT > RAT) in Climate Zones 5A, 6A, all B & C; fixed enthalpy > 28 Btu/lb (47kJ/kg) or fixed db OAT > 75°F (24°C) in Climate Zones 0A through 4A	3, 4, 5, 6, 9, 10, 11
Energy recovery	Sensible effectiveness	User defined	Heat exchanger sensible effectiveness at design heating and cooling conditions	3, 4, 9, 10, 11, 12
	Latent effectiveness	User defined	Heat exchanger latent effectiveness at design heating and cooling conditions	3, 4, 9, 10, 11, 12
	Economizer bypass	User defined	If ERV is bypassed or wheel rotation is slowed during economizer conditions (yes/no)	3, 4, 9, 10, 11, 12
	Economizer bypass active	Fixed	If there is a bypass, it will be active between 45°F and 75°F outside air temperature	3, 4, 9, 10, 11, 12
	Bypass SAT setpoint	User defined	If bypass, target SAT	3, 4, 9, 10, 11, 12
	Fan power reduction during bypass (W/cfm)	User defined	If ERV system includes bypass, static pressure setpoint and variable speed fan, fan power can be reduced during economizer conditions	3, 4, 9, 10, 11, 12
Demand control ventilation (DCV)	DCV application on/off	User defined	Percent of thermal block floor area under occupied standby controls, on/off only with occupancy sensor and no variable control	3, 4, 9, 10, 11, 12
	DCV application CO_2	User defined	Percentage of thermal block floor area under variable DCV control (CO_2); may include both variable and on/off controls	3, 4, 9, 10, 11, 12
DOAS	DOAS fan power W/cfm	User defined	Fan electrical input power in W/cfm of supply airflow	12
	DOAS supplemental heating and cooling	User defined	Heating source, cooling source, energy recovery and respective efficiencies	12
	Maximum SAT setpoint (cooling)	User defined	SAT setpoint if DOAS includes supplemental cooling	12
	Minimum SAT setpoint (heating)	User defined	SAT setpoint if DOAS includes supplemental heating	12

TABLE C409.6.1.10.2(1)—PROPOSED BUILDING SYSTEM PARAMETERS—continued				
CATEGORY	PARAMETER	FIXED OR USER DEFINED	REQUIRED	APPLICABLE SYSTEMS
Heating plant	Boiler efficiency	User defined	Boiler thermal efficiency	1, 6, 7, 9, 10, 11, 12
	Heating water loop configuration	User defined	Constant flow primary only; variable flow primary only; constant flow primary/variable flow secondary; variable flow primary and secondary	1, 6, 7, 9, 10, 11, 12
	Heating water primary pump power (W/gpm)	User defined	Heating water primary pump input W/gpm heating water flow	1, 6, 7, 9, 10, 11, 12
	Heating water secondary pump power (W/gpm)	User defined	Heating water secondary pump input W/gpm heating water flow (if primary/secondary)	1, 6, 7, 9, 10, 11, 12
	Heating water loop temperature	User defined	Heating water supply and return temperatures, °F	1, 6, 9, 10,11
	Heating water loop supply temperature reset	Fixed	Reset HWS by 27.3% of design delta-T (HWS-70°F space heating temperature setpoint) between 20°F and 50°F OAT	1, 6, 7, 9, 10, 11, 12
	Boiler type	Fixed	Noncondensing boiler where input thermal efficiency is less than 86%; condensing boiler otherwise	1, 6, 7, 9, 10, 11, 12
Chilled water plant	Chiller compressor type	User defined	Screw/scroll, centrifugal or reciprocating	6, 10, 11, 12
	Chiller condenser type	User defined	Air cooled or water cooled	6, 10, 11, 12
	Chiller full-load efficiency	User defined	Chiller COP	6, 10, 11, 12
	Chilled water loop configuration	User defined	Variable flow primary only, constant flow primary/variable flow secondary, variable flow primary and secondary	6, 10, 11,12
	Chilled water primary pump power (W/gpm)	User defined	Primary pump input W/gpm chilled water flow	6, 10, 11,12
	Chilled water secondary pump power (W/gpm)	User defined	Secondary pump input W/gpm chilled water flow (if primary/secondary)	6, 10, 11,12
	Chilled water temperature reset included	User defined	Yes/no	6, 10, 11,12
	Chilled water temperature reset schedule (if included)	Fixed	Outdoor air reset: CHW supply temperature of 44°F at 80°F (26.7°C) outdoor air db temperature and above, CHW supply temperature of 54°F at 60°F outdoor air db temperature and below, ramped linearly between	6, 10, 11,12
	Condenser water pump power (W/gpm)	User defined	Pump input W/gpm condenser water flow	6, 7, 8, 10, 11, 12
	Condenser water pump control	User defined	Constant speed or variable speed	6, 7, 8, 10, 11,12
	Heat rejection equipment efficiency	User defined	Gpm/hp tower fan	6, 7, 10, 11, 12
	Heat rejection fan control	User defined	Constant or variable speed	6, 7, 10, 11, 12
	Heat rejection approach and range	User defined	Design cooling tower approach and range temperature	6, 7, 10, 11, 12
Heat pump loop	Loop flow and heat pump control valve	Fixed	Two-position valve with VFD on pump; loop flow at 3 gpm/ton	7, 8
	Heat pump loop minimum and maximum temperature control	User defined	User input: restrict to minimum 20°F and maximum 40°F temperature difference	7
GLHP well field	—	Fixed	Bore depth = 250 ft Bore length 200 ft/ton for the greater of cooling or heating load Bore spacing = 15 ft Bore diameter = 5 in ¾" (19 mm) polyethylene pipe Ground and grout conductivity = 4.8 Btu × in/h × ft^2 × °F	8

For SI: 1 inch = 25.4 mm, 1 foot = 304.8 mm, °C = (°F – 32)/1.8, 1 British thermal unit per hour = 0.2931 W, 1 British thermal unit per pound = 2.33 kJ/kg, 1 cubic foot per minute = 0.4719 L/s, 1 cubic foot per minute/foot = 47.82 W, COP = (Btu/h × hp)/(2,550.7), 1 gallon per minute = 3.79 L/m.

CHW =Chilled Water, db = dry bulb, DOAS = Dedicated Outdoor Air System, GLHP = Ground Loop Heat Pump, HWS = Hot Water Supply, OAT = Outdoor Air Temperature, SAT = Supply Air Temperature, VFD = Variable Frequency Drive, wb = wet bulb.

a. Part-load fan power and pump power modified in accordance with Table C409.6.1.10.2(2).

TABLE C409.6.1.10.2(2)—FAN AND PUMP POWER CURVE COEFFICIENTS			
EQUATION TERM	FAN POWER COEFFICIENTS	PUMP POWER COEFFICIENTS	
	VSD + SP Reset	Ride Pump Curve	VSD + DP/Valve Reset
b	0.0408	0	0
x	0.088	3.2485	0.0205
x^2	-0.0729	-4.7443	0.4101
x^3	0.9437	2.5295	0.5753

C409.6.1.10.3 Demand control ventilation. Demand control ventilation (DCV) shall be modeled using a simplified approach that adjusts the design outdoor supply airflow rate based on the floor area of the *building* that is covered by DCV. The simplified method shall accommodate both variable DCV and on/off DCV, giving on/off DCV one third of the effective floor control area of the variable DCV. Outdoor air reduction coefficients shall be as stated in Table C409.6.1.10.3.

Exception: On/off DCV shall receive full effective area adjustment for R-1 and R-2 occupancies.

TABLE C409.6.1.10.3—DCV OUTDOOR AIR REDUCTION CURVE COEFFICIENTS				
EQUATION TERM	DCV OSA REDUCTION (*y*) AS A FUNCTION OF EFFECTIVE DCV CONTROL FLOOR AREA (*x*)			
	Office	School	Hotel, Motel, Multiple-Family, Dormitory	Retail
b	0	0	0	0
x	0.4053	0.2676	0.5882	0.4623
x^2	-0.8489	0.7753	-1.0712	-0.848
x^3	1.0092	-1.5165	1.3565	1.1925
x^4	-0.4168	0.7136	-0.6379	-0.5895

OSA = Outside Air.

C409.6.2 Simulation of the standard reference design. The *standard reference design* shall be configured and analyzed as specified in this section.

C409.6.2.1 Utility rates. Same as the *proposed design*.

C409.6.2.2 Thermal blocks. Same as the *proposed design*.

C409.6.2.3 Thermal zoning. Same as the *proposed design*.

C409.6.2.4 Occupancy type, schedule, density and heat gain. Same as the *proposed design*.

C409.6.2.5 Envelope components. Same as the *proposed design*

C409.6.2.6 Lighting. Same as the *proposed design*.

C409.6.2.7 Miscellaneous equipment. Same as the *proposed design*.

C409.6.2.8 Elevators. Not modeled. Same as the *proposed design*.

C409.6.2.9 Service water heating equipment. Not modeled. Same as the *proposed design*.

C409.6.2.10 On-site renewable energy systems. Not modeled. Same as the *proposed design*.

C409.6.2.11 HVAC equipment. The reference building design HVAC equipment consists of separate space conditioning systems as described in Tables C409.6.2.11(1) through C409.6.2.11(3) for the appropriate building use types. In these tables, "warm" refers to Climate Zones 0 through 2 and 3A, and "cold" refers to Climate Zones 3B, 3C and 4 through 8.

TABLE C409.6.2.11(1)—REFERENCE BUILDING DESIGN HVAC COMPLEX SYSTEMS

BUILDING TYPE PARAMETER	BUILDING TYPE			
	Large Office (warm)	**Large Office (cold)**	**School (warm)**	**School (cold)**
System type	VAV/RH Water-cooled chiller Electric reheat (PIU)	VAV/RH Water-cooled chiller Gas boiler	VAV/RH Water-cooled chiller Electric reheat (PIU)	VAV/RH Water-cooled chiller Gas boiler
Fan control	VSD (no SP reset)	VSD (no SP reset)	VSD (no SP reset)	VSD (no SP reset)
Main fan power [W/cfm (W × s/L)] proposed ≥ MERV 13	1.165 (2.468)	1.165 (2.468)	1.165 (2.468)	1.165 (2.468)
Main fan power [W/cfm (W × s/L)] proposed < MERV 13	1.066 (2.259)	1.066 (2.259)	1.066 (2.259)	1.066 (2.259)
Zonal fan power [W/cfm (W × s/L)]	0.35 (0.75)	NA	0.35 (0.75)	NA
Minimum zone airflow fraction	1.5 × Voz	1.5 × Voz	1.2 × Voz	1.2 × Voz
Heat/cool sizing factor	1.25/1.15	1.25/1.15	1.25/1.15	1.25/1.15
Outdoor air economizer	No	Yes except 4A	No	Yes except 4A
Occupied OSA (= proposed)	Sum(Voz)/0.75	Sum(Voz)/0.75	Sum(Voz)/0.65	Sum(Voz)/0.65
Energy recovery ventilator efficiency ERR; ERV bypass SAT setpoint	NA	NA	50% No bypass	50% 60°F except 4A
DCV	No	No	No	No
Cooling source	(2) Water-cooled centrifugal chillers	(2) Water-cooled centrifugal chillers	(2) Water-cooled screw chillers	(2) Water-cooled screw chillers
Cooling COP (net of fan)	Path B for profile	Path B for profile	Path B for profile	Path B for profile
Heating source (reheat)	Electric resistance	Gas boiler	Electric resistance	Gas boiler
Furnace or boiler efficiency	1.0	75% E_t	1.0	80% E_t
Condenser heat rejection	Axial fan open circuit cooling tower			
Cooling tower efficiency [gpm/fan hp (L/s × fan kW)]	38.2	38.2	38.2	38.2
Tower turndown [> 300 ton (1060 kW)]	50%	50%	50%	50%
Pump (constant flow/variable flow)	Constant flow; 10°F (5.6°C) range	Constant flow; 10°F (5.6°C) range	Constant flow; 10°F (5.6°C) range	Constant flow; 10°F (5.6°C) range
Tower approach	25.72 – (0.24 × wb), where wb is the 0.4% evaporation design wet-bulb temperature (°F)			
Cooling condenser pump power [W/gpm (W × s/L)]	19 (300)	19 (300)	19 (300)	19 (300)
Cooling primary pump power [W/gpm (W × s/L)]	9 (142)	9 (142)	9 (142)	9 (142)
Cooling secondary pump power [W/gpm (W × s/L)]	13 (205)	13 (205)	13 (205)	13 (205)
Cooling coil chilled water delta-T, °F (°C)	12 (6.7)	12 (6.7)	12 (6.7)	12 (6.7)
Design chilled water supply temperature, °F (°C)	44 (6.7)	44 (6.7)	44 (6.7)	44 (6.7)
Chilled water supply temperature (CHWST) reset setpoint vs. outside air temperature (OAT), °F (°C)	CHWST: 44-54/ OAT 80-60 (6.7-12.2/26.7-15.6)	CHWST: 44-54/ OAT 80-60 (6.7-12.2/26.7-15.6)	CHWST: 44-54/ OAT 80-60 (6.7-12.2/26.7-15.6)	CHWST: 44-54/ OAT 80-60 (6.7-12.2/26.7-15.6)
CHW cooling loop pumping control	2-way valves & pump VSD	2-way valves & pump VSD	2-way valves & pump VSD	2-way valves & pump VSD
Heating pump power [W/gpm (W × s/L)]	16.1 (254)	16.1 (254)	19 (300)	19 (300)
Heating oil HW dT, °F (°C)	50 (10)	50 (10)	50 (10)	50 (10)
Design hot water supply temperature (HWST), °F (°C)	180 (82.2)	180 (82.2)	180 (82.2)	180 (82.2)

TABLE C409.6.2.11(1)—REFERENCE BUILDING DESIGN HVAC COMPLEX SYSTEMS—continued				
BUILDING TYPE PARAMETER	**BUILDING TYPE**			
	Large Office (warm)	**Large Office (cold)**	**School (warm)**	**School (cold)**
HWST reset setpoint vs. OAT, °F (°C)	HWST: 180-150/ OAT 20-50 (82-65.6/-6.7-10)	HWST: 180-150/ OAT 20-50 (82-65.6/-6.7-10)	HWST: 180-150/ OAT 20-50 (82-65.6/-6.7-10)	HWST: 180-150/ OAT 20-50 (82-65.6/-6.7-10)
Heat loop pumping control	2-way valves & pump VSD	2-way valves & pump VSD	2-way valves & pump VSD	2-way valves & pump VSD

For SI: °C = (°F – 32)/1.8, 1 hp = 0.746 kW, 1 ton = 3.517 kW.
CHW = Chilled Water, ERR = Enthalpy Recovery Ratio, NA = Not Applicable, OSA = Outside Air, PIU = Parallel Powered Induction Unit, RH = Relative Humidity, SP = Static Pressure, Voz = Outdoor airflow to the zone, VSD = Variable Speed Drive.

TABLE C409.6.2.11(2)—TSPR REFERENCE BUILDING DESIGN HVAC SIMPLE SYSTEMS						
BUILDING TYPE PARAMETER	**BUILDING TYPE**					
	Medium Office (warm)	**Medium Office (cold)**	**Small Office (warm)**	**Small Office (cold)**	**Retail (warm)**	**Retail (cold)**
System type	Package VAV—electric reheat	Package VAV—hydronic reheat	PSZ-HP	PSZ-AC	PSZ-HP	PSZ-AC
Fan control	VSD (no SP reset)	VSD (no SP reset)	Constant volume	Constant volume	Constant volume	Constant volume
Main fan power [W/cfm (W × s/L)] proposed ≥ MERV 13	1.285 (2.723)	1.285 (2.723)	0.916 (1.941)	0.916 (1.941)	0.899 (1.905)	0.899 (1.905)
Main fan power [W/cfm (W × s/L)] proposed < MERV 13	1.176 (2.492)	1.176 (2.492)	0.850 (1.808)	0.850 (1.808)	0.835 (1.801)	0.835 (1.801)
Zonal fan power [W/cfm (W × s/L)]	0.35 (0.75)	NA	NA	NA	NA	NA
Minimum zone airflow fraction	30%	30%	NA	NA	NA	NA
Heat/cool sizing factor	1.25/1.15	1.25/1.15	1.25/1.15	1.25/1.15	1.25/1.15	1.25/1.15
Supplemental heating availability	NA	NA	< 40°F OAT	NA	< 40°F OAT	NA
Outdoor air economizer	No	Yes except 4A	No	Yes except 4A	No	Yes except 4A
Occupied OSA source	Packaged unit, occupied damper, all building use types					
Energy recovery ventilator	No	No	No	No	No	No
DCV	No	No	No	No	No	No
Cooling source	DX, multistage	DX, multistage	DX, 1 stage (heat pump)	DX, single stage	DX, 1 stage (heat pump)	DX, single stage
Cooling COP (net of fan)	3.40	3.40	3.00	3.00	3.40	3.50
Heating source	Electric resistance	Gas boiler	Heat pump	Furnace	Heat pump	Furnace
Heating COP (net of fan)/furnace or boiler efficiency	1.0	75% E_t	3.40	80% E_t	3.40	80% E_t

For SI: °C = (°F – 32)/1.8.
NA = Not Applicable, OSA = Outside Air, RH = Relative Humidity, SP = Static Pressure, VSD = Variable Speed Drive.

TABLE C409.6.2.11(3)—TSPR REFERENCE BUILDING DESIGN HVAC SIMPLE SYSTEMS				
BUILDING TYPE PARAMETER	**BUILDING TYPE**			
	Hotel (warm)	**Hotel (cold)**	**Multifamily (warm)**	**Multifamily (cold)**
System type	PTHP	PTAC	PTHP	PTAC
Fan control	Constant volume	Constant volume	Constant volume	Constant volume
Main fan power [W/cfm (W × s/L)]	0.300 (0.636)	0.300 (0.636)	0.300 (0.636)	0.300 (0.636)

TABLE C409.6.2.11(3)—TSPR REFERENCE BUILDING DESIGN HVAC SIMPLE SYSTEMS—continued				
BUILDING TYPE PARAMETER	BUILDING TYPE			
	Hotel (warm)	Hotel (cold)	Multifamily (warm)	Multifamily (cold)
Heat/cool sizing factor	1.25/1.15	1.25/1.15	1.25/1.15	1.25/1.15
Supplemental heating availability	< 40°F	NA	< 40°F	NA
Outdoor air economizer	No	No	No	No
Occupied OSA source	Packaged unit, occupied damper	Packaged unit, occupied damper	Packaged unit, occupied damper	Packaged unit, occupied damper
Energy recovery ventilator	No	No	No	No
DCV	No	No	No	No
Cooling source	DX, 1 stage (heat pump)	DX, 1 stage	DX, 1 stage (heat pump)	DX, 1 stage
Cooling COP (net of fan)	3.10	3.20	3.10	3.20
Heating source	PTHP	(2) Hydronic boiler	PTHP	(2) Hydronic boiler
Heating COP (net of fan)/furnace or boiler efficiency	3.10	75% E_t	3.10	75% E_t
Heating pump power [W/gpm (W × s/L)]	NA	19 (300)	NA	19 (300)
Heating coil heating water delta-T, °F (°C)	NA	50 (27.8)	NA	50 (27.8)
Design HWST, °F (°C)	NA	180 (82.2)	NA	180 (82.2)
HWST reset setpoint vs. OAT, °F (°C)	NA	HWST: 180-150/ OAT 20-50 (82-65.6/-6.7-10)	NA	HWST: 180-150/ OAT 20-50 (82-65.6/-6.7-10)
Heat loop pumping control	NA	2-way valves & ride pump curve	NA	2-way valves & ride pump curve

For SI: °C = (°F – 32)/1.8.
HWST = Hot Water Supply Temperature, NA = Not Applicable, OAT = Outdoor Air Temperature, OSA = Outside Air.

C409.7 Target design HVAC systems. Target system descriptions in Tables C409.7(1) through C409.7(3) are provided as reference. The target systems are used for developing mechanical performance factors and do not need to be programmed into TSPR software.

TABLE C409.7(1)—TARGET BUILDING DESIGN CRITERIA HVAC COMPLEX SYSTEMS				
BUILDING TYPE PARAMETER	BUILDING TYPE			
	Large Office (warm)	Large Office (cold)	School (warm)	School (cold)
System type	VAV/RH	VAV/RH	VAV/RH	VAV/RH
	Water-cooled chiller	Water-cooled chiller	Water-cooled chiller	Water-cooled chiller
	Electric reheat (PIU)	Gas boiler	Electric reheat (PIU)	Gas boiler
Fan control	VSD (no SP reset)	VSD (no SP reset)	VSD (no SP reset)	VSD (no SP reset)
Main fan power [W/cfm (W × s/L] Proposed ≥ MERV 13	1.127 (2.388)	1.127 (2.388)	1.127 (2.388)	1.127 (2.388)
Zonal fan power [W/CFM (W × s/L)]	0.35 (0.75)	NA	0.35 (0.75)	NA
Minimum zone airflow fraction	1.5 × Voz	1.5 × Voz	1.2 × Voz	1.2 × Voz
Heat/cool sizing factor	1.25/1.15	1.25/1.15	1.25/1.15	1.25/1.15
Outdoor air economizer	Yes except 0–1	Yes	Yes except 0–1	Yes
Occupied OSA (= proposed)	Sum(Voz)/0.75	Sum(Voz)/0.75	Sum(Voz)/0.65	Sum(Voz)/0.65
Energy recovery ventilator efficiency ERR	NA	NA	50%	50%
ERV bypass SAT setpoint	NA	NA	No bypass	60°F (15.6°C) except 4A
DCV	Yes	Yes	Yes	Yes
% Area variable control	15%	15%	70%	70%

TABLE C409.7(1)—TARGET BUILDING DESIGN CRITERIA HVAC COMPLEX SYSTEMS—continued

BUILDING TYPE PARAMETER	BUILDING TYPE			
	Large Office (warm)	Large Office (cold)	School (warm)	School (cold)
% Area on/off control	65%	65%	20%	20%
Cooling source	(2) Water-cooled centrif chillers	(2) Water-cooled centrif chillers	(2) Water-cooled screw chillers	(2) Water-cooled screw chillers
Cooling COP (net of fan)	ASHRAE 90.1 Appendix G, Table G3.5.3	ASHRAE 90.1 Appendix G, Table G3.5.3	ASHRAE 90.1 Appendix G, Table G3.5.3	ASHRAE 90.1 Appendix G, Table G3.5.3
Heating source (reheat)	Electric resistance	Gas boiler	Electric resistance	Gas boiler
Furnace or boiler efficiency	1.0	90% E_t	1.0	90% E_t
Condenser heat rejection	Cooling tower	Cooling tower	Cooling tower	Cooling tower
Cooling tower efficiency [gpm/hp (L/s × kW)]—See ASHRAE 90.1 Appendix G, Section G3.1.3.11	40.2 (3.40)	40.2 (3.40)	40.2 (3.40)	40.2 (3.40)
Tower turndown (> 300 ton (1060 kW))	50%	50%	50%	50%
Pump (constant flow/variable flow)	Constant flow; 10°F (5.6°C) range	Constant flow; 10°F (5.6°C) range	Constant flow; 10°F (5.6°C) range	Constant flow; 10°F (5.6°C) range
Tower approach	ASHRAE 90.1 Appendix G, Table G3.1.3.11	ASHRAE 90.1 Appendix G, Table G3.1.3.11	ASHRAE 90.1 Appendix G, Table G3.1.3.11	ASHRAE 90.1 Appendix G, Table G3.1.3.11
Cooling condenser pump power [W/gpm (W·s/L)]	19 (300)	19 (300)	19 (300)	19 (300)
Cooling primary pump power [W/gpm (W·s/L)]	9 (142)	9 (142)	9 (142)	9 (142)
Cooling secondary pump power [W/gpm (W·s/L)]	13 (205)	13 (205)	13 (205)	13 (205)
Cooling coil chilled water delta-T, °F (°C)	18 (10)	18 (10)	18 (10)	18 (10)
Design chilled water supply temperature, °F (°C)	42 (5.56)	42 (5.56)	42 (5.56)	42 (5.56)
Chilled water supply temperature (CHWST) reset setpoint vs. OAT, °F (°C)	CHWS 44-54/OAT 80-60 (6.7-12.2)/26.7-15.6)	CHWS 44-54/OAT 80-60 (6.7-12.2)/26.7-15.6)	CHWS 44-54/OAT 80-60 (6.7-12.2)/26.7-15.6)	CHWS 44-54/OAT 80-60 (6.7-12.2)/26.7-15.6)
CHW cooling loop pumping control	2-way valves & pump VSD	2-way valves & pump VSD	2-way valves & pump VSD	2-way valves & pump VSD
Heating pump power [W/gpm (W·s/L)]	16.1 (254)	16.1 (254)	19 (254)	19 (254)
Heating HW delta-T, °F (°C)	50 (27.78)	20 (11.11)	50 (27.78)	20 (11.11)
Design hot water supply temperature (HWST), °F (°C)	180 (82)	140 (60)	180 (82)	140 (60)
Hot water supply temperature (HWST) range vs. outside air temperature (OAT) range	HWST: 180-150/OAT 20-50 (82-65.6/-6.7-10)	HWST: 180-150/OAT 20-50 (82-65.6/-6.7-10)	HWST: 180-150/OAT 20-50 (82-65.6/-6.7-10)	HWST: 180-150/OAT 20-50 (82-65.6/-6.7-10)
Heat loop pumping control	2-way valves & pump VSD	2-way valves & pump VSD	2-way valves & pump VSD	2-way valves & pump VSD

For SI: °C = (°F – 32)/1.8.

CHW = Chilled Water, ERR = Enthalpy Recovery Ratio, NA = Not Applicable, OSA = Outside Air, PIU = Parallel Powered Induction Unit, RH = Relative Humidity, SP = Static Pressure, Voz = Outdoor airflow to the zone, VSD = Variable Speed Drive.

TABLE C409.7(2)—TARGET BUILDING DESIGN CRITERIA HVAC SIMPLE SYSTEMS

BUILDING TYPE PARAMETER	BUILDING TYPE					
	Medium Office(warm)	**Medium Office (cold)**	**Small Office (warm)**	**Small Office (cold)**	**Retail (warm)**	**Retail (cold)**
System type	Package VAV—electric reheat	Package VAV—hydronic reheat	PSZ-HP	PSZ-AC	PSZ-HP	PSZ-AC
Fan control	VSD (with SP reset)	VSD (with SP reset)	Constant volume	Constant volume	2-speed	2-speed
Main fan power [W/cfm(W × s/L)] proposed ≥ MERV 13	0.634 (1.343)	0.634 (1.343)	0.486 (1.03)	0.486 (1.03)	0.585 (1.245)	0.585 (1.245)
Zonal fan power [W/CFM (W × s/L)]	0.35 (5.53)	NA	NA	NA	NA	NA
Minimum zone airflow fraction	1.5 × Voz	1.5 × Voz	NA	NA	NA	NA
Heat/cool sizing factor	1.25/1.15	1.25/1.15	1.25/1.15	1.25/1.15	1.25/1.15	1.25/1.15
Supplemental heating availability	NA	NA	< 40°F (< 4.4°C) OAT	NA	< 40°F (< 4.4°C) OAT	NA
Outdoor air economizer	Yes except 0–1	Yes	Yes except 0–1	Yes	Yes except 0–1	Yes
Occupied OSA source	Packaged unit, occupied damper, all building use types					
Energy recovery ventilator	No	No	No	No	Yes in 0A, 1A, 2A, 3A	Yes all A, 6,7,8 CZ
ERR					50%	50%
DCV	Yes	Yes	No	No	Yes	Yes
% Area variable control	15%	15%			80%	80%
% Area on/off control	65%	65%			0%	0%
Cooling source	DX, multistage	DX, multistage	DX, 1 stage (heat pump)	DX, single stage	DX, 2 stage (heat pump)	DX, 2 stage
Cooling COP (net of fan)	3.83	3.83	3.82	3.8248	3.765	3.765
Heating source	Electric resistance	Gas boiler	Heat pump	Furnace	Heat pump	Furnace
Heating COP (net of fan)/ furnace or boiler efficiency	100%	81% E_t	3.81	81% E_t	3.536	81% E_t
Heating coil HW delta-T, °F (°C)	NA	20 (11.11)	NA	NA	NA	NA
Design HWST, °F (°C)	NA	140 (60)	NA	NA	NA	NA
HWST reset setpoint vs OAT, °F (°C)	NA	HWST: 180-150/OAT 20-50 (82-65.6/-6.7-10)	NA	NA	NA	NA
Heat loop pumping control	NA	2-way valves & ride pump curve	NA	NA	NA	NA
Heating pump power [W/gpm (W · s/L)]	NA	16.1	NA	NA	NA	NA

For SI: °C = (°F – 32)/1.8.

CHW = Chilled Water, ERR = Enthalpy Recovery Ratio, HWST = Hot Water Supply Temperature, NA = Not Applicable, OAT = Outside Air Temperature, OSA = Outside Air, SP = Static Pressure, Voz = Outdoor airflow to the zone, VSD = Variable Speed Drive.

TABLE C409.7(3)—TARGET BUILDING DESIGN CRITERIA HVAC SIMPLE SYSTEMS

BUILDING TYPE PARAMETER	BUILDING TYPE			
	Hotel (warm)	**Hotel (cold)**	**Multifamily (warm)**	**Multifamily (cold)**
System type	PTHP	PTAC with hydronic boiler	Split HP	Split AC
Fan control	Cycling	Cycling	Cycling	Cycling
Main fan power [W/cfm (W × s/L)]	0.300 (0.638)	0.300 (0.638)	0.246 (0.523)	0.271 (0.576)
Heat/cool sizing factor	1.25/1.15	1.25/1.15	1.25/1.15	1.25/1.15
Supplemental heating availability	< 40°F (< 4.4°C)	NA	< 40°F (< 4.4°C)	NA
Outdoor air economizer	Only CZ 2, 3	No	No	No
Occupied OSA source	DOAS	DOAS	DOAS	DOAS except 3C
Energy recovery ventilator	NA	NA	Yes	Yes except 3C
ERR	NA	NA	60%	60%
DCV	Yes	Yes	No	No
% Area variable control	70%	70%		
% Area variable control	0%	0%		
Cooling source	DX, 1 stage (heat pump)	DX, 1 stage	DX, 1 stage (heat pump)	DX, 1 stage
Cooling COP (net of fan)	3.83	3.83	3.823	3.6504
Heating source	Heat pump	(2) Hydronic boiler	Heat pump	Furnace
Heating COP (net of fan)/furnace or boiler efficiency	3.44	81% E_t	3.86	80% AFUE
Heating pump power (W/gpm (W·s/L))	NA	16.1	NA	NA
Heating coil heating water delta-T, °F (°C)	NA	20 (11.11)	NA	NA
Design HWST, °F (°C)	NA	140 (60)	NA	NA
HWST reset setpoint vs. OAT, °F (°C)	NA	HWST: 180-150/ OAT 20-50 (82-65.6/ -6.7-10)	NA	NA
Heat loop pumping control	NA	2-way valves & ride pump curve	NA	NA

For SI: °C = (°F – 32)/1.8.
DOAS = Dedicated Outdoor Air System, ERR = Enthalpy Recovery Ratio, HWST = Hot Water Supply Temperature, NA = Not Applicable, OAT = Outdoor Air Temperature, OSA = Outside Air.

CHAPTER 5 [CE] EXISTING BUILDINGS

User notes:

About this chapter: *Many buildings are renovated or altered in numerous ways that could affect the energy use of the building as a whole. Chapter 5 requires the application of certain parts of Chapter 4 in order to maintain, if not improve, the conservation of energy by the renovated or altered building.*

SECTION C501—GENERAL

C501.1 Scope. The provisions of this chapter shall control the *alteration*, *repair*, *addition* and *change of occupancy* of existing buildings and structures.

C501.1.1 Existing buildings. Except as specified in this chapter, this code shall not be used to require the removal, *alteration* or abandonment of, nor prevent the continued use and maintenance of, an existing building or building system lawfully in existence at the time of adoption of this code.

C501.2 Compliance. *Additions*, *alterations*, *repairs*, and changes of occupancy to, or relocation of, existing *buildings* and structures shall comply with Sections C502, C503, C504 and C505 of this code, as applicable, and with the provisions for *alterations*, *repairs*, *additions* and changes of occupancy or relocation, respectively, in the *International Building Code*, *International Existing Building Code, International Fire Code, International Fuel Gas Code, International Mechanical Code, International Plumbing Code, International Property Maintenance Code, International Private Sewage Disposal Code* and NFPA 70. Changes where unconditioned space is changed to conditioned space shall comply with Section C502.

Exception: *Additions*, *alterations*, *repairs* or changes of occupancy complying with ANSI/ASHRAE/IES 90.1.

C501.3 Maintenance. *Buildings* and structures, and parts thereof, shall be maintained in a safe and sanitary condition. Devices and systems required by this code shall be maintained in conformance to the code edition under which they were installed. The *owner* or the owner's authorized agent shall be responsible for the maintenance of buildings and structures. The requirements of this chapter shall not provide the basis for removal or abrogation of energy conservation, fire protection and safety systems and devices in existing structures.

C501.4 New and replacement materials. Except as otherwise required or permitted by this code, materials permitted by the applicable code for new construction shall be used. Like materials shall be permitted for *repairs*, provided that hazards to life, health or property are not created. Hazardous materials shall not be used where the code for new construction would not allow use of these materials in buildings of similar occupancy, purpose and location.

C501.5 Historic buildings. Provisions of this code relating to the construction, *repair, alteration*, restoration and movement of structures, and *change of occupancy* shall not be mandatory for *historic buildings* provided that a report has been submitted to the *code official* and signed by a *registered design professional*, or a representative of the State Historic Preservation Office or the historic preservation authority having jurisdiction, demonstrating that compliance with that provision would threaten, degrade or destroy the historic form, fabric or function of the *building*.

SECTION C502—ADDITIONS

C502.1 General. *Additions* to an existing building, building system or portion thereof shall conform to the provisions of this code as those provisions relate to new construction without requiring the unaltered portion of the existing building or building system to comply with this code. *Additions* shall not create an unsafe or hazardous condition or overload existing building systems. An *addition* shall be deemed to comply with this code if the *addition* alone complies or if the existing building and *addition* comply with this code as a single *building*.

C502.2 Change in space conditioning. Any nonconditioned or low-energy space that is altered to become *conditioned space* shall be required to comply with Section C502.

Exceptions:

1. Where the component performance alternative in Section C402.1.4 is used to comply with this section, the proposed UA shall be not greater than 110 percent of the target UA.
2. Where the simulated building performance option in Section C407 is used to comply with this section, the annual *energy cost* of the *proposed design* shall be not greater than 110 percent of the annual *energy cost* otherwise permitted by Section C407.2.

C502.3 Compliance. *Additions* shall comply with Sections C502.3.1 through C502.3.8.

C502.3.1 Vertical fenestration area. Additions shall comply with the following:

1. Where an addition has a new vertical fenestration area that results in a total building fenestration area less than or equal to that permitted by Section C402.5.1, the addition shall comply with Section C402.1.4, C402.5.3 or C407.

2. Where an addition with vertical fenestration that results in a total building fenestration area greater than Section C402.5.1 or an addition that exceeds the fenestration area greater than that permitted by Section C402.5.1, the fenestration shall comply with Section C402.5.1.1 for the addition only.
3. Where an addition has vertical fenestration that results in a total building vertical fenestration area exceeding that permitted by Section C402.5.1.1, the addition shall comply with Section C402.1.4 or C407.

C502.3.2 Skylight area. Skylights shall comply with the following:

1. Where an *addition* has new skylight area that results in a total building fenestration area less than or equal to that permitted by Section C402.5.1, the *addition* shall comply with Section C402.1.4 or C407.
2. Where an *addition* has new skylight area that results in a total building skylight area greater than permitted by Section C402.5.1 or where *additions* have skylight area greater than that permitted by Section C402.5.1, the skylight area shall comply with Section C402.5.1.2 for the *addition* only.
3. Where an *addition* has skylight area that results in a total building skylight area exceeding that permitted by Section C402.5.1.2, the *addition* shall comply with Section C402.1.4 or C407.

C502.3.3 Building mechanical systems. New mechanical systems and equipment that are part of the *addition* and serve the building heating, cooling and *ventilation* needs shall comply with Sections C403 and C408.

C502.3.4 Service water-heating systems. New service water-heating equipment, controls and service water-heating piping shall comply with Section C404.

C502.3.5 Pools and inground permanently installed spas. New pools and inground permanently installed spas shall comply with Section C404.8.

C502.3.6 Lighting power and systems. New lighting systems that are installed as part of the *addition* shall comply with Sections C405 and C408.

C502.3.6.1 Interior lighting power. The total interior lighting power for the *addition* shall comply with Section C405.3.2 for the *addition* alone, or the existing building and the *addition* shall comply as a single *building*.

C502.3.6.2 Exterior lighting power. The total exterior lighting power for the *addition* shall comply with Section C405.5.2 for the *addition* alone, or the existing building and the *addition* shall comply as a single *building*.

C502.3.7 Additional energy efficiency credit requirements. *Additions* shall comply with sufficient measures from Sections C406.2 and C406.3 to achieve not less than 50 percent of the number of required efficiency credits from Table C406.1.1(1) based on building occupancy group and *climate zone*. Where a project contains multiple occupancies, credits from Table C406.1.1(1) for each building occupancy shall be weighted by the gross floor area to determine the project weighted average energy credits required. Accessory occupancies shall be included with the primary occupancy group for purposes of this section. *Alterations* to the existing building that are not part of the *addition*, but are permitted with an *addition*, shall be permitted to be used to achieve the required credits.

Exceptions:

1. Buildings in Group U (Utility and Miscellaneous), Group S (Storage), Group F (Factory), Group H (High-Hazard).
2. *Additions* less than 1,000 square feet (93 m^2) and less than 50 percent of existing floor area.
3. *Additions* that do not include the *addition* or replacement of equipment covered by Tables C403.3.2(1) through C403.3.2(16) or Section C404.2.
4. *Additions* that do not increase *conditioned space*.
5. Where the *addition* alone or the existing building and *addition* together comply with Section C407.

C502.3.8 Renewable energy systems. *Additions* shall comply with Section C405.15 for the *addition* alone.

SECTION C503—ALTERATIONS

C503.1 General. *Alterations* to any *building* or structure shall comply with the requirements of Section C503. *Alterations* shall be such that the existing building or structure is not less conforming to the provisions of this code than the existing *building* or structure was prior to the *alteration*. *Alterations* to an existing *building*, *building* system or portion thereof shall conform to the provisions of this code as those provisions relate to new construction without requiring the unaltered portions of the existing *building* or *building* system to comply with this code. *Alterations* shall not create an unsafe or hazardous condition or overload existing building systems.

Exception: The following *alterations* need not comply with the requirements for new construction, provided that the energy use of the *building* is not increased:

1. Storm windows installed over existing *fenestration*.
2. Surface-applied window film installed on existing single-pane fenestration assemblies reducing solar heat gain, provided that the code does not require the glazing or *fenestration* to be replaced.
3. *Roof recover*.
4. *Roof replacement* where roof assembly insulation is integral to or located below the structural roof deck.

5. *Air barriers* shall not be required for *roof recover* and *roof replacement* where the *alterations* or renovations to the *building* do not include *alterations*, renovations or *repairs* to the remainder of the *building thermal envelope*.
6. An existing building undergoing *alterations* that complies with Section C407.

C503.2 Building thermal envelope. *Alterations* of existing *building thermal envelope* assemblies shall comply with this section. New *building thermal envelope* assemblies that are part of the *alteration* shall comply with Section C402. An area-weighted average *U-factor* for new and altered portions of the *building thermal envelope* shall be permitted to satisfy the *U-factor* requirements in Table C402.1.4. The existing *R-value* of insulation shall not be reduced or the *U-factor* of a *building thermal envelope* assembly be increased as part of a *building thermal envelope alteration* except where complying with Section C407.

Exception: Where the existing *building* exceeds the fenestration area limitations of Section C402.5.1 prior to alteration, the building is exempt from Section C402.5.1 provided that there is no increase in fenestration area.

C503.2.1 Roof, ceiling and attic alterations. Insulation complying with Sections C402.1 and C402.2.1, or an *approved* design that minimizes deviation from the insulation requirements, shall be provided for the following *alterations*:

1. An *alteration* of roof/ceiling construction other than *reroofing* where existing insulation located below the roof deck or on an attic floor above *conditioned space* does not comply with Table C402.1.2.
2. Roof replacement or a roof *alteration* that includes removing and replacing the roof covering, where the roof assembly includes insulation entirely above the roof deck.

   Exceptions: Where compliance with Section C402.1 cannot be met due to limiting conditions on an existing roof, an *approved* design shall be submitted with the following:

   1. *Construction documents* that include a report by a *registered design professional* or an *approved source* documenting details of the limiting conditions affecting compliance with the insulation requirements.
   2. *Construction documents* that include a roof design by a *registered design professional* or an *approved source* that minimizes deviation from the insulation requirements.
3. Conversion of unconditioned attic space into *conditioned space*.
4. Replacement of ceiling finishes exposing cavities or surfaces of the roof/ceiling construction.

C503.2.2 Vertical fenestration. The addition of *vertical fenestration* that results in a total building *fenestration* area less than or equal to that specified in Section C402.5.1 shall comply with Section C402.1.4, C402.5.3 or C407. The addition of *vertical* fenestration that results in a total building *fenestration* area greater than Section C402.5.1 shall comply with Section C402.5.1.1 for the space adjacent to the new fenestration only. *Alterations* that result in a total building *vertical fenestration* area exceeding that specified in Section C402.5.1.1 shall comply with Section C402.1.4 or C407. Provided that the vertical fenestration area is not changed, using the same vertical fenestration area in the *standard reference design* as the *building* prior to *alteration* shall be an alternative to using the vertical fenestration area specified in Table C407.4.1(1).

C503.2.2.1 Application to replacement fenestration products. Where some or all of an existing *fenestration* unit is replaced with a new fenestration product, including sash and glazing, the replacement fenestration unit shall meet the applicable requirements for *U-factor* and *SHGC* in Table C402.5.

Exception: An area-weighted average of the *U-factor* of replacement fenestration products being installed in the *building* for each fenestration product category listed in Table C402.5 shall be permitted to satisfy the *U-factor* requirements for each fenestration product category listed in Table C402.5. Individual fenestration products from different product categories listed in Table C402.5 shall not be combined in calculating the area-weighted average *U-factor*.

C503.2.3 Skylight area. New *skylight* area that results in a total building *skylight* area less than or equal to that specified in Section C402.5.1 shall comply with Section C402.1.4, C402.5 or C407. The addition of *skylight* area that results in a total building skylight area greater than Section C402.5.1 shall comply with Section C402.5.1.2 for the space adjacent to the new skylights. *Alterations* that result in a total building skylight area exceeding that specified in Section C402.5.1.2 shall comply with Section C402.1.4 or C407. Provided that the skylight area is not changed, using the same skylight area in the *standard reference design* as the *building* prior to *alteration* shall be an alternative to using the skylight area specified in Table C407.4.1(1).

C503.2.4 Above-grade wall alterations. *Above-grade wall alterations* shall comply with the following:

1. Where wall cavities are exposed, the cavity shall be filled with *cavity insulation* complying with Section C303.1.4. New cavities created shall be insulated in accordance with Section C402.1 or an *approved* design that minimizes deviation from the insulation requirements.
2. Where exterior wall coverings and *fenestration* are added or replaced for the full extent of any exterior wall assembly on one or more elevations of the *building*, insulation shall be provided where required in accordance with one of the following:
   - 2.1. An *R-value* of *continuous insulation* not less than that designated in Table C402.1.3 for the applicable *above-grade wall* type and existing cavity insulation *R-value*, if any;
   - 2.2. An *R-value* of not less than that required to bring the *above-grade wall* into compliance with Table C402.1.2; or,
   - 2.3. An *approved* design that minimizes deviation from the insulation requirements of Section C402.1.
3. Where Items 1 and 2 apply, the insulation shall be provided in accordance with Section C402.1.

Where any of the above requirements are applicable, the *above-grade wall alteration* shall comply with Sections 1402.2 and 1404.3 of the *International Building Code*.

C503.2.5 Floor alterations. Where an *alteration* to a floor or floor overhang exposes cavities or surfaces to which insulation can be applied, and the floor or floor overhang is part of the *building thermal envelope*, the floor or floor overhang shall be brought into compliance with Section C402.1 or an *approved* design that minimizes deviation from the insulation requirements. This requirement applies to floor *alterations* where the floor cavities or surfaces are exposed and unobstructed prior to construction.

C503.2.6 Below-grade wall alterations. Where unconditioned below-grade space is changed to *conditioned space*, walls enclosing such *conditioned space* shall be insulated where required in accordance with Section C402.1. Where the below-grade space is *conditioned space* and where walls enclosing such space are altered, they shall be insulated where required in accordance with Section C402.1.

C503.2.7 Air barrier. Altered *building thermal envelope* assemblies shall be provided with an *air barrier* in accordance with Section C402.6.1. Such *air barrier* need not be continuous with unaltered portions of the *building thermal envelope*. Testing requirements of Section C402.6.1.2 shall not be required.

C503.3 Heating and cooling systems. New heating, cooling and *duct systems* that are part of the *alteration* shall comply with Section C403.

C503.3.1 Economizers. New cooling systems that are part of *alteration* shall comply with Section C403.5.

C503.3.2 Mechanical system acceptance testing. Where an *alteration* requires compliance with Section C403 or any of its subsections, mechanical systems that serve the *alteration* shall comply with Sections C408.2.2, C408.2.3 and C408.2.5.

Exceptions:

1. *Buildings* with less than 10,000 square feet (929 m^2) and a combined heating, cooling and service water-heating capacity of less than 960,000 Btu/h (281 kW).
2. Systems included in Section C403.5 that serve individual *dwelling units* and *sleeping units*.

C503.3.3 Duct testing. *Ducts* and plenums designed to operate at static pressures not less than 3 inches water gauge (747 Pa) that serve an *alteration* shall be tested in accordance with this section where the *alteration* includes any of the following:

1. Twenty-five percent or more of the total length of the ducts in the system are relocated.
2. The total length of all ducts in the system is increased by 25 percent or more.

Ducts and plenums shall be leak tested in accordance with the SMACNA *HVAC Air Duct Leakage Test Manual* and shown to have a rate of *air leakage* (CL) less than or equal to 12.0 as determined in accordance with Equation 4-7 of Section C403.13.2.3. Documentation shall be available demonstrating that representative sections totaling not less than 25 percent of the *duct* area have been tested and that all tested sections comply with the requirements of this section.

C503.3.4 Controls. New heating and cooling equipment that is part of the *alteration* shall be provided with controls that comply with the control requirements in Sections C403.4 and C403.5 other than the requirements of Sections C403.4.3.3 and C403.4.4.

Exceptions:

1. Systems with *direct digital control* of individual *zones* reporting to a central control panel.
2. The replacement of individual components of multiple-zone VAV systems.

C503.3.5 System sizing. New heating and cooling equipment that is part of an *alteration* shall be sized in accordance with Section C403.3.1 based on the existing building features as modified by the *alteration*.

Exceptions:

1. Where it has been demonstrated to the *code official* that compliance with this section would result in heating or cooling equipment that is incompatible with the rest of the heating or cooling system.
2. Where it has been demonstrated to the *code official* that the additional capacity will be needed in the future.

C503.3.6 Replacement or added roof-mounted mechanical equipment. For roofs with insulation entirely above the roof deck and where existing roof-mounted mechanical equipment is replaced or new equipment is added, and the existing roof does not comply with the insulation requirements for new construction in accordance with Sections C402.1 and C402.2.1, curbs for added or replaced equipment shall be of a height necessary to accommodate the future addition of above-deck roof insulation to be installed in accordance with Section C503.2.1, Item 2. Alternatively, the curb height shall comply with Table C503.3.6. Curb height shall be the distance measured from the top of the curb to the top of the roof deck.

TABLE C503.3.6—ROOF-MOUNTED MECHANICAL EQUIPMENT CURB HEIGHTS

CLIMATE ZONE	CURB HEIGHT, MINIMUM
0, 1, 2 and 3	16 inches
4, 5 and 6	17 inches
7 and 8	18 inches

For SI: 1 inch = 25.4 mm.

C503.4 Service hot water systems. New service hot water systems that are part of the *alteration* shall comply with Section C404.

C503.4.1 Service hot water system acceptance testing. Where an *alteration* requires compliance with Section C404 or any of its subsections, service hot water systems that serve the *alteration* shall comply with Sections C408.2.3 and C408.2.5.

Exceptions:

1. *Buildings* with less than 10,000 square feet (929 m^2) and a combined heating, cooling and service water-heating capacity of less than 960,000 Btu/h (281 kW).
2. Systems included in Section C403.5 that serve individual *dwelling units* and *sleeping units*.

C503.5 Lighting systems. New lighting systems that are part of the *alteration* shall comply with Sections C503.5.1 and C503.5.2.

C503.5.1 Interior lighting and controls. *Alterations* to interior spaces, lighting or controls shall comply with the following:

1. Where an alteration of an interior space includes the addition or relocation of full height partitions, the space shall comply with Sections C405.2, C405.3 and C408.3.
2. Where the lighting within interior spaces is altered, those spaces shall comply with Sections C405.2, C405.3 and C408.3.
3. Where the lighting controls within interior spaces are altered, those spaces shall comply with Sections C405.2 and C408.3.

Exception: Compliance with Section C405.2.8 is not required for *alterations*.

C503.5.2 Exterior lighting and controls. *Alterations* to exterior lighting and controls shall comply with the following:

1. Where the connected exterior lighting power is increased by more than 400 watts, all exterior lighting, including lighting that is not proposed to be altered, shall comply with Section C405.5.
2. Where the combined power of added and replacement luminaires is more than 400 watts, all lighting that is added or altered shall be controlled in accordance with Sections C405.2 and C408.3.

   Exception: Individual luminaires less than 50 watts provided they pass functional tests verifying *automatic* shut off where daylight is present.
3. Where portions of exterior lighting controls are added or altered, those portions shall comply with Sections C405.2 and C408.3.

C503.6 Additional energy efficiency credit requirements for alterations. *Alterations* that are *substantial improvements* shall comply with measures from Sections C406.2, C406.3 or both to earn the number of required credits specified in Table C406.1.1(1) based on building occupancy group and *climate zone*. Where a project contains multiple occupancies, credits specified in Table C406.1.1(1) for each building occupancy shall be weighted by the gross *conditioned floor area* to determine the weighted average credits required. Accessory occupancies, other than Group F or H, shall be included with the primary occupancy group for the purposes of this section.

Exceptions:

1. *Alterations* that do not contain *conditioned space*.
2. Portions of *buildings* devoted to manufacturing or industrial use.
3. *Alterations* to *buildings* where the *building* after the *alteration* complies with Section C407.
4. *Alterations* that are permitted with an *addition* complying with Section C502.3.7.

SECTION C504—REPAIRS

C504.1 General. *Buildings* and structures, and parts thereof, shall be repaired in compliance with Section C501.3 and this section. Work on nondamaged components that is necessary for the required *repair* of damaged components shall be considered to be part of the *repair* and shall not be subject to the requirements for *alterations* in this chapter. Routine maintenance required by Section C501.3, ordinary *repairs* exempt from *permit* and abatement of wear due to normal service conditions shall not be subject to the requirements for *repairs* in this section.

Where a *building* was constructed to comply with ANSI/ASHRAE/IES 90.1, *repairs* shall comply with the standard and need not comply with Sections C402, C403, C404 and C405.

C504.2 Application. For the purposes of this code, the following shall be considered to be *repairs*:

1. Glass-only replacements in an existing sash and frame.
2. *Roof repairs*.
3. *Air barriers* shall not be required for *roof repair* where the *repairs* to the *building* do not include *alterations*, renovations or *repairs* to the remainder of the *building thermal envelope*.
4. Replacement of existing doors that separate *conditioned space* from the exterior shall not require the installation of a vestibule or revolving door, provided that an existing vestibule that separates a *conditioned space* from the exterior shall not be removed.
5. *Repairs* where only the bulb, the ballast or both within the existing luminaires in a space are replaced, provided that the replacement does not increase the installed interior lighting power.

SECTION C505—CHANGE OF OCCUPANCY OR USE

C505.1 General. Spaces undergoing a change in occupancy from Group F, H, S or U occupancy classification shall comply with Section C503. *Buildings* or portions of *buildings* undergoing a *change of occupancy* without *alterations* shall comply with Section C505.2.

Exception: Where the simulated building performance option in Section C407 is used to comply with this section, the annual *energy cost* of the *proposed design* shall be not greater than 110 percent of the annual *energy cost* otherwise permitted by Section C407.3.

C505.1.1 Alterations and change of occupancy. *Alterations* made concurrently with any *change of occupancy* shall be in accordance with Section C503.

C505.1.2 Portions of buildings. Where changes in occupancy and use are made to portions of an existing building, only those portions of the *building* shall be required to comply with Section C505.2.

C505.2 Energy use intensities. *Building thermal envelope*, space heating, cooling, *ventilation*, lighting and *service water heating* shall comply with Sections C505.2.1 through C505.2.4.

Exceptions:

1. Where it is demonstrated by analysis *approved* by the *code official* that the change will not increase *energy use intensity*.
2. Where the occupancy or use change is less than 5,000 square feet (465 m^2) in area.

C505.2.1 Building thermal envelope. Where a *change of occupancy* or use is made to a whole *building* that results in a fenestration area greater than the maximum fenestration area allowed by Section C402.5.1, the *building* shall comply with Section C402.1.4, with a proposed UA that shall be not greater than 110 percent of the target UA.

Exception: Where the *change of occupancy* or use is made to a portion of the *building*, the new occupancy is exempt from Section C402.5.1, provided that there is not an increase in fenestration area.

C505.2.2 Building mechanical systems. Where a *change of occupancy* or use results in the same or increased *energy use intensity* rank as specified in Table C505.2.2, the systems serving the *building* or space undergoing the change shall comply with Section C403.

TABLE C505.2.2—BUILDING MECHANICAL SYSTEMS

ENERGY USE INTENSITY RANK	*INTERNATIONAL BUILDING CODE* OCCUPANCY CLASSIFICATION AND USE
High	A-2, B (laboratories), I-2
Medium	A-1, A-3,[a] A-4, A-5, B,[b] E, I-1, I-3, I-4, M, R-4
Low	A-3 (places of religious worship), R-1, R-2, R-3,[c] S-1, S-2

a. Excluding places of religious worship.
b. Excluding laboratories.
c. Buildings three stories or less in height above grade plane shall comply with Section R505.

C505.2.3 Service water heating. Where a *change of occupancy* or use results in the same or increased *energy use intensity* rank as specified in Table C505.2.3, the *service water heating* systems serving the *building* or space undergoing the change shall comply with Section C404.

TABLE C505.2.3—SERVICE WATER HEATING

ENERGY USE INTENSITY RANK	*INTERNATIONAL BUILDING CODE* OCCUPANCY CLASSIFICATION AND USE
High	A-2, I-1, I-2, R-1
Low	All other occupancies and uses

C505.2.4 Lighting. Where a *change of occupancy* or use results in the same or increased *energy use intensity* rank as specified in Table C505.2.4, the lighting systems serving the *building* or space undergoing the change shall comply with Section C405 except for Sections C405.2.6 and C405.4.

TABLE C505.2.4—LIGHTING

ENERGY USE INTENSITY RANK	*INTERNATIONAL BUILDING CODE* OCCUPANCY CLASSIFICATION AND USE
High	B (laboratories), B (outpatient healthcare), I-2, M
Medium	A-2, A-3 (courtrooms), B,[a] I-1, I-3, I-4, R-1, R-2, R-3,[b] R-4, S-1, S-2
Low	A-1, A-3,[c] A-4, E

a. Excluding laboratories and outpatient healthcare.
b. Buildings three stories or less in height above grade plane shall comply with Section R505.
c. Excluding courtrooms.

CHAPTER 6 [CE] REFERENCED STANDARDS

User notes:

About this chapter: *Chapter 6 lists the full title, edition year and address of the promulgator for all standards that are referenced in the code. The section numbers in which the standards are referenced are also listed.*

This chapter lists the standards that are referenced in various sections of this document. The standards are listed herein by the promulgating agency of the standard, the standard identification, the effective date and title, and the section or sections of this document that reference the standard. The application of the referenced standards shall be as specified in Section C110.

AERC *Attachments Energy Rating Council, 355 Lexington Ave 15th Floor, New York, NY 10017*

AERC 1—2017: Procedures for Determining Energy Performance Properties of Fenestration Attachments
C406.3.4

AHAM *Association of Home Appliance Manufacturers, 1111 19th Street NW, Suite 402, Washington, DC 20036*

ANSI/AHAM RAC-1—2020: Room Air Conditioners
Table C403.3.2(4)

AHRI *Air-Conditioning, Heating, & Refrigeration Institute, 2311 Wilson Blvd, Suite 400, Arlington, VA 22201*

210/240—2023 (2020): Performance Rating of Unitary Air-conditioning and Air-source Heat Pump Equipment
Table C403.3.2(1), Table C403.3.2(2), Table C403.3.2(8), Table C403.3.2(9)

310/380—2017 (CSA-C744-17): Packaged Terminal Air Conditioners and Heat Pumps
Table C403.3.2(4)

340/360—2022: Performance Rating of Commercial and Industrial Unitary Air-conditioning and Heat Pump Equipment
Table C403.3.2(1), Table C403.3.2(2)

365 (I-P)—2009: Commercial and Industrial Unitary Air-conditioning Condensing Units
Table C403.3.2(1)

390 (I-P)—2003: Performance Rating of Single Package Vertical Air-conditioners and Heat Pumps
Table C403.3.2(4)

400 (I-P)—2015: Performance Rating of Liquid to Liquid Heat Exchangers
C403.3.2

440 (I-P)—2019: Performance Rating of Fan Coils
C403.13.3

460—2005: Performance Rating of Remote Mechanical-draft Air-cooled Refrigerant Condensers
Table C403.3.2(7)

550/590 (I-P)—2022: Performance Rating of Water-chilling and Heat Pump Water-heating Packages Using the Vapor Compression Cycle
Table C403.3.2(3), Table C403.3.2(15)

560—2000: Absorption Water Chilling and Water Heating Packages
Table C403.3.2(3), Table C403.3.2(15)

840 (I-P)—1998: Performance Rating of Unit Ventilators
C403.13.3

910 (I-P)—2014: Performance Rating of Indoor Pool Dehumidifiers
Table C403.3.2(11)

920 (I-P)—2020: Performance Rating of Direct Expansion-Dedicated Outdoor Air System Units (with Addendum 1)
Table C403.3.2(12), Table C403.3.2(13)

1200 (I-P)—2022: Performance Rating of Commercial Refrigerated Display Merchandisers and Storage Cabinets
Table C403.12.1

1230—2021: Performance Rating of Variable Refrigerant Flow (VRF) Multi-split Air-Conditioning and Heat Pump Equipment
Table C403.3.2(8), Table C403.3.2(9)

1250 (I-P)—2020: Standard for Performance Rating in Walk-in Coolers and Freezers
Table C403.12.2.1(3)

1360 (I-P)—2017: Performance Rating of Computer and Data Processing Room Air Conditioners
Table C403.3.2(10), Table C403.3.2(16)

1380 (I-P)—2019: Demand Response through Variable Capacity HVAC Systems in Residential and Small Commercial Applications
C406.3

ANSI/AHRI 1300—2013 (R2023) (I-P): Performance Rating of Commercial Heat Pump Water Heaters
C406.2.3.1.2

ASHRAE/ANSI/AHRI/ISO 13256-1—1998 (R2012): Water-to-Air and Brine-to-Air Heat Pumps—Testing and Rating for Performance
Table C403.3.2(14)

ASHRAE/ANSI/AHRI/ISO 13256-2—1998 (R2012): Water-to-Water and Brine-to-Water Heat Pumps—Testing and Rating for Performance
Table C403.3.2(14)

AISI *American Iron and Steel Institute, 25 Massachusetts Avenue, NW, Suite 800, Washington, DC 20001*

AISI S250—22: North American Standard for Thermal Transmittance of Building Envelopes with Cold-Formed Steel Framing, with Supplement 1, dated 2022
C402.1.2.1.6

AMCA *Air Movement and Control Association International, 30 West University Drive, Arlington Heights, IL 60004-1806*

208—18: Calculation of the Fan Energy Index
C403.8.3, C403.9.1

500D—18: Laboratory Methods for Testing Dampers for Rating
C403.7.7

ANSI/AMCA 220—21: Laboratory Methods of Testing Air Curtain Units for Aerodynamic Performance Rating
C402.6.6

ANSI/AMCA 230—22: Laboratory Methods of Testing Air Circulating Fans for Rating and Certification
C403.9

ANSI *American National Standards Institute, 25 West 43rd Street, 4th Floor, New York, NY 10036*

ANSI Z21.47—2016/CSA 2.3—2021: Gas-Fired Central Furnaces
Table C403.3.2(5)

ANSI Z83.8—2016/CSA 2.6—2016: Gas Unit Heater, Gas Packaged Heaters, Gas Utility Heaters and Gas-Fired Duct Furnaces
Table C403.3.2(5)

ANSI/CTA-2045-A—2018: Modular Communications Interface for Energy Management
C406.3

ANSI/CTA-2045-B—2018: Modular Communications Interface for Energy Management
C406.3, C406.3.7

ANSI/NEMA WD 6—2016: Wiring Devices—Dimensional Specifications
C405.9.2

APSP *Pool & Hot Tub Alliance (formerly the Association of Pool and Spa Professionals), 2111 Eisenhower Avenue, Suite 580, Alexandria, VA 22314*

14—2019: American National Standard for Portable Electric Spa Energy Efficiency
C404.9

ASABE *American Society of Agricultural and Biological Engineers, 2950 Niles Road, St. Joseph, MI 49085*

S640—July 2017(R2022): Quantities and Units of Electromagnetic Radiation for Plants (Photosynthetic Organisms)
C202

ASHRAE *ASHRAE, 180 Technology Parkway NW, Peachtree Corners, GA 3009*

55—2020: Thermal Environmental Conditions for Human Occupancy
Table C407.4.1(1),

62.1—2019: Ventilation for Acceptable Indoor Air Quality
C403.6.1, C406.3.3, Table C409.6.1.10.2(1)

90.1—2022: Energy Standard for Buildings Except Low-rise Residential Buildings

C101.3, C401.2.2, C402.1.2, Table C402.1.2, C402.1.2.1, C402.1.3, Table C403.3.2(5), Table C403.3.2(15), C406.2, C406.2, C406.2.1.1, C409.6.1.3.2, C409.6.1.5, C409.6.1.6, Table C409.6.1.10.2(1), Table C409.7(1), C501.2, C501.3

90.4—2022: Energy Standard for Data Centers

C403.1.2, C405.9.1, C405.9.2, C406.2.2.3

140—2020: Method of Test for Evaluating Building Performance Simulation Software (with Addenda A and B)

C407.5.1.2, C409.5.1

ANSI/ASHRAE/ACCA Standard 183—2007 (RA2020): Peak Cooling and Heating Load Calculations in Buildings, Except Low-rise Residential Buildings

C403.1.1

ANSI/ASHRAE/ASHE Standard 170—2021: Ventilation of Health Care Facilities

C409.2.1

ASHRAE Standard 51—16/ANSI/AMCA Standard 210—16: Laboratory Methods of Testing Fans for Certified Aerodynamic Performance Rating

Table C403.8.5

ASHRAE—2020: 2020 ASHRAE Handbook—HVAC Systems and Equipment

C403.1.1

ISO/AHRI/ASHRAE 13256-1—1998 (R2012): Water-to-Air and Brine-to-Air Heat Pumps—Testing and Rating for Performance

Table C403.3.2(14)

ISO/AHRI/ASHRAE 13256-2—1998 (R2012): Water-to-Water and Brine-to-Water Heat Pumps—Testing and Rating for Performance

Table C403.3.2(14)

ASME *American Society of Mechanical Engineers, Two Park Avenue, New York, NY 10016-5990*

ASME A17.1—2022/CSA B44—: Safety Code for Elevators and Escalators

C405.10.2

ASTM *ASTM International, 100 Barr Harbor Drive, P.O. Box C700, West Conshohocken, PA 19428-2959*

C90—21: Specification for Load-bearing Concrete Masonry Units

Table C402.1.3

C835—06(2020): Standard Test Method for Total Hemispherical Emittance of Surfaces up to 1400°C

C402.3

C1363—19: Standard Test Method for Thermal Performance of Building Materials and Envelope Assemblies by Means of a Hot Box Apparatus

C303.1.4.1, Table C402.1.2, C402.1.2.1, C402.1.2.1.7, Table C402.1.2.1.7, C402.2.7

C1371—15: Standard Test Method for Determination of Emittance of Materials Near Room Temperature Using Portable Emissometers

C402.3, Table C402.4

C1549—16: Standard Test Method for Determination of Solar Reflectance Near Ambient Temperature Using a Portable Solar Reflectometer

C402.3, Table C402.4

D1003—21: Standard Test Method for Haze and Luminous Transmittance of Transparent Plastics

C402.6.2.3.2

D8052/D8052M—22: Standard Test Method for Quantification of Air Leakage in Low-Sloped Membrane Roof Assemblies

C402.6.2.3.2

E283/E283M—19: Standard Test Method for Determining the Rate of Air Leakage Through Exterior Windows, Curtain Walls and Doors Under Specified Pressure Differences Across the Specimen

C402.6.1.2.1, C402.6.2.3.2, Table C402.6.3

E408—13(2019): Standard Test Methods for Total Normal Emittance of Surfaces Using Inspection-meter Techniques

C402.3, Table C402.4

E779—19: Standard Test Method for Determining Air Leakage Rate by Fan Pressurization

C402.6.2.1, C402.6.2.2

E903—20: Standard Test Method Solar Absorptance, Reflectance and Transmittance of Materials Using Integrating Spheres

C402.3, Table C402.4

E1186—22: Standard Practices for Air Leakage Site Detection in Building Envelopes and Air Barrier Systems
C402.6.2

E1677—19: Standard Specification for Air Barrier (AB) Material or Assemblies for Low-Rise Framed Building Walls
C402.6.2.3.2

E1827—22: Standard Test Methods for Determining Airtightness of Building Using an Orifice Blower Door
C402.6.2.1, C402.6.2.2

E1918—21: Standard Test Method for Measuring Solar Reflectance of Horizontal or Low-sloped Surfaces in the Field
Table C402.4

E1980—11(2019): Standard Practice for Calculating Solar Reflectance Index of Horizontal and Low-sloped Opaque Surfaces
Table C402.4

E2178—21a: Standard Test Method for Determining Air Leakage Rate and Calculation of Air Permeance of Building Materials
C402.6.2.3.1

E2357—23: Standard Test Method for Determining Air Leakage of Air Barriers Assemblies
C402.6.2.3.2

E3158—18: Standard Test Method for Measuring the Air Leakage Rate of a Large or Multizone Building
C402.6.2.1

F1281—17(2021): Standard Specification for Cross-linked Polyethylene/Aluminum/Cross-linked Polyethylene (PEX-AL_PEX) Pressure Pipe
Table C404.5.2.1

F1361—21: Standard Test Method for Performance of Open Vat Fryers
Table C406.2.6.2(1)

F1484—18: Standard Test Method for Performance of Steam Cookers
Table C406.2.6.2(2)

F1495—20: Standard Specification for Combination Oven Electric or Gas Fired
Table C406.2.6.2(4)

F1496—13(2019): Standard Test Method for Performance of Convection Ovens
Table C406.2.6.2(4)

F1696—20: Standard Test Method for Energy Performance of Stationary-Rack, Door-Type Commercial Dishwashing Machines
Table C406.2.6.2(3)

F1920—20: Standard Test Method for Performance of Rack Conveyor Commercial Dishwashing Machines
Table C406.2.6.2(3)

F2093—18: Standard Test Method for Performance of Rack Ovens
Table C406.2.6.2(4)

F2144—21: Standard Test Method for Performance of Large Open Vat Fryers
Table C406.2.6.2(1)

F2861—20: Standard Test Method for Enhanced Performance of Combination Oven in Various Modes
Table C406.2.6.2(4)

CRRC *Cool Roof Rating Council, 2435 North Lombard Street, Portland, OR 97217*

ANSI/CRRC S100—2021: Standard Test Methods for Determining Radiative Properties of Materials
Table C402.4, C402.4.1

CSA *CSA Group, 8501 East Pleasant Valley Road, Cleveland, OH 44131-551*

AAMA/WDMA/CSA 101/I.S.2/A440—22: North American Fenestration Standard/Specification for Windows, Doors and Skylights
Table C402.6.3

CAN/CSA C439—18: Laboratory Methods of Test for Rating the Performance of Heat/Energy-Recovery Ventilators
C403.7.4.1

CSA B55.1—20: Test Method for Measuring Efficiency and Pressure Loss of Drain Water Heat Recovery Units
C404.7, C406.2.3.6

CSA B55.2—20: Drain Water Heat Recovery Units
C404.7

CTI *Cooling Technology Institute, P. O. Box 681807, Houston, TX 77268*

ATC-105—2019: Acceptance Test Code for Water Cooling Towers
Table C403.3.2(7)

ATC-105DS—2019: Acceptance Test Code for Dry Fluid Coolers
Table C403.3.2(7)

ATC-105S—2021: Acceptance Test Code for Closed Circuit Cooling Towers
Table C403.3.2(7)

ATC-106—2011: Acceptance Test for Mechanical Draft Evaporative Vapor Condensers
Table C403.3.2(7)

CTI STD-201 RS—2021: Performance Rating of Evaporative Heat Rejection Equipment
Table C403.3.2(7)

DASMA *Door & Access Systems Manufacturers Association, International, 1300 Sumner Avenue, Cleveland, OH 44115-2851*

ANSI/DASMA 105—2020: Test Method for Thermal Transmittance and Air Infiltration of Garage Doors and Rolling Doors
C303.1.3, Table C402.6.3

DOE *US Department of Energy, c/o Superintendent of Documents, 1000 Independence Avenue SW, Washington, DC 20585*

10 CFR 50: Domestic Licensing of Production and Utilization Facilities
C403.17

10 CFR, Part 430—2015: Energy Conservation Program for Consumer Products: Test Procedures and Certification and Enforcement Requirement for Plumbing Products; and Certification and Enforcement Requirements for Residential Appliances; Final Rule
Table C403.3.2(1), Table C403.3.2(2), Table C403.3.2(4), Table C403.3.2(5), Table C403.3.2(6), Table C403.3.2(14), C403.15, Table C404.2, C406.2.3.1.2, Table C406.2.3.5

10 CFR, Part 430, Appendix U: Uniform Test Method for Measuring the Energy Consumption of Ceiling Fans
C403.8.5, Table C403.9

10 CFR, Part 431—2015: Energy Efficiency Program for Certain Commercial and Industrial Equipment: Test Procedures and Efficiency Standards; Final Rules
Table C403.3.2(6), C403.8.4, C403.12, C403.12.1, Table C403.12.1, C403.12.2, Table C403.12.2.1(2), C404.2, Table C404.2, C405.7, Table C405.7, C405.8, Table C405.8(1), Table C405.8(2), Table C405.8(3), Table C405.8(4)

FGIA *Fenestration & Glazing Industry Alliance (formerly, American Architectural Manufacturers Association), 1900 E. Golf Road, Suite 1250, Schaumburg, IL 60173-4268*

AAMA/WDMA/CSA 101/I.S.2/A440—22: North American Fenestration Standard/Specification for Windows, Doors, and Skylights
Table C402.6.3

Green-e *Green-e, c/o Center for Resource Solutions1012 Torney Ave., 2nd Floor, San Francisco, CA 94129*

Green-e, Version 1.0, July 7, 2017: Green-e Framework for Renewable Energy Certification
C405.15.4

ICC *International Code Council, Inc., 200 Massachusetts Ave, NW, Suite 250, Washington, DC 20001*

IBC—24: International Building Code®
C201.3, C202, C303.1.1, C303.1.2, C303.2, C402.1.5, C402.2.7, C402.6.4, C405.2, C405.5.1, C405.15.2.1, C501.2, C501.2, C503.2.4

ICC 500—20: Standard for the Design and Construction of Storm Shelters
C402.5.2

IFC—24: International Fire Code®
C201.3, C501.2

IFGC—24: International Fuel Gas Code®
C201.3, C501.2

IMC—24: International Mechanical Code®
C201.3, C402.1.5, C403.2.2, C403.6, C403.6.1, C403.6.6, C403.7.1, C403.7.2, C403.7.4.2, C403.7.5, C403.7.7, C403.8.6.1, C403.13.1, C403.13.2, C403.13.2.1, C403.13.2.2, C406.2.2.5, C406.3.3, Table C407.4.1(1), C408.2.2.1, C501.2

IPC—24: International Plumbing Code®
C201.3, C501.2

IPMC—24: International Property Maintenance Code®
C501.2

IPSDC—24: International Private Sewage Disposal Code®
C501.2

IEC *IEC Regional Centre for North America, IEC International Electrotechnical Commission, 446 Main Street, 16th Floor, Worcester, MA 01608*

IEC 62746-10-1—2018: Systems interface between customer energy management system and the power management system – Part 10-1: Open automated demand response
C406.3

IEEE *Institute of Electrical and Electronic Engineers, Inc., 3 Park Avenue, 17th Floor, New York, NY 10016-5997*

515.1—2012: IEEE Standard for the Testing, Design, Installation, and Maintenance of Electrical Resistance Trace Heating for Commercial Applications
C404.6.2

1547—2018a: IEEE Standard for Interconnection and Interoperability of Distributed Energy Resources with Associated Electric Power Systems Interfaces
C405.16

IES Illuminating Engineering Society, 120 Wall Street, 17th Floor, New York, NY 10005-4001

ANSI/ASHRAE/IES 90.1—2022: Energy Standard for Buildings, Except Low-Rise Residential Buildings
C101.3, C401.2, C401.2.2, Table C402.1.2, C402.1.2.1, C402.1.3, Table C402.1.3, Table C403.3.2(5), Table C403.3.2(15), C406.2, C406.2.1.1, C409.6.1.3.2, C409.6.1.3.2, C409.6.1.5, C409.6.1.6, Table C409.6.1.10.2(1), Table C409.7(1), C501.2, C501.3

ANSI/IES RP-2—2020: Recommended Practice: Lighting Retail Spaces
C406.2.5

ANSI/IES RP-3—2020: Recommended Practice: Lighting Educational Facilities
C406.2.5

ANSI/IES RP-4—2020: Recommended Practice: Lighting Library Spaces
C406.2.5

ANSI/IES RP-6—2020: Recommended Practice: Lighting Sports and Recreational Areas
C406.2.5

ANSI/IES RP-7—2020: Recommended Practice: Lighting Industrial Facilities
C406.2.5

ANSI/IES RP-8—2021: Recommended Practice: Lighting Roadway and Parking Facilities
C406.2.5

ANSI/IES RP-9—2020: Recommended Practice: Lighting Hospitality Spaces
C406.2.5

ANSI/IES RP-10—2020: Recommended Practice: Lighting Common Applications
C406.2.5

ANSI/IES RP-27—2020: Recommended Practice: Photobiological Safety for Lighting Systems
C406.2.5

ANSI/IES RP-29—2020: Recommended Practice: Lighting Hospital and Healthcare Facilities
C406.2.5

ANSI/IES RP-30—2020: Recommended Practice: Lighting Museums
C406.2.5

ANSI/IES RP-41—2020: Recommended Practice: Lighting Theaters and Worship Spaces
C406.2.5

ANSI/IES/ALARP-11—2020: Recommended Practice: Lighting for Interior and Exterior Residential Environments
C406.2.5

ISO *International Organization for Standardization, Chemin de Blandonnet 8, CP 401 - 1214 Vernier, Geneva, Switzerland*

ISO 9050—2003: Glass in Building: Determination of Light Transmittance, Solar Direct Transmittance, Total Solar Energy Transmittance, Ultraviolet Transmittance and Related Glazing Factors

C402.3

ISO 25745-2—2015: Energy Performance of Lifts, Escalators and Moving Walks—Part 2: Energy calculation and classification for lifts (elevators)

C406.2.6.1

ISO 27327-1—2009: Fans—Air Curtain Units—Laboratory Methods of Testing for Aerodynamic Performance Rating

C402.6.6

ISO/AHRI/ASHRAE 13256-1—1998 (R2012): Water-to-Air and Brine-to-Air Heat Pumps—Testing and Rating for Performance

Table C403.3.2(14)

ISO/AHRI/ASHRAE 13256-2—1998 (R2012): Water-to-Water and Brine-to-Water Heat Pumps—Testing and Rating for Performance

Table C403.3.2(14)

NEMA *National Electrical Manufacturers Association, 1300 17th Street N #900, Arlington, VA 22209*

ANSI/NEMA MG1—2021: Motors and Generators

C202

OS 4—2016: Requirements for Air-Sealed Boxes for Electrical and Communication Applications

C402.6.1.2.2.1

NFPA *National Fire Protection Association, 1 Batterymarch Park, Quincy, MA 02169-7471*

70—23: National Electrical Code

C405.12.1, C501.2

NFRC *National Fenestration Rating Council, Inc., 6305 Ivy Lane, Suite 140, Greenbelt, MD 20770*

100—2023: Procedure for Determining Fenestration Products *U*-factors

C303.1.3, Table C402.1.2, C402.1.2.1.7, Table C402.1.4, C402.2.1.2, C402.5.1.1

200—2023: Procedure for Determining Fenestration Product Solar Heat Gain Coefficient and Visible Transmittance at Normal Incidence

C303.1.3, C402.5.1.1

203—2023: Procedure for Determining Visible Transmittance of Tubular Daylighting Devices

C303.1.3

300—2023: Test Method for Determining the Solar Optical Properties of Glazing Materials and Systems

C402.3

400—2023: Procedure for Determining Fenestration Product Air Leakage

Table C402.6.3

OpenADR *OpenADR Alliance, 111 Deerwood Road, Suite 200, San Roman, CA 94583*

OpenADR 2.0a and 2.0b—2019: Profile Specification Distributed Energy Resources

C406.3

RESNET *Residential Energy Services Network, Inc., P.O. Box 4561, Oceanside, CA 92052-4561*

ANSI/RESNET/ICC 380—2022: Standard for Testing Airtightness of Building, Dwelling Unit and Sleeping Unit Enclosures; Airtightness of Heating and Cooling Air Distribution Systems, and Airflow of Mechanical Ventilation Systems

C402.6.2.2

SMACNA *Sheet Metal and Air Conditioning Contractors' National Association, Inc., 4021 Lafayette Center Drive Chantilly, VA 20151-1219*

ANSI/SMACNA 016, 2nd edition—2012: HVAC Air Duct Leakage Test Manual Second Edition (ANSI/SMACNA 016—2012)

C403.13.2.3, C503.3.3

UL *UL LLC, 333 Pfingsten Road, Northbrook, IL 60062-2096*

710—2012: Exhaust Hoods for Commercial Cooking Equipment—with Revisions through February 2021

C403.7.5

727—2018: Oil-fired Central Furnaces

Table C403.3.2(5)

731—1995: Oil-fired Unit Heaters—with Revisions through October 2015

Table C403.3.2(5)

1741—2021: Inverters, Converters, Controllers and Interconnection System Equipment for Use with Distributed Energy Resources

C405.16

1784—2015: Air Leakage Tests of Door Assemblies—with Revisions through February 2015

C402.6.4

US-FTC *United States-Federal Trade Commission, 600 Pennsylvania Avenue NW, Washington, DC 20580*

CFR Title 16 (2015): *R*-Value Rule

C303.1.4

WDMA *Window & Door Manufacturers Association, 2001 K Street NW, 3rd Floor North, Washington, DC 20006*

AAMA/WDMA/CSA 101/I.S.2/A440—22: North American Fenestration Standard/Specification for windows, doors and skylights

Table C402.6.3

APPENDIX CA BOARD OF APPEALS—COMMERCIAL

The provisions contained in this appendix are not mandatory unless specifically referenced in the adopting ordinance.

User notes:

About this appendix: *Appendix CA provides criteria for Board of Appeals members. Also provided are procedures by which the Board of Appeals should conduct its business.*

SECTION CA101—GENERAL

CA101.1 Scope. A board of appeals shall be established within the jurisdiction for the purpose of hearing applications for modification of the requirements of this code pursuant to the provisions of Section C109. The board shall be established and operated in accordance with this section, and shall be authorized to hear evidence from appellants and the *code official* pertaining to the application and intent of this code for the purpose of issuing orders pursuant to these provisions.

CA101.2 Application for appeal. Any person shall have the right to appeal a decision of the *code official* to the board. An application for appeal shall be based on a claim that the intent of this code or the rules legally adopted hereunder have been incorrectly interpreted, the provisions of this code do not fully apply or an equally good or better form of construction is proposed. The application shall be filed on a form obtained from the *code official* within 20 days after the notice was served.

CA101.2.1 Limitation of authority. The board shall not have authority to waive requirements of this code or interpret the administration of this code.

CA101.2.2 Stays of enforcement. Appeals of notice and orders, other than Imminent Danger notices, shall stay the enforcement of the notice and order until the appeal is heard by the board.

CA101.3 Membership of board. The board shall consist of five voting members appointed by the chief appointing authority of the jurisdiction. Each member shall serve for **[INSERT NUMBER OF YEARS]** years or until a successor has been appointed. The board member's terms shall be staggered at intervals, so as to provide continuity. The *code official* shall be an ex officio member of said board but shall not vote on any matter before the board.

CA101.3.1 Qualifications. The board shall consist of five individuals, who are qualified by experience and training to pass on matters pertaining to building construction and are not employees of the jurisdiction.

CA101.3.2 Alternate members. The chief appointing authority is authorized to appoint two alternate members who shall be called by the board chairperson to hear appeals during the absence or disqualification of a member. Alternate members shall possess the qualifications required for board membership, and shall be appointed for the same term or until a successor has been appointed.

CA101.3.3 Vacancies. Vacancies shall be filled for an unexpired term in the same manner in which original appointments are required to be made.

CA101.3.4 Chairperson. The board shall annually select one of its members to serve as chairperson.

CA101.3.5 Secretary. The chief appointing authority shall designate a qualified clerk to serve as secretary to the board. The secretary shall file a detailed record of all proceedings which shall set forth the reasons for the board's decision, the vote of each member, the absence of a member and any failure of a member to vote.

CA101.3.6 Conflict of interest. A member with any personal, professional or financial interest in a matter before the board shall declare such interest and refrain from participating in discussions, deliberations and voting on such matters.

CA101.3.7 Compensation of members. Compensation of members shall be determined by law.

CA101.3.8 Removal from the board. A member shall be removed from the board prior to the end of their terms only for cause. Any member with continued absence from regular meeting of the board may be removed at the discretion of the chief appointing authority.

CA101.4 Rules and procedures. The board shall establish policies and procedures necessary to carry out its duties consistent with the provisions of this code and applicable state law. The procedures shall not require compliance with strict rules of evidence, but shall mandate that only relevant information be presented.

CA101.5 Notice of meeting. The board shall meet upon notice from the chairperson, within 10 days of the filing of an appeal or at stated periodic intervals.

CA101.5.1 Open hearing. All hearings before the board shall be open to the public. The appellant, the appellant's representative, the *code official* and any person whose interests are affected shall be given an opportunity to be heard.

CA101.5.2 Quorum. Three members of the board shall constitute a quorum.

CA101.5.3 Postponed hearing. When five members are not present to hear an appeal, either the appellant or the appellant's representative shall have the right to request a postponement of the hearing.

CA101.6 Legal counsel. The jurisdiction shall furnish legal counsel to the board to provide members with general legal advice concerning matters before them for consideration. Members shall be represented by legal counsel at the jurisdiction's expense in all matters arising from service within the scope of their duties.

CA101.7 Board decision. The board shall only modify or reverse the decision of the *code official* by a concurring vote of three or more members.

CA101.7.1 Resolution. The decision of the board shall be by resolution. Every decision shall be promptly filed in writing in the office of the *code official* within three days and shall be open to the public for inspection. A certified copy shall be furnished to the appellant or the appellant's representative and to the *code official*.

CA101.7.2 Administration. The *code official* shall take immediate action in accordance with the decision of the board.

CA101.8 Court review. Any person, whether or not a previous party of the appeal, shall have the right to apply to the appropriate court for a writ of certiorari to correct errors of law. Application for review shall be made in the manner and time required by law following the filing of the decision in the office of the chief administrative officer.

SOLAR-READY ZONE—COMMERCIAL

The provisions contained in this appendix are not mandatory unless specifically referenced in the adopting ordinance.

User notes:

About this appendix: *Appendix CB is intended to encourage the installation of renewable energy systems by preparing buildings for the future installation of solar energy equipment, piping and wiring.*

SECTION CB101—SCOPE

CB101.1 General. These provisions shall be applicable for new construction where solar-ready provisions are required.

SECTION CB102—GENERAL DEFINITION

SOLAR-READY ZONE. A section or sections of the roof or building overhang designated and reserved for the future installation of a solar photovoltaic or solar thermal system.

SECTION CB103—SOLAR-READY ZONE

CB103.1 General. A solar-ready zone shall be located on the roof of buildings that are five stories or less in height above grade plane, and are oriented between 110 degrees and 270 degrees of true north or have *low slope* roofs. Solar-ready zones shall comply with Sections CB103.2 through CB103.8.

Exceptions:

1. A *building* with a permanently installed, on-site renewable energy system.
2. A *building* with a solar-ready zone that is shaded for more than 70 percent of daylight hours annually.
3. A *building* where the licensed design professional certifies that the incident solar radiation available to the *building* is not suitable for a solar-ready zone.
4. A *building* where the licensed design professional certifies that the solar zone area required by Section CB103.3 cannot be met because of extensive rooftop equipment, skylights, *vegetative roof* areas or other obstructions.

CB103.2 Construction document requirements for a solar-ready zone. *Construction documents* shall indicate the solar-ready zone.

CB103.3 Solar-ready zone area. The total solar-ready zone area shall be not less than 40 percent of the roof area calculated as the horizontally projected gross roof area less the area covered by skylights, occupied roof decks, *vegetative roof* areas and mandatory access or set back areas as required by the *International Fire Code*. The solar-ready zone shall be a single area or smaller, separated sub-zone areas. Each sub-zone shall be not less than 5 feet (1524 mm) in width in the narrowest dimension.

CB103.4 Obstructions. Solar ready zones shall be free from obstructions, including pipes, vents, *ducts*, HVAC equipment, skylights and roof-mounted equipment.

CB103.5 Roof loads and documentation. A collateral dead load of not less than 5 pounds per square foot (5 psf) (24.41 kg/m^2) shall be included in the gravity and lateral design calculations for the solar-ready zone. The structural design loads for roof dead load and roof live load shall be indicated on the *construction documents*.

CB103.6 Interconnection pathway. *Construction documents* shall indicate pathways for routing of conduit or piping from the solar-ready zone to the electrical service panel or service hot water system.

CB103.7 Electrical service reserved space. The main electrical service panel shall have a reserved space to allow installation of a dual-pole circuit breaker for future solar electric and shall be labeled "For Future Solar Electric." The reserved space shall be positioned at the end of the panel that is opposite from the panel supply conductor connection.

CB103.8 Construction documentation certificate. A permanent certificate, indicating the solar-ready zone and other requirements of this section, shall be posted near the electrical distribution panel, *water heater* or other conspicuous location by the builder or *registered design professional*.

APPENDIX CC

ZERO ENERGY COMMERCIAL BUILDING PROVISIONS

The provisions contained in this appendix are not mandatory unless specifically referenced in the adopting ordinance.

User notes:

About this appendix: *Appendix CC provides a model for applying new renewable energy generation when new buildings add electric load to the grid. This renewable energy will avoid the additional emissions that would otherwise occur from conventional power generation.*

SECTION CC101—GENERAL

CC101.1 Purpose. The purpose of this appendix is to supplement the *International Energy Conservation Code* and require renewable energy systems of adequate capacity to achieve net zero operational energy.

CC101.2 Scope. This appendix applies to new buildings that are addressed by the *International Energy Conservation Code.*

Exceptions:

1. Detached one- and two-family dwellings and townhouses as well as Group R-2 buildings three stories or less in height above grade plane, manufactured homes (mobile dwellings), and manufactured houses (modular dwellings).
2. Buildings that use neither electricity nor fossil fuel.

SECTION CC102—DEFINITIONS

CC102.1 Definitions. The definitions contained in this section supplement or modify the definitions in the *International Energy Conservation Code.*

ADJUSTED OFF-SITE RENEWABLE ENERGY. The amount of energy production from off-site renewable energy systems that may be used to offset building energy.

BUILDING ENERGY. All energy consumed at the *building site* as measured at the site boundary. Contributions from on-site or off-site renewable energy systems shall not be considered when determining the building energy.

DIRECT ACCESS TO WHOLESALE MARKET. An agreement by the *owner* and a renewable energy developer to purchase renewable energy from the wholesale market.

DIRECT OWNERSHIP. An off-site renewable energy system under the ownership or control of the building project *owner.*

GREEN RETAIL PRICING. A program by the retail electricity provider to provide 100 percent renewable energy to the building project *owner.*

MINIMUM RENEWABLE ENERGY REQUIREMENT. The minimum amount of on-site or adjusted off-site renewable energy needed to comply with this appendix.

OFF-SITE RENEWABLE ENERGY SYSTEM. A renewable energy system that serves the building project and is not an on-site renewable energy system, including contracted purchases of renewable energy and renewable energy certificates (RECs).

ON-SITE RENEWABLE ENERGY SYSTEM. Renewable energy systems located on any of the following:

1. The building.
2. The property on which the building is located.
3. A property that shares a boundary with and is under the same ownership or control as the property on which the building is located.
4. A property that is under the same ownership or control as the property on which the building is located and is separated only by a public right-of-way from the building served by the renewable energy system.

RENEWABLE ENERGY INVESTMENT FUND (REIF). A fund established by a jurisdiction to accept payment from building project owners to construct or acquire interests in qualifying renewable energy systems, together with their associated RECS, on the building project owner's behalf.

RENEWABLE ENERGY SYSTEM. Photovoltaic, solar thermal, geothermal energy extracted from hot fluid or steam, wind, or other *approved* systems used to generate renewable energy.

SEMIHEATED SPACE. An *enclosed space* within a *building* that is heated by a heating system whose output capacity is greater than or equal to 3.4 Btu/h × ft^2 of floor area but is not a *conditioned space.*

SECTION CC103—MINIMUM RENEWABLE ENERGY

Scan for Changes 5b1de92

CC103.1 Renewable energy. On-site renewable energy systems shall be installed, or adjusted off-site renewable energy shall be procured to meet the minimum renewable energy requirement in accordance with Equation CC-1.

Equation CC-1 $RE_{on\text{-}site} + RE_{off\text{-}site} \geq RE_{min}$

where:

$RE_{on\text{-}site}$ = Annual site energy production from on-site renewable energy systems, including installed on-site renewable energy systems used for compliance with Sections C405.13.1 and C406.

$RE_{off\text{-}site}$ = Adjusted annual site energy production from off-site renewable energy systems that is permitted to be credited against the minimum renewable energy requirement. This includes off-site renewable energy purchased for compliance with Section C405.13.2.

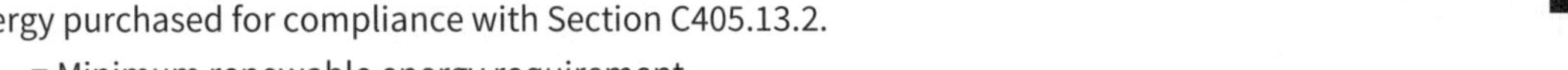

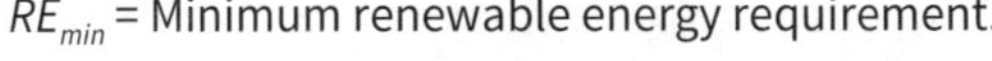

RE_{min} = Minimum renewable energy requirement.

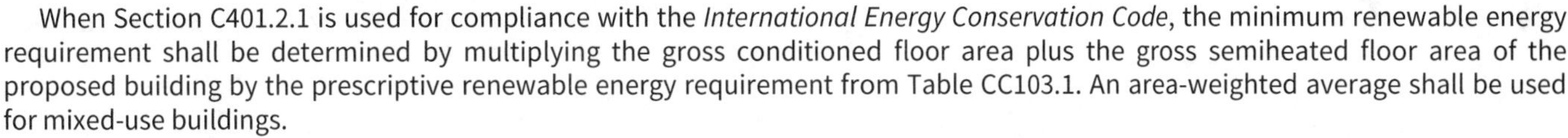

When Section C401.2.1 is used for compliance with the *International Energy Conservation Code*, the minimum renewable energy requirement shall be determined by multiplying the gross conditioned floor area plus the gross semiheated floor area of the proposed building by the prescriptive renewable energy requirement from Table CC103.1. An area-weighted average shall be used for mixed-use buildings.

When Section C401.2.1, Item 2 or Section C401.2.2 is used for compliance with the *International Energy Conservation Code*, the minimum renewable energy requirement shall be equal to the building energy as determined from energy simulations.

TABLE CC103.1—PRESCRIPTIVE RENEWABLE ENERGY REQUIREMENT FOR BUILDING TYPES AND CLIMATES (kWh/ft²/yr)

CLIMATE ZONE	BUILDING AREA TYPE											
	Multifamily (R-2)	Healthcare/ Hospital (I-2)	Hotel/ Motel (R-2)	Office (B)	Restaurant (A-2)	Retail (M)	School (E)	Warehouse (S)	Grocery Store (M)	Laboratory (B)	Assembly (A)	All Others
0A	13	35	23	10	129	17	16	3	27	41	5	17
0B	12	34	22	10	123	17	15	3	26	40	5	16
1A	11	32	20	9	113	14	13	3	24	36	4	15
1B	11	32	20	9	118	15	14	3	24	37	5	15
2A	11	32	20	8	114	13	12	3	22	34	4	14
2B	11	30	18	8	108	12	11	3	22	33	4	13
3A	11	30	18	8	117	13	11	3	21	31	4	13
3B	10	29	18	8	110	12	10	3	20	31	4	13
3C	9	28	18	7	100	10	9	2	18	27	3	12
4A	12	31	18	8	123	15	11	6	21	32	4	14
4B	11	29	18	7	113	12	10	4	20	30	4	13
4C	10	28	17	7	111	13	10	4	18	28	3	13
5A	12	31	19	8	133	17	11	8	22	34	4	15
5B	11	29	18	8	125	14	11	5	21	31	4	14
5C	10	29	17	7	116	13	10	4	18	27	3	13
6A	14	33	20	10	151	20	13	11	26	39	5	17
6B	13	33	19	8	137	17	11	7	22	34	4	16
7	14	37	21	9	164	20	13	10	25	37	5	18
8	15	40	22	11	190	23	16	10	28	43	5	20

For SI: 1 kilowatt hour per square foot = 10.76 kWh/m².

CC103.2 Calculation of on-site renewable energy. The annual energy production from on-site renewable energy systems shall be determined using *approved* software.

TABLE CC103.2—PROCUREMENT FACTORS FOR RENEWABLE ENERGY SYSTEM COMPLIANCE ALTERNATIVES

ON-SITE RENEWABLE ENERGY	PROCUREMENT FACTOR	
	Unbundled RECs	Other Procurement Methods
7.5 W/ft² of roof area or more or where one or more of Exceptions 1, 2 and 3 to Section C405.15.1 are satisfied.	0.20	1.0

TABLE CC103.2—PROCUREMENT FACTORS FOR RENEWABLE ENERGY SYSTEM COMPLIANCE ALTERNATIVES—continued		
ON-SITE RENEWABLE ENERGY	**PROCUREMENT FACTOR**	
Less than 7.5 W/ft^2 of roof area and none among Exceptions 1, 2 and 3 to Section C405.15.1 is satisfied.	0.20	0.75
For SI: 1 watt per square foot = W/0.0929 m^2. W = Watts.		

CC103.2.1 Renewable energy certificates. Renewable energy certificates (RECs) associated with the on-site renewable energy system shall be assigned to the initial and subsequent building owner(s) for a cumulative period of not less than 15 years. The building owner(s) are permitted to transfer RECs to building tenants occupying the *building*.

CC103.3 Off-site renewable energy. Off-site energy shall comply with Sections CC103.3.1 and CC103.3.2.

CC103.3.1 Off-site procurement methods. One or more of the following off-site renewable energy procurement methods shall be used to comply with Section CC103.1:

1. Community renewables energy facility.
2. *Renewable energy investment fund.*
3. *Financial renewable energy power purchase agreement.*
4. *Direct ownership.*
5. *Direct access to wholesale market.*
6. Green retail pricing.
7. Unbundled Renewable Energy Certificates (RECs).
8. *Physical renewable energy power purchase agreement.*

CC103.3.2 Requirements for all procurement methods. Off-site renewable energy systems and procurement methods used to comply with Section CC103.1 shall comply with all of the following:

1. The building *owner* shall sign a legally binding contract or other *approved* agreement to procure qualifying off-site renewable energy.
2. The procurement contract shall have duration of not less than 15 years and shall be structured to survive a partial or full transfer of ownership of the property.
3. RECs associated with the procured *off-site renewable energy* shall comply with the following requirements:
    3.1. The RECs shall be retained or retired by or on behalf of the property *owner* or tenant for a period of not less than 15 years.
    3.2. The RECs shall be created within a 12-month period of use of the REC.
    3.3. The RECs shall be from a generating asset constructed not more than 5 years before the issuance of the certificate of occupancy.
4. The generating source shall be a renewable energy system.
5. The generation source shall be located where the energy can be delivered to the *building site* by any of the following:
    5.1. Direct connection to the off-site renewable energy facility.
    5.2. The local utility or distribution entity.
    5.3. An interconnected electrical network where energy delivery capacity between the generator and the *building* site is available.
6. Records on power sent to or purchased by the building shall be retained by the building owner and made available for inspection by the code official upon request.

CC103.3.3 Adjusted off-site renewable energy. The process for calculating the adjusted *off-site renewable energy* is shown in Equation CC-2.

Equation CC-2 $RE_{off\text{-}site} = PF_{NonRecs} \times RE_{NonRecs} + 0.20 \times RE_{Recs}$

where:

$RE_{off\text{-}site}$ = Adjusted off-site renewable energy.

$PF_{NonRecs}$ = The renewable energy procurement factor for off-site renewable energy other than RECs, in accordance with Section CC103.3.3.1.

$RE_{NonRecs}$ = Annual energy production for renewable energy procurement methods other than RECs.

RE_{Recs} = Annual energy production associated with unbundled RECs.

CC103.3.3.1 Procurement factors. The procurement factors for renewable energy system compliance alternatives shall be as specified in Table CC103.2.

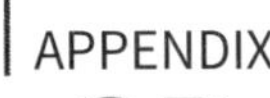

APPENDIX CD THE 2030 GLIDE PATH

The provisions contained in this appendix are not mandatory unless specifically referenced in the adopting ordinance.

User notes:

About this appendix: *This voluntary appendix is suited for adopting authorities that wish to extend beyond the mandatory provisions of this code toward zero net energy goals. Appendix CD is intended to be adopted by jurisdictions that will require new construction to operate at zero net energy by the year 2030. It reduces the net annual energy use of buildings by approximately one-third in comparison with buildings constructed in compliance with the 2021 IECC. It is assumed that the 2027 and 2030 editions will also reduce energy use by one-third each.*

SECTION CD101—COMPLIANCE

CD101.1 Reserved.

CD101.2 Simulated Building Performance compliance. Where compliance is demonstrated using the Simulated Building Performance option of Section C401.2.1, the percentage of annual *energy cost* (PAEC), applied to the *standard reference design* referenced in Equation 4-32, shall be multiplied by 0.97.

CD101.3 On-site renewable electricity systems. In addition to any renewable energy generation equipment provided to comply with Section C406.3, buildings shall install equipment for *on-site renewable energy* generation with a direct current (DC) nameplate capacity rating of not less than that computed using Equation CD-2.

Equation CD-1 $AA = CA + SNA/3$

where:

AA = Adjusted area, in ft^2 (m^2).

CA = Conditioned area, in ft^2 (m^2).

SNA = Semi-heated and nonconditioned area, in ft^2 (m^2).

Equation CD-2 $REQ = AA \times CF$

where:

REQ = Required on-site capacity, in DC watts.

AA = Adjusted area from Equation CD-1, in ft^2 (m^2).

CF = Capacity factor from Table CD101.3, in watts/ft^2 (m^2).

Exceptions:

1. Any required renewable energy generation capacity in excess of 10 watts per square foot (108 W/m^2) of net available roof area is permitted to be provided using an off-site renewable energy system in accordance with Section CD101.4. For the purposes of this section, net available roof area is the gross roof area minus the roof area occupied by any combination of skylights, mechanical equipment, vegetated areas, required access pathways, vehicle parking and occupied roof terrace area.
2. The following buildings are permitted to provide off-site renewable energy generation in accordance with Section CD101.4 in lieu of all or part of the *on-site renewable energy* generation capacity required by Section CD101.3:
   - 2.1. Any *building* where more than 50 percent of roof area would be shaded from direct-beam sunlight by existing natural objects or by structures that are not part of the *building* for more than 2,500 annual hours between 8:00 a.m. and 4:00 p.m.
   - 2.2. Any *building* with gross *conditioned floor area* less than 1,000 square feet (93 m^2).
   - 2.3. Any *building* whose primary roof slope is 2 units vertical in 12 units horizontal (17 percent slope) or greater.
3. Alternate forms of renewable energy generation capacity are permitted where the annual energy generation is not less than that produced by the required solar capacity, and where annual energy generation is calculated using an *approved* methodology.
4. All or part of the required renewable energy generation capacity is permitted to be replaced by other efficiency measures provided that such measures will reduce the annual energy consumption of the *building* by an amount no less than that which would otherwise be produced annually by the required renewable energy capacity, as calculated using the Simulated Building Performance compliance path in Section C407 and an *approved* calculation methodology for solar production.

TABLE CD101.3—ON-SITE RENEWABLE ELECTRICITY	
CLIMATE ZONE	**CAPACITY FACTOR**
1A, 2B, 3B, 3C, 4B and 5B	2.0 W/ft^2
0A, 0B, 1B, 2A, 3A and 6B	2.3 W/ft^2
4A, 4C, 5A, 5C, 6A, 7 and 8	2.6 W/ft^2
For SI: 1 watt per square foot = 10.76 W/m^2.	

CD101.4 Off-site renewable energy. *Buildings* that qualify for one or more of the exceptions to Section CD101.3 and that do not have on-site renewable energy systems sufficiently sized to fully comply with Section CD101.3 shall procure off-site renewable energy in accordance with Sections CD101.4.1 through CD101.4.3. Such procured energy shall provide not less than the total annual required off-site renewable energy determined in accordance with Equation CD-4 and shall be provided in addition to any renewable energy provided to comply with Section C406.3.

Equation CD-3 $DEF = REQ - INSTL$

where:

DEF = Renewable capacity deficit, in DC watts.

REQ = Required on-site capacity in DC watts, from Equation CD-2.

$INSTL$ = Installed on-site capacity, in DC watts.

Equation CD-4 $OFF = 4.4 \times DEF$

where:

OFF = Off-site renewable energy to be procured, in kWh/year.

CD101.4.1 Off-site procurement. The *building owner* shall procure and be credited for the total amount of off-site renewable energy required by Equation CD-4. Procured off-site renewable energy shall comply with the requirements applicable to not less than one of the following:

1. Community renewables energy facility.
2. Financial renewable energy power purchase agreement.
3. Physical renewable energy power purchase agreement.
4. Direct ownership.
5. Renewable Energy Investment Fund.
6. Green retail tariff.

CD101.4.2 Off-site contract. The renewable energy shall be delivered or credited to the *building site* under an energy contract with a duration of not less than 10 years. The contract shall be structured to survive a partial or full transfer of ownership of the building property. The total required off-site renewable energy shall be procured in equal installments over the duration of the off-site contract.

CD101.4.3 Renewable energy certificate (REC) documentation. The property *owner* or *owner's* authorized agent shall demonstrate that where RECs are associated with on-site and off-site renewable energy production required by Sections CD101.3 and CD101.4, the following criteria shall be met:

1. The RECs shall be retained and retired by or on behalf of the property *owner* or tenant for a period of not less than 10 years or the duration of the contract in Section CD101.4.2, whichever is less.
2. The RECs shall be created within a 12-month period of the use of the REC.
3. The RECs represent a generating asset constructed not more than 5 years before the issuance of the certificate of occupancy.

APPENDIX CE
REQUIRED HVAC TOTAL SYSTEM PERFORMANCE RATIO (TSPR)

The provisions contained in this appendix are not mandatory unless specifically referenced in the adopting ordinance.

User notes:

About this appendix: *Appendix CE can be adopted for stretch codes and utility incentive certification that requires Total System Performance Ratio (TSPR) analysis where it is applicable and requires a higher level of performance, saving 5 percent versus minimum efficiency systems.*

SECTION CE101—GENERAL

Scan for Changes

7043cc0

CE101.1 Required HVAC total system performance ratio (TSPR). For jurisdictions that wish to adopt a stretch code or HVAC incentive system, make the following changes to Section C403.

CE101.2 (Replace Section C403.1 with the following) General. Mechanical systems and equipment serving the building heating, cooling, ventilating or refrigerating needs shall comply with one of the following:

1. Sections C403.1.1 and C403.2 through C403.17 and where applicable, Section C409.
2. *Data centers* shall comply with Sections C403.1.1, C403.1.2 and C403.6 through C403.17.

CE101.3 HVAC total system performance ratio (TSPR). (Add the following three exceptions to Section C409.2.)

Exceptions:

1. Buildings with *conditioned floor area* less than 5,000 square feet (465 m^2).
2. *Alterations* to existing buildings that do not substantially replace the entire HVAC system and are not serving initial build-out construction.
3. HVAC systems meeting or exceeding all the requirements of the applicable target design HVAC system described in Tables C409.7(1) through C409.7(3).

TABLE CE101.3—MECHANICAL PERFORMANCE FACTORS

BUILDING TYPE	OCCUPANCY GROUP	CLIMATE ZONE																		
		0A	0B	1A	1B	2A	2B	3A	3B	3C	4A	4B	4C	5A	5B	5C	6A	6B	7	8
Office (small and medium)[a]	B	0.68	0.68	0.67	0.67	0.65	0.62	0.67	0.65	0.61	0.76	0.67	0.74	0.80	0.73	0.76	0.82	0.79	0.83	0.85
Office (large)[a]	B	0.79	0.79	0.80	0.80	0.75	0.78	0.68	0.77	0.73	0.64	0.72	0.60	0.67	0.68	0.60	0.69	0.67	0.67	0.67
Retail	M	0.57	0.54	0.48	0.52	0.44	0.44	0.41	0.48	0.38	0.43	0.54	0.65	0.44	0.65	0.64	0.48	0.43	0.42	0.36
Hotel/motel	R-1	0.59	0.59	0.60	0.60	0.59	0.65	0.58	0.67	0.69	0.43	0.56	0.49	0.36	0.45	0.48	0.33	0.36	0.29	0.25
Multifamily/ dormitory	R-2	0.61	0.60	0.64	0.60	0.62	0.61	0.56	0.68	0.52	0.50	0.48	0.42	0.51	0.45	0.36	0.52	0.48	0.48	0.45
School/ education and libraries	E (A-3)	0.78	0.77	0.76	0.75	0.71	0.68	0.67	0.68	0.64	0.69	0.68	0.65	0.78	0.69	0.58	0.85	0.76	0.79	0.73

For SI: 1 square foot = 0.0929 m^2.

a. Large office = gross conditioned floor area greater than 150,000 square feet or greater than five floors; all other offices are small or medium.

ENERGY CREDITS

The provisions contained in this appendix are not mandatory unless specifically referenced in the adopting ordinance.

User notes:

About this appendix: *Appendix CF can be adopted by authorities having jurisdiction seeking stretch codes building on the methodology of Section C406.*

SECTION CF101—GENERAL

CF101.1 Purpose. The purpose of this appendix is to supplement the *International Energy Conservation Code* and requires projects to comply with Advanced Energy Credit Package requirements.

CF101.2 Scope. This appendix applies to all buildings that, in accordance with Section C406.1, are required to comply with Section C406.1.1 or C406.1.2.

SECTION CF102—ADVANCED ENERGY CREDIT PACKAGE

CF102.1 Advanced Energy Credit Package requirements. The requirements of this section supersede the requirements of Section C406.1.1. Projects shall comply with measures from Section C406.2 to achieve the minimum number of required efficiency credits from Table CF102.1(1) based on building occupancy group and *climate zone*. Projects with multiple occupancies, unconditioned parking garages and *buildings* with separate shell-and-core and build-out construction permits shall comply as follows:

Where a project contains multiple occupancies, credits in Table CF102.1(1) from each building occupancy shall be weighted by the gross floor area to determine the weighted average project energy credits required. Accessory occupancies shall be included with the primary occupancy group for purposes of Section C406 and this appendix.

Exceptions:

1. Unconditioned parking garages that achieve 50 percent of the credits required for use in Groups S-1 and S-2 in Table CF102.1(1).
2. Portions of buildings devoted to manufacturing or industrial use.
3. Where a *building* achieves more renewable and load management credits in Section C406.3 than are required in Section C406.1.2, surplus credits shall be permitted to reduce required energy efficiency credits as follows:

$$EEC_{red} = EEC_{tbl} - \{\text{the lesser of: } [SRLM_{lim}, SLRM_{adj} \times (RLM_{ach} - RLM_{req})]\}$$

where:

EEC_{red} = Reduced required energy efficiency credits.

EEC_{tbl} = Required energy efficiency credits from Table C406.1.1(1).

$SRLM_{lim}$ = Surplus renewable and load management credit limit from Table C406.1.1(2).

$SRLM_{adj}$ = 1.0 for all-electric or all-renewable buildings (excluding emergency generation); 0.7 for buildings with fossil fuel equipment (excluding emergency generation).

RLM_{ach} = Achieved renewable and load management credits from Section C406.3.

RLM_{req} = Required renewable and load management credits from Section C406.1.2.

TABLE CF102.1(1)—ENERGY CREDIT REQUIREMENTS BY BUILDING OCCUPANCY GROUP

BUILDING OCCUPANCY GROUPS	CLIMATE ZONE																		
	0A	0B	1A	1B	2A	2B	3A	3B	3C	4A	4B	4C	5A	5B	5C	6A	6B	7	8
R-2, R-4 and I-1	179	174	188	197	200	200	200	200	200	200	200	200	193	200	200	200	200	200	200
I-2	78	75	73	71	80	90	100	85	90	97	83	90	99	90	96	107	106	130	117
R-1	106	100	110	105	109	122	123	125	131	137	129	136	157	139	147	171	158	180	176
B	114	110	112	115	108	107	116	111	114	126	118	123	135	125	125	152	142	153	141
A-2	83	81	82	82	86	86	108	91	97	126	99	111	147	117	113	160	143	163	151
M	113	113	121	118	123	127	116	116	133	109	100	92	99	134	125	171	146	150	137
E	91	95	91	100	96	100	105	104	101	113	110	110	120	117	122	131	132	126	131
S-1 and S-2	108	106	111	109	109	108	89	106	108	134	100	130	200	143	123	200	190	189	148
All other	54	53	55	56	57	60	61	60	63	68	60	65	73	68	69	84	79	84	78

TABLE CF102.1(2)—LIMIT TO ENERGY EFFICIENCY CREDIT CARRYOVER FROM RENEWABLE AND LOAD MANAGEMENT CREDITS

BUILDING OCCUPANCY GROUPS	CLIMATE ZONE																		
	0A	0B	1A	1B	2A	2B	3A	3B	3C	4A	4B	4C	5A	5B	5C	6A	6B	7	8
R-2, R-4 and I-1	100	100	114	110	113	91	95	115	101	73	102	99	54	73	101	45	50	66	62
I-2	30	25	26	20	28	33	38	31	33	37	30	32	41	41	50	53	56	75	80
R-1	20	8	20	5	26	22	20	28	30	19	26	23	24	28	28	27	30	43	54
B	25	19	18	20	15	15	15	24	25	31	36	32	37	40	43	42	40	51	66
A-2	9	5	5	5	5	5	5	5	5	9	5	5	21	9	5	32	19	49	61
M	5	5	5	5	5	5	5	5	5	5	5	5	5	5	5	5	5	5	10
E	24	24	31	29	29	28	19	33	39	31	43	33	34	37	33	31	33	46	54
S-1 and S-2	5	5	5	5	5	5	5	5	5	5	5	5	37	19	5	49	41	51	56
All other	5	5	5	5	5	5	5	5	15	5	6	8	5	11	15	5	5	9	20

SECTION CF103—BUILDINGS WITHOUT HEAT PUMPS

CF103.1 Buildings without heat pumps. The number of efficiency credits required by Section C406.1.1 shall be multiplied by 1.25 for the following:

1. Buildings using purchased energy that is not electricity for space heating or service water heating.
2. Buildings with electric storage water heaters that are not heat pumps.
3. Buildings with total heat pump space heating capacity less than the space heating load at heating design conditions calculated in accordance with Section C403.1.1.

Exceptions:

1. Portions of buildings devoted to manufacturing or industrial use.
2. Buildings complying with all of the following:
   - 2.1. The building's peak heating load calculated in accordance with Section C403.1.1 is greater than the building's peak cooling load calculated in accordance with Section C403.1.1.
   - 2.2. The building's total heat pump space heating capacity is not less than 50 percent of the building's space heating load at heating design conditions calculated in accordance with Section C403.1.1.
   - 2.3. Any energy source other than electricity or on-site renewable energy is used for space heating only where a heat pump cannot provide the necessary heating energy to satisfy the thermostat setting.
   - 2.4. Electric resistance heat is used only in accordance with Section C403.4.1.1.
3. Low-energy buildings complying with Section C402.1.1.1.
4. Portions of buildings in Utility and Miscellaneous Group U, Storage Group S, Factory Group F or High-Hazard Group H.
5. Buildings located in Climate Zones 0A, 0B, 1A, 1B, 2A and 2B.

SECTION CF104—EXISTING BUILDINGS

CF104.1 Additions not served by heat pumps. The number of efficiency credits required by Section C502.3.7 shall by multiplied by 1.25 for the following:

1. Additions using *purchased energy* that is not electricity for space heating or *service water heating*.
2. Additions served by electric storage water heaters that are not heat pumps.
3. Additions served by total heat pump space heating capacity less than the peak space heating load at heating design conditions calculated in accordance with Section C403.1.1.

Exceptions: Additions complying with all of the following:

1. The *addition's* peak heating load calculated in accordance with Section C403.1.1 is greater than the *addition's* peak cooling load calculated in accordance with Section C403.1.1.
2. The *addition's* total heat pump space heating capacity serving the *addition* is not less than 50 percent of the *addition's* space heating load at heating design conditions calculated in accordance with Section C403.1.1.
3. Any energy source other than electricity or *on-site renewable energy* is used for space heating serving the *addition* only where a heat pump cannot provide the necessary heating energy to satisfy the *thermostat* setting.
4. Electric resistance heat serving the *addition* is used only in accordance with Section C403.4.1.1.

APPENDIX CG ELECTRIC VEHICLE CHARGING INFRASTRUCTURE

The provisions contained in this appendix are not mandatory unless specifically referenced in the adopting ordinance.

User notes:

About this appendix: *Appendix CG can be adopted by authorities having jurisdiction seeking electric vehicle charging infrastructure requirements.*

SECTION CG101—ELECTRIC VEHICLE POWER TRANSFER

CG101.1 Definitions.

AUTOMOBILE PARKING SPACE. A space within a building or private or public parking lot, exclusive of driveways, ramps, columns, office and work areas, for the parking of an automobile.

ELECTRIC VEHICLE (EV). An automotive-type vehicle for on-road use, such as passenger automobiles, buses, trucks, vans, neighborhood electric vehicles and electric motorcycles, primarily powered by an electric motor that draws current from a building electrical service, electric vehicle supply equipment (EVSE), a rechargeable storage battery, a fuel cell, a photovoltaic array or another source of electric current.

ELECTRIC VEHICLE CAPABLE SPACE (EV CAPABLE SPACE). A designated automobile parking space that is provided with electrical infrastructure such as, but not limited to, raceways, cables, electrical capacity, a panelboard or other electrical distribution equipment space necessary for the future installation of an EVSE

ELECTRIC VEHICLE READY SPACE (EV READY SPACE). An automobile parking space that is provided with a branch circuit and an outlet, junction box or receptacle that will support an installed EVSE

ELECTRIC VEHICLE SUPPLY EQUIPMENT (EVSE). Equipment for plug-in power transfer, including ungrounded, grounded and equipment grounding conductors; electric vehicle connectors; attached plugs; any personal protection system; and all other fittings, devices, power outlets or apparatus installed specifically for the purpose of transferring energy between the premises wiring and the electric vehicle.

ELECTRIC VEHICLE SUPPLY EQUIPMENT INSTALLED SPACE (EVSE SPACE). An automobile parking space that is provided with a dedicated EVSE connection.

CG101.2 Electric vehicle power transfer infrastructure. Parking facilities shall be provided with electric vehicle power transfer infrastructure in accordance with Sections CG101.2.1 through CG101.2.6.

CG101.2.1 Quantity. The number of required electric vehicle (EV) spaces, *EV capable spaces* and *EV ready spaces* shall be determined in accordance with this section and Table CG101.2.1 based on the total number of *automobile parking spaces* and shall be rounded up to the nearest whole number. For R-2 buildings, the Table CG101.2.1 requirements shall be based on the total number of *dwelling units* or the total number of *automobile parking spaces*, whichever is less.

1. Where more than one parking facility is provided on a *building site*, the number of required *automobile parking spaces* required to have EV power transfer infrastructure shall be calculated separately for each parking facility.
2. Where one shared parking facility serves multiple building occupancies, the required number of spaces shall be determined proportionally based on the floor area of each building occupancy.
3. Installed electric vehicle supply equipment installed spaces (*EVSE spaces*) that exceed the minimum requirements of this section may be used to meet the minimum requirements for *EV ready spaces* and *EV capable spaces*.
4. Installed *EV ready spaces* that exceed the minimum requirements of this section may be used to meet the minimum requirements for *EV capable spaces*.
5. Where the number of *EV ready spaces* allocated for R-2 occupancies is equal to the number of *dwelling units* or to the number of *automobile parking spaces* allocated to R-2 occupancies, whichever is less, requirements for *EVSE spaces* for R-2 occupancies shall not apply.
6. Requirements for a Group S-2 parking garage shall be determined by the occupancies served by that parking garage. Where new automobile spaces do not serve specific occupancies, the values for Group S-2 parking garage in Table CG101.2.1 shall be used.

Exception: Parking facilities serving occupancies other than R2 with fewer than 10 *automobile parking spaces*.

TABLE CG101.2.1—REQUIRED EV POWER TRANSFER INFRASTRUCTURE

OCCUPANCY	EVSE SPACES	EV READY SPACES	EV CAPABLE SPACES
Group A	10%	0%	10%
Group B	15%	0%	30%
Group E	15%	0%	30%

TABLE CG101.2.1—REQUIRED EV POWER TRANSFER INFRASTRUCTURE—continued			
OCCUPANCY	EVSE SPACES	EV READY SPACES	EV CAPABLE SPACES
Group F	2%	0%	5%
Group H	1%	0%	0%
Group I	15%	0%	30%
Group M	15%	0%	30%
Group R-1	20%	5%	75%
Group R-2	20%	5%	75%
Groups R-3 and R-4	2%	0%	5%
Group S exclusive of parking garages	1%	0%	0%
Group S-2 parking garages	15%	0%	30%

CG101.2.2 EV capable spaces. Each *EV capable space* used to meet the requirements of Section CG101.2.1 shall comply with the following:

1. A continuous raceway or cable assembly shall be installed between an enclosure or outlet located within 3 feet (914 mm) of the *EV capable space* and electrical distribution equipment.
2. Installed raceway or cable assembly shall be sized and rated to supply a minimum circuit capacity in accordance with Section CG101.2.5.
3. The electrical distribution equipment to which the raceway or cable assembly connects shall have dedicated overcurrent protection device space and electrical capacity to supply a calculated load in accordance with Section CG101.2.5.
4. The enclosure or outlet and the electrical distribution equipment directory shall be marked: "For electric vehicle supply equipment (EVSE)."

CG101.2.3 EV ready spaces. Each branch circuit serving *EV ready spaces* used to meet the requirements of Section CG101.2.1 shall comply with the following:

1. Terminate at an outlet or enclosure located within 3 feet (914 mm) of each *EV ready space* it serves.
2. Have a minimum system and circuit capacity in accordance with Section CG101.2.5.
3. The electrical distribution equipment directory shall designate the branch circuit as "For electric vehicle supply equipment (EVSE)" and the outlet or enclosure shall be marked "For electric vehicle supply equipment (EVSE)."

CG101.2.4 EVSE spaces. An installed EVSE with multiple output connections shall be permitted to serve multiple *EVSE spaces*. Each EVSE installed to meet the requirements of Section CG101.2.1, serving either a single *EVSE space* or multiple *EVSE spaces*, shall comply with the following:

1. Have a minimum system and circuit capacity in accordance with Section CG101.2.5.
2. Have a nameplate rating not less than 6.2 kW.
3. Be located within 3 feet (914 mm) of each *EVSE space* it serves.
4. Be installed in accordance with Section CG101.2.6.

CG101.2.5 System and circuit capacity. The system and circuit capacity shall comply with Sections CG101.2.5.1 and CG101.2.5.2.

CG101.2.5.1 System capacity. The electrical distribution equipment supplying the branch circuit(s) serving each *EV capable space*, *EV ready space* and *EVSE space* shall comply with one of the following:

1. Have a calculated load of 7.2 kVA or the nameplate rating of the equipment, whichever is larger, for each *EV capable space*, *EV ready space* and *EVSE space*.
2. Meets the requirements of Section CG101.2.5.3.1.

CG101.2.5.2 Circuit capacity. The branch circuit serving each *EV capable space*, *EV ready space* and *EVSE space* shall comply with one of the following:

1. Have a rated capacity not less than 50 amperes or the nameplate rating of the equipment, whichever is larger.
2. Meets the requirements of Section CG101.2.5.3.2.

CG101.2.5.3 System and circuit capacity management. Where system and circuit capacity management is selected in Section CG101.2.5.1 or CG101.2.5.2, the installation shall comply with Sections CG101.2.5.3.1 and CG101.2.5.3.2.

CG101.2.5.3.1 System capacity management. The maximum equipment load on the electrical distribution equipment supplying the branch circuits(s) serving *EV capable spaces*, *EV ready spaces* and *EVSE spaces* controlled by an energy management system shall be the maximum load permitted by the energy management system, but not less than 3.3 kVA per space.

CG101.2.5.3.2 Circuit capacity management. Each branch circuit serving multiple *EVSE spaces*, *EV ready spaces* or *EV capable spaces* controlled by an energy management system shall comply with one of the following:

1. Have a minimum capacity of 25 amperes per space.

2. Have a minimum capacity of 20 amperes per space for R-2 occupancies where all *automobile parking spaces* are *EV ready spaces* or *EVSE spaces*.

CG101.2.6 EVSE installation. *EVSE* shall be installed in accordance with NFPA 70 and shall be *listed* and *labeled* in accordance with UL 2202 or UL 2594. *EVSE* shall be accessible in accordance with Section 1107 of the *International Building Code*.

SECTION CG102—REFERENCED STANDARDS

CG102.1 General. See Table CG102.1 for standards that are referenced in various sections of this appendix. Standards are listed by the standard identification with the effective date, standard title, and the section or sections of this appendix that reference the standard.

TABLE CG102.1—REFERENCED STANDARDS

STANDARD ACRONYM	STANDARD NAME	SECTIONS HEREIN REFERENCED
UL 2202—2009	*Electric Vehicle (EV) Charging System—with revisions through February 2018*	CG101.2.6
UL 2594—2016	*Standard for Electric Vehicle Supply Equipment*	CG101.2.6

APPENDIX CH
ELECTRIC-READY COMMERCIAL BUILDING PROVISIONS

The provisions contained in this appendix are not mandatory unless specifically referenced in the adopting ordinance.

User notes:

About this appendix: *Appendix CH can be adopted where authorities having jurisdiction seek new building to be electric ready.*

SECTION CH101—GENERAL

CH101.1 Intent. The intent of this appendix is to amend the *International Energy Conservation Code* to reduce future retrofit costs by requiring *commercial buildings* with combustion equipment to install the electrical infrastructure for electric equipment.

CH101.2 Scope. The provisions in this appendix are applicable to *commercial buildings*. New construction shall comply with Section CH103.

SECTION CH102—DEFINITIONS

APPLIANCE. A device or apparatus that is manufactured and designed to utilize energy and for which this code provides specific requirements.

COMBUSTION EQUIPMENT. Any equipment or appliance used for space heating, service water heating, cooking, clothes drying or lighting that uses a fossil fuel.

COMMERCIAL COOKING APPLIANCES. Commercial cooking appliances used in a commercial food service establishment for heating or cooking food and which produce grease vapors, steam, fumes, smoke or odors that are required to be removed through a local exhaust ventilation system. Such appliances include deep fat fryers, upright broilers, griddles, broilers, steam-jacketed kettles, hot-top ranges, under-fired broilers (charbroilers), ovens, barbecues, rotisseries and similar appliances.

SECTION CH103—NEW COMMERCIAL BUILDING

CH103.1 Additional electric infrastructure. Electric infrastructure in *buildings* that contain combustion equipment shall be installed in accordance with this section.

CH103.1.1 Combustion space heating. Spaces containing combustion equipment for space heating shall comply with Sections CH103.1.1.1, CH103.1.1.2 and CH103.1.1.3.

TABLE CH103.1.1—ALTERNATE ELECTRIC SPACE HEATING EQUIPMENT CONVERSION FACTORS (VA/kBtu/h)

99.6% HEATING DESIGN TEMPERATURE		P_s
Greater Than (°F)	Not Greater Than	VA/kBtu/h
50	N/A	N/A
45	50	94
40	45	100
35	40	107
30	35	115
25	30	124
20	25	135
15	20	149
10	15	164
5	10	184
0	5	210
-5	0	243
-10	-5	289
-15	-10	293

For SI: °C = [(° F) – 32]/1.8, 1 British thermal unit per hour = 0.2931 kW.

CH103.1.1.1 Designated exterior locations for future electric space-heating equipment. Spaces containing combustion equipment for space heating shall be provided with designated exterior location(s) shown on the plans and of sufficient size for outdoor space-heating heat pump equipment, with a chase that is sized to accommodate refrigerant lines between the exterior location and the interior location of the space-heating equipment, and with natural drainage for condensate from heating operation or a condensate drain located within 3 feet (914 mm) of the location of the future exterior space-heating heat pump equipment.

CH103.1.1.2 Dedicated branch circuits for future electric space-heating equipment. Spaces containing combustion space-heating equipment with a capacity not more than 65,000 Btu/h (19 kW) shall be provided with a dedicated 240-volt branch circuit with ampacity of not less than 50. The branch circuit shall terminate within 6 feet (1829 mm) of the space heating equipment and be in a location with ready *access*. Both ends of the branch circuit shall be labeled with the words "For Future Electric Space Heating Equipment" and be electrically isolated. Spaces containing combustion equipment for space heating with a capacity of not less than 65,000 Btu/h (19 kW) shall be provided with a dedicated branch circuit rated and sized in accordance with Section CH103.1.1.3, and terminating in a junction box within 3 feet (914 mm) of the location the space heating equipment in a location with ready *access*. Both ends of the branch circuit shall be labeled "For Future Electric Space Heating Equipment."

Exceptions:

1. Where a branch circuit provides electricity to the space heating combustion equipment and is rated and sized in accordance with Section CH103.1.1.3.
2. Where a branch circuit provides electricity to space cooling equipment and is rated and sized in accordance with Section CH103.1.1.3.
3. Where future electric space heating equipment would require three-phase power and the space containing combustion equipment for space heating is provided with an electrical panel with a label stating "For Future Electric Space Heating Equipment" and a bus bar rated and sized in accordance with Section CH103.1.1.3.
4. Buildings where the 99.6 percent design heating temperature is not less than 50°F (10°C).

CH103.1.1.3 Additional space heating electric infrastructure sizing. Electric infrastructure for future electric space heating equipment shall be sized to accommodate not less than one of the following:

1. An electrical capacity not less than the nameplate space heating combustion equipment heating capacity multiplied by the value in Table CH103.1.1, in accordance with Equation CH-1.

   Equation CH-1 $VA_S = Q_{com} \times P_S$

   VA_s = The required electrical capacity of the electrical infrastructure in volt-amps.

   Q_{com} = The nameplate heating capacity of the combustion equipment in kBtu/h.

   P_s = The VA per kBtu/h from Table CH103.1.1 in VA/kBtu/h.

2. An electrical capacity not less than the peak space heating load of the building areas served by the space heating combustion equipment, calculated in accordance with Section C403.1.1, multiplied by the value for the 99.6 percent design heating temperature in Table CH103.1.1, in accordance with Equation CH-2.

   Equation CH-2 $VA_S = Q_{design} \times P_S$

   VA_s = The required electrical capacity of the electrical infrastructure in volt-amps.

   Q_{design} = The 99.6 percent design heating load of the spaces served by the combustion equipment in kBtu/h.

   P_s = The VA per kBtu/h from Table CH103.1.1 in VA/kBtu/h.

3. An *approved* alternate design that uses no energy source other than electricity or *on-site renewable energy*.

CH103.1.2 Combustion service water heating Spaces containing combustion equipment for *service water heating* shall comply with Sections CH103.1.2.1, CH103.1.2.2 and CH103.1.2.3.

TABLE CH103.1.2—ALTERNATE ELECTRIC WATER HEATING EQUIPMENT CONVERSION FACTORS (VA/kBtu/h)

99.6% HEATING DESIGN TEMPERATURE		P_w
Greater Than (°F)	Not More Than	VA/kBtu/h
55	60	118
50	55	123
45	50	129
40	45	136
35	40	144
30	35	152
25	30	162
20	25	173
15	20	185

TABLE CH103.1.2—ALTERNATE ELECTRIC WATER HEATING EQUIPMENT CONVERSION FACTORS (VA/kBtu/h)—continued

99.6% HEATING DESIGN TEMPERATURE		P_w
Greater Than (°F)	Not More Than	VA/kBtu/h
10	15	293
5	10	293
0	5	293
Less than 0°F		293

For SI: °C = [(°F) – 32]/1.8, 1 British thermal unit per hour = 0.2931 kW.

CH103.1.2.1 Combustion service water heating electrical infrastructure. For each piece of combustion equipment for water heating with an input capacity of not more than 75,000 Btu/h (22 kW), the following electrical infrastructure is required:

1. An individual 240-volt branch circuit with an ampacity of not less than 30 shall be provided and terminate within 6 feet (1829 mm) of the water heater and shall be in a location with ready access.
2. The branch circuit overcurrent protection device and the termination of the branch circuit shall be labeled "For future electric water heater."
3. The space for containing the future water heater shall include the space occupied by the combustion equipment and shall have a height of not less than 7 feet (2134 mm), a width of not less than 3 feet (914 mm), a depth of not less than 3 feet (914 mm) and with a volume of not less than 700 cubic feet (20 m^3).

Exception: Where the space containing the water heater provides for air circulation sufficient for the operation of a heat pump water heater, the minimum room volume shall not be required.

CH103.1.2.2 Designated locations for future electric heat pump water heating equipment. Designated locations for future electric heat pump water heating equipment shall be in accordance with one of the following:

1. Designated exterior location(s) shown on the plans, of sufficient size for outdoor water heating heat pump equipment and with a chase that is sized to accommodate refrigerant lines between the exterior location and the interior location of the water heating equipment.
2. An interior location with a minimum volume the greater of 700 cubic feet (19 822 L) or 7 cubic feet (198 L) per 1,000 Btu/h (293 W) combustion equipment water heating capacity. The interior location shall include the space occupied by the combustion equipment.
3. An interior location with sufficient airflow to exhaust cool air from future water heating heat pump equipment provided by not fewer than one 16-inch (406 mm) by 24-inch (610 mm) grill to a heated space and one 8-inch (203 mm) *duct* of not more than 10 feet (3048 mm) in length for cool exhaust air.

CH103.1.2.3 Dedicated branch circuits for future electric heat pump water heating equipment. Spaces containing combustion equipment for water heating with a capacity of greater than 75,000 Btu/h (21 980 W) shall be provided with a dedicated branch circuit rated and sized in accordance with Section CH103.1.2.4 and terminating in a junction box within 3 feet (914 mm) of the location the water heating equipment in a location with ready *access*. Both ends of the branch circuit shall be labeled "For Future Electric Water Heating Equipment."

Exception: Where future electric water heating equipment would require three-phase power and the main electrical service panel has a reserved space for a bus bar rated and sized in accordance with Section CH103.1.2.4 and labeled "For Future Electric Water Heating Equipment."

CH103.1.2.4 Additional water heating electric infrastructure sizing. Electric infrastructure water heating equipment with a capacity of greater than 75,000 Btu/h (21 980 W) shall be sized to accommodate one of the following:

1. An electrical capacity not less than the combustion equipment water heating capacity multiplied by the value in Table CH103.1.2 plus electrical capacity to serve recirculating loads as shown in Equation CH-3.

   $$VA_W = (Q_{capacity} \times P_W) + [Q_{recirc} \times 293(VA(\text{Btu/h}))]$$

2. An alternate design that complies with this code, is *approved* by the authority having jurisdiction and uses no energy source other than electricity or *on-site renewable energy.*

CH103.1.3 Combustion cooking. Spaces containing combustion equipment for cooking shall comply with Section CH103.1.3.1 or CH103.1.3.2.

CH103.1.3.1 Commercial cooking. Spaces containing commercial cooking appliances shall be provided with a dedicated branch circuit with a minimum electrical capacity in accordance with Table CH103.1.3.1 based on the appliance in the space. The branch circuit shall terminate within 3 feet (914 mm) of the appliance in a location with ready *access*. Both ends of the branch circuit shall be labeled with the words "For Future Electric Cooking Equipment" and be electrically isolated.

TABLE CH103.1.3.1—COMMERCIAL COOKING MINIMUM BRANCH CIRCUIT CAPACITY	
COMMERCIAL COOKING APPLIANCE	**MINIMUM BRANCH CIRCUIT CAPACITY**
Range	469 VA/kBtu/h
Steamer	114 VA/kBtu/h
Fryer	200 VA/kBtu/h
Oven	266 VA/kBtu/h
Griddle	195 VA/kBtu/h
All other commercial cooking appliances	114 VA/kBtu/h
For SI: 1 British thermal unit per hour = 0.2931 kW.	

CH103.1.3.2 All other cooking. Spaces containing all other cooking equipment not designated as commercial cooking appliances shall be provided with a dedicated branch circuit in compliance with NFPA 70 Section 422.10. The branch circuit shall terminate within 6 feet (1829 mm) of fossil fuel ranges, cooktops and ovens and be in a location with ready *access*. Both ends of the branch circuit shall be labeled with the words "For Future Electric Cooking Equipment" and be electrically isolated.

CH103.1.4 Combustion clothes drying. Spaces containing combustion equipment for clothes drying shall comply with Section CH103.1.4.1 or CH103.1.4.2.

CH103.1.4.1 Commercial drying. Spaces containing clothes drying equipment and end uses for commercial laundry applications shall be provided with conduit that is continuous between a junction box located within 3 feet (914 mm) of the equipment and an electrical panel. The junction box, conduit and bus bar in the electrical panel shall be rated and sized to accommodate a branch circuit with sufficient capacity for equivalent electric equipment with equivalent equipment capacity. The electrical junction box and electrical panel shall have labels stating, "For Future Electric Clothes Drying Equipment."

CH103.1.4.2 Residential drying. Spaces containing clothes drying equipment, appliances and end uses serving multiple *dwelling units* or sleeping areas with a capacity less than or equal to 9.2 cubic feet (0.26 m^3) shall be provided with a dedicated 240-volt branch circuit with a minimum capacity of 30 amperes, shall terminate within 6 feet (1829 mm) of fossil fuel clothes dryers and shall be in a location with ready *access*. Both ends of the branch circuit shall be labeled with the words "For Future Electric Clothes Drying Equipment" and be electrically isolated.

CH103.1.5 On-site transformers. *Enclosed spaces* and underground vaults containing on-site electric transformers on the *building* side of the electric utility meter shall have sufficient space to accommodate transformers sized to serve the additional electric loads identified in Sections CH103.1.1, CH103.1.2, CH103.1.3 and CH103.1.4.

CH103.2 Hydronic heating design requirements. For all hydronic space heating systems, the design entering water temperature for coils, radiant panels, radiant floor systems, radiators, baseboard heaters and any other device that uses hot water to provide heat to a space shall be not more than 130°F (54°C).

CH103.3 Construction documentation. The *construction documents* shall provide details for additional electric infrastructure, including branch circuits, conduit, prewiring, panel capacity and electrical service capacity, as well as interior and exterior spaces designated for future electric equipment.

APPENDIX CI DEMAND RESPONSIVE CONTROLS

The provisions contained in this appendix are not mandatory unless specifically referenced in the adopting ordinance.

User notes:

About this appendix: *Appendix CI can by adopted by authorities having jurisdiction seeking demand responsive controls to be integrated into heating and cooling systems, water heating systems and lighting systems.*

SECTION CI101—DEMAND RESPONSIVE HEATING AND COOLING SYSTEMS

CI101.1 Demand responsive controls. Electric heating and cooling systems shall be provided with demand responsive controls capable of executing the following actions in response to a *demand response signal*:

1. Automatically increasing the zone operating cooling setpoint by the following values: 1°F (0.5°C), 2°F (1°C), 3°F (1.5°C) and 4°F (2°C).
2. Automatically decreasing the zone operating heating setpoint by the following values: 1°F (0.5°C), 2°F (1°C), 3°F (1.5°C) and 4°F (2°C).

Where a *demand response signal* is not available, the heating and cooling system controls shall be capable of performing all other functions. Where *thermostats* are controlled by direct digital control including, but not limited to, an energy management system, the system shall be capable of *demand responsive control* and capable of adjusting all thermal setpoints to comply. The demand responsive controls shall comply with either Section CI101.1.1 or CI101.1.2.

Exceptions:

1. Group I occupancies.
2. Group H occupancies.
3. Controls serving data center systems.
4. Occupancies or applications requiring precision in indoor temperature control as approved by the code official.
5. Buildings that comply with Load Management measure G02 in Section C406.3.3.
6. Buildings with energy storage capacity for not less than a 25 percent load reduction at peak load for a period of not less than 3 hours.

CI101.1.1 Air conditioners and heat pumps with two or more stages of control and cooling capacity of less than 65,000 Btu/h. *Thermostats* for air conditioners and heat pumps with two or more stages of control and a cooling capacity less than 65,000 Btu/h (19 kW) shall be provided with a *demand responsive control* that complies with the communication and performance requirements of AHRI 1380.

CI101.1.2 All other heating and cooling systems. *Thermostats* for heating and cooling systems shall be provided with a *demand responsive control* that complies with one of the following:

1. Certified OpenADR 2.0a VEN, as specified under Clause 11, Conformance.
2. Certified OpenADR 2.0b VEN, as specified under Clause 11, Conformance.
3. Certified by the manufacturer as being capable of responding to a *demand response signal* from a certified OpenADR 2.0b VEN by automatically implementing the control functions requested by the VEN for the equipment it controls.
4. IEC 62746-10-1.
5. The communication protocol required by a controlling entity, such as a utility or service provider, to participate in an automated demand response program.
6. The physical configuration and communication protocol of ANSI/CTA 2045-A or ANSI/CTA 2045-B.

SECTION CI102—DEMAND RESPONSIVE WATER HEATING

CI102.1 Demand responsive water heating. Electric storage water heaters with a rated water storage volume of 40 gallons (151 L) to 120 gallons (454 L) and a nameplate input rating equal to or less than 12 kW shall be provided with demand responsive controls in accordance with Table CI102.1.

Exceptions:

1. Water heaters that provide a hot water delivery temperature of 180°F (82°C) or greater.
2. Water heaters that comply with Section IV, Part HLW or Section X of the ASME *Boiler and Pressure Vessel Code.*
3. Water heaters that use three-phase electric power.

TABLE CI102.1—DEMAND RESPONSIVE CONTROLS FOR WATER HEATING		
EQUIPMENT TYPE	CONTROLS	
	Manufactured before 7/1/2025	Manufactured on or after 7/1/2025
Electric storage water heaters	AHRI Standard 1430 or ANSI/CTA-2045-B Level 1 and also capable of initiating water heating to meet the temperature setpoint in response to a demand response signal	AHRI Standard 1430

SECTION CI103—DEMAND RESPONSIVE LIGHTING CONTROLS

CI103.1 Demand responsive lighting controls. Interior general lighting in Group B, E, M and S occupancies shall have demand responsive controls complying with Section C405.2.8.1 in not less than 75 percent of the interior floor area.

Exceptions:

1. Where the combined interior floor area of Group B, E, M and S occupancies is less than 10,000 square feet (929 m^2).
2. Buildings where a *demand response signal* is not available from a controlling entity other than the *owner*.
3. Parking garages.
4. Ambulatory care facilities.
5. Outpatient clinics.
6. Physician or dental offices.

SECTION CI104—REFERENCED STANDARDS

CI104.1 General. See Table CI104.1 for standards that are referenced in various sections of this appendix. Standards are listed by the standard identification with the effective date, standard title, and the section or sections of this appendix that reference the standard.

TABLE CI104.1—REFERENCED STANDARDS		
STANDARD ACRONYM	STANDARD NAME	SECTIONS HEREIN REFERENCED
AHRI 1430 (I-P)—2022	*Demand Flexible Electric Storage Water Heaters (with Addendum 1)*	Table CI102.1
ASME BPVC	*Boiler and Pressure Vessel Code*	CI102.1

APPENDIX CJ

ELECTRICAL ENERGY STORAGE SYSTEM

The provisions contained in this appendix are not mandatory unless specifically referenced in the adopting ordinance.

User notes:

About this appendix: *This voluntary appendix provides requirements for electric energy storage readiness provisions.*

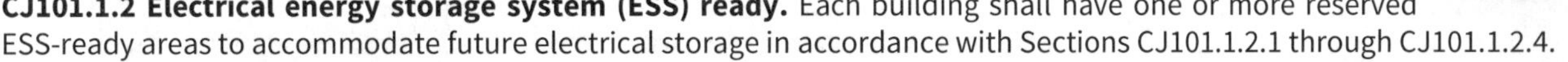

SECTION CJ101—ELECTRICAL ENERGY STORAGE SYSTEM

CJ101.1 Electrical energy storage system. Buildings shall comply with Section CJ101.1.1 or CJ101.1.2.

CJ101.1.1 Electrical energy storage system (ESS) capacity. Each building shall have one or more ESS with a total rated energy capacity and rated power capacity as follows:

1. ESS-rated energy capacity (kWh) ≥ 1.0 × installed on-site renewable electric energy system rated power (kWDC).
2. ESS-rated power capacity (kW) ≥ 0.25 × installed on-site renewable electric energy system rated power (kWDC).

Where installed, DC-coupled battery systems shall meet the requirements for rated energy capacity alone.

CJ101.1.2 Electrical energy storage system (ESS) ready. Each building shall have one or more reserved ESS-ready areas to accommodate future electrical storage in accordance with Sections CJ101.1.2.1 through CJ101.1.2.4.

CJ101.1.2.1 ESS-ready location. Each ESS-ready area shall be located in accordance with Section 1207 of the *International Fire Code*.

CJ101.1.2.2 ESS-ready minimum area requirements. Each ESS-ready area shall be sized in accordance with the spacing requirements of Section 1207 of the *International Fire Code* and the UL 9540 or UL 9540A designated rating of the planned system. Where rated to UL 9540A, the area shall be sized in accordance with the manufacturer's instructions.

CJ101.1.2.3 Electrical distribution equipment. The on-site electrical distribution equipment shall have sufficient capacity, rating and space to allow the installation of overcurrent devices and circuit wiring in accordance with NFPA 70 for future electrical ESS complying with the capacity criteria of Section CJ101.1.2.4.

CJ101.1.2.4 ESS-ready minimum system capacity. Compliance with ESS-ready requirements in Sections CJ101.1.2.1 through CJ101.1.2.3 shall be based on a minimum total energy capacity and minimum rated power capacity as follows:

1. ESS-rated energy capacity (kWh) ≥ gross conditioned floor area of the three largest floors (ft^2) × 0.0008 kWh/ft^2.
2. ESS-rated power capacity (kW) ≥ gross conditioned floor area of the three largest floors (ft^2) × 0.0002 kW/ft^2.

SECTION CJ102—REFERENCED STANDARDS

CJ102.1 General. See Table CJ102.1 for standards that are referenced in various sections of this appendix. Standards are listed by the standard identification with the effective date, standard title, and the section or sections of this appendix that reference the standard.

TABLE CJ102.1—REFERENCED STANDARDS

STANDARD ACRONYM	STANDARD NAME	SECTIONS HEREIN REFERENCED
ANSI/CAN/UL 9540—2020	*Energy Storage Systems and Equipment*	CJ101.1.2.2
ANSI/CAN/UL 9540A—2019	*Test Method for Evaluating Runaway Fire Propagation Energy Storage Systems*	CJ101.1.2.2

INDEX

ALL-ELECTRIC COMMERCIAL BUILDING PROVISIONS

Resources are related information that is not part of the code.

User notes:

About this resource: *This resource provides code compliance pathways for commercial buildings intended to result in all-electric buildings for adopting jurisdictions or individual projects.*

ICC Council Policy-49 Note: *In considering whether to adopt the content in this resource, jurisdictions in the United States should note that federal law might be found to preempt the provisions it prescribes. See the Public Health and Welfare Act, 42 U.S.C. § 6297: Effect on other law. Whether the content of this resource or a modification thereof is subject to preemption may depend on court decisions or whether a waiver has been issued by the US Department of Energy pursuant to 42 U.S.C. § 6297(d).*

SECTION CRA101—GENERAL

CRA101.1 Intent. The intent of this resource is to amend the *International Energy Conservation Code* to reduce greenhouse gas emissions from *buildings* and improve the safety and health for *commercial building* occupants by requiring new *all-electric buildings* and efficient electrification of *existing buildings*.

CRA101.2 Scope. The provisions in this resource are applicable to *commercial buildings*. New construction shall comply with Section CRA103. *Additions, alterations, repairs* and *changes of occupancy* to *existing buildings* shall comply with Chapter 5 and Section CRA104.

SECTION CRA102—DEFINITIONS

ALL-ELECTRIC BUILDING. A *building* using no *purchased energy* other than electricity when utility power is available.

APPLIANCE. A device or apparatus that is manufactured and designed to utilize energy and for which this code provides specific requirements.

COMBUSTION EQUIPMENT. Any equipment or appliance used for space heating, *service water heating*, cooking, clothes drying, humidification or lighting that uses fuel gas or fuel oil.

PURCHASED ENERGY. Energy or power purchased for consumption and delivered to the building site.

SUBSTANTIAL IMPROVEMENT. Any repair, reconstruction, rehabilitation, alteration, *addition* or other improvement of a building or structure, the cost of which is equal to or greater than 50 percent of the market value of the structure before the improvement. Where the structure has sustained substantial damage as defined in the *International Building Code*, any repairs are considered substantial improvement regardless of the actual repair work performed. Substantial improvement does not include the following:

1. Improvement of a building required to correct health, sanitary or safety code violations ordered by the code official.
2. Alteration of a historic building where the alteration will not affect the building's designation as a historic building.

SECTION CRA103—NEW COMMERCIAL BUILDINGS

CRA103.1 Application. New *commercial buildings* shall be all-electric buildings and comply with Section C401.2.1 or C401.2.2.

1. *Purchased energy* other than electricity shall be permitted where it has been demonstrated to the *code official* that the *building* is required by an applicable law or regulation to provide space heating with an emergency power system or a standby power system.
2. *Purchased energy* shall be permitted for an emergency power system or a standby power system.

CRA103.2 Electric resistance heating equipment. The sole use of electric resistance equipment and *appliances* for space and water heating shall be prohibited other than for *buildings* or portions of *buildings* that comply with not less than one of Sections CRA103.2.1 through CRA103.2.8.

CRA103.2.1 Low space heating capacity. Electric resistance *appliances* or equipment shall be permitted in *buildings* or areas of *buildings* not served by a mechanical cooling system and with a total space heating capacity not greater than 4.0 Btu/h (1.2 watts) per square foot of *conditioned space*.

CRA103.2.2 Small systems. *Buildings* in which electric resistance *appliances* or equipment comprise less than 5 percent of the total system heating capacity or serve less than 5 percent of the *conditioned floor area*.

CRA103.2.3 Specific conditions. Portions of *buildings* or specific equipment and *appliances,* subject to approval, that require electric resistance heating and cannot practicably be served by electric heat pumps.

CRA103.2.4 Kitchen makeup air. Makeup air for commercial kitchen exhaust systems required to be tempered by Section 508.1.1 of the *International Mechanical Code* is permitted to be heated by electric resistance.

CRA103.2.5 Freeze protection. The use of electric resistance heat for freeze protection shall comply with Sections CRA103.2.5.1 and CRA103.2.5.2.

CRA103.2.5.1 Low indoor design conditions. Space heating systems sized for spaces with indoor design conditions of not greater than 40°F (4.5°C) and intended for freeze protection, including temporary systems in unfinished spaces, shall be permitted to use electric resistance. The *building thermal envelope* of any such space shall be insulated in accordance with Section C402.1.

CRA103.2.5.2 Freeze protection system. Freeze protection systems shall comply with Section C403.14.4.

CRA103.2.6 Preheating of outdoor air. Hydronic systems without energy recovery ventilation and that do not use freeze protection fluids shall be permitted to utilize electric resistance to temper air to not more than 40°F (4.5°C). All systems with energy recovery ventilation shall be permitted to utilize electric resistance to preheat outdoor air to defrost or temper air entering the energy recovery device and shall comply with one of the following:

1. Where the space is mechanically humidified or has a *process application* that will maintain the space above 30 percent relative humidity when the outdoor temperature is not greater than 25°F (-4°C) and the system recovers latent energy, the outdoor air shall not be preheated to greater than 25°F (-4°C).
2. For sensible-only heat recovery exchangers, outdoor air shall not be preheated to greater than 25°F (-4°C).
3. For all other systems, outdoor air shall not be preheated to greater than 5°F (-15°C).

CRA103.2.7 Small buildings. *Buildings* with a *conditioned floor area* of not more than 250 square feet (23.2 m^2) and not served by a mechanical space cooling system shall be permitted to use electric resistance *appliances* or equipment for space heating.

CRA103.2.8 Supplemental heat. Electric resistance heat shall be permitted as supplemental heat where installed with heat pumps sized in accordance with Section CRA103.3 and where operated only when a heat pump cannot provide the necessary heating energy to satisfy the *thermostat* setting.

CRA103.3 Heat pump sizing for space heating. Heat pump space heating systems shall be sized to meet the building heating load at the greater of 0°F (-18°C) or the 99 percent annual heating dry-bulb for the nearest weather station provided in the ASHRAE *Handbook of Fundamentals*. The heat pump space heating system shall not require the use of supplemental electric heat at or above this temperature other than for defrosting. Lower capacity heat pumps that operate in conjunction with thermal storage shall be permitted if the system meets the requirements of this section.

CRA103.4 Heat pump sizing for water heating. Heat pump *service heating systems* shall be sized to meet not less than the *building service water heating* load at the greater of 15°F (-9.5°C) or the 99 percent annual heating dry-bulb for the nearest weather station provided in the latest edition of the ASHRAE *Handbook of Fundamentals*. Supplemental electric heat shall not be required at or above this temperature other than for temperature maintenance in recirculating systems and defrosting.

CRA103.5 Heating outside a building. Systems for heating outside a *building* shall comply with Section C403.14.1.

CRA103.6 Low capacity cooling equipment. Air conditioners with capacities less than 240,000 Btu/h (70 kW) shall be electric heat pump equipment sized and configured to provide both space cooling and space heating.

SECTION CRA104—EXISTING COMMERCIAL BUILDINGS

CRA104.1 Combustion equipment in additions. *Additions* shall use no *purchased energy* other than electricity and new equipment installed to serve additions shall not use *purchased energy* other than electricity. Where existing systems using *purchased energy* other than electricity serve an *addition,* the *existing building* and *addition* together shall not use more *purchased energy* other than electricity than the *existing building* alone.

CRA104.2 Substantial improvement. Buildings undergoing *substantial improvements* shall be all-electric buildings, comply with Section C402.5 and meet a site EUI by building type in accordance with ASHRAE Standard 100 Table 7-2a.

Exception: Compliance with ASHRAE Standard 100 shall not be required where *Group R* occupancies achieve an ERI score of 80 or below without *on-site renewable energy* included, in accordance with ANSI/RESNET/ICC 301, for each *dwelling unit*.

CRA104.3 Cooling equipment. New and replacement air conditioners shall be electric heat pump equipment sized and configured to provide both space cooling and space heating. Any existing space heating systems, other than existing heat pump equipment, that serve the same *zone* as the new equipment shall be configured as supplementary heat in accordance with Section CRA104.6.

CRA104.4 Service water heating equipment. Where *water heaters* are added or replaced, they shall not use *purchased energy* other than electricity.

CRA104.5 Furnace replacement. Newly installed warm air furnaces provided for space heating shall be permitted only as supplementary heat controlled in accordance with Section CRA104.6.

CRA104.6 Heat pump supplementary heat. Heat pumps having combustion equipment or electric resistance equipment for supplementary space or *service water heating* shall have controls that limit supplemental heat operation to only those times when one of the following applies:

1. The heat pump is operating in defrost mode.
2. The vapor compression cycle malfunctions.
3. For space heating systems, the *thermostat* malfunctions.
4. For space heating systems, the vapor compression cycle cannot provide the necessary heating energy to satisfy the *thermostat* setting.
5. The outdoor air temperature is less than the design temperature determined in accordance with Section CRA103.3.

6. For *service water heating*, the heat pump water heater cannot maintain an output water temperature of 120°F (49°C).
7. For temperature maintenance in *service water heating* systems.

New supplementary space and *service water heating* systems for heat pump equipment shall not be permitted to have a heating output capacity greater than the heating output capacity of the heat pump equipment.

SECTION CRA105—REFERENCED STANDARDS

CRA105.1 General. See Table CRA105.1 for standards that are referenced in various sections of this resource. Standards are listed by the standard identification with the effective date, the standard title, and the section or sections of this resource that reference this standard.

TABLE CRA105.1—REFERENCED STANDARDS

STANDARD ACRONYM	STANDARD NAME	SECTION HEREIN REFERENCED
ASHRAE 100—2018	*Energy Efficiency in Existing Buildings*	CRA104.2
ASHRAE—2017	*2017 ASHRAE Handbook of Fundamentals*	CRA103.3
ANSI/RESNET/ICC 301—2022	*Standard for the Calculation and Labeling of Energy Performance of Dwelling and Sleeping Units using an Energy Rating Index*—includes Addendum A, Approved July 28, 2022, and Addendum B, Approved October 12, 2022	CRA104.2

THE 2030 GLIDE PATH (PRESCRIPTIVE)

Resources are related information that is not part of the code.

User notes:

About this resource: *This resource is intended to be adopted by jurisdictions that will require new construction to operate at net zero energy by the year 2030. It reduces the net annual energy use of buildings by approximately one-third in comparison with buildings constructed in compliance with the 2021 IECC. It is assumed that the 2027 and 2030 editions will also reduce energy use by one-third each.*

ICC Council Policy-49 Note: *This resource is an accompaniment to the performance pathway included within Appendix CD and is intended for adopting authorities that wish to extend beyond the mandatory provisions of this code toward Zero Net Energy goals. For jurisdictions in the United States, compliance options for this prescriptive path may be limited if using only minimum efficiency mechanical and service water heating equipment. Adopting authorities may need to consider alternative means to expand methods for compliance under these conditions (see Section C104.1). Adopting authorities should be aware of potential preemption issues based on the Energy Policy and Conservation Act when evaluating whether to adopt the content in this resource. See the Public Health and Welfare Act, 42 U.S.C. § 6297: Effect on other law. Whether the content in this resource or a modification thereof is subject to preemption may depend on court decisions or whether a waiver has been issued by the US Department of Energy pursuant to 42 U.S.C. § 6297(d).*

SECTION CRB101—COMPLIANCE

CRB101.1 Prescriptive compliance. Where compliance is demonstrated using the Prescriptive Compliance option in Section C401.2.1, the number of additional efficiency credits required by Section C406.1 shall be 1.4 times the number that is required by Section C406.1.1.

Exception: Where a building achieves more renewable and load management credits in Section C406.3 than are required in Section C406.1.2, surplus credits shall be permitted to reduce required energy efficiency credits as follows:

$$EEC_{red} = EEC_{tbl} - \{\text{the lesser of: } [SRLM_{lim}, SLRM_{adj} \times (RLM_{ach} - RLM_{req})]\}$$

where:

EEC_{red} = Reduced required energy efficiency credits.

EEC_{tbl} = Required energy efficiency credits from Table C406.1.1(1).

$SRLM_{lim}$ = Surplus renewable and load management credit limit from Table CRB101.1.

$SRLM_{adj}$ = 1.0 for all-electric or all-renewable buildings (excluding emergency generation); 0.7 for buildings with fossil fuel equipment (excluding emergency generation).

RLM_{ach} = Achieved renewable and load management credits from Section C406.3.

RLM_{req} = Required renewable and load management credits from Section C406.1.2.

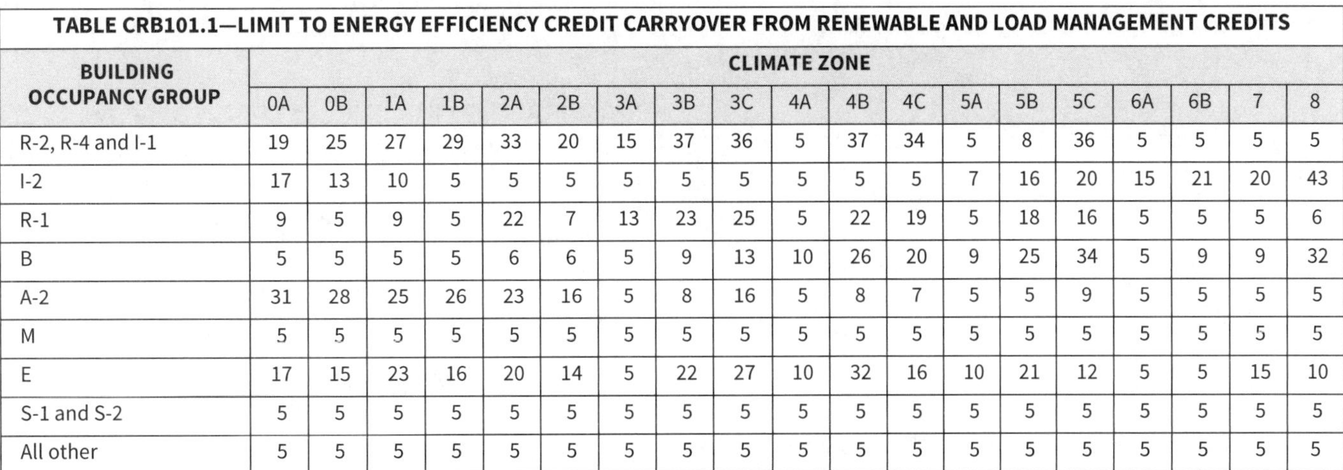

TABLE CRB101.1—LIMIT TO ENERGY EFFICIENCY CREDIT CARRYOVER FROM RENEWABLE AND LOAD MANAGEMENT CREDITS

BUILDING OCCUPANCY GROUP	CLIMATE ZONE																		
	0A	0B	1A	1B	2A	2B	3A	3B	3C	4A	4B	4C	5A	5B	5C	6A	6B	7	8
R-2, R-4 and I-1	19	25	27	29	33	20	15	37	36	5	37	34	5	8	36	5	5	5	5
I-2	17	13	10	5	5	5	5	5	5	5	5	5	7	16	20	15	21	20	43
R-1	9	5	9	5	22	7	13	23	25	5	22	19	5	18	16	5	5	5	6
B	5	5	5	5	6	6	5	9	13	10	26	20	9	25	34	5	9	9	32
A-2	31	28	25	26	23	16	5	8	16	5	8	7	5	5	9	5	5	5	5
M	5	5	5	5	5	5	5	5	5	5	5	5	5	5	5	5	5	5	5
E	17	15	23	16	20	14	5	22	27	10	32	16	10	21	12	5	5	15	10
S-1 and S-2	5	5	5	5	5	5	5	5	5	5	5	5	5	5	5	5	5	5	5
All other	5	5	5	5	5	5	5	5	5	5	5	5	5	5	5	5	5	5	5

IECC—RESIDENTIAL PROVISIONS

CONTENTS

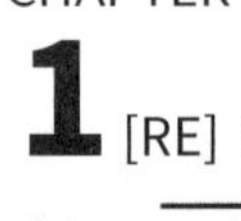

CHAPTER 1 [RE] SCOPE AND ADMINISTRATION

User notes:

About this chapter: *Chapter 1 establishes the limits of applicability of this code and describes how the code is to be applied and enforced. Chapter 1 is in two parts: Part 1—Scope and Application (Sections R101 and R102) and Part 2—Administration and Enforcement (Sections R103– R110). Section R101 identifies which buildings and structures come under its purview and references other I-Codes as applicable. Standards and codes are scoped to the extent referenced (see Section R102.4).*

This code is intended to be adopted as a legally enforceable document, and it cannot be effective without adequate provisions for its administration and enforcement. The provisions of Chapter 1 establish the authority and duties of the code official appointed by the authority having jurisdiction and also establish the rights and privileges of the design professional, contractor and property owner.

QR code use: *A QR code is placed at the beginning of any section that has undergone technical revision. To see those revisions, scan the QR code with a smart device or enter the 7-digit code beneath the QR code at the end of the following URL: qr.iccsafe.org/ (see Formatting Changes to the 2024 International Codes for more information).*

PART 1—SCOPE AND APPLICATION

SECTION R101—SCOPE AND GENERAL REQUIREMENTS

R101.1 Title. This code shall be known as the *Energy Conservation Code* of **[NAME OF JURISDICTION]** and shall be cited as such. It is referred to herein as "this code."

R101.2 Scope. This code applies to the design and construction of detached one- and two-family dwellings and multiple single-family dwellings (townhouses) and Group R-2, R-3 and R-4 buildings three stories or less in height above *grade plane*.

R101.2.1 Appendices. Provisions in the appendices shall not apply unless specifically adopted.

R101.3 Intent. The IECC—Residential Provisions provide market-driven, enforceable requirements for the design and construction of residential buildings, providing minimum efficiency requirements for buildings that result in the maximum level of energy efficiency that is safe, technologically feasible, and life cycle cost-effective, considering economic feasibility, including potential costs and savings for consumers and building owners, and return on investment. Additionally, the code provides jurisdictions with optional supplemental requirements, including requirements that lead to achievement of zero energy buildings, presently, and, through glidepaths that achieve zero energy buildings by 2030 and on additional timelines sought by governments, and achievement of additional policy goals as identified by the Energy and Carbon Advisory Council and approved by the Board of Directors. The code may include nonmandatory appendices incorporating additional energy efficiency and greenhouse gas reduction resources developed by the International Code Council and others. Requirements contained in the code will include, but not be limited to, prescriptive- and performance-based pathways. The code will aim to simplify code requirements to facilitate the code's use and compliance rate. The code is updated on a 3-year cycle with each subsequent edition providing increased energy savings over the prior edition. The IECC residential provisions shall include an update to Chapter 11 of the *International Residential Code*. This code is intended to provide flexibility to permit the use of innovative approaches and techniques to achieve this intent. This code is not intended to abridge safety, health or environmental requirements contained in other applicable codes or ordinances.

R101.4 Compliance. *Residential buildings* shall meet the provisions of IECC—Residential Provisions. *Commercial buildings* shall meet the provisions of IECC—Commercial Provisions.

R101.4.1 Compliance materials. The *code official* shall be permitted to approve specific computer software, worksheets, compliance manuals and other similar materials that meet the intent of this code.

SECTION R102—APPLICABILITY

R102.1 Applicability. Where, in any specific case, different sections of this code specify different materials, methods of construction or other requirements, the most restrictive shall govern. Where there is a conflict between a general requirement and a specific requirement, the specific requirement shall govern.

R102.1.1 Mixed residential and commercial buildings. Where a *building* includes both *residential building* and *commercial building* portions, each portion shall be separately considered and meet the applicable provisions of the IECC—Commercial Provisions or IECC—Residential Provisions.

R102.2 Other laws. The provisions of this code shall not be deemed to nullify any provisions of local, state or federal law.

R102.3 Application of references. References to chapter or section numbers, or to provisions not specifically identified by number, shall be construed to refer to such chapter, section or provision of this code.

R102.4 Referenced codes and standards. The codes and standards referenced in this code shall be those indicated in Chapter 6, and such codes and standards shall be considered as part of the requirements of this code to the prescribed extent of each such reference and as further regulated in Sections R102.4.1 and R102.4.2.

R102.4.1 Conflicts. Where conflicts occur between provisions of this code and referenced codes and standards, the provisions of this code shall apply.

R102.4.2 Provisions in referenced codes and standards. Where the extent of the reference to a referenced code or standard includes subject matter that is within the scope of this code, the provisions of this code, as applicable, shall take precedence over the provisions in the referenced code or standard.

R102.5 Partial invalidity. If a portion of this code is held to be illegal or void, such a decision shall not affect the validity of the remainder of this code.

PART 2—ADMINISTRATION AND ENFORCEMENT

SECTION R103—CODE COMPLIANCE AGENCY

R103.1 Creation of enforcement agency. The [INSERT NAME OF DEPARTMENT] is hereby created and the official in charge thereof shall be known as the authority having jurisdiction (AHJ). The function of the agency shall be the implementation, administration and enforcement of the provisions of this code.

R103.2 Appointment. The AHJ shall be appointed by the chief appointing authority of the jurisdiction.

R103.3 Deputies. In accordance with the prescribed procedures of this jurisdiction and with the concurrence of the appointing authority, the AHJ shall have the authority to appoint a deputy AHJ, other related technical officers, inspectors and other employees. Such employees shall have powers as delegated by the AHJ.

SECTION R104—ALTERNATIVE MATERIALS, DESIGN AND METHODS OF CONSTRUCTION AND EQUIPMENT

R104.1 General. The provisions of this code are not intended to prevent the installation of any material or to prohibit any design or method of construction not specifically prescribed by this code, provided that any such alternative has been *approved*. The *code official* shall have the authority to approve an alternative material, design or method of construction upon the written application of the owner or the owner's authorized agent. The *code official* shall first find that the proposed design is satisfactory and complies with the intent of the provisions of this code, and that the material, method or work offered is, for the purpose intended, not less than the equivalent of that prescribed in this code for strength, effectiveness, fire resistance, durability, energy conservation and safety. The *code official* shall respond to the applicant, in writing, stating the reasons why the alternative was *approved* or was not *approved*.

R104.1.1 Above code programs. The *code official* or other AHJ shall be permitted to deem a national, state or local energy-efficiency program to exceed the energy efficiency required by this code. *Buildings approved* in writing by such an energy-efficiency program shall be considered to be in compliance with this code where such buildings also meet the requirements identified in Table R405.2 and the proposed total *building thermal envelope* thermal conductance (TC) shall be less than or equal to the total *building thermal envelope* TC using the prescriptive *U*-factors and *F*-factors from Table R402.1.2 multiplied by 1.08 in Climate Zones 0, 1 and 2, and by 1.15 in Climate Zones 3 through 8, in accordance with Equation 1-1. The area-weighted maximum *fenestration solar heat gain coefficients* (SHGC) permitted in Climate Zones 0 through 3 shall be 0.30.

Equation 1-1 For Climate Zones 0–2: $TC_{Proposed\ design} \leq 1.08 \times TC_{Prescriptive\ reference\ design}$
For Climate Zones 3–8: $TC_{Proposed\ design} \leq 1.15 \times TC_{Prescriptive\ reference\ design}$

SECTION R105—CONSTRUCTION DOCUMENTS

R105.1 General. *Construction documents*, technical reports and other supporting data shall be submitted in one or more sets, or in a digital format where allowed by the code *official*, with each application for a permit. The *construction documents* and technical reports shall be prepared by a registered design professional where required by the statutes of the jurisdiction in which the project is to be constructed. Where special conditions exist, the *code official* is authorized to require necessary *construction documents* to be prepared by a registered design professional.

Exception: The *code official* is authorized to waive the requirements for *construction documents* or other supporting data if the *code official* determines they are not necessary to confirm compliance with this code.

R105.2 Information on construction documents. *Construction documents* shall be drawn to scale on suitable material. Electronic media documents are permitted to be submitted where *approved* by the *code official*. *Construction documents* shall be of sufficient clarity to indicate the location, nature and extent of the work proposed, and show in sufficient detail pertinent data and features of the *building*, systems and equipment as herein governed. Details shall include the following as applicable:

1. Energy compliance path.
2. Insulation materials and their *R*-values.
3. *Fenestration U*-factors and *solar heat gain coefficients* (SHGC).

4. Area-weighted *U-factor* and *solar heat gain coefficients* (SHGC) calculations.
5. Mechanical system design criteria.
6. Mechanical and service water-heating systems and equipment types, sizes and efficiencies.
7. Equipment and system controls.
8. *Duct* sealing, *duct* and pipe insulation and location.
9. Air sealing details.

R105.2.1 Building thermal envelope depiction. The *building thermal envelope* shall be represented on the *construction documents*.

R105.2.2 Solar-ready system. Where a *solar-ready zone* is provided, the *construction documents* shall indicate details for a dedicated roof area for the *solar-ready zone*, roof dead load, roof live load, ground snow load and the routing of conduit or prewiring from the *solar-ready zone* to an electrical service panel or plumbing from the *solar-ready zone* to a *service water heating* system.

R105.3 Examination of documents. The *code official* shall examine or cause to be examined the accompanying *construction documents* and shall ascertain whether the construction indicated and described is in accordance with the requirements of this code and other pertinent laws or ordinances. The *code official* is authorized to utilize a registered design professional, or other *approved* entity not affiliated with the *building* design or construction, in conducting the review of the plans and specifications for compliance with the code.

R105.3.1 Approval of construction documents. When the *code official* issues a permit where *construction documents* are required, the *construction documents* shall be endorsed in writing and stamped "Reviewed for Code Compliance." Such *approved construction documents* shall not be changed, modified or altered without authorization from the *code official*. Work shall be done in accordance with the *approved construction documents*.

One set of *construction documents* so reviewed shall be retained by the *code official*. The other set shall be returned to the applicant, kept at the site of work and shall be open to inspection by the *code official* or a duly authorized representative.

R105.3.2 Previous approvals. This code shall not require changes in the *construction documents*, construction or designated occupancy of a structure for which a lawful permit has been heretofore issued or otherwise lawfully authorized, and the construction of which has been pursued in good faith within 180 days after the effective date of this code and has not been abandoned.

R105.3.3 Phased approval. The *code official* shall have the authority to issue a permit for the construction of part of an energy conservation system before the *construction documents* for the entire system have been submitted or *approved*, provided adequate information and detailed statements have been filed complying with all pertinent requirements of this code. The holders of such permit shall proceed at their own risk without assurance that the permit for the entire energy conservation system will be granted.

R105.4 Amended construction documents. Work shall be installed in accordance with the *approved construction documents*, and any changes made during construction that are not in compliance with the *approved construction documents* shall be resubmitted for approval as an amended set of *construction documents*.

R105.5 Retention of construction documents. One set of *approved construction documents* shall be retained by the *code official* for a period of not less than 180 days from date of completion of the permitted work, or as required by state or local laws.

SECTION R106—FEES

R106.1 Payment of fees. A permit shall not be valid until the fees prescribed by law have been paid. Nor shall an amendment to a permit be released until the additional fee, if any, has been paid.

R106.2 Schedule of permit fees. Where a permit is required, a fee for each permit shall be paid as required, in accordance with the schedule as established by the applicable governing authority.

R106.3 Permit valuation. The applicant for a permit shall provide an estimated value of the work for which the permit is being issued at the time of application. Such estimated valuations shall include the total value of the work, including materials and labor. Where, in the opinion of the *code official*, the valuation is underestimated, the permit shall be denied unless the applicant can show detailed estimates acceptable to the *code official*. The final valuation shall be *approved* by the *code official*.

R106.4 Work commencing before permit issuance. Any person who commences any work before obtaining the necessary permits shall be subject to an additional fee established by the *code official* that shall be in addition to the required permit fees.

R106.5 Related fees. The payment of the fee for the construction, *alteration*, removal or demolition of work done in connection to or concurrently with the work or activity authorized by a permit shall not relieve the applicant or holder of the permit from the payment of other fees that are prescribed by law.

R106.6 Refunds. The *code official* is authorized to establish a refund policy.

SECTION R107—INSPECTIONS

R107.1 General. Construction or work for which a permit is required shall be subject to inspection by the *code official* or his or her designated agent, and such construction or work shall remain visible and able to be accessed for inspection purposes until *approved*. It shall be the duty of the permit applicant to cause the work to remain visible and able to be accessed for inspection purposes. Neither the *code official* nor the jurisdiction shall be liable for expense entailed in the removal or replacement of any material, product, system or building component required to allow inspection to validate compliance with this code.

R107.2 Required inspections. The *code official* or his or her designated agent, upon notification, shall make the inspections set forth in Sections R107.2.1 through R107.2.7.

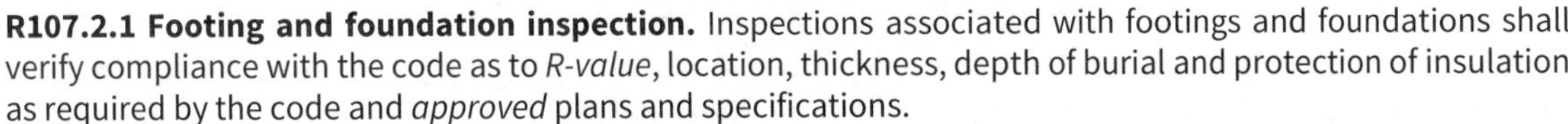

R107.2.1 Footing and foundation inspection. Inspections associated with footings and foundations shall verify compliance with the code as to *R-value*, location, thickness, depth of burial and protection of insulation as required by the code and *approved* plans and specifications.

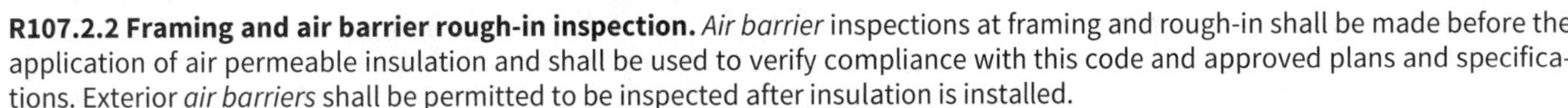

R107.2.2 Framing and air barrier rough-in inspection. *Air barrier* inspections at framing and rough-in shall be made before the application of air permeable insulation and shall be used to verify compliance with this code and approved plans and specifications. Exterior *air barriers* shall be permitted to be inspected after insulation is installed.

R107.2.3 Plumbing rough-in inspection. Inspections at plumbing rough-in shall verify compliance as required by the code and *approved* plans and specifications as to types of insulation and corresponding *R*-values and protection, and required controls. Where a *solar-ready zone* is provided for a solar thermal system, inspections shall verify pathways for routing of plumbing from *solar-ready zone* to *service water heating* system.

R107.2.4 Mechanical rough-in inspection. Inspections at mechanical rough-in shall verify compliance as required by the code and *approved* plans and specifications as to installed HVAC equipment type and size, required controls, system insulation and corresponding *R-value*, system air leakage control, programmable thermostats, *dampers*, whole-house *ventilation*, and minimum fan efficiency.

Exception: Systems serving multiple *dwelling units* shall be inspected in accordance with Section C107.2.4.

R107.2.5 Electrical rough-in inspection. Inspections at electrical rough-in shall verify compliance as required by the code and the *approved* plans and specifications as to the locations, distribution and capacity of the electrical system. Where the *solar-ready zone* is installed for electricity generation, inspections shall verify conduit or prewiring from *solar-ready zone* to electrical panel.

R107.2.6 Insulation and fenestration rough-in inspection. Inspections at insulation and *fenestration* rough-in shall be made before the application of interior finish and shall be used to verify compliance with this code as to types of insulation, corresponding *R*-values and their correct location and proper installation; and *fenestration* properties such as *U*-factors, SHGC and proper installation.

R107.2.7 Final inspection. The *building* shall have a final inspection and shall not be occupied until *approved*. The final inspection shall include verification of the installation of all required *building* systems, equipment and controls and their proper operation and the required number of high-efficacy lamps and fixtures.

R107.3 Reinspection. A *building* shall be reinspected where determined necessary by the *code official*.

R107.4 Approved third-party inspection agencies. The *code official* is authorized to accept reports of third-party inspection agencies not affiliated with the *building* design or construction, provided that such agencies are *approved* as to qualifications and reliability relevant to the *building* components and systems that they are inspecting or testing, and approval is granted prior to issuance of the building permit.

R107.4.1 Authorization of approved third-party inspection agency. An *approved* third-party inspection agency shall provide all requested information for the *code official* to determine that the agency meets the applicable requirements specified in Sections R107.4.1.1 through R107.4.1.3 and to authorize its work in the jurisdiction.

R107.4.1.1 Independence. An *approved* third-party inspection agency shall be an independent business identity. The agency shall perform its duties in accordance with the scope of delegated responsibilities established by the *code official*. The agency shall disclose to the *code official* any conflicts of interest, including where fees for service are derived. The agency shall acknowledge in writing that it is authorized to work only within the scope of delegated responsibilities.

R107.4.1.2 Equipment. An *approved* third-party inspection agency shall have adequate equipment to perform inspections and tests required by the *code official* and this code. All testing equipment shall be periodically calibrated as required by the manufacturer, testing standards used in this code or certifications held by the *approved* third-party inspection agency.

R107.4.1.3 Personnel. Personnel assigned by an *approved* third-party inspection agency to perform inspections and testing shall be trained or credentialed, and documentation of training or credentials shall be available to the *code official* upon request.

R107.4.1.4 Delegated authority. Where *approved*, a third-party inspection agency shall have the authority to perform delegated inspections and determine compliance or noncompliance of work with *approved construction documents*.

R107.4.2 Approved third-party inspection agency reporting. An *approved* third-party inspection agency shall keep records of delegated inspections, tests and compliance documentation required by this code. The agency shall submit reports of delegated inspections and tests to the *code official* and to the owner or owner's representative. Reports shall indicate the compliance determination for the inspected or tested work based on *approved construction documents*. A final report documenting required delegated inspections and tests, and correction of any discrepancies noted in the inspections or tests, shall be submitted with other required compliance documentation at a time required by the *code official*.

R107.5 Inspection requests. It shall be the duty of the holder of the permit or their duly authorized agent to notify the *code official* when work is ready for inspection. It shall be the duty of the permit holder to provide *access to* and means for inspections of such work that are required by this code.

R107.6 Reinspection and testing. Where any work or installation does not pass an initial test or inspection, the necessary corrections shall be made to achieve compliance with this code. The work or installation shall then be resubmitted to the *code official* for inspection and testing.

SECTION R108—NOTICE OF APPROVAL

R108.1 Approval. After the prescribed tests and inspections indicate that the work complies in all respects with this code, a notice of approval shall be issued by the *code official*.

R108.2 Revocation. The *code official* is authorized to, in writing, suspend or revoke a notice of approval issued under the provisions of this code wherever the certificate is issued in error, or on the basis of incorrect information supplied, or where it is determined that the *building* or structure, premise, or portion thereof is in violation of any ordinance or regulation or any of the provisions of this code.

SECTION R109—MEANS OF APPEALS

R109.1 General. In order to hear and decide appeals of orders, decisions or determinations made by the *code official* relative to the application and interpretation of this code, there shall be and is hereby created a board of appeals. The board of appeals shall be appointed by the applicable governing authority and shall hold office at its pleasure. The board shall adopt rules of procedure for conducting its business and shall render all decisions and findings in writing to the appellant with a duplicate copy to the *code official*.

R109.2 Limitations on authority. An application for appeal shall be based on a claim that the true intent of this code or the rules legally adopted thereunder have been incorrectly interpreted, the provisions of this code do not fully apply or an equivalent or better form of construction is proposed. The board shall not have authority to waive requirements of this code.

R109.3 Qualifications. The board of appeals shall consist of members who are qualified by experience and training on matters pertaining to the provisions of this code and are not employees of the jurisdiction.

R109.4 Administration. The *code official* shall take action in accordance with the decision of the board.

SECTION R110—STOP WORK ORDER

R110.1 Authority. Where the *code official* finds any work regulated by this code being performed in a manner contrary to the provisions of this code or in a dangerous or unsafe manner, the *code official* is authorized to issue a stop work order.

R110.2 Issuance. The stop work order shall be in writing and shall be given to the owner of the property, the owner's authorized agent or the person performing the work. Upon issuance of a stop work order, the cited work shall immediately cease. The stop work order shall state the reason for the order and the conditions under which the cited work is authorized to resume.

R110.3 Emergencies. Where an emergency exists, the *code official* shall not be required to give a written notice prior to stopping the work.

R110.4 Failure to comply. Any person who shall continue any work after having been served with a stop work order, except such work as that person is directed to perform to remove a violation or unsafe condition, shall be subject to fines established by the AHJ.

CHAPTER 2 [RE] DEFINITIONS

User notes:

About this chapter: *Codes, by their very nature, are technical documents. Every word, term and punctuation mark can add to or change the meaning of a technical requirement. It is necessary to maintain a consensus on the specific meaning of each term contained in the code. Chapter 2 performs this function by stating clearly what specific terms mean for the purpose of the code.*

SECTION R201—GENERAL

R201.1 Scope. Unless stated otherwise, the following words and terms in this code shall have the meanings indicated in this chapter.

R201.2 Interchangeability. Words used in the present tense include the future; words in the masculine gender include the feminine and neuter; the singular number includes the plural and the plural includes the singular.

R201.3 Terms defined in other codes. Terms that are not defined in this code but are defined in the *International Building Code*, *International Fire Code*, *International Fuel Gas Code*, *International Mechanical Code*, *International Plumbing Code* or the *International Residential Code* shall have the meanings ascribed to them in those codes.

R201.4 Terms not defined. Terms not defined by this chapter shall have ordinarily accepted meanings such as the context implies.

SECTION R202—GENERAL DEFINITIONS

ABOVE-GRADE WALL. A wall more than 50 percent above grade and enclosing *conditioned space*. This includes between-floor spandrels, peripheral edges of floors, roof and basement knee walls, dormer walls, gable end walls, walls enclosing a mansard roof and *skylight* shafts.

ACCESS (TO). That which enables a device, appliance or equipment to be reached by *ready access* or by a means that first requires the removal or movement of a panel or similar obstruction.

ADDITION. An extension or increase in the floor area, number of stories or height of a *building* or structure.

AIR BARRIER. One or more materials joined together in a continuous manner to restrict or prevent the passage of air through the *building thermal envelope* and its assemblies.

AIR-HANDLING UNIT. A blower or fan used for the purpose of distributing supply air to a room, space or area.

ALTERATION. Any construction, retrofit or renovation to an existing structure other than *repair* or *addition*. Also, a change in a building, electrical, gas, mechanical or plumbing system that involves an extension, *addition* or change to the arrangement, type or purpose of the original installation.

APPROVED. Acceptable to the *code official.*

APPROVED AGENCY. An established and recognized agency that is regularly engaged in conducting tests furnishing inspection services, or furnishing product certification, where such agency has been *approved* by the *code official.*

APPROVED SOURCE. An independent person, firm or corporation approved by the *code official*, who is competent and experienced in the application of engineering principles to materials, methods or system analyses.

AUTOMATIC. Self-acting, operating by its own mechanism when actuated by some impersonal influence, as, for example, a change in current strength, pressure, temperature or mechanical configuration (see "*Manual*").

AUTOMATIC SHUTOFF CONTROL. A device capable of automatically turning loads off without *manual* intervention. *Automatic shutoff controls* include devices such as, but not limited to, occupancy sensors, vacancy sensors, door switches, programmable time switches (i.e., timeclocks), or count-down timers.

BALANCED VENTILATION SYSTEM. A ventilation system that simultaneously supplies outdoor air to and exhausts air from a space, where the mechanical supply airflow rate and the mechanical exhaust airflow rate are each within 10 percent of the average of the two airflow rates.

BASEMENT WALL. A wall 50 percent or more below grade and enclosing *conditioned space.*

BIODIESEL BLEND. A homogeneous mixture of hydrocarbon oils and mono alkyl esters of long chain fatty acids.

BUILDING. Any structure used or intended for supporting or sheltering any use or occupancy, including any mechanical systems, service water-heating systems and electric power and lighting systems located on the *building site* and supporting the building.

BUILDING SITE. A contiguous area of land that is under the ownership or control of one entity.

BUILDING THERMAL ENVELOPE. The *basement walls, exterior walls,* floors, ceiling, roofs and any other *building* element assemblies that enclose *conditioned space* or provide a boundary between *conditioned space* and exempt or unconditioned space.

CAVITY INSULATION. Insulating material located between framing members.

CIRCULATING HOT WATER SYSTEM. A specifically designed water distribution system where one or more pumps are operated in the service hot water piping to circulate heated water from the water-heating equipment to fixtures and back to the water-heating equipment.

CLIMATE ZONE. A geographical region based on climatic criteria as specified in this code.

CODE OFFICIAL. The officer or other designated authority charged with the administration and enforcement of this code or a duly authorized representative.

COMMERCIAL BUILDING. For this code, all buildings that are not included in the definition of "*Residential building*."

COMMON AREAS. All conditioned spaces within Group R occupancy buildings that are not *dwelling units* or *sleeping units*.

CONDITIONED FLOOR AREA. The horizontal projection of the floors associated with the *conditioned space*.

CONDITIONED SPACE. An area, room or space that is enclosed within the *building thermal envelope* and that is directly or indirectly heated or cooled. Spaces are indirectly heated or cooled where they communicate through openings with conditioned spaces, where they are separated from conditioned spaces by uninsulated walls, floors or ceilings, or where they contain uninsulated *ducts*, piping or other sources of heating or cooling.

CONSTRUCTION DOCUMENTS. Written, graphic and pictorial documents prepared or assembled for describing the design, location and physical characteristics of the elements of a project necessary for obtaining a building permit.

CONTINUOUS AIR BARRIER. A combination of materials and assemblies that restrict or prevent the passage of air through the *building thermal envelope*.

CONTINUOUS INSULATION (ci). Insulating material that is continuous across all structural members without thermal bridges other than fasteners and service openings. It is installed on the interior or exterior, or is integral to any opaque surface, of the *building thermal* envelope.

CONTINUOUS PILOT. A pilot which, once placed in operation, is intended to remain ignited continuously until it is manually interrupted.

CRAWL SPACE WALL. The opaque portion of a wall that encloses a crawl space and is partially or totally below grade.

CURTAIN WALL. *Fenestration* products used to create an external nonload-bearing wall that is designed to separate the exterior and interior environments.

DAMPER. A manually or automatically controlled device to regulate draft or the rate of flow of air or combustion gases.

DEMAND RECIRCULATION WATER SYSTEM. A water distribution system where one or more pumps prime the service hot water piping with heated water upon demand for hot water.

DEMAND RESPONSE SIGNAL. A signal that indicates a price or a request to modify electricity consumption for a limited time period.

DEMAND RESPONSIVE CONTROL. A control capable of receiving and automatically responding to a *demand response signal*.

DIMMER. A control device that is capable of continuously varying the light output and energy use of light sources.

DISTRIBUTION SYSTEM EFFICIENCY (DSE). A system efficiency factor that adjusts for the energy losses associated with delivery of energy from the equipment to the source of the load.

DUCT. A tube or conduit utilized for conveying air. The air passages of self-contained systems are not to be construed as air *ducts*.

DUCT SYSTEM. A system that consists of *space conditioning equipment* and *ductwork*, and includes any apparatus installed in connection therewith.

DUCTWORK. The assemblies of connected *ducts*, plenums, boots, fittings, *dampers*, supply registers, return grilles, and filter grilles through which air is supplied to or returned from the space to be heated, cooled, or ventilated. Supply *ductwork* delivers air to the spaces from the *space conditioning equipment*. Return *ductwork* conveys air from the spaces back to the space *conditioning equipment*. *Ventilation ductwork* conveys air to or from any space.

DWELLING UNIT. A single unit providing complete independent living facilities for one or more persons, including permanent provisions for living, sleeping, eating, cooking and sanitation.

EMITTANCE. The ratio of the radiant heat flux emitted by a specimen measured on a scale from 0 to 1, where a value of 1 indicates perfect release of thermal radiation.

ENCLOSED REFLECTIVE AIRSPACE. An unventilated cavity with a low-*emittance* surface bounded on all sides by building components.

ENERGY ANALYSIS. A method for estimating the annual energy use of the *proposed design* and *standard reference design* based on estimates of energy use.

ENERGY COST. The total estimated annual cost for purchased energy for the *building* functions regulated by this code, including applicable demand charges.

ENERGY RATING INDEX (ERI). A numerical integer value that represents the relative energy performance of a *rated design* or constructed *dwelling unit* as compared with the energy performance of the *ERI Reference Design*, where an ERI value of 100 represents the energy performance of the *ERI Reference Design* and an ERI value of 0 represents a *rated design* or constructed *dwelling unit* with zero net energy performance.

ENERGY SIMULATION TOOL. An *approved* software program or calculation-based methodology that projects the annual energy use of a *building*.

ERI REFERENCE DESIGN. A version of the *rated design* that meets the minimum requirements of the 2006 *International Energy Conservation Code*.

EXISTING BUILDING. A *building* erected prior to the date of adoption of the appropriate code, or one for which a legal building permit has been issued.

EXTERIOR WALL. Walls including both *above-grade walls* and *basement walls.*

FENESTRATION. Products classified as either *vertical fenestration* or *skylights.*

Skylights. Glass or other transparent or translucent glazing material installed at a slope of less than 60 degrees (1.05 rad) from horizontal including unit skylights, tubular daylighting devices, and glazing materials in solariums, sunrooms, roofs and sloped walls.

Vertical fenestration. Windows that are fixed or operable, opaque doors, glazed doors, glazed block and combination opaque/glazed doors composed of glass or other transparent or translucent glazing materials and installed at a slope of not less than 60 degrees (1.05 rad) from horizontal.

FENESTRATION PRODUCT, SITE-BUILT. A *fenestration* designed to be made up of field-glazed or field-assembled units using specific factory cut or otherwise factory-formed framing and glazing units. Examples of site-built fenestration include storefront systems, *curtain walls* and atrium roof systems.

F-FACTOR (THERMAL TRANSMITTANCE). The perimeter heat loss factor for slab-on-grade floors (Btu/h × ft × °F) [W/(m × K)].

FUEL GAS. A natural gas, manufactured gas, liquified petroleum gas or a mixture of these.

FUEL OIL. Kerosene or any hydrocarbon oil having a flash point not less than 100°F (38°C).

GRADE PLANE. A reference plane representing the average of the finished ground level adjoining the *building* at all *exterior walls*. Where the finished ground level slopes away from the *exterior wall*, the reference plane is established by the lowest points within the area between the *building* and the lot line or, where the lot line is more than 6 feet (1829 mm) from the *building* between the structure and a point 6 feet (1829 mm) from the *building*.

HEAT EXCHANGER. A device that transfers heat from one medium to another.

HEATED SLAB. Slab-on-grade construction in which the heating elements, hydronic tubing, or hot air distribution system is in contact with, or placed within or under, the slab.

HISTORIC BUILDING. Any *building* or structure that is one or more of the following:

1. Listed, or certified as eligible for listing by the State Historic Preservation Officer or the Keeper of the National Register of Historic Places, in the National Register of Historic Places.
2. Designated as historic under an applicable state or local law.
3. Certified as a contributing resource within a National Register-listed, state-designated or locally designated historic district.

INFILTRATION. The uncontrolled inward air leakage into a *building* caused by the pressure effects of wind or the effect of differences in the indoor and outdoor air density or both.

INSULATED SIDING. A type of *continuous insulation* with manufacturer-installed insulating material as an integral part of the cladding product having an *R-value* of not less than R-2.

INTERMITTENT IGNITION. Type of ignition that is energized when an appliance is called on to operate and that remains continuously energized during each period of main burner operation and where the ignition is deenergized when the main burner operating cycle is completed.

INTERRUPTED IGNITION. Type of ignition that is energized prior to the admission of fuel to the main burner and that is deenergized when the main flame is established.

KNEE WALL. An *above-grade wall* assembly, or wall defined by vertical truss members, of any height that separates *conditioned space* from unconditioned buffer spaces, such as ventilated attics and entry porch roofs, rather than ambient outdoors.

LABELED. Equipment, materials or products to which have been affixed a label, seal, symbol or other identifying mark of a nationally recognized testing laboratory, *approved* agency or other organization concerned with product evaluation that maintains periodic inspection of the production of such *labeled* items and whose labeling indicates either that the equipment, material or product meets identified standards or has been tested and found suitable for a specified purpose.

LIQUID FUEL. A *fuel oil* or *biodiesel blend.*

LISTED. Equipment, materials, products or services included in a list published by an organization acceptable to the *code official* and concerned with evaluation of products or services that maintains periodic inspection of production of *listed* equipment or materials or periodic evaluation of services and whose listing states either that the equipment, material, product or service meets identified standards or has been tested and found suitable for a specified purpose.

LIVING SPACE. Space within a *dwelling unit* utilized for living, sleeping, eating, cooking, bathing, washing and sanitation purposes.

LOW SLOPE. A roof slope less than 2 units vertical in 12 units horizontal (17 percent slope).

MANUAL. Capable of being operated by personal intervention (see "*Automatic*").

OCCUPANT SENSOR CONTROL. An *automatic* control device that detects the presence or absence of people within an area and causes lighting, equipment or appliances to be regulated accordingly.

OCCUPIABLE SPACE. An enclosed space intended for human activities, excluding those spaces intended primarily for other purposes, such as storage rooms and equipment rooms, that are only intended to be occupied occasionally and for short periods of time.

ON-DEMAND PILOT. A pilot that, once placed into operation, is intended to remain ignited for a predetermined period of time following an *automatic* or *manual* operation of the main burner gas valve, after which the pilot is automatically extinguished when no *automatic* or *manual* operation of the main burner gas valve occurs during the predetermined period of time.

ON-SITE RENEWABLE ENERGY. Energy from *renewable energy resources* harvested at the *building site*.

OPAQUE DOOR. A door that is not less than 50 percent opaque in surface area.

PLENUM. An enclosed portion of the *building* structure, other than an *occupiable space* being conditioned, that is designed to allow air movement and thereby serve as part of the supply or return *ductwork*.

PROPOSED DESIGN. A description of the proposed *dwelling unit* used to estimate annual energy use for determining compliance based on *simulated building performance*.

RADIANT BARRIER. A material having a low emittance surface of 0.1 or less installed in building assemblies.

RATED DESIGN. A description of the proposed *dwelling unit* used to determine the *energy rating index*.

READY ACCESS (TO). That which enables a device, appliance or equipment to be directly reached without requiring the removal or movement of any panel or similar obstruction.

REFLECTIVE INSULATION. A material with a surface *emittance* of 0.1 or less in an assembly consisting of one or more *enclosed reflective airspaces*.

RENEWABLE ENERGY CERTIFICATE (REC). A market-based instrument that represents and conveys the environmental attributes of 1 megawatt hour of renewable electricity generation and could be sold separately from the underlying physical electricity associated with *renewable energy resources*; also known as an energy attribute certificate (EAC).

RENEWABLE ENERGY RESOURCES. Energy derived from solar radiation, wind, waves, tides, landfill gas, biogas, biomass or extracted from hot fluid or steam heated within the earth.

REPAIR. The reconstruction or renewal of any part of an existing *building* for the purpose of its maintenance or to correct damage.

REROOFING. The process of recovering or replacing an existing roof covering. See "*Roof recover*" and "*Roof replacement*."

RESIDENTIAL BUILDING. For this code, includes detached one- and two-family dwellings and townhouses as well as *Group R*-2, *R*-3 and *R*-4 buildings three stories or less in height above *grade plane*.

ROOF ASSEMBLY. A system designed to provide weather protection and resistance to design loads. The system consists of a roof covering and roof deck or a single component serving as both the roof covering and the roof deck. A *roof assembly* includes the roof covering, underlayment and roof deck and can also include a thermal barrier, an ignition barrier, insulation or a vapor retarder.

ROOF RECOVER. The process of installing an additional roof covering over an existing roof covering without removing the existing roof covering.

ROOF REPAIR. Reconstruction or renewal of any part of an existing roof for the purposes of its maintenance.

ROOF REPLACEMENT. An *alteration* that includes the removal of all existing layers of *roof assembly* materials down to the roof deck and the installation of replacement materials above the existing roof deck.

***R*-VALUE (THERMAL RESISTANCE).** The inverse of the time rate of heat flow through a body from one of its bounding surfaces to the other surface for a unit temperature difference between the two surfaces, under steady state conditions, per unit area (h × ft^2 × °F/Btu) [(m^2 × K)/W].

SERVICE WATER HEATING. Supply of hot water for purposes other than comfort heating.

SIMULATED BUILDING PERFORMANCE. A process in which the proposed building design is compared to a *standard reference design* for the purposes of estimating relative energy use to determine code compliance.

SLEEPING UNIT. A single unit that provides rooms or spaces for one or more persons, includes permanent provisions for sleeping and can include provisions for living, eating and either sanitation or kitchen facilities but not both. Such rooms and spaces that are part of a *dwelling unit* are not *sleeping units*.

SOLAR HEAT GAIN COEFFICIENT (SHGC). The ratio of the solar heat gain entering the space through the *fenestration* assembly to the incident solar radiation. Solar heat gain includes directly transmitted solar heat and absorbed solar radiation that is then reradiated, conducted or convected into the space.

SOLAR-READY ZONE. A section or sections of the roof or building overhang designated and reserved for the future installation of a solar photovoltaic or solar thermal system.

SPACE CONDITIONING. The treatment of air so as to control the temperature, humidity, filtration or distribution of the air to meet the requirements of a *conditioned space*.

SPACE CONDITIONING EQUIPMENT. The *heat exchangers*, *air-handling units*, filter boxes and any apparatus installed in connection therewith used to provide *space conditioning*.

STANDARD REFERENCE DESIGN. A version of the *proposed design* that meets the minimum requirements of this code and is used to determine the maximum annual energy use requirement for compliance based on simulated building performance.

STEEP SLOPE. A roof slope 2 units vertical in 12 units horizontal (17 percent slope) or greater.

SUBSTANTIAL IMPROVEMENT. Any *repair*, reconstruction, rehabilitation, *alteration*, *addition* or other improvement of a *building* or structure, the cost of which equals or is more than 50 percent of the market value of the structure before the improvement. Where the structure has sustained substantial damage as defined in the *International Building Code*, any repairs are considered *substantial improvement* regardless of the actual *repair* work performed. *Substantial improvement* does not include the following:

1. Improvement of a *building* ordered by the code official to correct health, sanitary or safety code violations.
2. *Alteration* of a historic building where the *alteration* will not affect the designation as a historic building.

SUNROOM. A one-story structure attached to a dwelling with a glazing area in excess of 40 percent of the gross area of the structure's *exterior walls* and roof.

TESTING UNIT ENCLOSURE AREA. The sum of the area of ceiling, floors, and walls separating a *dwelling unit's* or *sleeping unit's conditioned space* from the exterior or from adjacent conditioned or unconditioned spaces. Wall height shall be measured from the finished floor of the *dwelling unit* or *sleeping unit* to the underside of the floor above.

THERMAL DISTRIBUTION EFFICIENCY (TDE). The resistance to changes in air heat as air is conveyed through a distance of air *duct*. TDE is a heat loss calculation evaluating the difference in the heat of the air between the air *duct* inlet and outlet caused by differences in temperatures between the air in the *duct* and the *duct* material. TDE is expressed as a percent difference between the inlet and outlet heat in the *duct*.

THERMAL ISOLATION. Physical and *space conditioning* separation from *conditioned spaces*. The *conditioned spaces* shall be controlled as separate *zones* for heating and cooling or conditioned by separate equipment.

THERMOSTAT. An *automatic* control device used to maintain temperature at a fixed or adjustable setpoint.

***U*-FACTOR (THERMAL TRANSMITTANCE).** The coefficient of heat transmission (air to air) through a building component or assembly, equal to the time rate of heat flow per unit area and unit temperature difference between the warm side and cold side air films (Btu/h $\times$ ft^2 $\times$ °F) [W/(m^2 $\times$ K)].

VENTILATION. The natural or mechanical process of supplying conditioned or unconditioned air to, or removing such air from, any space.

VENTILATION AIR. That portion of supply air that comes from outside (outdoors) plus any recirculated air that has been treated to maintain the desired quality of air within a designated space.

VISIBLE TRANSMITTANCE (VT). The ratio of visible light entering the space through the *fenestration* product assembly to the incident visible light. *Visible Transmittance* includes the effects of glazing material and frame and is expressed as a number between 0 and 1.

WHOLE-HOUSE MECHANICAL VENTILATION SYSTEM An exhaust system, supply system, or combination thereof that is designed to mechanically exchange indoor air with *outdoor air* when operating continuously or through a programmed intermittent schedule to satisfy the whole house *ventilation* rates.

WORK AREA. That portion or portions of a *building* consisting of all reconfigured spaces as indicated on the *construction documents*. *Work area* excludes other portions of the *building* where incidental work entailed by the intended work must be performed and portions of the *building* where work not initially intended by the owner is specifically required by this code.

ZONE. A space or group of spaces within a *building* with heating or cooling requirements that are sufficiently similar so that desired conditions can be maintained throughout using a single controlling device.

CHAPTER

3 [RE] GENERAL REQUIREMENTS

User notes:

About this chapter: *Chapter 3 addresses broadly applicable requirements that would not be at home in other chapters having more specific coverage of subject matter. This chapter establishes climate zone by US counties and territories and includes methodology for determining climate zones elsewhere. It also contains product rating, marking and installation requirements for materials such as insulation, windows, doors and siding.*

SECTION R301—CLIMATE ZONES

R301.1 General. *Climate zones* from Figure R301.1 or Table R301.1 shall be used for determining the applicable requirements from Chapter 4. Locations not indicated in Table R301.1 shall be assigned a *climate zone* in accordance with Section R301.3.

FIGURE R301.1—CLIMATE ZONES

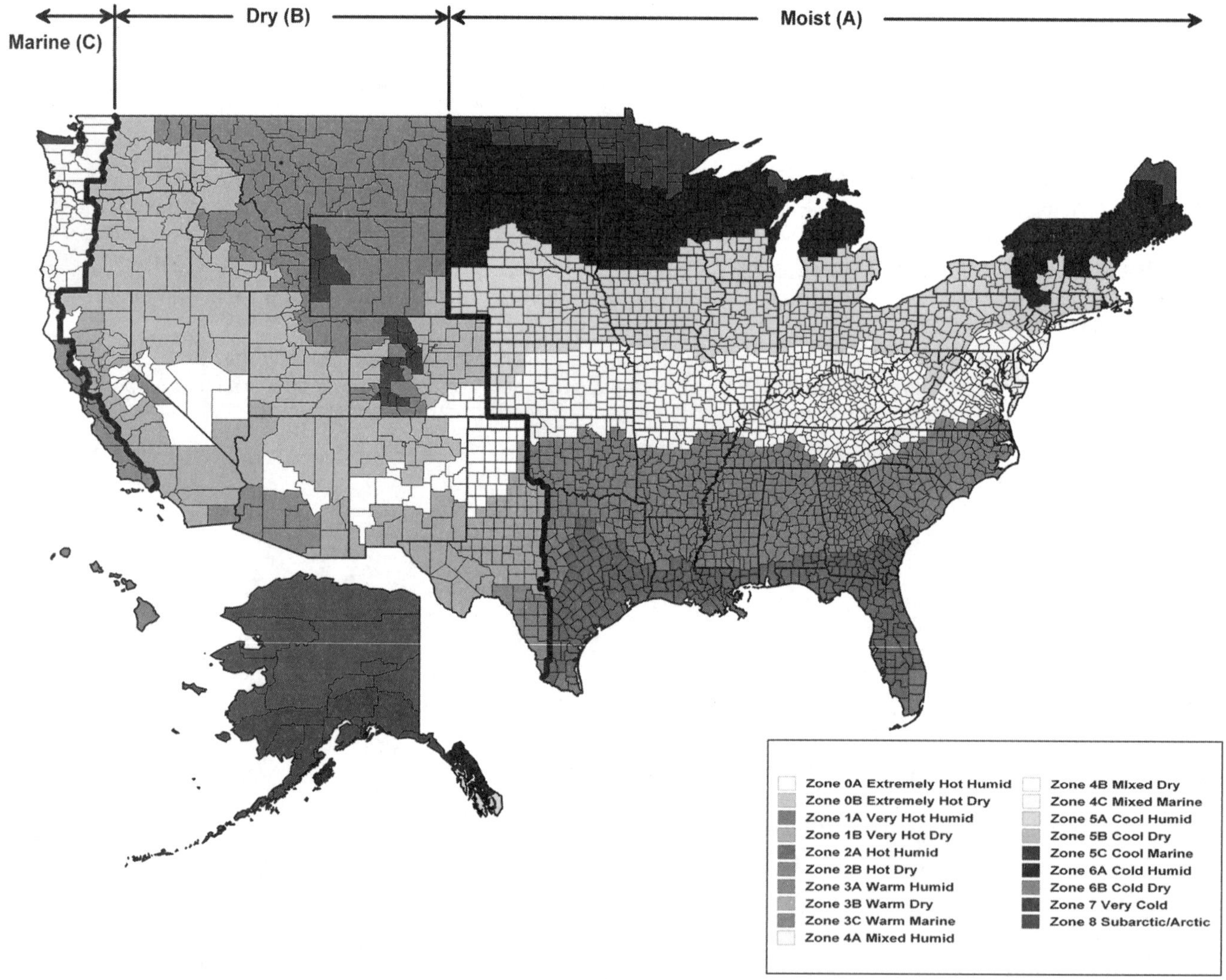

TABLE R301.1—CLIMATE ZONES, MOISTURE REGIMES AND WARM HUMID DESIGNATIONS BY STATE, COUNTY AND TERRITORY[a]

US STATES
ALABAMA
3A Autauga*
2A Baldwin*
3A Barbour*
3A Bibb
3A Blount
3A Bullock*
3A Butler*
3A Calhoun
3A Chambers
3A Cherokee
3A Chilton
3A Choctaw*
3A Clarke*
3A Clay
3A Cleburne
2A Coffee*
3A Colbert
3A Conecuh*
3A Coosa
2A Covington*
3A Crenshaw*
3A Cullman
2A Dale*
3A Dallas*
3A DeKalb
3A Elmore*
2A Escambia*
3A Etowah
3A Fayette
3A Franklin
2A Geneva*
3A Greene
3A Hale
2A Henry*
2A Houston*
3A Jackson
3A Jefferson
3A Lamar
3A Lauderdale
3A Lawrence
3A Lee
3A Limestone
3A Lowndes*
3A Macon*
3A Madison
3A Marengo*
3A Marion
3A Marshall
2A Mobile*
3A Monroe*
3A Montgomery*
3A Morgan
3A Perry*
3A Pickens
3A Pike*
3A Randolph
3A Russell*
3A Shelby
3A St. Clair
3A Sumter
3A Talladega
3A Tallapoosa
3A Tuscaloosa
3A Walker
3A Washington*
3A Wilcox*
3A Winston
ALASKA
7 Aleutians East
7 Aleutians West
7 Anchorage
7 Bethel
7 Bristol Bay
8 Denali
7 Dillingham
8 Fairbanks North Star
6A Haines
6A Juneau
7 Kenai Peninsula
5C Ketchikan Gateway
6A Kodiak Island
7 Lake and Peninsula
7 Matanuska-Susitna
8 Nome
8 North Slope
8 Northwest Arctic
5C Prince of Wales Outer Ketchikan
5C Sitka
6A Skagway-Hoonah-Angoon
8 Southeast Fairbanks

TABLE R301.1—CLIMATE ZONES, MOISTURE REGIMES AND WARM HUMID DESIGNATIONS BY STATE, COUNTY AND TERRITORY[a]—continued

US STATES—continued
ALASKA *(continued)*
7 Valdez-Cordova
8 Wade Hampton
6A Wrangell-Petersburg
7 Yakutat
8 Yukon-Koyukuk
ARIZONA
5B Apache
3B Cochise
5B Coconino
4B Gila
3B Graham
3B Greenlee
2B La Paz
2B Maricopa
3B Mohave
5B Navajo
2B Pima
2B Pinal
3B Santa Cruz
4B Yavapai
2B Yuma
ARKANSAS
3A Arkansas
3A Ashley
4A Baxter
4A Benton
4A Boone
3A Bradley
3A Calhoun
4A Carroll
3A Chicot
3A Clark
3A Clay
3A Cleburne
3A Cleveland
3A Columbia*
3A Conway
3A Craighead
3A Crawford
3A Crittenden
3A Cross
3A Dallas
3A Desha
3A Drew
3A Faulkner
3A Franklin
4A Fulton
3A Garland
3A Grant
3A Greene
3A Hempstead*
3A Hot Spring
3A Howard
3A Independence
4A Izard
3A Jackson
3A Jefferson
3A Johnson
3A Lafayette*
3A Lawrence
3A Lee
3A Lincoln
3A Little River*
3A Logan
3A Lonoke
4A Madison
4A Marion
3A Miller*
3A Mississippi
3A Monroe
3A Montgomery
3A Nevada
4A Newton
3A Ouachita
3A Perry
3A Phillips
3A Pike
3A Poinsett
3A Polk
3A Pope
3A Prairie
3A Pulaski
3A Randolph
3A Saline
3A Scott
4A Searcy
3A Sebastian
3A Sevier*
3A Sharp
3A St. Francis

TABLE R301.1—CLIMATE ZONES, MOISTURE REGIMES AND WARM HUMID DESIGNATIONS BY STATE, COUNTY AND TERRITORY[a]—continued

US STATES—continued
ARKANSAS *(continued)*
4A Stone
3A Union*
3A Van Buren
4A Washington
3A White
3A Woodruff
3A Yell
CALIFORNIA
3C Alameda
6B Alpine
4B Amador
3B Butte
4B Calaveras
3B Colusa
3B Contra Costa
4C Del Norte
4B El Dorado
3B Fresno
3B Glenn
4C Humboldt
2B Imperial
4B Inyo
3B Kern
3B Kings
4B Lake
5B Lassen
3B Los Angeles
3B Madera
3C Marin
4B Mariposa
3C Mendocino
3B Merced
5B Modoc
6B Mono
3C Monterey
3C Napa
5B Nevada
3B Orange
3B Placer
5B Plumas
3B Riverside
3B Sacramento
3C San Benito
3B San Bernardino
3B San Diego
3C San Francisco
3B San Joaquin
3C San Luis Obispo
3C San Mateo
3C Santa Barbara
3C Santa Clara
3C Santa Cruz
3B Shasta
5B Sierra
5B Siskiyou
3B Solano
3C Sonoma
3B Stanislaus
3B Sutter
3B Tehama
4B Trinity
3B Tulare
4B Tuolumne
3C Ventura
3B Yolo
3B Yuba
COLORADO
5B Adams
6B Alamosa
5B Arapahoe
6B Archuleta
4B Baca
4B Bent
5B Boulder
5B Broomfield
6B Chaffee
5B Cheyenne
7 Clear Creek
6B Conejos
6B Costilla
5B Crowley
5B Custer
5B Delta
5B Denver
6B Dolores
5B Douglas
6B Eagle
5B Elbert
5B El Paso
5B Fremont

TABLE R301.1—CLIMATE ZONES, MOISTURE REGIMES AND WARM HUMID DESIGNATIONS BY STATE, COUNTY AND TERRITORY[a]—continued

US STATES—continued
COLORADO *(continued)*
5B Garfield
5B Gilpin
7 Grand
7 Gunnison
7 Hinsdale
5B Huerfano
7 Jackson
5B Jefferson
5B Kiowa
5B Kit Carson
7 Lake
5B La Plata
5B Larimer
4B Las Animas
5B Lincoln
5B Logan
5B Mesa
7 Mineral
6B Moffat
5B Montezuma
5B Montrose
5B Morgan
4B Otero
6B Ouray
7 Park
5B Phillips
7 Pitkin
4B Prowers
5B Pueblo
6B Rio Blanco
7 Rio Grande
7 Routt
6B Saguache
7 San Juan
6B San Miguel
5B Sedgwick
7 Summit
5B Teller
5B Washington
5B Weld
5B Yuma
CONNECTICUT
5A (all)

DELAWARE
4A (all)
DISTRICT OF COLUMBIA
4A (all)
FLORIDA
2A Alachua*
2A Baker*
2A Bay*
2A Bradford*
2A Brevard*
1A Broward*
2A Calhoun*
2A Charlotte*
2A Citrus*
2A Clay*
2A Collier*
2A Columbia*
2A DeSoto*
2A Dixie*
2A Duval*
2A Escambia*
2A Flagler*
2A Franklin*
2A Gadsden*
2A Gilchrist*
2A Glades*
2A Gulf*
2A Hamilton*
2A Hardee*
2A Hendry*
2A Hernando*
2A Highlands*
2A Hillsborough*
2A Holmes*
2A Indian River*
2A Jackson*
2A Jefferson*
2A Lafayette*
2A Lake*
2A Lee*
2A Leon*
2A Levy*
2A Liberty*
2A Madison*
2A Manatee*

TABLE R301.1—CLIMATE ZONES, MOISTURE REGIMES AND WARM HUMID DESIGNATIONS BY STATE, COUNTY AND TERRITORY[a]—continued

US STATES—continued
FLORIDA *(continued)*
2A Marion*
2A Martin*
1A Miami-Dade*
1A Monroe*
2A Nassau*
2A Okaloosa*
2A Okeechobee*
2A Orange*
2A Osceola*
1A Palm Beach*
2A Pasco*
2A Pinellas*
2A Polk*
2A Putnam*
2A Santa Rosa*
2A Sarasota*
2A Seminole*
2A St. Johns*
2A St. Lucie*
2A Sumter*
2A Suwannee*
2A Taylor*
2A Union*
2A Volusia*
2A Wakulla*
2A Walton*
2A Washington*
GEORGIA
2A Appling*
2A Atkinson*
2A Bacon*
2A Baker*
3A Baldwin
3A Banks
3A Barrow
3A Bartow
3A Ben Hill*
2A Berrien*
3A Bibb
3A Bleckley*
2A Brantley*
2A Brooks*
2A Bryan*
3A Bulloch*
3A Burke
3A Butts
2A Calhoun*
2A Camden*
3A Candler*
3A Carroll
3A Catoosa
2A Charlton*
2A Chatham*
3A Chattahoochee*
3A Chattooga
3A Cherokee
3A Clarke
3A Clay*
3A Clayton
2A Clinch*
3A Cobb
2A Coffee*
2A Colquitt*
3A Columbia
2A Cook*
3A Coweta
3A Crawford
3A Crisp*
3A Dade
3A Dawson
2A Decatur*
3A DeKalb
3A Dodge*
3A Dooly*
2A Dougherty*
3A Douglas
2A Early*
2A Echols*
2A Effingham*
3A Elbert
3A Emanuel*
2A Evans*
3A Fannin
3A Fayette
3A Floyd
3A Forsyth
3A Franklin
3A Fulton
3A Gilmer
3A Glascock

TABLE R301.1—CLIMATE ZONES, MOISTURE REGIMES AND WARM HUMID DESIGNATIONS BY STATE, COUNTY AND TERRITORY[a]—continued

US STATES—continued
GEORGIA *(continued)*
2A Glynn*
3A Gordon
2A Grady*
3A Greene
3A Gwinnett
3A Habersham
3A Hall
3A Hancock
3A Haralson
3A Harris
3A Hart
3A Heard
3A Henry
3A Houston*
3A Irwin*
3A Jackson
3A Jasper
2A Jeff Davis*
3A Jefferson
3A Jenkins*
3A Johnson*
3A Jones
3A Lamar
2A Lanier*
3A Laurens*
3A Lee*
2A Liberty*
3A Lincoln
2A Long*
2A Lowndes*
3A Lumpkin
3A Macon*
3A Madison
3A Marion*
3A McDuffie
2A McIntosh*
3A Meriwether
2A Miller*
2A Mitchell*
3A Monroe
3A Montgomery*
3A Morgan
3A Murray
3A Muscogee
3A Newton
3A Oconee
3A Oglethorpe
3A Paulding
3A Peach*
3A Pickens
2A Pierce*
3A Pike
3A Polk
3A Pulaski*
3A Putnam
3A Quitman*
3A Rabun
3A Randolph*
3A Richmond
3A Rockdale
3A Schley*
3A Screven*
2A Seminole*
3A Spalding
3A Stephens
3A Stewart*
3A Sumter*
3A Talbot
3A Taliaferro
2A Tattnall*
3A Taylor*
3A Telfair*
3A Terrell*
2A Thomas*
2A Tift*
2A Toombs*
3A Towns
3A Treutlen*
3A Troup
3A Turner*
3A Twiggs*
3A Union
3A Upson
3A Walker
3A Walton
2A Ware*
3A Warren
3A Washington
2A Wayne*
3A Webster*

TABLE R301.1—CLIMATE ZONES, MOISTURE REGIMES AND WARM HUMID DESIGNATIONS BY STATE, COUNTY AND TERRITORY[a]—continued

US STATES—continued
GEORGIA *(continued)*
3A Wheeler*
3A White
3A Whitfield
3A Wilcox*
3A Wilkes
3A Wilkinson
2A Worth*
HAWAII
1A (all)*
IDAHO
5B Ada
6B Adams
6B Bannock
6B Bear Lake
5B Benewah
6B Bingham
6B Blaine
6B Boise
6B Bonner
6B Bonneville
6B Boundary
6B Butte
6B Camas
5B Canyon
6B Caribou
5B Cassia
6B Clark
5B Clearwater
6B Custer
5B Elmore
6B Franklin
6B Fremont
5B Gem
5B Gooding
5B Idaho
6B Jefferson
5B Jerome
5B Kootenai
5B Latah
6B Lemhi
5B Lewis
5B Lincoln
6B Madison
5B Minidoka
5B Nez Perce
6B Oneida
5B Owyhee
5B Payette
5B Power
5B Shoshone
6B Teton
5B Twin Falls
6B Valley
5B Washington
ILLINOIS
5A Adams
4A Alexander
4A Bond
5A Boone
5A Brown
5A Bureau
4A Calhoun
5A Carroll
5A Cass
5A Champaign
4A Christian
4A Clark
4A Clay
4A Clinton
4A Coles
5A Cook
4A Crawford
4A Cumberland
5A DeKalb
5A De Witt
5A Douglas
5A DuPage
5A Edgar
4A Edwards
4A Effingham
4A Fayette
5A Ford
4A Franklin
5A Fulton
4A Gallatin
4A Greene
5A Grundy
4A Hamilton
5A Hancock
4A Hardin

TABLE R301.1—CLIMATE ZONES, MOISTURE REGIMES AND WARM HUMID DESIGNATIONS BY STATE, COUNTY AND TERRITORY[a]—continued

US STATES—continued
ILLINOIS *(continued)*
5A Henderson
5A Henry
5A Iroquois
4A Jackson
4A Jasper
4A Jefferson
4A Jersey
5A Jo Daviess
4A Johnson
5A Kane
5A Kankakee
5A Kendall
5A Knox
5A Lake
5A La Salle
4A Lawrence
5A Lee
5A Livingston
5A Logan
5A Macon
4A Macoupin
4A Madison
4A Marion
5A Marshall
5A Mason
4A Massac
5A McDonough
5A McHenry
5A McLean
5A Menard
5A Mercer
4A Monroe
4A Montgomery
5A Morgan
5A Moultrie
5A Ogle
5A Peoria
4A Perry
5A Piatt
5A Pike
4A Pope
4A Pulaski
5A Putnam
4A Randolph
4A Richland
5A Rock Island
4A Saline
5A Sangamon
5A Schuyler
5A Scott
4A Shelby
5A Stark
4A St. Clair
5A Stephenson
5A Tazewell
4A Union
5A Vermilion
4A Wabash
5A Warren
4A Washington
4A Wayne
4A White
5A Whiteside
5A Will
4A Williamson
5A Winnebago
5A Woodford
INDIANA
5A Adams
5A Allen
4A Bartholomew
5A Benton
5A Blackford
5A Boone
4A Brown
5A Carroll
5A Cass
4A Clark
4A Clay
5A Clinton
4A Crawford
4A Daviess
4A Dearborn
4A Decatur
5A De Kalb
5A Delaware
4A Dubois
5A Elkhart
4A Fayette
4A Floyd

TABLE R301.1—CLIMATE ZONES, MOISTURE REGIMES AND WARM HUMID DESIGNATIONS BY STATE, COUNTY AND TERRITORY[a]—continued

US STATES—continued
INDIANA *(continued)*
5A Fountain
4A Franklin
5A Fulton
4A Gibson
5A Grant
4A Greene
5A Hamilton
5A Hancock
4A Harrison
4A Hendricks
5A Henry
5A Howard
5A Huntington
4A Jackson
5A Jasper
5A Jay
4A Jefferson
4A Jennings
4A Johnson
4A Knox
5A Kosciusko
5A LaGrange
5A Lake
5A LaPorte
4A Lawrence
5A Madison
4A Marion
5A Marshall
4A Martin
5A Miami
4A Monroe
5A Montgomery
4A Morgan
5A Newton
5A Noble
4A Ohio
4A Orange
4A Owen
5A Parke
4A Perry
4A Pike
5A Porter
4A Posey
5A Pulaski
4A Putnam
5A Randolph
4A Ripley
4A Rush
4A Scott
4A Shelby
4A Spencer
5A Starke
5A Steuben
5A St. Joseph
4A Sullivan
4A Switzerland
5A Tippecanoe
5A Tipton
4A Union
4A Vanderburgh
5A Vermillion
4A Vigo
5A Wabash
5A Warren
4A Warrick
4A Washington
5A Wayne
5A Wells
5A White
5A Whitley
IOWA
5A Adair
5A Adams
5A Allamakee
5A Appanoose
5A Audubon
5A Benton
6A Black Hawk
5A Boone
5A Bremer
5A Buchanan
5A Buena Vista
5A Butler
5A Calhoun
5A Carroll
5A Cass
5A Cedar
6A Cerro Gordo
5A Cherokee
5A Chickasaw

TABLE R301.1—CLIMATE ZONES, MOISTURE REGIMES AND WARM HUMID DESIGNATIONS BY STATE, COUNTY AND TERRITORY[a]—continued

US STATES—continued
IOWA *(continued)*
5A Clarke
6A Clay
5A Clayton
5A Clinton
5A Crawford
5A Dallas
5A Davis
5A Decatur
5A Delaware
5A Des Moines
6A Dickinson
5A Dubuque
6A Emmet
5A Fayette
5A Floyd
5A Franklin
5A Fremont
5A Greene
5A Grundy
5A Guthrie
5A Hamilton
6A Hancock
5A Hardin
5A Harrison
5A Henry
5A Howard
5A Humboldt
5A Ida
5A Iowa
5A Jackson
5A Jasper
5A Jefferson
5A Johnson
5A Jones
5A Keokuk
6A Kossuth
5A Lee
5A Linn
5A Louisa
5A Lucas
6A Lyon
5A Madison
5A Mahaska
5A Marion
5A Marshall
5A Mills
6A Mitchell
5A Monona
5A Monroe
5A Montgomery
5A Muscatine
6A O'Brien
6A Osceola
5A Page
6A Palo Alto
5A Plymouth
5A Pocahontas
5A Polk
5A Pottawattamie
5A Poweshiek
5A Ringgold
5A Sac
5A Scott
5A Shelby
6A Sioux
5A Story
5A Tama
5A Taylor
5A Union
5A Van Buren
5A Wapello
5A Warren
5A Washington
5A Wayne
5A Webster
6A Winnebago
5A Winneshiek
5A Woodbury
6A Worth
5A Wright
KANSAS
4A Allen
4A Anderson
4A Atchison
4A Barber
4A Barton
4A Bourbon
4A Brown
4A Butler
4A Chase

TABLE R301.1—CLIMATE ZONES, MOISTURE REGIMES AND WARM HUMID DESIGNATIONS BY STATE, COUNTY AND TERRITORY[a]—continued

US STATES—continued
KANSAS *(continued)*
4A Chautauqua
4A Cherokee
5A Cheyenne
4A Clark
4A Clay
4A Cloud
4A Coffey
4A Comanche
4A Cowley
4A Crawford
5A Decatur
4A Dickinson
4A Doniphan
4A Douglas
4A Edwards
4A Elk
4A Ellis
4A Ellsworth
4A Finney
4A Ford
4A Franklin
4A Geary
5A Gove
4A Graham
4A Grant
4A Gray
5A Greeley
4A Greenwood
4A Hamilton
4A Harper
4A Harvey
4A Haskell
4A Hodgeman
4A Jackson
4A Jefferson
5A Jewell
4A Johnson
4A Kearny
4A Kingman
4A Kiowa
4A Labette
4A Lane
4A Leavenworth
4A Lincoln
4A Linn
5A Logan
4A Lyon
4A Marion
4A Marshall
4A McPherson
4A Meade
4A Miami
4A Mitchell
4A Montgomery
4A Morris
4A Morton
4A Nemaha
4A Neosho
4A Ness
5A Norton
4A Osage
4A Osborne
4A Ottawa
4A Pawnee
5A Phillips
4A Pottawatomie
4A Pratt
5A Rawlins
4A Reno
5A Republic
4A Rice
4A Riley
4A Rooks
4A Rush
4A Russell
4A Saline
5A Scott
4A Sedgwick
4A Seward
4A Shawnee
5A Sheridan
5A Sherman
5A Smith
4A Stafford
4A Stanton
4A Stevens
4A Sumner
5A Thomas
4A Trego
4A Wabaunsee

TABLE R301.1—CLIMATE ZONES, MOISTURE REGIMES AND WARM HUMID DESIGNATIONS BY STATE, COUNTY AND TERRITORY[a]—continued

US STATES—continued
KANSAS *(continued)*
5A Wallace
4A Washington
5A Wichita
4A Wilson
4A Woodson
4A Wyandotte
KENTUCKY
4A (all)
LOUISIANA
2A Acadia*
2A Allen*
2A Ascension*
2A Assumption*
2A Avoyelles*
2A Beauregard*
3A Bienville*
3A Bossier*
3A Caddo*
2A Calcasieu*
3A Caldwell*
2A Cameron*
3A Catahoula*
3A Claiborne*
3A Concordia*
3A De Soto*
2A East Baton Rouge*
3A East Carroll
2A East Feliciana*
2A Evangeline*
3A Franklin*
3A Grant*
2A Iberia*
2A Iberville*
3A Jackson*
2A Jefferson*
2A Jefferson Davis*
2A Lafayette*
2A Lafourche*
3A La Salle*
3A Lincoln*
2A Livingston*
3A Madison*
3A Morehouse
3A Natchitoches*
2A Orleans*
3A Ouachita*
2A Plaquemines*
2A Pointe Coupee*
2A Rapides*
3A Red River*
3A Richland*
3A Sabine*
2A St. Bernard*
2A St. Charles*
2A St. Helena*
2A St. James*
2A St. John the Baptist*
2A St. Landry*
2A St. Martin*
2A St. Mary*
2A St. Tammany*
2A Tangipahoa*
3A Tensas*
2A Terrebonne*
3A Union*
2A Vermilion*
3A Vernon*
2A Washington*
3A Webster*
2A West Baton Rouge*
3A West Carroll
2A West Feliciana*
3A Winn*
MAINE
6A Androscoggin
7 Aroostook
6A Cumberland
6A Franklin
6A Hancock
6A Kennebec
6A Knox
6A Lincoln
6A Oxford
6A Penobscot
6A Piscataquis
6A Sagadahoc
6A Somerset
6A Waldo
6A Washington
6A York

TABLE R301.1—CLIMATE ZONES, MOISTURE REGIMES AND WARM HUMID DESIGNATIONS BY STATE, COUNTY AND TERRITORY[a]—continued

US STATES—continued
MARYLAND
5A Allegany
4A Anne Arundel
4A Baltimore
4A Baltimore (city)
4A Calvert
4A Caroline
4A Carroll
4A Cecil
4A Charles
4A Dorchester
4A Frederick
5A Garrett
4A Harford
4A Howard
4A Kent
4A Montgomery
4A Prince George's
4A Queen Anne's
4A Somerset
4A St. Mary's
4A Talbot
4A Washington
4A Wicomico
4A Worcester
MASSACHUSETTS
5A (all)
MICHIGAN
6A Alcona
6A Alger
5A Allegan
6A Alpena
6A Antrim
6A Arenac
6A Baraga
5A Barry
5A Bay
6A Benzie
5A Berrien
5A Branch
5A Calhoun
5A Cass
6A Charlevoix
6A Cheboygan
6A Chippewa
6A Clare
5A Clinton
6A Crawford
6A Delta
6A Dickinson
5A Eaton
6A Emmet
5A Genesee
6A Gladwin
6A Gogebic
6A Grand Traverse
5A Gratiot
5A Hillsdale
6A Houghton
5A Huron
5A Ingham
5A Ionia
6A Iosco
6A Iron
6A Isabella
5A Jackson
5A Kalamazoo
6A Kalkaska
5A Kent
7 Keweenaw
6A Lake
5A Lapeer
6A Leelanau
5A Lenawee
5A Livingston
6A Luce
6A Mackinac
5A Macomb
6A Manistee
7 Marquette
6A Mason
6A Mecosta
6A Menominee
5A Midland
6A Missaukee
5A Monroe
5A Montcalm
6A Montmorency
5A Muskegon
6A Newaygo
5A Oakland

TABLE R301.1—CLIMATE ZONES, MOISTURE REGIMES AND WARM HUMID DESIGNATIONS BY STATE, COUNTY AND TERRITORY[a]—continued

US STATES—continued
MICHIGAN *(continued)*
6A Oceana
6A Ogemaw
6A Ontonagon
6A Osceola
6A Oscoda
6A Otsego
5A Ottawa
6A Presque Isle
6A Roscommon
5A Saginaw
5A Sanilac
6A Schoolcraft
5A Shiawassee
5A St. Clair
5A St. Joseph
5A Tuscola
5A Van Buren
5A Washtenaw
5A Wayne
6A Wexford
MINNESOTA
7 Aitkin
6A Anoka
6A Becker
7 Beltrami
6A Benton
6A Big Stone
6A Blue Earth
6A Brown
7 Carlton
6A Carver
7 Cass
6A Chippewa
6A Chisago
6A Clay
7 Clearwater
7 Cook
6A Cottonwood
7 Crow Wing
6A Dakota
6A Dodge
6A Douglas
6A Faribault
5A Fillmore
6A Freeborn
6A Goodhue
6A Grant
6A Hennepin
5A Houston
7 Hubbard
6A Isanti
7 Itasca
6A Jackson
6A Kanabec
6A Kandiyohi
7 Kittson
7 Koochiching
6A Lac qui Parle
7 Lake
7 Lake of the Woods
6A Le Sueur
6A Lincoln
6A Lyon
7 Mahnomen
7 Marshall
6A Martin
6A McLeod
6A Meeker
6A Mille Lacs
6A Morrison
6A Mower
6A Murray
6A Nicollet
6A Nobles
7 Norman
6A Olmsted
6A Otter Tail
7 Pennington
7 Pine
6A Pipestone
7 Polk
6A Pope
6A Ramsey
7 Red Lake
6A Redwood
6A Renville
6A Rice
6A Rock
7 Roseau
6A Scott

TABLE R301.1—CLIMATE ZONES, MOISTURE REGIMES AND WARM HUMID DESIGNATIONS BY STATE, COUNTY AND TERRITORY[a]—continued

US STATES—continued
MINNESOTA *(continued)*
6A Sherburne
6A Sibley
6A Stearns
6A Steele
6A Stevens
7 St. Louis
6A Swift
6A Todd
6A Traverse
6A Wabasha
7 Wadena
6A Waseca
6A Washington
6A Watonwan
6A Wilkin
5A Winona
6A Wright
6A Yellow Medicine
MISSISSIPPI
3A Adams*
3A Alcorn
3A Amite*
3A Attala
3A Benton
3A Bolivar
3A Calhoun
3A Carroll
3A Chickasaw
3A Choctaw
3A Claiborne*
3A Clarke
3A Clay
3A Coahoma
3A Copiah*
3A Covington*
3A DeSoto
3A Forrest*
3A Franklin*
2A George*
3A Greene*
3A Grenada
2A Hancock*
2A Harrison*
3A Hinds*
3A Holmes
3A Humphreys
3A Issaquena
3A Itawamba
2A Jackson*
3A Jasper
3A Jefferson*
3A Jefferson Davis*
3A Jones*
3A Kemper
3A Lafayette
3A Lamar*
3A Lauderdale
3A Lawrence*
3A Leake
3A Lee
3A Leflore
3A Lincoln*
3A Lowndes
3A Madison
3A Marion*
3A Marshall
3A Monroe
3A Montgomery
3A Neshoba
3A Newton
3A Noxubee
3A Oktibbeha
3A Panola
2A Pearl River*
3A Perry*
3A Pike*
3A Pontotoc
3A Prentiss
3A Quitman
3A Rankin*
3A Scott
3A Sharkey
3A Simpson*
3A Smith*
2A Stone*
3A Sunflower
3A Tallahatchie
3A Tate
3A Tippah
3A Tishomingo

TABLE R301.1—CLIMATE ZONES, MOISTURE REGIMES AND WARM HUMID DESIGNATIONS BY STATE, COUNTY AND TERRITORY[a]—continued

US STATES—continued
MISSISSIPPI *(continued)*
3A Tunica
3A Union
3A Walthall*
3A Warren*
3A Washington
3A Wayne*
3A Webster
3A Wilkinson*
3A Winston
3A Yalobusha
3A Yazoo
MISSOURI
5A Adair
5A Andrew
5A Atchison
4A Audrain
4A Barry
4A Barton
4A Bates
4A Benton
4A Bollinger
4A Boone
4A Buchanan
4A Butler
4A Caldwell
4A Callaway
4A Camden
4A Cape Girardeau
4A Carroll
4A Carter
4A Cass
4A Cedar
4A Chariton
4A Christian
5A Clark
4A Clay
4A Clinton
4A Cole
4A Cooper
4A Crawford
4A Dade
4A Dallas
5A Daviess
5A DeKalb
4A Dent
4A Douglas
3A Dunklin
4A Franklin
4A Gasconade
5A Gentry
4A Greene
5A Grundy
5A Harrison
4A Henry
4A Hickory
5A Holt
4A Howard
4A Howell
4A Iron
4A Jackson
4A Jasper
4A Jefferson
4A Johnson
5A Knox
4A Laclede
4A Lafayette
4A Lawrence
5A Lewis
4A Lincoln
5A Linn
5A Livingston
5A Macon
4A Madison
4A Maries
5A Marion
4A McDonald
5A Mercer
4A Miller
4A Mississippi
4A Moniteau
4A Monroe
4A Montgomery
4A Morgan
4A New Madrid
4A Newton
5A Nodaway
4A Oregon
4A Osage
4A Ozark
3A Pemiscot

TABLE R301.1—CLIMATE ZONES, MOISTURE REGIMES AND WARM HUMID DESIGNATIONS BY STATE, COUNTY AND TERRITORY[a]—continued

US STATES—continued
MISSOURI *(continued)*
4A Perry
4A Pettis
4A Phelps
5A Pike
4A Platte
4A Polk
4A Pulaski
5A Putnam
5A Ralls
4A Randolph
4A Ray
4A Reynolds
4A Ripley
4A Saline
5A Schuyler
5A Scotland
4A Scott
4A Shannon
5A Shelby
4A St. Charles
4A St. Clair
4A St. Francois
4A St. Louis
4A St. Louis (city)
4A Ste. Genevieve
4A Stoddard
4A Stone
5A Sullivan
4A Taney
4A Texas
4A Vernon
4A Warren
4A Washington
4A Wayne
4A Webster
5A Worth
4A Wright
MONTANA
6B (all)
NEBRASKA
5A (all)
NEVADA
4B Carson City (city)
5B Churchill
3B Clark
4B Douglas
5B Elko
4B Esmeralda
5B Eureka
5B Humboldt
5B Lander
4B Lincoln
4B Lyon
4B Mineral
4B Nye
5B Pershing
5B Storey
5B Washoe
5B White Pine
NEW HAMPSHIRE
6A Belknap
6A Carroll
5A Cheshire
6A Coos
6A Grafton
5A Hillsborough
5A Merrimack
5A Rockingham
5A Strafford
6A Sullivan
NEW JERSEY
4A Atlantic
5A Bergen
4A Burlington
4A Camden
4A Cape May
4A Cumberland
4A Essex
4A Gloucester
4A Hudson
5A Hunterdon
4A Mercer
4A Middlesex
4A Monmouth
5A Morris
4A Ocean
5A Passaic
4A Salem
5A Somerset
5A Sussex

TABLE R301.1—CLIMATE ZONES, MOISTURE REGIMES AND WARM HUMID DESIGNATIONS BY STATE, COUNTY AND TERRITORY[a]—continued

US STATES—continued
NEW JERSEY *(continued)*
4A Union
5A Warren
NEW MEXICO
4B Bernalillo
4B Catron
3B Chaves
4B Cibola
5B Colfax
4B Curry
4B DeBaca
3B Doña Ana
3B Eddy
4B Grant
4B Guadalupe
5B Harding
3B Hidalgo
3B Lea
4B Lincoln
5B Los Alamos
3B Luna
5B McKinley
5B Mora
3B Otero
4B Quay
5B Rio Arriba
4B Roosevelt
5B Sandoval
5B San Juan
5B San Miguel
5B Santa Fe
3B Sierra
4B Socorro
5B Taos
5B Torrance
4B Union
4B Valencia
NEW YORK
5A Albany
5A Allegany
4A Bronx
5A Broome
5A Cattaraugus
5A Cayuga
5A Chautauqua
5A Chemung
6A Chenango
6A Clinton
5A Columbia
5A Cortland
6A Delaware
5A Dutchess
5A Erie
6A Essex
6A Franklin
6A Fulton
5A Genesee
5A Greene
6A Hamilton
6A Herkimer
6A Jefferson
4A Kings
6A Lewis
5A Livingston
6A Madison
5A Monroe
6A Montgomery
4A Nassau
4A New York
5A Niagara
6A Oneida
5A Onondaga
5A Ontario
5A Orange
5A Orleans
5A Oswego
6A Otsego
5A Putnam
4A Queens
5A Rensselaer
4A Richmond
5A Rockland
5A Saratoga
5A Schenectady
5A Schoharie
5A Schuyler
5A Seneca
5A Steuben
6A St. Lawrence
4A Suffolk
6A Sullivan

TABLE R301.1—CLIMATE ZONES, MOISTURE REGIMES AND WARM HUMID DESIGNATIONS BY STATE, COUNTY AND TERRITORY[a]—continued

US STATES—continued
NEW YORK *(continued)*
5A Tioga
5A Tompkins
6A Ulster
6A Warren
5A Washington
5A Wayne
4A Westchester
5A Wyoming
5A Yates
NORTH CAROLINA
3A Alamance
3A Alexander
5A Alleghany
3A Anson
5A Ashe
5A Avery
3A Beaufort
3A Bertie
3A Bladen
3A Brunswick*
4A Buncombe
4A Burke
3A Cabarrus
4A Caldwell
3A Camden
3A Carteret*
3A Caswell
3A Catawba
3A Chatham
3A Cherokee
3A Chowan
3A Clay
3A Cleveland
3A Columbus*
3A Craven
3A Cumberland
3A Currituck
3A Dare
3A Davidson
3A Davie
3A Duplin
3A Durham
3A Edgecombe
3A Forsyth
3A Franklin
3A Gaston
3A Gates
4A Graham
3A Granville
3A Greene
3A Guilford
3A Halifax
3A Harnett
4A Haywood
4A Henderson
3A Hertford
3A Hoke
3A Hyde
3A Iredell
4A Jackson
3A Johnston
3A Jones
3A Lee
3A Lenoir
3A Lincoln
4A Macon
4A Madison
3A Martin
4A McDowell
3A Mecklenburg
4A Mitchell
3A Montgomery
3A Moore
3A Nash
3A New Hanover*
3A Northampton
3A Onslow*
3A Orange
3A Pamlico
3A Pasquotank
3A Pender*
3A Perquimans
3A Person
3A Pitt
3A Polk
3A Randolph
3A Richmond
3A Robeson
3A Rockingham
3A Rowan

TABLE R301.1—CLIMATE ZONES, MOISTURE REGIMES AND WARM HUMID DESIGNATIONS BY STATE, COUNTY AND TERRITORY[a]—continued

US STATES—continued
NORTH CAROLINA *(continued)*
3A Rutherford
3A Sampson
3A Scotland
3A Stanly
4A Stokes
4A Surry
4A Swain
4A Transylvania
3A Tyrrell
3A Union
3A Vance
3A Wake
3A Warren
3A Washington
5A Watauga
3A Wayne
3A Wilkes
3A Wilson
4A Yadkin
5A Yancey
NORTH DAKOTA
6A Adams
6A Barnes
7 Benson
6A Billings
7 Bottineau
6A Bowman
7 Burke
6A Burleigh
6A Cass
7 Cavalier
6A Dickey
7 Divide
6A Dunn
6A Eddy
6A Emmons
6A Foster
6A Golden Valley
7 Grand Forks
6A Grant
6A Griggs
6A Hettinger
6A Kidder
6A LaMoure
6A Logan
7 McHenry
6A McIntosh
6A McKenzie
6A McLean
6A Mercer
6A Morton
6A Mountrail
7 Nelson
6A Oliver
7 Pembina
7 Pierce
7 Ramsey
6A Ransom
7 Renville
6A Richland
7 Rolette
6A Sargent
6A Sheridan
6A Sioux
6A Slope
6A Stark
6A Steele
6A Stutsman
7 Towner
6A Traill
7 Walsh
7 Ward
6A Wells
6A Williams
OHIO
4A Adams
5A Allen
5A Ashland
5A Ashtabula
4A Athens
5A Auglaize
5A Belmont
4A Brown
4A Butler
5A Carroll
5A Champaign
5A Clark
4A Clermont
4A Clinton
5A Columbiana

TABLE R301.1—CLIMATE ZONES, MOISTURE REGIMES AND WARM HUMID DESIGNATIONS BY STATE, COUNTY AND TERRITORY[a]—continued

US STATES—continued
OHIO *(continued)*
5A Coshocton
5A Crawford
5A Cuyahoga
5A Darke
5A Defiance
5A Delaware
5A Erie
5A Fairfield
4A Fayette
4A Franklin
5A Fulton
4A Gallia
5A Geauga
4A Greene
5A Guernsey
4A Hamilton
5A Hancock
5A Hardin
5A Harrison
5A Henry
4A Highland
4A Hocking
5A Holmes
5A Huron
4A Jackson
5A Jefferson
5A Knox
5A Lake
4A Lawrence
5A Licking
5A Logan
5A Lorain
5A Lucas
4A Madison
5A Mahoning
5A Marion
5A Medina
4A Meigs
5A Mercer
5A Miami
5A Monroe
5A Montgomery
5A Morgan
5A Morrow
5A Muskingum
5A Noble
5A Ottawa
5A Paulding
5A Perry
4A Pickaway
4A Pike
5A Portage
5A Preble
5A Putnam
5A Richland
4A Ross
5A Sandusky
4A Scioto
5A Seneca
5A Shelby
5A Stark
5A Summit
5A Trumbull
5A Tuscarawas
5A Union
5A Van Wert
4A Vinton
4A Warren
4A Washington
5A Wayne
5A Williams
5A Wood
5A Wyandot
OKLAHOMA
3A Adair
4A Alfalfa
3A Atoka
4B Beaver
3A Beckham
3A Blaine
3A Bryan
3A Caddo
3A Canadian
3A Carter
3A Cherokee
3A Choctaw
4B Cimarron
3A Cleveland
3A Coal
3A Comanche

TABLE R301.1—CLIMATE ZONES, MOISTURE REGIMES AND WARM HUMID DESIGNATIONS BY STATE, COUNTY AND TERRITORY[a]—continued

US STATES—continued
OKLAHOMA *(continued)*
3A Cotton
4A Craig
3A Creek
3A Custer
4A Delaware
3A Dewey
4A Ellis
4A Garfield
3A Garvin
3A Grady
4A Grant
3A Greer
3A Harmon
4A Harper
3A Haskell
3A Hughes
3A Jackson
3A Jefferson
3A Johnston
4A Kay
3A Kingfisher
3A Kiowa
3A Latimer
3A Le Flore
3A Lincoln
3A Logan
3A Love
4A Major
3A Marshall
3A Mayes
3A McClain
3A McCurtain
3A McIntosh
3A Murray
3A Muskogee
3A Noble
4A Nowata
3A Okfuskee
3A Oklahoma
3A Okmulgee
4A Osage
4A Ottawa
3A Pawnee
3A Payne
3A Pittsburg
3A Pontotoc
3A Pottawatomie
3A Pushmataha
3A Roger Mills
3A Rogers
3A Seminole
3A Sequoyah
3A Stephens
4B Texas
3A Tillman
3A Tulsa
3A Wagoner
4A Washington
3A Washita
4A Woods
4A Woodward
OREGON
5B Baker
4C Benton
4C Clackamas
4C Clatsop
4C Columbia
4C Coos
5B Crook
4C Curry
5B Deschutes
4C Douglas
5B Gilliam
5B Grant
5B Harney
5B Hood River
4C Jackson
5B Jefferson
4C Josephine
5B Klamath
5B Lake
4C Lane
4C Lincoln
4C Linn
5B Malheur
4C Marion
5B Morrow
4C Multnomah
4C Polk
5B Sherman

TABLE R301.1—CLIMATE ZONES, MOISTURE REGIMES AND WARM HUMID DESIGNATIONS BY STATE, COUNTY AND TERRITORY[a]—continued

US STATES—continued
OREGON *(continued)*
4C Tillamook
5B Umatilla
5B Union
5B Wallowa
5B Wasco
4C Washington
5B Wheeler
4C Yamhill
PENNSYLVANIA
4A Adams
5A Allegheny
5A Armstrong
5A Beaver
5A Bedford
4A Berks
5A Blair
5A Bradford
4A Bucks
5A Butler
5A Cambria
5A Cameron
5A Carbon
5A Centre
4A Chester
5A Clarion
5A Clearfield
5A Clinton
5A Columbia
5A Crawford
4A Cumberland
4A Dauphin
4A Delaware
5A Elk
5A Erie
5A Fayette
5A Forest
4A Franklin
5A Fulton
5A Greene
5A Huntingdon
5A Indiana
5A Jefferson
5A Juniata
5A Lackawanna
4A Lancaster
5A Lawrence
4A Lebanon
5A Lehigh
5A Luzerne
5A Lycoming
5A McKean
5A Mercer
5A Mifflin
5A Monroe
4A Montgomery
5A Montour
5A Northampton
5A Northumberland
4A Perry
4A Philadelphia
5A Pike
5A Potter
5A Schuylkill
5A Snyder
5A Somerset
5A Sullivan
5A Susquehanna
5A Tioga
5A Union
5A Venango
5A Warren
5A Washington
5A Wayne
5A Westmoreland
5A Wyoming
4A York
RHODE ISLAND
5A (all)
SOUTH CAROLINA
3A Abbeville
3A Aiken
3A Allendale*
3A Anderson
3A Bamberg*
3A Barnwell*
2A Beaufort*
3A Berkeley*
3A Calhoun
3A Charleston*
3A Cherokee

TABLE R301.1—CLIMATE ZONES, MOISTURE REGIMES AND WARM HUMID DESIGNATIONS BY STATE, COUNTY AND TERRITORY[a]—continued

US STATES—continued
SOUTH CAROLINA *(continued)*
3A Chester
3A Chesterfield
3A Clarendon
3A Colleton*
3A Darlington
3A Dillon
3A Dorchester*
3A Edgefield
3A Fairfield
3A Florence
3A Georgetown*
3A Greenville
3A Greenwood
3A Hampton*
3A Horry*
2A Jasper*
3A Kershaw
3A Lancaster
3A Laurens
3A Lee
3A Lexington
3A Marion
3A Marlboro
3A McCormick
3A Newberry
3A Oconee
3A Orangeburg
3A Pickens
3A Richland
3A Saluda
3A Spartanburg
3A Sumter
3A Union
3A Williamsburg
3A York
SOUTH DAKOTA
6A Aurora
6A Beadle
5A Bennett
5A Bon Homme
6A Brookings
6A Brown
5A Brule
6A Buffalo
6A Butte
6A Campbell
5A Charles Mix
6A Clark
5A Clay
6A Codington
6A Corson
6A Custer
6A Davison
6A Day
6A Deuel
6A Dewey
5A Douglas
6A Edmunds
6A Fall River
6A Faulk
6A Grant
5A Gregory
5A Haakon
6A Hamlin
6A Hand
6A Hanson
6A Harding
6A Hughes
5A Hutchinson
6A Hyde
5A Jackson
6A Jerauld
5A Jones
6A Kingsbury
6A Lake
6A Lawrence
6A Lincoln
5A Lyman
6A Marshall
6A McCook
6A McPherson
6A Meade
5A Mellette
6A Miner
6A Minnehaha
6A Moody
6A Pennington
6A Perkins
6A Potter
6A Roberts

TABLE R301.1—CLIMATE ZONES, MOISTURE REGIMES AND WARM HUMID DESIGNATIONS BY STATE, COUNTY AND TERRITORY[a]—continued

US STATES—continued
SOUTH DAKOTA *(continued)*
6A Sanborn
6A Shannon
6A Spink
5A Stanley
6A Sully
5A Todd
5A Tripp
6A Turner
5A Union
6A Walworth
5A Yankton
6A Ziebach
TENNESSEE
4A Anderson
3A Bedford
4A Benton
4A Bledsoe
4A Blount
4A Bradley
4A Campbell
4A Cannon
4A Carroll
4A Carter
4A Cheatham
3A Chester
4A Claiborne
4A Clay
4A Cocke
3A Coffee
3A Crockett
4A Cumberland
3A Davidson
3A Decatur
4A DeKalb
4A Dickson
3A Dyer
3A Fayette
4A Fentress
3A Franklin
3A Gibson
3A Giles
4A Grainger
4A Greene
3A Grundy
4A Hamblen
3A Hamilton
4A Hancock
3A Hardeman
3A Hardin
4A Hawkins
3A Haywood
3A Henderson
4A Henry
3A Hickman
4A Houston
4A Humphreys
4A Jackson
4A Jefferson
4A Johnson
4A Knox
4A Lake
3A Lauderdale
3A Lawrence
3A Lewis
3A Lincoln
4A Loudon
4A Macon
3A Madison
3A Marion
3A Marshall
3A Maury
4A McMinn
3A McNairy
4A Meigs
4A Monroe
4A Montgomery
3A Moore
4A Morgan
4A Obion
4A Overton
3A Perry
4A Pickett
4A Polk
4A Putnam
4A Rhea
4A Roane
4A Robertson
3A Rutherford
4A Scott
4A Sequatchie

TABLE R301.1—CLIMATE ZONES, MOISTURE REGIMES AND WARM HUMID DESIGNATIONS BY STATE, COUNTY AND TERRITORY[a]—continued

US STATES—continued
TENNESSEE *(continued)*
4A Sevier
3A Shelby
4A Smith
4A Stewart
4A Sullivan
4A Sumner
3A Tipton
4A Trousdale
4A Unicoi
4A Union
4A Van Buren
4A Warren
4A Washington
3A Wayne
4A Weakley
4A White
3A Williamson
4A Wilson
TEXAS
2A Anderson*
3B Andrews
2A Angelina*
2A Aransas*
3A Archer
4B Armstrong
2A Atascosa*
2A Austin*
4B Bailey
2B Bandera
2A Bastrop*
3B Baylor
2A Bee*
2A Bell*
2A Bexar*
3A Blanco*
3B Borden
2A Bosque*
3A Bowie*
2A Brazoria*
2A Brazos*
3B Brewster
4B Briscoe
2A Brooks*
3A Brown*
2A Burleson*
3A Burnet*
2A Caldwell*
2A Calhoun*
3B Callahan
1A Cameron*
3A Camp*
4B Carson
3A Cass*
4B Castro
2A Chambers*
2A Cherokee*
3B Childress
3A Clay
4B Cochran
3B Coke
3B Coleman
3A Collin*
3B Collingsworth
2A Colorado*
2A Comal*
3A Comanche*
3B Concho
3A Cooke
2A Coryell*
3B Cottle
3B Crane
3B Crockett
3B Crosby
3B Culberson
4B Dallam
2A Dallas*
3B Dawson
4B Deaf Smith
3A Delta
3A Denton*
2A DeWitt*
3B Dickens
2B Dimmit
4B Donley
2A Duval*
3A Eastland
3B Ector
2B Edwards
2A Ellis*
3B El Paso

TABLE R301.1—CLIMATE ZONES, MOISTURE REGIMES AND WARM HUMID DESIGNATIONS BY STATE, COUNTY AND TERRITORY[a]—continued

US STATES—continued
TEXAS *(continued)*
3A Erath*
2A Falls*
3A Fannin
2A Fayette*
3B Fisher
4B Floyd
3B Foard
2A Fort Bend*
3A Franklin*
2A Freestone*
2B Frio
3B Gaines
2A Galveston*
3B Garza
3A Gillespie*
3B Glasscock
2A Goliad*
2A Gonzales*
4B Gray
3A Grayson
3A Gregg*
2A Grimes*
2A Guadalupe*
4B Hale
3B Hall
3A Hamilton*
4B Hansford
3B Hardeman
2A Hardin*
2A Harris*
3A Harrison*
4B Hartley
3B Haskell
2A Hays*
3B Hemphill
3A Henderson*
1A Hidalgo*
2A Hill*
4B Hockley
3A Hood*
3A Hopkins*
2A Houston*
3B Howard
3B Hudspeth
3A Hunt*
4B Hutchinson
3B Irion
3A Jack
2A Jackson*
2A Jasper*
3B Jeff Davis
2A Jefferson*
2A Jim Hogg*
2A Jim Wells*
2A Johnson*
3B Jones
2A Karnes*
3A Kaufman*
3A Kendall*
2A Kenedy*
3B Kent
3B Kerr
3B Kimble
3B King
2B Kinney
2A Kleberg*
3B Knox
3A Lamar*
4B Lamb
3A Lampasas*
2B La Salle
2A Lavaca*
2A Lee*
2A Leon*
2A Liberty*
2A Limestone*
4B Lipscomb
2A Live Oak*
3A Llano*
3B Loving
3B Lubbock
3B Lynn
2A Madison*
3A Marion*
3B Martin
3B Mason
2A Matagorda*
2B Maverick
3B McCulloch
2A McLennan*

TABLE R301.1—CLIMATE ZONES, MOISTURE REGIMES AND WARM HUMID DESIGNATIONS BY STATE, COUNTY AND TERRITORY[a]—continued

US STATES—continued
TEXAS *(continued)*
2A McMullen*
2B Medina
3B Menard
3B Midland
2A Milam*
3A Mills*
3B Mitchell
3A Montague
2A Montgomery*
4B Moore
3A Morris*
3B Motley
3A Nacogdoches*
2A Navarro*
2A Newton*
3B Nolan
2A Nueces*
4B Ochiltree
4B Oldham
2A Orange*
3A Palo Pinto*
3A Panola*
3A Parker*
4B Parmer
3B Pecos
2A Polk*
4B Potter
3B Presidio
3A Rains*
4B Randall
3B Reagan
2B Real
3A Red River*
3B Reeves
2A Refugio*
4B Roberts
2A Robertson*
3A Rockwall*
3B Runnels
3A Rusk*
3A Sabine*
3A San Augustine*
2A San Jacinto*
2A San Patricio*
3A San Saba*
3B Schleicher
3B Scurry
3B Shackelford
3A Shelby*
4B Sherman
3A Smith*
3A Somervell*
2A Starr*
3A Stephens
3B Sterling
3B Stonewall
3B Sutton
4B Swisher
2A Tarrant*
3B Taylor
3B Terrell
3B Terry
3B Throckmorton
3A Titus*
3B Tom Green
2A Travis*
2A Trinity*
2A Tyler*
3A Upshur*
3B Upton
2B Uvalde
2B Val Verde
3A Van Zandt*
2A Victoria*
2A Walker*
2A Waller*
3B Ward
2A Washington*
2B Webb
2A Wharton*
3B Wheeler
3A Wichita
3B Wilbarger
1A Willacy*
2A Williamson*
2A Wilson*
3B Winkler
3A Wise
3A Wood*
4B Yoakum

TABLE R301.1—CLIMATE ZONES, MOISTURE REGIMES AND WARM HUMID DESIGNATIONS BY STATE, COUNTY AND TERRITORY[a]—continued

US STATES—continued
TEXAS *(continued)*
3A Young
2B Zapata
2B Zavala
UTAH
5B Beaver
5B Box Elder
5B Cache
5B Carbon
6B Daggett
5B Davis
6B Duchesne
5B Emery
5B Garfield
5B Grand
5B Iron
5B Juab
5B Kane
5B Millard
6B Morgan
5B Piute
6B Rich
5B Salt Lake
5B San Juan
5B Sanpete
5B Sevier
6B Summit
5B Tooele
6B Uintah
5B Utah
6B Wasatch
3B Washington
5B Wayne
5B Weber
VERMONT
6A (all)
VIRGINIA
4A (all except as follows:)
5A Alleghany
5A Bath
3A Brunswick
3A Chesapeake
5A Clifton Forge
5A Covington
3A Emporia
3A Franklin
3A Greensville
3A Halifax
3A Hampton
5A Highland
3A Isle of Wight
3A Mecklenburg
3A Newport News
3A Norfolk
3A Pittsylvania
3A Portsmouth
3A South Boston
3A Southampton
3A Suffolk
3A Surry
3A Sussex
3A Virginia Beach
WASHINGTON
5B Adams
5B Asotin
5B Benton
5B Chelan
5C Clallam
4C Clark
5B Columbia
4C Cowlitz
5B Douglas
6B Ferry
5B Franklin
5B Garfield
5B Grant
4C Grays Harbor
5C Island
4C Jefferson
4C King
5C Kitsap
5B Kittitas
5B Klickitat
4C Lewis
5B Lincoln
4C Mason
5B Okanogan
4C Pacific
6B Pend Oreille
4C Pierce
5C San Juan

TABLE R301.1—CLIMATE ZONES, MOISTURE REGIMES AND WARM HUMID DESIGNATIONS BY STATE, COUNTY AND TERRITORY[a]—continued

US STATES—continued
WASHINGTON *(continued)*
4C Skagit
5B Skamania
4C Snohomish
5B Spokane
6B Stevens
4C Thurston
4C Wahkiakum
5B Walla Walla
4C Whatcom
5B Whitman
5B Yakima
WEST VIRGINIA
5A Barbour
4A Berkeley
4A Boone
4A Braxton
5A Brooke
4A Cabell
4A Calhoun
4A Clay
4A Doddridge
4A Fayette
4A Gilmer
5A Grant
4A Greenbrier
5A Hampshire
5A Hancock
5A Hardy
5A Harrison
4A Jackson
4A Jefferson
4A Kanawha
4A Lewis
4A Lincoln
4A Logan
5A Marion
5A Marshall
4A Mason
4A McDowell
4A Mercer
5A Mineral
4A Mingo
5A Monongalia
4A Monroe
4A Morgan
4A Nicholas
5A Ohio
5A Pendleton
4A Pleasants
5A Pocahontas
5A Preston
4A Putnam
4A Raleigh
5A Randolph
4A Ritchie
4A Roane
4A Summers
5A Taylor
5A Tucker
4A Tyler
4A Upshur
4A Wayne
4A Webster
5A Wetzel
4A Wirt
4A Wood
4A Wyoming
WISCONSIN
5A Adams
6A Ashland
6A Barron
6A Bayfield
6A Brown
6A Buffalo
6A Burnett
5A Calumet
6A Chippewa
6A Clark
5A Columbia
5A Crawford
5A Dane
5A Dodge
6A Door
6A Douglas
6A Dunn
6A Eau Claire
6A Florence
5A Fond du Lac
6A Forest
5A Grant

TABLE R301.1—CLIMATE ZONES, MOISTURE REGIMES AND WARM HUMID DESIGNATIONS BY STATE, COUNTY AND TERRITORY[a]—continued

US STATES—continued
WISCONSIN *(continued)*
5A Green
5A Green Lake
5A Iowa
6A Iron
6A Jackson
5A Jefferson
5A Juneau
5A Kenosha
6A Kewaunee
5A La Crosse
5A Lafayette
6A Langlade
6A Lincoln
6A Manitowoc
6A Marathon
6A Marinette
6A Marquette
6A Menominee
5A Milwaukee
5A Monroe
6A Oconto
6A Oneida
5A Outagamie
5A Ozaukee
6A Pepin
6A Pierce
6A Polk
6A Portage
6A Price
5A Racine
5A Richland
5A Rock
6A Rusk
5A Sauk
6A Sawyer
6A Shawano
6A Sheboygan
6A St. Croix
6A Taylor
6A Trempealeau
5A Vernon
6A Vilas
5A Walworth
6A Washburn
5A Washington
5A Waukesha
6A Waupaca
5A Waushara
5A Winnebago
6A Wood
WYOMING
6B Albany
6B Big Horn
6B Campbell
6B Carbon
6B Converse
6B Crook
6B Fremont
5B Goshen
6B Hot Springs
6B Johnson
5B Laramie
7 Lincoln
6B Natrona
6B Niobrara
6B Park
5B Platte
6B Sheridan
7 Sublette
6B Sweetwater
7 Teton
6B Uinta
6B Washakie
6B Weston
US TERRITORIES
AMERICAN SAMOA
1A (all)*
GUAM
1A (all)*
NORTHERN MARIANA ISLANDS
1A (all)*
PUERTO RICO
1A (all except as follows:)*
2B Barraquitas
2B Cayey
VIRGIN ISLANDS
1A (all)*

a. Key: A – Moist, B – Dry, C – Marine. Absence of moisture designation indicates moisture regime is irrelevant. Asterisk (*) indicates a Warm Humid location.

R301.2 Warm Humid counties. In Table R301.1, Warm Humid counties are identified by an asterisk.

R301.3 Climate zone definitions. To determine the *climate zones* for locations not listed in this code, use the following information to determine *climate zone* numbers and letters in accordance with Items 1 through 5.

1. Determine the thermal *climate zone*, 0 through 8, from Table R301.3 using the heating (HDD) and cooling degree-days (CDD) for the location.
2. Determine the moisture zone (Marine, Dry or Humid) in accordance with Items 2.1 through 2.3.
   - 2.1. If monthly average temperature and precipitation data are available, use the Marine, Dry and Humid definitions to determine the moisture zone (C, B or A).
   - 2.2. If annual average temperature information (including degree-days) and annual precipitation (i.e., annual mean) are available, use Items 2.2.1 through 2.2.3 to determine the moisture zone. If the moisture zone is not Marine, then use the Dry definition to determine whether Dry or Humid.
     - 2.2.1. If thermal *climate zone* is 3 and CDD50°F ≤ 4,500 (CDD10°C ≤ 2500), *climate zone* is Marine (3C).
     - 2.2.2. If thermal *climate zone* is 4 and CDD50°F ≤ 2,700 (CDD10°C ≤ 1500), *climate zone* is Marine (4C).
     - 2.2.3. If thermal *climate zone* is 5 and CDD50°F ≤ 1,800 (CDD10°C ≤ 1000), *climate zone* is Marine (5C).
   - 2.3. If only degree-day information is available, use Items 2.3.1 through 2.3.3 to determine the moisture zone. If the moisture zone is not Marine, then it is not possible to assign Humid or Dry moisture zone for this location.
     - 2.3.1. If thermal *climate zone* is 3 and CDD50°F ≤ 4,500 (CDD10°C ≤ 2500), *climate zone* is Marine (3C).
     - 2.3.2. If thermal *climate zone* is 4 and CDD50°F ≤ 2,700 (CDD10°C ≤ 1500), *climate zone* is Marine (4C).
     - 2.3.3. If thermal *climate zone* is 5 and CDD50°F ≤ 1,800 (CDD10°C ≤ 1000), *climate zone* is Marine (5C).
3. Marine (C) Zone definition: Locations meeting all the criteria in Items 3.1 through 3.4.
   - 3.1. Mean temperature of coldest month between 27°F (-3°C) and 65°F (18°C).
   - 3.2. Warmest month mean < 72°F (22°C).
   - 3.3. Not fewer than four months with mean temperatures over 50°F (10°C).
   - 3.4. Dry season in summer. The month with the heaviest precipitation in the cold season has at least three times as much precipitation as the month with the least precipitation in the rest of the year. The cold season is October through March in the Northern Hemisphere and April through September in the Southern Hemisphere.
4. Dry (B) definition: Locations meeting the criteria in Items 4.1 through 4.4.
   - 4.1. Not Marine (C).
   - 4.2. If 70 percent or more of the precipitation, *P*, occurs during the high sun period, defined as April through September in the Northern Hemisphere and October through March in the Southern Hemisphere, then the dry/humid threshold is in accordance with Equation 3-1.

     Equation 3-1 $P < 0.44 \times (T - 7)$
     $[P < 20.0 \times (T + 14) \text{ in SI units}]$

     where:

     P = Annual precipitation, inches (mm).

     T = Annual mean temperature, °F (°C).
   - 4.3. If between 30 and 70 percent of the precipitation, *P*, occurs during the high sun period, defined as April through September in the Northern Hemisphere and October through March in the Southern Hemisphere, then the dry/humid threshold is in accordance with Equation 3-2.

     Equation 3-2 $P < 0.44 \times (T - 19.5)$
     $[P < 20.0 \times (T + 7) \text{ in SI units}]$

     where:

     P = Annual precipitation, inches (mm).

     T = Annual mean temperature, °F (°C).
   - 4.4. If 30 percent or less of the precipitation, *P*, occurs during the high sun period, defined as April through September in the Northern Hemisphere and October through March in the Southern Hemisphere, then the dry/humid threshold is in accordance with Equation 3-3.

     Equation 3-3 $P < 0.44 \times (T - 32)$
     $[P < 20.0 \times T \text{ in SI units}]$

     where:

     P = Annual precipitation, inches (mm).

     T = Annual mean temperature, °F (°C).
5. Humid (A) definition: Locations that are not Marine (C) or Dry (B).

TABLE R301.3—THERMAL CLIMATE ZONE DEFINITIONS

ZONE NUMBER	THERMAL CRITERIA	
	IP Units	SI Units
0	10,800 < CDD50°F	6000 < CDD10°C
1	9,000 < CDD50°F < 10,800	5000 < CDD10°C < 6000
2	6,300 < CDD50°F ≤ 9,000	3500 < CDD10°C ≤ 5000
3	CDD50°F ≤ 6,300 AND HDD65°F ≤ 3,600	CDD10°C < 3500 AND HDD18°C ≤ 2000
4	CDD50°F ≤ 6,300 AND 3,600 < HDD65°F ≤ 5,400	CDD10°C < 3500 AND 2000 < HDD18°C ≤ 3000
5	CDD50°F < 6,300 AND 5,400 < HDD65°F ≤ 7,200	CDD10°C < 3500 AND 3000 < HDD18°C ≤ 4000
6	7,200 < HDD65°F ≤ 9,000	4000 < HDD18°C ≤ 5000
7	9,000 < HDD65°F ≤ 12,600	5000 < HDD18°C ≤ 7000
8	12,600 < HDD65°F	7000 < HDD18°C

For SI: °C = [(°F) – 32]/1.8.

R301.4 Tropical climate region. The tropical region shall be defined as:

1. Hawaii, Puerto Rico, Guam, American Samoa, US Virgin Islands, Commonwealth of Northern Mariana Islands; and
2. Islands in the area between the Tropic of Cancer and the Tropic of Capricorn.

SECTION R302—DESIGN CONDITIONS

R302.1 Interior design conditions. The interior design temperatures used for heating and cooling load calculations shall be a maximum of 72°F (22°C) for heating and minimum of 75°F (24°C) for cooling.

SECTION R303—MATERIALS, SYSTEMS AND EQUIPMENT

R303.1 Identification. Materials, systems and equipment shall be identified in a manner that will allow a determination of compliance with the applicable provisions of this code.

R303.1.1 Building thermal envelope insulation. An *R-value* identification mark shall be applied by the manufacturer to each piece of *building thermal envelope* insulation that is 12 inches (305 mm) or greater in width. Alternatively, the insulation installers shall provide a certification that indicates the type, manufacturer and *R-value* of insulation installed in each element of the *building thermal envelope*. For blown-in or sprayed fiberglass and cellulose insulation, the initial installed thickness, settled thickness, settled *R-value*, installed density, coverage area and number of bags installed shall be indicated on the certification. For sprayed polyurethane foam (SPF) insulation, the installed thickness of the areas covered and the *R*-value of the installed thickness shall be indicated on the certification. For *reflective insulation*, the number of reflective sheets, the number and thickness of the enclosed reflective airspaces and the *R-value* for the installed assembly determined in accordance with Section R303.1.6 shall be *listed* on the certification. For *insulated siding*, the *R-value* shall be on a label on the product's package and shall be indicated on the certification. The insulation installer shall sign, date and post the certification in a conspicuous location on the job site.

Exception: For roof insulation installed above the deck, the *R-value* shall be *labeled* as required by the material standards specified in Table 1508.2 of the *International Building Code* or Table R906.2 of the *International Residential Code*, as applicable.

R303.1.1.1 Blown-in or sprayed roof and ceiling insulation. The thickness of blown-in or sprayed fiberglass and cellulose roof and ceiling insulation shall be written in inches (mm) on markers that are installed at not less than one for every 300 square feet (28 m^2) throughout the attic space. The markers shall be affixed to the trusses or joists and marked with the minimum initial installed thickness with numbers not less than 1 inch (25 mm) in height. Each marker shall face the attic access opening. The thickness and installed *R-value* of sprayed polyurethane foam insulation shall be indicated on the certification provided by the insulation installer.

R303.1.2 Insulation mark installation. Insulating materials shall be installed such that the manufacturer's *R-value* mark is readily observable at inspection. For insulation materials that are installed without an observable manufacturer's *R-value* mark, such as blown or draped products, an insulation certificate complying with Section R303.1.1 shall be left immediately after installation by the installer, in a conspicuous location within the *building*, to certify the installed *R-value* of the insulation material.

Exception: For roof insulation installed above the deck, the *R-value* shall be *labeled* as specified by the material standards in Table 1508.2 of the *International Building Code* or Table R906.2 of the *International Residential Code*, as applicable.

R303.1.3 Fenestration product rating. *U*-factors of *fenestration* products such as windows, doors and *skylights* shall be determined in accordance with NFRC 100.

Exception: Where required, garage door *U*-factors shall be determined in accordance with either NFRC 100 or ANSI/DASMA 105.

U-factors shall be determined by an accredited, independent laboratory, and *labeled* and certified by the manufacturer.

Products lacking such a *labeled U-factor* shall be assigned a default *U-factor* from Table R303.1.3(1) or Table R303.1.3(2). The *solar heat gain coefficient* (SHGC) and *visible transmittance* (VT) of glazed *fenestration* products such as windows, glazed doors and *skylights* shall be determined in accordance with NFRC 200 by an accredited, independent laboratory, and *labeled* and certified by the manufacturer. Products lacking such a *labeled* SHGC or VT shall be assigned a default SHGC or VT from Table R303.1.3(3).

TABLE R303.1.3(1)—DEFAULT GLAZED WINDOW, GLASS DOOR AND SKYLIGHT *U*-FACTORS

FRAME TYPE	WINDOW AND GLASS DOOR		SKYLIGHT	
	Single pane	Double pane	Single	Double
Metal	1.20	0.80	2.00	1.30
Metal with thermal break	1.10	0.65	1.90	1.10
Nonmetal or metal clad	0.95	0.55	1.75	1.05
Glazed block	0.60			

TABLE R303.1.3(2)—DEFAULT OPAQUE DOOR *U*-FACTORS

DOOR TYPE	OPAQUE *U*-FACTOR
Uninsulated metal	1.20
Insulated metal	0.60
Wood	0.50
Insulated, nonmetal edge, not exceeding 45% glazing, any glazing double pane	0.35

TABLE R303.1.3(3)—DEFAULT GLAZED FENESTRATION SHGC AND VT

	SINGLE GLAZED		DOUBLE GLAZED		GLAZED BLOCK
	Clear	Tinted	Clear	Tinted	
SHGC	0.8	0.7	0.7	0.6	0.6
VT	0.6	0.3	0.6	0.3	0.6

R303.1.4 Insulation product rating. The thermal resistance, *R-value*, of insulation shall be determined in accordance with Part 460 of US-FTC CFR Title 16 in units of h × ft^2 × °F/Btu at a mean temperature of 75°F (24°C).

R303.1.4.1 Insulated siding. The thermal resistance, *R-value*, of *insulated siding* shall be determined in accordance with ASTM C1363. Installation for testing shall be in accordance with the manufacturer's instructions.

R303.1.5 Air-impermeable insulation. Insulation having an air permeability not greater than 0.004 cubic feet per minute per square foot [0.002 L/(s × m^2)] under pressure differential of 0.3 inch water gauge (75 Pa) when tested in accordance with ASTM E2178 shall be determined air-impermeable insulation.

R303.1.6 Airspaces. Where the *R-value* of an enclosed reflective airspace or enclosed nonreflective airspace is used for compliance with this code, the airspace shall be enclosed in a cavity bounded on all sides by building components and constructed to minimize airflow into and out of the enclosed airspace. Airflow shall be deemed minimized where one of the following conditions occur:

1. The enclosed airspace is unventilated.
2. The enclosed airspace is bounded on one or more sides by an anchored masonry veneer, constructed in accordance with Chapter 7 of the *International Residential Code*, and vented by veneer weep holes located only at the bottom portion of the airspace and spaced not less than 15 inches (381 mm) on center with the top of the cavity airspace closed.

Exception: For ventilated cavities, the effect of the *ventilation* of airspaces located on the exterior side of the continuous *air barrier* and adjacent to and behind the exterior wall covering material shall be determined in accordance with ASTM C1363, modified with an airflow entering the bottom and exiting the top of the airspace at an air movement rate of not less than 70 millimeters per second.

R303.2 Installation. Materials, systems and equipment shall be installed in accordance with the manufacturer's instructions and the *International Building Code* or the *International Residential Code*, as applicable.

R303.2.1 Protection of exposed foundation insulation. Insulation applied to the exterior of *basement walls*, *crawl space walls* and the perimeter of slab-on-grade floors shall have a rigid, opaque and weather-resistant protective covering to prevent the degradation of the insulation's thermal performance. The protective covering shall cover the exposed exterior insulation and extend not less than 6 inches (153 mm) below grade.

R303.2.2 Radiant barrier. Where installed, radiant barriers shall comply with the requirements of ASTM C1313/C1313M and shall be installed in accordance with ASTM C1743.

R303.3 Maintenance information. Maintenance instructions shall be furnished for equipment and systems that require preventive maintenance. Required regular maintenance actions shall be clearly stated and incorporated on a readily visible label. The label shall include the title or publication number for the operation and maintenance manual for that particular model and type of product.

CHAPTER
4 [RE]

RESIDENTIAL ENERGY EFFICIENCY

User notes:

About this chapter: *Chapter 4 presents the paths and options for compliance with the energy efficiency provisions. Chapter 4 contains energy efficiency provisions for the building envelope, mechanical and water heating systems, lighting and additional efficiency requirements. A performance alternative, energy rating alternative, and tropical regional alternative are also provided to allow for energy code compliance other than by the prescriptive method.*

SECTION R401—GENERAL

R401.1 Scope. This chapter applies to residential buildings.

R401.2 Application. Residential buildings shall comply with Section R401.2.1, R401.2.2, R401.2.3 or R401.2.4.

Exception: *Additions*, *alterations*, *repairs* and *changes of occupancy* to *existing buildings* complying with Chapter 5.

R401.2.1 Prescriptive Compliance Option. The Prescriptive Compliance Option requires compliance with Sections R401 through R404 and R408.

R401.2.2 Simulated Building Performance Option. The Simulated Building Performance Option requires compliance with Section R405.

R401.2.3 Energy Rating Index Option. The *Energy Rating Index* (ERI) Option requires compliance with Section R406.

R401.2.4 Tropical Climate Region Option. The Tropical Climate Region Option requires compliance with Section R407.

R401.3 Certificate. A permanent certificate shall be completed by the builder or other *approved* party and posted on a wall in the space where the furnace is located, a utility room or an *approved* location inside the *building*. Where located on an electrical panel, the certificate shall not cover or obstruct the visibility of the circuit directory *label*, service disconnect *label* or other required labels. The certificate shall indicate the following:

1. The predominant *R*-values of insulation installed in or on ceilings, roofs, walls, foundation components such as slabs, *basement walls*, *crawl space walls* and floors and *ducts* outside *conditioned spaces.*
2. *U*-factors of *fenestration* and the *solar heat gain coefficient* (SHGC) of *fenestration*. Where there is more than one value for any component of the *building thermal envelope*, the certificate shall indicate both the value covering the largest area and the area weighted average value if available.
3. The results from any required *duct system* and *building thermal envelope* air leakage testing performed on the *building*.
4. The types, sizes and efficiencies of heating, cooling and service water-heating equipment. Where a gas-fired unvented room heater, electric furnace or baseboard electric heater is installed in the residence, the certificate shall indicate "gas-fired unvented room heater," "electric furnace" or "baseboard electric heater," as appropriate. An efficiency shall not be indicated for gas-fired unvented room heaters, electric furnaces and electric baseboard heaters.
5. Where on-site photovoltaic panel systems have been installed, the array capacity, inverter efficiency, panel tilt and orientation shall be noted on the certificate.
6. For *buildings* where an *Energy Rating Index* score is determined in accordance with Section R406, the *Energy Rating Index* score, both with and without any on-site generation, shall be listed on the certificate.
7. The code edition under which the structure was permitted, the compliance path used and, where applicable, the additional efficiency measures selected for compliance with Section R408.
8. The location and dimensions of a *solar-ready zone* where one is provided.

SECTION R402—BUILDING THERMAL ENVELOPE

R402.1 General. The *building thermal envelope* shall comply with the requirements of one of the following:

1. Sections R402.1.1 through R402.1.4 and Section R402.1.6.
2. Sections R402.1.1, R402.1.5 and R402.1.6.

Exceptions:

1. The following low-energy *buildings*, or portions thereof, separated from the remainder of the building by *building thermal envelope* assemblies complying with this section shall be exempt from the *building thermal envelope* provisions of Section R402.
   - 1.1. Those with a peak design rate of energy usage less than 3.4 Btu/h × ft^2 (10.7 W/m^2) or 1.0 watt/ft^2 of floor area for space-conditioning purposes.
   - 1.2. Those that do not contain *conditioned space*.
2. Log homes designed in accordance with ICC 400.

R402.1.1 Vapor retarder. Wall assemblies in the *building thermal envelope* shall comply with the vapor retarder requirements of Section R702.7 of the *International Residential Code* or Section 1404.3 of the *International Building Code*, as applicable.

R402.1.2 Insulation and fenestration criteria. The *building thermal envelope* shall meet the requirements of Table R402.1.2, based on the *climate zone* specified in Chapter 3. Assemblies shall have a *U-factor* or *F-factor* equal to or less than that specified in Table R402.1.2. *Fenestration* shall have a *U-factor* and glazed fenestration SHGC equal to or less than that specified in Table R402.1.2.

TABLE R402.1.2—MAXIMUM ASSEMBLY *U*-FACTORS[a] AND FENESTRATION REQUIREMENTS

CLIMATE ZONE	0	1	2	3	4 EXCEPT MARINE	5 AND MARINE 4	6	7 AND 8
Vertical fenestration *U*-factor	0.50	0.50	0.40	0.30	0.30	0.28[d]	0.28[d]	0.27[d]
Skylight *U*-factor	0.60	0.60	0.60	0.53	0.53	0.50	0.50	0.50
Glazed vertical fenestration SHGC	0.25	0.25	0.25	0.25	0.40	NR	NR	NR
Skylight SHGC	0.28	0.28	0.28	0.28	0.40	NR	NR	NR
Ceiling *U*-factor	0.035	0.035	0.030	0.030	0.026	0.026	0.026	0.026
Insulation entirely above roof deck	0.039	0.039	0.039	0.039	0.032	0.032	0.032	0.028
Wood-framed wall *U*-factor	0.084	0.084	0.084	0.060	0.045	0.045	0.045	0.045
Mass wall *U*-factor[b]	0.197	0.197	0.165	0.098	0.098	0.082	0.060	0.057
Floor *U*-factor	0.064	0.064	0.064	0.047	0.047	0.033	0.033	0.028
Basement wall *U*-factor	0.360	0.360	0.360	0.091[c]	0.059	0.050	0.050	0.050
Unheated slab *F*-factor[e]	0.73	0.73	0.73	0.54	0.51	0.51	0.48	0.48
Heated slab *F*-factor[e]	0.74	0.74	0.74	0.66	0.66	0.66	0.66	0.66
Crawl space wall *U*-factor	0.477	0.477	0.477	0.136	0.065	0.055	0.055	0.055

For SI: 1 foot = 304.8 mm.

a. Nonfenestration *U*-factors and *F*-factors shall be obtained from measurement, calculation, an *approved source*, or Appendix RF where such appendix is adopted or *approved*.

b. Mass walls shall be in accordance with Section R402.2.6. Where more than half the insulation is on the interior, the mass wall *U*-factors shall not exceed 0.17 in Climate Zones 0 and 1, 0.14 in Climate Zone 2, 0.12 in Climate Zone 3, 0.087 in Climate Zone 4 except Marine, 0.065 in Climate Zone 5 and Marine 4, and 0.057 in Climate Zones 6 through 8.

c. In Warm Humid locations as defined by Figure R301.1 and Table R301.1, the basement wall *U*-factor shall not exceed 0.360.

d. A maximum *U*-factor of 0.30 shall apply in Marine Climate Zone 4 and Climate Zones 5 through 8 to vertical fenestration products installed in buildings located either:
  1. Above 4,000 feet in elevation above sea level, or
  2. In windborne debris regions where protection of openings is required by Section R301.2.1.2 of the *International Residential Code*.

e. *F*-factors for slabs shall correspond to the *R*-values of Table R402.1.3 and the installation conditions of Section R402.2.10.1.

R402.1.3 *R*-value alternative. Assemblies with an *R-value* of insulation materials equal to or greater than that specified in Table R402.1.3 shall be an alternative to the *U-factor* or *F-factor* in Table R402.1.2

TABLE R402.1.3—INSULATION MINIMUM *R*-VALUES AND FENESTRATION REQUIREMENTS BY COMPONENT[a]

CLIMATE ZONE	0	1	2	3	4 EXCEPT MARINE	5 AND MARINE 4	6	7 AND 8
Vertical fenestration *U*-factor	0.50	0.50	0.40	0.30	0.30	0.28[g]	0.28[g]	0.27[g]
Skylight *U*-factor	0.60	0.60	0.60	0.53	0.53	0.50	0.50	0.50
Glazed vertical fenestration SHGC	0.25	0.25	0.25	0.25	0.40	NR	NR	NR
Skylight SHGC	0.28	0.28	0.28	0.28	0.40	NR	NR	NR
Ceiling *R*-value	30	30	38	38	49	49	49	49
Insulation entirely above roof deck	25ci	25ci	25ci	25ci	30ci	30ci	30ci	35ci
Wood-framed wall *R*-value[e]	13 or 0&10ci	13 or 0&10ci	13 or 0&10ci	20 or 13&5ci or 0&15ci	30 or 20&5ci or 13&10ci or 0&20ci	30 or 20&5ci or 13&10ci or 0&20ci	30 or 20&5ci or 13&10ci or 0&20ci	30 or 20&5ci or 13&10ci or 0&20ci
Mass wall *R*-value[f]	3/4	3/4	4/6	8/13	8/13	13/17	15/20	19/21

TABLE R402.1.3—INSULATION MINIMUM *R*-VALUES AND FENESTRATION REQUIREMENTS BY COMPONENT[a]—continued

CLIMATE ZONE	0	1	2	3	4 EXCEPT MARINE	5 AND MARINE 4	6	7 AND 8
Floor *R*-value[h]	13 or 7+5ci or 10ci	13 or 7+5ci or 10ci	13 or 7+5ci or 10ci	19 or 13+5ci or 15ci	19 or 13+5ci or 15ci	30 or 19+7.5ci or 20ci	30 or 19+7.5ci or 20ci	38 or 19+10ci or 25ci
Basement wall *R*-value[b, e]	0	0	0	5ci or 13[d]	10ci or 13	15ci or 19 or 13&5ci	15ci or 19 or 13&5ci	15ci or 19 or 13&5ci
Unheated slab *R*-value & depth[c]	0	0	0	10ci, 2 ft	10ci, 3 ft	10ci, 3 ft	10ci, 4 ft	10ci, 4 ft
Heated slab *R*-value & depth[c]	R-5ci edge and R-5 full slab	R-5ci edge and R-5 full slab	R-5ci edge and R-5 full slab	R-10ci, 2 ft and R-5 full slab	R-10ci, 3 ft and R-5 full slab	R-10ci, 3 ft and R-5 full slab	R-10ci, 4 ft and R-5 full slab	R-10ci, 4 ft and R-5 full slab
Crawl space wall *R*-value[b, e]	0	0	0	5ci or 13[d]	10ci or 13	15ci or 19 or 13&5ci	15ci or 19 or 13&5ci	15ci or 19 or 13&5ci

For SI: 1 foot = 304.8 mm.

NR = Not Required, ci = Continuous Insulation.

a. *R*-values are minimums. *U*-factors and SHGC are maximums. Where insulation is installed in a cavity that is less than the label or design thickness of the insulation, the installed *R*-value of the insulation shall be not less than the *R*-value specified in the table.

b. "5ci or 13" means R-5 continuous insulation (ci) on the interior or exterior surface of the wall or R-13 cavity insulation on the interior side of the wall. "10ci or 13" means R-10 continuous insulation (ci) on the interior or exterior surface of the wall or R-13 cavity insulation on the interior side of the wall. "15ci or 19 or 13&5ci" means R-15 continuous insulation (ci) on the interior or exterior surface of the wall; or R-19 cavity insulation on the interior side of the wall; or R-13 cavity insulation on the interior of the wall in addition to R-5 continuous insulation on the interior or exterior surface of the wall.

c. Slab insulation shall be installed in accordance with Section R402.2.10.1.

d. Basement wal*l* insulation is not required in Warm Humid locations as defined by Figure R301.1 and Table R301.1.

e. The first value is cavity insulation; the second value is continuous insulation. Therefore, as an example, "13&5" means R-13 cavity insulation plus R-5 continuous insulation.

f. Mass walls shall be in accordance with Section R402.2.6. The second *R*-value applies where more than half of the insulation is on the interior of the mass wall.

g. A maximum *U*-factor of 0.30 shall apply in Marine Climate Zone 4 and Climate Zones 5 through 8 to vertical fenestration products installed in buildings located either:

1. Above 4,000 feet in elevation.
2. In windborne debris regions where protection of openings is required by Section R301.2.1.2 of the *International Residential Code*.

h. "30 or 19+7.5ci or 20ci" means R-30 cavity insulation alone or R-19 cavity insulation with R-7.5 continuous insulation or R-20 continuous insulation alone.

R402.1.4 *R*-value computation. *Cavity insulation* alone shall be used to determine compliance with the *cavity insulation R-value* requirements in Table R402.1.3. Where *cavity insulation* is installed in multiple layers, the *R*-values of the *cavity insulation* layers shall be summed to determine compliance with the *cavity insulation R-value* requirements. The manufacturer's settled *R-value* shall be used for blown-in insulation. *Continuous insulation* (ci) alone shall be used to determine compliance with the *continuous insulation R-value* requirements in Table R402.1.3. Where *continuous insulation* is installed in multiple layers, the *R*-values of the *continuous insulation* layers shall be summed to determine compliance with the *continuous insulation R-value* requirements. *Cavity insulation R-values* shall not be used to determine compliance with the *continuous insulation R-value* requirements in Table R402.1.3. Computed *R-values* shall not include an *R-value* for other building materials or air films. Where *insulated siding* is used for the purpose of complying with the *continuous insulation* requirements of Table R402.1.3, the manufacturer's *labeled R-value* for the *insulated siding* shall be reduced by R-0.6.

R402.1.5 Component performance alternative. Where the proposed total *building thermal envelope* thermal conductance (TC_p) is less than or equal to the total *building thermal envelope* thermal conductance (TC_r) using factors in Table R402.1.2, the *building* shall be considered to be in compliance with Table R402.1.2. The total thermal conductance (TC) shall be determined in accordance with Equation 4-1. Proposed *U*-factors and slab-on-grade *F*-factors shall be taken from ANSI/ASHRAE/IES Standard 90.1 Appendix A or determined using a method consistent with the ASHRAE *Handbook of Fundamentals* and shall include the thermal bridging effects of framing materials. In addition to TC compliance, the SHGC requirements of Table R402.1.2 and the maximum *fenestration U-factors* of Section R402.6 shall be met.

Equation 4-1 $TC_p \leq TC_r$

where:

$TC_p = U_pA + F_pP$

$TC_r = U_rA + F_rP$

U_pA = the sum of proposed *U*-factors times the assembly areas in the proposed building.

F_pP = the sum of the proposed *F*-factors times the slab-on-grade perimeter lengths in the proposed building.

U_rA = the sum of *U*-factors in Table R402.1.2 times the same assembly areas as in the proposed building.

F_rP = the sum of *F*-factors in Table R402.1.2 times the same slab-on-grade perimeter lengths as in the proposed building.

Exception: For Climate Zones 0, 1 and 2, the value of F_rP shall equal the value of F_pP.

R402.1.6 Rooms containing fuel-burning appliances. In Climate Zones 3 through 8, where open combustion air *ducts* provide combustion air to open combustion fuel-burning appliances, the appliances and combustion air opening shall be located outside the *building thermal envelope* or enclosed in a room that is isolated from inside the *building thermal envelope*. Such rooms shall be sealed and insulated in accordance with the *building thermal envelope* requirements of Table R402.1.3, where the walls, floors and ceilings shall meet a minimum of the *basement wall R-value* requirement. The door into the room shall be fully gasketed and any water lines and *ducts* in the room insulated in accordance with Section R403. The combustion air *duct* shall be insulated where it passes through *conditioned space* to an *R-value* of not less than R-8.

Exceptions:

1. Direct vent appliances with both intake and exhaust pipes installed continuous to the outside.
2. Fireplaces and stoves complying with Section R402.5.2 and Section R1006 of the *International Residential Code*.

R402.2 Specific insulation requirements. In addition to the requirements of Section R402.1, insulation shall meet the specific requirements of Sections R402.2.1 through R402.2.13.

R402.2.1 Ceilings with attics. Where Section R402.1.3 requires R-38 insulation in the ceiling or attic, installing R-30 over 100 percent of the ceiling or attic area requiring insulation shall satisfy the requirement for R-38 insulation wherever the full height of uncompressed R-30 insulation extends over the wall top plate at the eaves. Where Section R402.1.3 requires R-49 insulation in the ceiling or attic, installing R-38 over 100 percent of the ceiling or attic area requiring insulation shall satisfy the requirement for R-49 insulation wherever the full height of uncompressed R-38 insulation extends over the wall top plate at the eaves. This reduction shall not apply to the insulation and *fenestration* criteria in Section R402.1.2 and the component performance alternative in Section R402.1.5.

R402.2.2 Ceilings without attics. Where Section R402.1.3 requires insulation *R*-values greater than R-30 in the interstitial space above a ceiling and below the structural roof deck, and the design of the roof/ceiling assembly does not allow sufficient space for the required insulation, the minimum required insulation *R-value* for such roof/ceiling assemblies shall be R-30. Insulation shall extend over the top of the wall plate to the outer edge of such plate and shall not be compressed. This reduction of insulation from the requirements of Section R402.1.3 shall be limited to 500 square feet (46 m^2) or 20 percent of the total insulated ceiling area, whichever is less. This reduction shall not apply to the component performance alternative in Section R402.1.5.

R402.2.3 Attic knee wall. Wood attic *knee wall* assemblies that separate *conditioned space* from unconditioned attic spaces shall comply with Table R402.1.3 for wood-framed walls. Steel attic knee wall assemblies shall comply with Section R402.2.7. Such *knee walls* shall have an *air barrier* between conditioned and unconditioned space.

R402.2.3.1 Roof truss framing separating conditioned and unconditioned space. Where wood vertical roof truss framing members are used to separate *conditioned space* and unconditioned space, they shall comply with Table R402.1.3 for wood-framed walls. Steel frame vertical roof truss framing members used to separate *conditioned space* and unconditioned space shall comply with Section R402.2.7.

R402.2.4 Eave baffle. For air-permeable insulation in vented attics, a baffle shall be installed adjacent to soffit and eave vents. Baffles shall maintain a net free area opening equal to or greater than the size of the vent. The baffle shall extend over the top of the attic insulation. The baffle shall be permitted to be any solid material. The baffle shall be installed to the outer edge of the *exterior wall* top plate so as to provide maximum space for attic insulation coverage over the top plate. Where soffit venting is not continuous, baffles shall be installed continuously to prevent *ventilation air* in the eave soffit from bypassing the baffle.

R402.2.5 Access hatches and doors. Access hatches and doors from conditioned to unconditioned spaces such as attics and crawl spaces shall be insulated to the same *R-value* required by Table R402.1.3 for the wall or ceiling in which they are installed.

Exceptions:

1. Vertical doors providing access from *conditioned spaces* to unconditioned spaces that comply with the *fenestration* requirements of Table R402.1.3 based on the applicable *climate zone* specified in Chapter 3.
2. Horizontal pull-down, stair-type access hatches in ceiling assemblies that provide access from conditioned to unconditioned spaces in Climate Zones 0 through 4 shall not be required to comply with the insulation level of the surrounding surfaces provided the hatch meets all of the following:
   - 2.1. The average *U-factor* of the hatch shall be less than or equal to U-0.10 or have an average insulation *R-value* of R-10 or greater.
   - 2.2. Not less than 75 percent of the panel area shall have an insulation *R-value* of R-13 or greater.
   - 2.3. The net area of the framed opening shall be less than or equal to 13.5 square feet (1.25 m^2).
   - 2.4. The perimeter of the hatch edge shall be weather-stripped.

The reduction shall not apply to the component performance alternative in Section R402.1.5.

R402.2.5.1 Access hatches and door insulation installation and retention. Vertical or horizontal access hatches and doors from *conditioned spaces* to unconditioned spaces such as attics and crawl spaces shall be weather-stripped. Access that prevents damaging or compressing the insulation shall be provided to all equipment. Where loose-fill insulation is installed, a wood-framed or equivalent baffle, retainer or dam shall be installed to prevent loose-fill insulation from spilling into *living space* from higher to lower sections of the attic and from attics covering *conditioned spaces* to unconditioned spaces. The baffle or retainer shall provide a permanent means of maintaining the installed *R-value* of the loose-fill insulation.

R402.2.6 Mass walls. Mass walls where used as a component of the *building thermal envelope* shall be one of the following:

1. Above-ground walls of concrete block, concrete, insulated concrete form, masonry cavity, brick but not brick veneer, adobe, compressed earth block, rammed earth, solid timber, mass timber or solid logs.
2. Any wall having a heat capacity greater than or equal to 6 Btu/ft^2 × °F (123 kJ/m^2 × K).

R402.2.7 Steel-frame ceilings, walls and floors. Steel-frame ceilings, walls, and floors shall comply with the *U-factor* requirements of Table R402.1.2. The calculation of the *U-factor* for steel-framed ceilings and walls in a *building thermal envelope* assembly shall be determined in accordance with AISI S250, modified as follows:

1. Where the steel-framed wall contains no *cavity insulation*, and uses *continuous insulation* to satisfy the *U-factor* maximum, the steel-framed wall member spacing is permitted to be installed at any on-center spacing.
2. Where the steel-framed wall contains framing spaced at 24 inches (610 mm) on center with a 23 percent framing factor or framing spaced at 16 inches (400 mm) on center with a 25 percent framing factor, the next lower framing member spacing input values shall be used when calculating using AISI S250.
3. Where the steel-framed wall contains less than 23 percent framing factors AISI S250 shall be used without any modifications.
4. Where the steel-framed wall contains other than standard C-shaped framing members the AISI S250 calculation option for other than standard C-shaped framing is permitted to be used.

R402.2.8 Floors. Floor *insulation* shall be installed in accordance with all of the following:

1. Table R402.1.2 or R402.1.3 and manufacturer's instructions.
2. Floor framing members that are part of the *building thermal envelope* shall be air sealed to maintain a *continuous air barrier*.
3. One of the following methods:
    - 3.1. *Cavity insulation* shall be installed to maintain permanent contact with the underside of the subfloor decking.
    - 3.2. *Cavity insulation* shall be installed to maintain contact with the top side of sheathing separating the cavity and the unconditioned space below. Insulation shall extend from the bottom to the top of all perimeter floor framing members.
    - 3.3. A combination of *cavity insulation* and *continuous insulation* shall be installed such that the *cavity insulation* maintains contact with the top side of the *continuous insulation* and the *continuous insulation* maintains contact with the underside of the structural floor system. Insulation shall extend from the bottom to the top of all perimeter floor framing members.
    - 3.4. *Continuous insulation* shall be installed to maintain contact with the underside of the structural floor system. Insulation shall extend from the bottom to the top of all perimeter floor framing members.

R402.2.9 Basement walls. *Basement walls* shall be insulated in accordance with Table R402.1.3.

Exception: *Basement walls* associated with unconditioned basements where the following requirements are met:

1. The floor overhead, including the underside stairway stringer leading to the basement, is insulated in accordance with Section R402.1.3 and applicable provisions of Sections R402.2 and R402.2.8.
2. There are no uninsulated ductwork, domestic hot water piping, or hydronic heating surfaces exposed to the basement.
3. There are no HVAC supply or return diffusers serving the basement.
4. The walls surrounding the stairway and adjacent to *conditioned space* are insulated in accordance with Section R402.1.3 and applicable provisions of Section R402.2.
5. The door(s) leading to the basement from *conditioned spaces* are insulated in accordance with Section R402.1.3 and applicable provisions of Section R402.2, and weather-stripped in accordance with Section R402.5.
6. The *building thermal envelope* separating the basement from adjacent *conditioned spaces* complies with Section R402.5.

R402.2.9.1 Basement wall insulation installation. Where *basement walls* are insulated, the insulation shall be installed from the top of the *basement wall* down to 10 feet (3048 mm) below grade or to the basement floor, whichever is less, or in accordance with the *proposed design* or the *rated design*, as applicable.

R402.2.10 Slab-on-grade floors. Slab-on-grade floors with a floor surface within 24 inches (610 mm) above or below grade shall be insulated in accordance with Section R402.2.10.1 or R402.2.10.2.

Exception: Slab-edge insulation is not required in jurisdictions designated by the *code official* as having a very heavy termite infestation probability.

R402.2.10.1 Slab-on-grade floor insulation installation. For buildings complying with Section R401.2.1, the slab edge *continuous insulation* shall extend downward from the top of the slab on the outside or inside of the foundation wall. Insulation located below grade shall extend the vertical distance provided in Table R402.1.3, but need not exceed the footing depth in accordance with Section R403.1.4 of the *International Residential Code*. Where a proposed design includes insulation extending away from the *building*, it shall be protected by pavement or by not less than 10 inches (254 mm) of soil. The top edge of the insulation installed between the *exterior wall* and the edge of the interior slab shall be permitted to be cut at a 45-degree (0.79 rad) angle away from the *exterior wall*. Full-slab insulation shall be continuous under the entire area of the slab-on-grade floor, except at structural column locations and service penetrations. Slab edge insulation required at the *heated slab* perimeter shall not be required to extend below the bottom of the *heated slab* and shall be continuous with the full slab insulation.

R402.2.10.2 Alternative slab-on-grade insulation configurations. For *buildings* complying with Section R405 or R406, slab-on-grade insulation shall be installed in accordance with the *proposed design* or *rated design*.

R402.2.11 Crawl space walls. *Crawl space walls* shall be insulated in accordance with Section R402.2.11.1 or R402.2.11.2.

Exception: *Crawl space walls* associated with a crawl space that is vented to the outdoors and the floor overhead is insulated in accordance with Table R402.1.3 and Section R402.2.8.

R402.2.11.1 Crawl space wall insulation installations. Where installed, *crawl space wall* insulation shall be secured to the wall and extend downward from the sill plate to not less than the top of the foundation wall footing.

Exception: Where the *crawl space wall* insulation is installed on the interior side of the wall and the crawl space floor is more than 24 inches (610 mm) below the exterior grade, the crawl space wall insulation shall be permitted to extend downward from the sill plate at the top of the foundation wall to not less than the interior floor of the crawl space.

Exposed earth in crawl space foundations shall be covered with a continuous Class I vapor retarder in accordance with the *International Building Code* or *International Residential Code*, as applicable. Joints of the vapor retarder shall overlap by 6 inches (153 mm) and be sealed or taped. The edges of the vapor retarder shall extend not less than 6 inches (153 mm) up stem walls and shall be attached to the stem walls.

R402.2.11.2 Alternative crawl space wall insulation configurations. For *buildings* complying with Section R405 or R406, *crawl space wall* insulation shall be installed in accordance with the *proposed design* or *rated design*.

R402.2.12 Masonry veneer. Insulation shall not be required on the horizontal portion of a foundation that supports a masonry veneer.

R402.2.13 Sunroom and heated garage insulation. *Sunrooms* enclosing *conditioned space* and heated garages shall meet the insulation requirements of this code.

Exception: For *sunrooms* and heated garages provided with *thermal isolation*, and enclosing *conditioned space*, the following exceptions to the insulation requirements of this code shall apply:

1. The minimum ceiling insulation *R*-values shall be R-19 in *Climate Zones* 0 through 4 and R-24 in *Climate Zones* 5 through 8.
2. The minimum wall insulation *R-value* shall be R-13 in all *climate zones*. Walls separating a *sunroom* or heated garage with *thermal isolation* from *conditioned space* shall comply with the *building thermal envelope* requirements of this code.

R402.3 Radiant barriers. Where installed, radiant barriers shall be installed in accordance with ASTM C1743.

R402.4 Fenestration. In addition to the requirements of Section R402, *fenestration* shall comply with Sections R402.4.1 through R402.4.5.

R402.4.1 *U*-factor. An area-weighted average of *fenestration* products shall be permitted to satisfy the *U*-factor requirements.

R402.4.2 Glazed fenestration SHGC. An area-weighted average of *fenestration* products more than 50 percent glazed shall be permitted to satisfy the SHGC requirements.

Dynamic glazing shall be permitted to satisfy the SHGC requirements of Table R402.1.2 provided that the ratio of the higher to lower *labeled SHGC* is greater than or equal to 2.4, and the dynamic glazing is automatically controlled to modulate the amount of solar gain into the space in multiple steps. Dynamic glazing shall be considered separately from other *fenestration*, and area-weighted averaging with other *fenestration* that is not dynamic glazing shall be prohibited.

Exception: Dynamic glazing shall not be required to comply with this section where both the lower and higher *labeled* SHGC comply with the requirements of Table R402.1.2.

R402.4.3 Glazed fenestration exemption. Not greater than 15 square feet (1.4 m^2) of glazed *fenestration* per *dwelling unit* shall be exempt from the *U-factor* and SHGC requirements in Section R402.1.2. This exemption shall not apply to the component performance alternative in Section R402.1.5.

R402.4.4 Opaque door exemption. One side-hinged *opaque door* assembly not greater than 24 square feet (2.22 m^2) in area shall be exempt from the *U-factor* requirement in Section R402.1.2. This exemption shall not apply to the component performance alternative in Section R402.1.5.

R402.4.5 Sunroom and heated garage fenestration. *Sunrooms* and heated garages enclosing *conditioned space* shall comply with the *fenestration* requirements of this code.

Exception: In Climate Zones 2 through 8, for *sunrooms* and heated garages with *thermal isolation* and enclosing *conditioned space*, the *fenestration U-factor* shall not exceed 0.45 and the skylight *U-factor* shall not exceed 0.70.

New *fenestration* separating a *sunroom* or heated garage with *thermal isolation* from *conditioned space* shall comply with the *building thermal envelope* requirements of this code.

R402.5 Air leakage. The *building thermal envelope* shall be constructed to limit air leakage in accordance with the requirements of Sections R402.5.1 through R402.5.4.

R402.5.1 Building thermal envelope. The *building thermal envelope* shall comply with Sections R402.5.1.1 through R402.5.1.3. The sealing methods between dissimilar materials shall allow for differential expansion and contraction.

R402.5.1.1 Installation. The components of the *building thermal envelope* as indicated in Table R402.5.1.1 shall be installed in accordance with the manufacturer's instructions and the criteria indicated in Table R402.5.1.1, as applicable to the method of construction. Where required by the *code official*, an *approved* third party shall inspect all components and verify compliance.

TABLE R402.5.1.1—AIR BARRIER, AIR SEALING AND INSULATION INSTALLATION[a]		
COMPONENT	**AIR BARRIER, AIR SEALING CRITERIA**	**INSULATION INSTALLATION CRITERIA**
General requirements	A continuous air barrier shall be installed in the building thermal envelope. Breaks or joints in the air barrier shall be sealed.	Air-permeable insulation shall not be used as a sealing material.
Ceiling/attic	An air barrier shall be installed in any dropped ceiling or soffit to separate it from unconditioned space. Access openings, drop down stairs or knee wall doors to unconditioned attic spaces shall be sealed with gasketing materials that allow for repeated entrance over time.	The insulation in any dropped ceiling/soffit shall be aligned with the air barrier. Access hatches and doors shall be installed and insulated in accordance with Section R402.2.5. Eave baffles shall be installed in accordance with Section R402.2.4.
Walls	The junction of the foundation and sill plate shall be sealed. The junction of the top plate and the top of exterior walls shall be sealed. Knee walls shall be sealed.	Cavities within corners and headers of frame walls shall be insulated by completely filling the cavity with a material having a thermal resistance, *R*-value, of not less than R-3 per inch. Exterior building thermal envelope insulation for framed walls shall be installed in substantial contact and continuous alignment with the air barrier.
Knee wall	Knee walls shall have an air barrier between conditioned and unconditioned space	Insulation installed in a knee wall assembly shall be installed in accordance with Section R402.2.3. Air-permeable insulation shall be enclosed inside an air barrier assembly.
Windows, skylights and doors	The rough opening gap between framing and the frames of skylights, windows and doors, shall be sealed in accordance with fenestration manufacturer's instructions.	Insulation shall not be required in the rough opening gap except as required by the fenestration manufacturer's instructions.
Rim joists	Rim joists shall include an air barrier. The junctions of the rim board to the sill plate and the rim board and the subfloor shall be air sealed.	Rim joists shall be insulated so that the insulation maintains permanent contact with the exterior rim board.[b]
Floors, including cantilevered floors and floors above garages	Floor framing members that are part of the building thermal envelope shall be air sealed to maintain a continuous air barrier. Air permeable floor cavity insulation shall be enclosed.	Floor insulation shall be installed in accordance with the requirements of Section R402.2.8.
Basement, crawl space and slab foundations	Exposed earth in unvented crawl spaces shall be covered with a Class I vapor retarder/air barrier in accordance with Section R402.2.11. Penetrations through concrete foundation walls and slabs shall be air sealed. Class 1 vapor retarders shall not be used as an air barrier on below-grade walls and shall be installed in accordance with Section R702.7 of the *International Residential Code*.	Crawl space insulation, where provided instead of floor insulation, shall be installed in accordance with Section R402.2.11. Conditioned basement foundation wall insulation shall be installed in accordance with Section R402.2.9.1. Slab-on-grade floor insulation shall be installed in accordance with Section R402.2.10.
Shafts, penetrations	Duct and flue shafts to exterior or unconditioned space shall be sealed. Utility penetrations of the air barrier shall be caulked, gasketed or otherwise sealed and shall allow for expansion, contraction of materials and mechanical vibration.	Insulation shall be fitted tightly around utilities passing through shafts and penetrations in the building thermal envelope to maintain required *R*-value.
Narrow cavities	Narrow cavities of 1 inch or less that are not able to be insulated shall be air sealed.	Batts to be installed in narrow cavities shall be cut to fit or narrow cavities shall be filled with insulation that on installation readily conforms to the available cavity space.
Garage separation	Air sealing shall be provided between the garage and conditioned spaces.	Insulated portions of the garage separation assembly shall be installed in accordance with Sections R303 and R402.2.8.
Recessed lighting	Recessed light fixtures installed in the building thermal envelope shall be air sealed in accordance with Section R402.5.4.	Recessed light fixtures installed in the building thermal envelope shall be airtight and IC rated, and shall be buried in or surrounded with insulation.

TABLE R402.5.1.1—AIR BARRIER, AIR SEALING AND INSULATION INSTALLATION[a]—continued		
COMPONENT	**AIR BARRIER, AIR SEALING CRITERIA**	**INSULATION INSTALLATION CRITERIA**
Plumbing, wiring or other obstructions	All holes created by wiring, plumbing or other obstructions in the air barrier assembly shall be air sealed.	Insulation shall be installed to fill the available space and surround wiring, plumbing, or other obstructions, unless the required *R*-value can be met by installing insulation and air barrier systems completely to the exterior side of the obstructions.
Showers, tubs and fireplaces adjacent to the building thermal envelope	An air barrier shall separate insulation in the building thermal envelope from the shower, tub or fireplace assemblies.	Exterior framed walls adjacent to showers, tubs and fireplaces shall be insulated.
Electrical, communication and other equipment boxes, housings and enclosures	Boxes, housing and enclosures that penetrate the air barrier shall be caulked, taped, gasketed or otherwise sealed to the air barrier element being penetrated. All concealed openings into the box, housing or enclosure shall be sealed. Alternatively, air-sealed boxes shall be installed in accordance with Section R402.5.5.	Boxes, housing and enclosures shall be buried in or surrounded by insulation.
HVAC register boots	HVAC supply and return register boots shall be sealed to the subfloor, wall covering or ceiling penetrated by the boot.	HVAC supply and return register boots located within a building thermal envelope assembly shall be buried in or surrounded by insulation.
Concealed sprinklers	Where required to be sealed, concealed fire sprinklers shall only be sealed in a manner that is recommended by the manufacturer. Caulking or other adhesive sealants shall not be used to fill voids between fire sprinkler cover plates and walls or ceilings.	—
Common walls or double walls separating attached single-family dwellings or townhouses	An interior air barrier shall be provided. Air sealing at the intersections with building thermal envelope shall be provided. Where installed in a fire-resistance-rated wall assembly, air sealing materials shall comply with one of the following: 1. Be in accordance with an approved design for the fire-resistance-rated assembly. 2. Be supported by approved data that shows the assembly as installed complies with the required fire-resistance rating.	Insulation materials recognized in the approved common wall or double-wall design and installed in accordance with the approved design shall be permitted to be used.

a. Inspection of log walls shall be in accordance with the provisions of ICC 400.

b. Insulation full enclosure is not required in unconditioned/ventilated attic spaces and at rim joists.

R402.5.1.2 Air leakage testing. The *building* or each *dwelling unit* or *sleeping unit* in the *building* shall be tested for air leakage. Testing shall be conducted in accordance with ANSI/RESNET/ICC 380, ASTM E779, ASTM E1827 or ASTM E3158 and reported at a pressure differential of 0.2 inch water gauge (50 Pa). Where required by the *code official*, testing shall be conducted by an *approved* third party. A written report of the results of the test shall be signed by the party conducting the test and provided to the *code official*. Testing shall be performed at any time after creation of all penetrations of the *building thermal envelope* have been sealed.

During testing:

1. Exterior windows and doors, fireplace and stove doors shall be closed, but not sealed, beyond the intended weatherstripping or other *infiltration* control measures.
2. *Dampers* including exhaust, intake, makeup air, backdraft and flue *dampers* shall be closed, but not sealed beyond intended *infiltration* control measures.
3. Interior doors, where installed at the time of the test, shall be open.
4. Exterior or interior terminations for continuous *ventilation* systems shall be sealed.
5. Heating and cooling systems, where installed at the time of the test, shall be turned off.
6. Supply and return registers, where installed at the time of the test, shall be fully open.

   Exceptions:

   1. For heated, attached private garages and heated, detached private garages accessory to one- and two-family dwellings and townhouses not more than three stories above *grade plane* in height, *building thermal envelope* tightness and insulation installation shall be considered acceptable where the items in Table R402.5.1.1, applicable to the method of construction, are field verified. Where required by the *code official*, an *approved* third party independent from the installer shall inspect both *air barrier* and insulation installa-

tion criteria. Heated, attached private garage space and heated, detached private garage space shall be thermally isolated from all other habitable, *conditioned spaces* in accordance with Sections R402.2.13 and R402.4.5, as applicable.

2. Where tested in accordance with Section R402.5.1.2.1, testing of each *dwelling unit* or *sleeping unit* is not required.

R402.5.1.2.1 Unit sampling. For *buildings* with eight or more *dwelling units* or *sleeping units*, seven or 20 percent of the *dwelling units* or *sleeping units,* whichever is greater, shall be tested. Tested units shall include a top-floor unit, a ground-floor unit, a middle-floor unit and the *dwelling unit* or *sleeping unit* with the largest *testing unit enclosure area*. Where the air leakage rate of a tested unit is greater than the maximum permitted rate, corrective actions shall be taken and the unit retested until it passes. For each tested *dwelling unit* or *sleeping unit* with an air leakage rate greater than the maximum permitted rate, three additional units, including the corrected unit, shall be tested. Where *buildings* have fewer than eight *dwelling units* or *sleeping units,* each unit shall be tested.

R402.5.1.3 Maximum air leakage rate. Where tested in accordance with Section R402.5.1.2, the air leakage rate for *buildings*, *dwelling units* or *sleeping units* shall be as follows:

1. Where complying with Section R401.2.1, the *building* or the *dwelling units* or *sleeping units* in the *building* shall have an air leakage rate not greater than 4.0 air changes per hour in Climate Zones 0, 1 and 2; 3.0 air changes per hour in Climate Zones 3 through 5; and 2.5 air changes per hour in Climate Zones 6 through 8.
2. Where complying with Section R401.2.2 or R401.2.3, the *building* or the *dwelling units* or *sleeping units* in the *building* shall have an air leakage rate not greater than 4.0 air changes per hour, or 0.22 cubic feet per minute per square foot [1.1 L/(s × m^2)] of the *building thermal envelope* area or the dwelling *testing unit enclosure area*, as applicable.

Exceptions:

1. Where *dwelling units* or *sleeping units* are attached or located in an R-2 occupancy, and are tested without simultaneously testing adjacent *dwelling units* or *sleeping units*, the air leakage rate is permitted to be not greater than 0.27 cubic feet per minute per square foot [1.4 L/(s × m^2)] of the *testing unit enclosure area*. Where adjacent *dwelling units* are simultaneously tested in accordance with ASTM E779, the air leakage rate is permitted to be not greater than 0.27 cubic feet per minute per square foot [1.4 L/(s × m^2)] of the *testing unit enclosure area* that separates *conditioned space* from the exterior.
2. Where *buildings* have 1,500 square feet (139.4 m^2) or less of *conditioned floor area*, the air leakage rate is permitted to be not greater than 0.27 cubic feet per minute per square foot [1.4 L/(s × m^2)].

R402.5.2 Fireplaces. New wood-burning fireplaces shall have tight-fitting flue *dampers* or doors, and outdoor combustion air. Where using tight-fitting doors on factory-built fireplaces *listed* and *labeled* in accordance with UL 127, the doors shall be tested and *listed* for the fireplace.

R402.5.3 Fenestration air leakage. Windows, *skylights* and sliding glass doors shall have an air *infiltration* rate of not greater than 0.3 cubic feet per minute per square foot (1.5 L/s/m^2), and for swinging doors, not greater than 0.5 cubic feet per minute per square foot (2.6 L/s/m^2), when tested in accordance with NFRC 400 or AAMA/WDMA/CSA 101/I.S.2/A440 by an accredited, independent laboratory and *listed* and *labeled* by the manufacturer.

Exception: Site-built windows, *skylights* and doors.

R402.5.4 Recessed lighting. Recessed luminaires installed in the *building thermal envelope* shall be sealed to limit air leakage between conditioned and *unconditioned spaces*. Recessed luminaires shall be IC-rated and *labeled* as having an air leakage rate of not greater than 2.0 cfm (0.944 L/s) when tested in accordance with ASTM E283 at a pressure differential of 1.57 psf (75 Pa). Recessed luminaires shall be sealed with a gasket or caulked between the housing and the interior wall or ceiling covering.

R402.5.5 Air-sealed electrical and communication outlet boxes. Air-sealed electrical and communication outlet boxes that penetrated the *air barrier* of the *building thermal envelope* shall be caulked, taped, gasketed or otherwise sealed to the *air barrier* element being penetrated. Air-sealed boxes shall be buried in or surrounded by insulation. Air-sealed boxes shall be tested and marked in accordance with NEMA OS 4. Air-sealed boxes shall be installed in accordance with the manufacturer's instructions.

R402.6 Maximum fenestration *U*-factor and SHGC. The area-weighted average maximum *fenestration U-factor* permitted using tradeoffs from Section R402.1.5 or R405 shall be 0.48 in Climate Zones 4 and 5 and 0.40 in Climate Zones 6 through 8 for vertical *fenestration*, and 0.75 in Climate Zones 4 through 8 for skylights. The area-weighted average maximum *fenestration SHGC* permitted using tradeoffs from Section R405 in Climate Zones 0 through 3 shall be 0.40.

Exception: The maximum *U-factor* and *SHGC* for *fenestration* shall not be required in storm shelters complying with ICC 500.

SECTION R403—SYSTEMS

R403.1 Controls. Not less than one *thermostat* shall be provided for each separate heating and cooling system.

R403.1.1 Programmable thermostat. The *thermostat* controlling the primary heating or cooling system of the *dwelling unit* shall be capable of controlling the heating and cooling system on a daily schedule to maintain different temperature set points at different times of day and different days of the week. This *thermostat* shall include the capability to set back or temporarily operate the system to maintain *zone* temperatures of not less than 55°F (13°C) to not greater than 85°F (29°C). The *thermostat* shall be programmed initially by the manufacturer with a heating temperature setpoint of not greater than 70°F (21°C) and a cooling temperature setpoint of not less than 78°F (26°C).

R403.1.2 Heat pump supplementary heat. Heat pumps having supplementary electric-resistance, *fuel gas* or *liquid fuel* heating systems shall have controls that are configured to prevent supplemental heat operation when the capacity of the heat pump compressor can meet the heating load. Supplemental heat operation shall be limited to only where one of the following applies:

1. The vapor compression cycle cannot provide the necessary heating energy to satisfy the *thermostat* setting.
2. The heat pump is operating in defrost mode.
3. The vapor compression cycle malfunctions.
4. The *thermostat* malfunctions.

R403.2 Hot water boiler temperature reset. Other than where equipped with tankless domestic water heating coils, the manufacturer shall equip each gas, *liquid fuel* and electric boiler with *automatic* means of adjusting the water temperature supplied by the boiler so that incremental change of the inferred heat load will cause an incremental change in the temperature of the water supplied by the boiler. This can be accomplished with outdoor reset, indoor reset or water temperature sensing.

R403.3 Duct systems. *Duct systems* shall be installed in accordance with Sections R403.3.1 through R403.3.9.

Exception: *Ventilation ductwork* that is not integrated with *duct systems* serving heating or cooling systems.

R403.3.1 Duct system design. *Duct systems* serving one or two *dwelling units* or *sleeping units* shall be designed and sized in accordance with ANSI/ACCA Manual D. *Duct systems* serving more than two *dwelling units* or *sleeping units* shall be sized in accordance with the ASHRAE *Handbook of Fundamentals*, ANSI/ACCA Manual D or other equivalent computation procedure.

R403.3.2 Building cavities. *Building* framing cavities shall not be used as ductwork or plenums.

R403.3.3 Ductwork located outside conditioned space. Supply and return *ductwork* located outside *conditioned space* shall be insulated to an *R-value* of not less than R-8 for *ducts* 3 inches (76 mm) in diameter and larger and not less than R-6 for *ducts* smaller than 3 inches (76 mm) in diameter. *Ductwork* buried beneath a *building* shall be insulated as required per this section or have an equivalent *thermal distribution efficiency*. Underground *ductwork* utilizing the *thermal distribution efficiency* method shall be *listed* and *labeled* to indicate the *R-value* equivalency.

R403.3.4 Duct systems located in conditioned space. For *duct systems* to be considered inside a *conditioned space,* the *space conditioning equipment* shall be located completely on the conditioned side of the *building thermal envelope*. The *ductwork* shall comply with the following, as applicable:

1. The *ductwork* shall be located completely on the conditioned side of the *building thermal envelope*.
2. *Ductwork* in ventilated attic spaces or unvented attics with vapor diffusion ports shall be buried within ceiling insulation in accordance with Section R403.3.5 and shall comply with the following:
   - 2.1. The ductwork leakage, as measured either by a rough-in test of the supply and return *ductwork* or a post-construction *duct system* leakage test to outside the *building thermal envelope* in accordance with Section R403.3.7, is not greater than 1.5 cubic feet per minute (42.5 L/min) per 100 square feet (9.29 m^2) of *conditioned floor area* served by the *duct system*.
   - 2.2. The ceiling insulation *R-value* installed against and above the insulated *ductwork* is greater than or equal to the proposed ceiling insulation *R-value*, less the *R-value* of the insulation on the *ductwork*.
3. *Ductwork* contained within wall or floor assemblies separating unconditioned from *conditioned space* shall comply with the following:
   - 3.1. A *continuous air barrier* shall be installed as part of the building assembly between the *ductwork* and the unconditioned space.
   - 3.2. *Ductwork* shall be installed in accordance with Section R403.3.3.

     Exception: Where the building assembly cavities containing *ductwork* have been air sealed in accordance with Section R402.5.1 and insulated in accordance with Item 3.3, *duct* insulation is not required.
   - 3.3. Not less than R-10 insulation, or not less than 50 percent of the required insulation *R-value* specified in Table R402.1.3, whichever is greater, shall be located between the *ductwork* and the unconditioned space.
   - 3.4. Segments of *ductwork* contained within these building assemblies shall not be considered completely inside *conditioned space* for compliance with Section R405 or R406.

R403.3.5 Ductwork buried within ceiling insulation. Where supply and return *ductwork* is partially or completely buried in ceiling insulation, such *ductwork* shall comply with the following:

1. The supply and return *ductwork* shall be insulated with not less than R-8 insulation.
2. At all points along the *ductwork*, the sum of the ceiling insulation *R-value* against and above the top of the *ductwork*, and against and below the bottom of the *ductwork*, shall be not less than R-19, excluding the *R-value* of the duct insulation.
3. In Climate Zones 0A, 1A, 2A and 3A, the supply *ductwork* shall be completely buried within ceiling insulation, insulated to an *R-value* of not less than R-13 and in compliance with the vapor retarder requirements of Section 604.11 of the *International Mechanical Code* or Section M1601.4.6 of the *International Residential Code*, as applicable.

   Exception: Sections of the supply *ductwork* that are less than 3 feet (914 mm) from the supply outlet.
4. In Climate Zones 0A, 1A, 2A and 3A where installed in an unvented attic with vapor diffusion ports, the supply *ductwork* shall be completely buried within the insulation in the ceiling assembly at the floor of the attic, insulated to an *R*-value of not less than R-8 and in compliance with the vapor retarder requirements of Section 604.11 of the *International Mechanical Code* or Section M1601.4.6 of the *International Residential Code*, as applicable.

   Exception: Sections of the supply *ductwork* that are less than 3 feet (914 mm) from the supply outlet.

   4.1. Air permeable insulation installed in unvented attics shall comply with Section R806.5 of the *International Residential Code*.

R403.3.5.1 Effective *R*-value of deeply buried ducts. Where complying using Section R405, sections of *ductwork* that are installed in accordance with Section R403.3.5 surrounded with blown-in attic insulation having an *R-value* of R-30 or greater and located such that the top of the *ductwork* is not less than 3.5 inches (89 mm) below the top of the insulation shall be considered as having an effective duct insulation *R-value* of R-25.

R403.3.6 Sealing. *Ductwork*, *air-handling units* and filter boxes shall be sealed. Joints and seams shall comply with the *International Mechanical Code* or the *International Residential Code*, as applicable.

R403.3.6.1 Sealed air-handling unit. *Air-handling units* shall have a manufacturer's designation for an air leakage of not greater than 2 percent of the design airflow rate when tested in accordance with ASHRAE 193.

R403.3.7 Duct system testing. Each *duct system* shall be tested for air leakage in accordance with ANSI/RESNET/ICC 380 or ASTM E1554. Total leakage shall be measured with a pressure differential of 0.1 inch water gauge (25 Pa) across the *duct system* and shall include the measured leakage from the supply and return *ductwork*. A written report of the test results shall be signed by the party conducting the test and provided to the *code official*. *Duct system* leakage testing at either rough-in or post construction shall be permitted with or without the installation of registers or grilles. Where installed, registers and grilles shall be sealed during the test. Where registers and grilles are not installed, the face of the register boots shall be sealed during the test.

Exceptions:

1. Testing shall not be required for *duct systems* serving *ventilation* systems that are not integrated with *duct systems* serving heating or cooling systems.
2. Testing shall not be required where there is not more than 10 feet (3048 mm) of total *ductwork* external to the *space conditioning equipment* and both the following are met:

   2.1. The *duct system* is located entirely within *conditioned space*.

   2.2. The *ductwork* does not include *plenums* constructed of building cavities or gypsum board.
3. Where the *space conditioning equipment* is not installed, testing shall be permitted. The total measured leakage of the supply and return *ductwork* shall be less than or equal to 3.0 cubic feet per minute (85 L/min) per 100 square feet (9.29 m^2) of *conditioned floor area*.
4. Where tested in accordance with Section R403.3.9, testing of each *duct system* is not required.

R403.3.8 Duct system leakage. The total measured *duct system* leakage shall not be greater than the values in Table R403.3.8, based on the *conditioned floor area*, number of ducted returns, and location of the *duct system*. For *buildings* complying with Section R405 or R406, where *duct system* leakage to outside is tested in accordance with ANSI/RESNET/ICC 380 or ASTM E1554, the leakage to outside value shall not be used for compliance with this section, but shall be permitted to be used in the calculation procedures of Section R405 and R406.

TABLE R403.3.8—MAXIMUM TOTAL DUCT SYSTEM LEAKAGE

EQUIPMENT AND DUCT CONFIGURATION	DUCT SYSTEMS SERVING MORE THAN 1,000 FT^2 OF CONDITIONED FLOOR AREA		DUCT SYSTEMS SERVING 1,000 FT^2 OR LESS OF CONDITIONED FLOOR AREA
	cfm/100 ft^2		cfm
	Number of ducted returns[a]		
	< 3	≥ 3	Any
Space conditioning equipment is not installed[b, c]	3	4	30
All components of the duct system are installed[c]	4	6	40

TABLE R403.3.8—MAXIMUM TOTAL DUCT SYSTEM LEAKAGE—continued			
EQUIPMENT AND DUCT CONFIGURATION	DUCT SYSTEMS SERVING MORE THAN 1,000 FT² OF CONDITIONED FLOOR AREA		DUCT SYSTEMS SERVING 1,000 FT² OR LESS OF CONDITIONED FLOOR AREA
	cfm/100 ft²		cfm
	Number of ducted returns[a]		
	< 3	≥ 3	Any
Space conditioning equipment is not installed, but the ductwork is located entirely in conditioned space[c, d]	6	8	60
All components of the duct system are installed and entirely located in conditioned space[c]	8	12	80

For SI: 1 cubic foot per minute per square foot = 0.0033 LPM/m², 1 cubic foot per minute = 28.3 LPM.

a. A ducted return is a duct made of sheet metal or flexible duct that connects one or more return grilles to the return-side inlet of the air-handling unit. Any other method to convey air from return or transfer grilles to the air-handling unit does not constitute a ducted return for the purpose of determining maximum total duct system leakage allowance.

b. Duct system testing is permitted where space conditioning equipment is not installed, provided that the return ductwork is installed and the measured leakage from the supply and return ductwork is included.

c. For duct systems to be considered inside a conditioned space, where the ductwork is located in ventilated attic spaces or unvented attics with vapor diffusion ports, duct system leakage to outside must comply with Item 2.1 of Section R403.3.4.

d. Prior to the issuance of a certificate of occupancy, where the air-handling unit is not verified as being located in conditioned space, the total duct system leakage must be retested.

R403.3.9 Unit sampling. For *buildings* with eight or more *dwelling units* or *sleeping units,* the *duct systems* in the greater of seven or 20 percent of the *dwelling units* or *sleeping units* shall be tested, including a top floor unit, a ground floor unit, a middle floor unit and the unit with the largest *conditioned floor area*. Where buildings have fewer than eight *dwelling units* or *sleeping units*, the *duct systems* in each unit shall be tested. Where the leakage of a *duct system* is greater than the maximum permitted *duct system* leakage, corrective actions shall be made to the *duct system* and the *duct system* shall be system retested until it passes. For each tested *dwelling unit* or *sleeping unit* that has a greater total duct system leakage than the maximum permitted *duct system* leakage, an additional three *dwelling units* or *sleeping units*, including the corrected unit, shall be tested.

R403.4 Mechanical system piping insulation. Mechanical system piping capable of carrying fluids greater than 105°F (41°C) or less than 55°F (13°C) shall be insulated to an *R-value* of not less than R-3.

R403.4.1 Protection of piping insulation. Piping insulation exposed to weather shall be protected from damage, including that caused by sunlight, moisture, physical contact and wind. The protection shall provide shielding from solar radiation that can cause degradation of the material and shall be removable not less than 6 feet (1828 mm) from the equipment for maintenance. Adhesive tape shall be prohibited.

R403.5 Service hot water systems. Energy conservation measures for service hot water systems shall be in accordance with Sections R403.5.1 through R403.5.3.

R403.5.1 Heated water circulation and temperature maintenance systems. Heated water circulation systems shall be in accordance with Section R403.5.1.1. Heat trace temperature maintenance systems shall be in accordance with Section R403.5.1.2. *Automatic* controls, temperature sensors and pumps shall be in a location with access. *Manual* controls shall be in a location with *ready access*.

R403.5.1.1 Circulation systems. Heated water circulation systems shall be provided with a circulation pump. Gravity and thermosyphon circulation systems shall be prohibited. Controls for *circulating hot water system* pumps shall automatically turn off the pump when the water in the circulation loop is at the desired temperature and when there is no demand for hot water. The system return pipe shall be a dedicated return pipe or a cold water supply pipe. Where a cold water supply pipe is used as the return pipe, a temperature sensor connected to the controls shall be located on the hot water supply not more than two feet (305 mm) from the connection to the cold water supply pipe. The controls shall limit the temperature of the water entering the cold water piping to not greater than 104°F (40°C).

R403.5.1.1.1 Demand recirculation water systems. *Demand recirculation water systems* shall have controls that start the pump upon receiving a signal from the action of a user of a fixture or appliance, sensing the presence of a user of a fixture or sensing the flow of hot or tempered water to a fixture fitting or appliance. The controls shall limit pump operation by:

1. Shutting off the pump when the temperature sensor detects one of the following:
    1.1. An increase in the water temperature of not more than 10°F (5.6°C) above the initial temperature of the water in the pipe.
    1.2. The temperature of the water in the pipe reaches 104°F (40°C).
2. Limiting pump operation to a maximum of 5 minutes following activation.
3. Not activating the pump for at least 5 minutes following shutoff or when the temperature of the water in the pipe exceeds 104°F (40°C).

R403.5.1.2 Heat trace systems. Electric heat trace systems shall comply with IEEE 515.1 or UL 515. Controls for such systems shall automatically adjust the energy input to the heat tracing to maintain the desired water temperature in the piping in accordance with the times when heated water is used in the occupancy.

R403.5.2 Hot water pipe insulation. Insulation for service hot water piping shall comply with Table R403.5.2 and be applied to the following:

1. Piping $^{3}/_{4}$ inch (19.1 mm) and larger in nominal diameter located inside the *conditioned space*.
2. Piping located outside the *conditioned space*.
3. Piping from the water heater to a distribution manifold.
4. Piping located under a floor slab.
5. Buried piping.
6. Supply and return piping in circulating hot water systems.

Exception: Cold water returns in *demand recirculation water systems*.

TABLE R403.5.2—MINIMUM PIPE INSULATION THICKNESS

FLUID OPERATING TEMPERATURE RANGE AND USAGE (°F)	INSULATION CONDUCTIVITY		MINIMUM PIPE INSULATION THICKNESS (inches)
	Conductivity Btu × in/(h × ft² × °F)[a]	Mean rating temperature (°F)	
141–200	0.25–0.29	125	1.0
105–140	0.21–0.28	100	1.0

For SI: 1 inch = 25.4 mm, °C = (°F – 32)/1.8.

a. For insulation outside the stated conductivity range listed in this table, the minimum thickness (T) listed in this table shall be determined as follows:

$T = r[(1 + t/r)^{K/k} - 1]$

where:

T = Minimum insulation thickness.

r = Actual outside radius of pipe.

t = Insulation thickness listed in the table for applicable fluid temperature and pipe size (1 inch).

K = Conductivity of alternate material at mean rating temperature indicated for the applicable fluid temperature (Btu × in/h × ft² × °F).

k = The upper value of the conductivity range listed in this table for the applicable fluid temperature.

R403.5.3 Drain water heat recovery units. Where installed, drain water heat recovery units shall comply with CSA B55.2. Drain water heat recovery units shall be tested in accordance with CSA B55.1. Potable water-side pressure loss of drain water heat recovery units shall be less than 3 psi (20.7 kPa) for individual units connected to one or two showers. Potable water-side pressure loss of drain water heat recovery units shall be less than 2 psi (13.8 kPa) for individual units connected to three or more showers.

R403.6 Mechanical ventilation. The *buildings* and *dwelling units* complying with Section R402.5.1.1 shall be provided with mechanical *ventilation* that complies with the requirements of Section M1505 of the *International Residential Code* or the *International Mechanical Code*, as applicable, or with other *approved* means of *ventilation*. Outdoor air intakes and exhausts shall have *automatic* or gravity *dampers* that close when the *ventilation* system is not operating.

R403.6.1 Heat or energy recovery ventilation. *Dwelling units* shall be provided with a heat recovery or energy recovery *ventilation* system in Climate Zones 6, 7 and 8. The system shall be a balanced *ventilation* system with a sensible recovery efficiency (SRE) of not less than 65 percent at 32°F (0°C) at an airflow greater than or equal to the design airflow. The SRE shall be determined from a *listed* value or from interpolation of *listed* values.

R403.6.2 Fan efficacy for whole-house mechanical ventilation systems and outdoor air ventilation systems. Fans used to provide whole-dwelling mechanical *ventilation* shall meet the efficacy requirements of Table R403.6.2 at one or more rating points. Fans shall be tested in accordance with the test procedure referenced by Table R403.6.2 and *listed*. The airflow shall be reported in the product listing or on the label. Fan efficacy shall be reported in the product listing or shall be derived from the input power and airflow values reported in the product listing or on the label. Fan efficacy for fully ducted HRV, ERV, balanced *ventilation* systems and in-line fans shall be determined at a static pressure of not less than 0.2 inch water gauge (50 Pa). Fan efficacy for ducted range hoods, bathroom and utility room fans shall be determined at a static pressure of not less than 0.1 inch water gauge (25 Pa).

TABLE R403.6.2—FAN EFFICACY FOR WHOLE-HOUSE MECHANICAL VENTILATION SYSTEMS AND OUTDOOR AIR VENTILATION SYSTEMS[a]

SYSTEM TYPE	AIRFLOW RATE (CFM)	MINIMUM EFFICACY (CFM/WATT)	TEST PROCEDURE
HRV or ERV	Any	1.2[a]	CAN/CSA C439
Balanced ventilation system without heat or energy recovery	Any	1.2[a]	ANSI/AMCA 210-ANSI/ASHRAE 51
Range hood	Any	2.8	

TABLE R403.6.2—WHOLE-DWELLING MECHANICAL VENTILATION SYSTEM FAN EFFICACY[a]—continued			
SYSTEM TYPE	AIRFLOW RATE (CFM)	MINIMUM EFFICACY (CFM/WATT)	TEST PROCEDURE
In-line supply or exhaust fan	Any	3.8	ANSI/AMCA 210-ANSI/ASHRAE 51
Other exhaust fan	< 90	2.8	
	≥ 90 and < 200	3.5	
	≥ 200	4.0	
Air-handling unit that is integrated to tested and listed HVAC equipment	Any	1.2	Outdoor airflow as specified. Air-handling unit fan power determined in accordance with the applicable US Department of Energy Code of Federal Regulations DOE10 CFR 430 or other approved test method.

For SI: 1 cubic foot per minute = 28.3 L/min.
a. For balanced ventilation systems, HRVs and ERVs, determine the efficacy as the outdoor airflow divided by the total fan power.

R403.6.3 Testing. Mechanical *ventilation* systems shall be tested and verified to provide the minimum *ventilation* flow rates required by Section R403.6, in accordance with ANSI/RESNET/ICC 380. Where required by the *code official*, testing shall be conducted by an *approved* third party. A written report of the results of the test shall be signed by the party conducting the test and provided to the *code official*.

Exceptions:

1. Kitchen range hoods that are ducted to the outside with ducting having a diameter of 6 inches (152 mm) or larger, a length of 10 feet (3028 mm) or less, and not more than two 90-degree (1.57 rad) elbows or equivalent shall not require testing.
2. A third-party test shall not be required where the ventilation system has an integrated diagnostic tool used for airflow measurement, and a user interface that communicates the installed airflow rate.
3. Where tested in accordance with Section R403.6.4, testing of each mechanical ventilation system is not required.

R403.6.4 Unit sampling. For *buildings* with eight or more *dwelling units* or *sleeping units,* the mechanical *ventilation* systems in the greater of seven units or 20 percent of the total units shall be tested. Tested systems shall include systems in a top floor unit, systems in a ground floor unit, systems in a middle floor unit, and the systems in the *dwelling unit* or *sleeping unit* with the largest *conditioned floor area*. Where *buildings* have fewer than eight *dwelling units* or *sleeping units*, the mechanical *ventilation* systems in each unit shall be tested. Where the *ventilation* flow rate of a mechanical *ventilation* system is less than the minimum permitted rate, corrective actions shall be taken and the system retested until it passes. For each tested *dwelling unit* or *sleeping unit* system with a *ventilation* flow rate lower than the minimum permitted, three additional systems, including the corrected system, shall be tested.

R403.6.5 Intermittent exhaust control for bathrooms and toilet rooms. Where an exhaust system serving a bathroom or toilet room is designed for intermittent operation, the exhaust system controls shall include one or more of the following:

1. A timer control with one or more delay setpoints that automatically turns off exhaust fans when the selected setpoint is reached. Not fewer than one delay-off setpoint shall be 30 minutes or less.
2. An *occupant sensor control* with one or more delay setpoints that automatically turns off exhaust fans in accordance with the selected delay setpoint after all occupants have vacated the space. Not fewer than one delay-off setpoint shall be 30 minutes or less.
3. A humidity control with an adjustable setpoint ranging between 50 percent or more and 80 percent or less relative humidity that automatically turns off exhaust fans when the selected setpoint is reached.
4. A contaminant control that responds to a particle or gaseous concentration and automatically turns off exhaust fans when a design setpoint is reached.

Manual off functionality shall not be used in lieu of the minimum setpoint functionality required by this section.

Exception: Bathroom and toilet room exhaust systems serving as an integral component of an outdoor air *ventilation* system or a whole-house mechanical *ventilation* system.

R403.7 Equipment sizing and efficiency rating. Heating and cooling *equipment* shall be sized in accordance with ACCA Manual S based on *building* loads calculated in accordance with ACCA Manual J or other *approved* heating and cooling calculation methodologies. New or replacement heating and cooling equipment shall have an efficiency rating equal to or greater than the minimum required by federal law for the geographic location where the equipment is installed.

R403.7.1 Electric-resistance space heating. Detached one- and two-family dwellings and townhouses in Climate Zones 4 through 8 using electric-resistance space heating shall limit the total installed heating capacity of all electric-resistance space heating serving the *dwelling unit* to not more than 2.0 kW or shall install a heat pump in the largest space that is not used as a bedroom.

R403.8 Systems serving multiple dwelling units. Except for systems complying with Section R403.9, systems serving multiple *dwelling units* shall comply with Sections C403 and C404 of the *International Energy Conservation Code*—Commercial Provisions instead of Section R403.

R403.9 Mechanical systems located outside of the building thermal envelope. Mechanical systems providing heat outside of the *building thermal envelope* of a *building* shall comply with Sections R403.9.1 through R403.9.4.

R403.9.1 Heating outside a building. Systems installed to provide heat outside a *building* shall be radiant systems. Such heating systems shall be controlled by an occupancy-sensing device or a timer switch, so that the system is automatically de-energized when occupants are not present.

R403.9.2 Snow melt and ice system controls. Snow- and ice-melting systems, supplied through energy service to the *building*, shall include *automatic* controls capable of shutting off the system when the pavement temperature is greater than 50°F (10°C) and precipitation is not falling, and an *automatic* or *manual* control that will allow shutoff when the outdoor temperature is greater than 40°F (4.8°C).

R403.9.3 Roof and gutter deicing controls. Roof and gutter deicing systems, including but not limited to self-regulating cable, shall include *automatic* controls that are configured to shut off the system when the outdoor temperature is above 40°F (4.4°C) and shall include one of the following:

1. A moisture sensor configured to shut off the system in the absence of moisture.
2. A daylight sensor or other means configured to shut off the system between sunset and sunrise.

R403.9.4 Freeze protection system controls. Freeze protection systems, such as heat tracing of outdoor piping and *heat exchangers*, including self-regulating heat tracing, shall include *automatic* controls configured to shut off the systems when outdoor air temperatures are above 40°F (4.4°C) or when the conditions of the protected fluid will prevent freezing.

R403.10 Energy consumption of pools and spas. The energy consumption of pools and permanent spas shall be controlled by the requirements in Sections R403.10.1 through R403.10.3.

R403.10.1 Heaters. The electric power to heaters shall be controlled by an on-off switch that is an integral part of the heater mounted on the exterior of the heater in a location with *ready access*, or external to and within 3 feet (914 mm) of the heater. Operation of such switch shall not change the setting of the heater *thermostat*. Such switches shall be in addition to a circuit breaker for the power to the heater. Gas-fired heaters shall not be equipped with continuously burning ignition pilots.

R403.10.2 Time switches. Time switches or other control methods that can automatically turn heaters and pump motors off and on according to a preset schedule shall be installed for heaters and pump motors. Heaters and pump motors that have built-in time switches shall be in compliance with this section.

Exceptions:

1. Where public health standards require 24-hour pump operation.
2. Pumps that operate *on-site renewable energy* and waste-heat-recovery pool heating systems.

R403.10.3 Covers. Outdoor heated pools and outdoor permanent spas shall be provided with a vapor-retardant cover or other *approved* vapor-retardant means.

Exception: Where more than 75 percent of the energy for heating, computed over an operation season of not fewer than 3 calendar months, is from a heat pump or an *on-site renewable energy* system, covers or other vapor-retardant means shall not be required.

R403.11 Portable spas. The energy consumption of electric-powered portable spas shall be controlled by the requirements of APSP 14.

R403.12 Residential pools and permanent residential spas. Where installed, the energy consumption of residential swimming pools and permanent residential spas shall be controlled in accordance with the requirements of APSP 15.

R403.13 Gas fireplaces. Gas fireplace systems shall not be equipped with a *continuous pilot* and shall be equipped with an *on-demand pilot*, *intermittent ignition* or *interrupted ignition,* as defined by ANSI Z21.20.

Exception: Gas-fired appliances using pilots within a *listed* combustion safety device.

R403.13.1 Gas fireplace efficiency. Vented gas fireplace heaters shall have a fireplace efficiency (FE) rating not less than 50 percent as determined in accordance with CSA P.4.1, and shall be *listed* and *labeled* in accordance with CSA/ANSI Z21.88. Vented gas fireplaces (decorative appliances) shall be *listed* and *labeled* in accordance with CSA/ANSI Z21.50.

SECTION R404—ELECTRICAL POWER, LIGHTING AND RENEWABLE ENERGY SYSTEMS

R404.1 Lighting equipment. All permanently installed luminaires shall be capable of operation with an efficacy of not less than 45 lumens per watt or shall contain lamps capable of operation with an efficacy of not less than 65 lumens per watt.

Exceptions:

1. Appliance lamps.
2. Antimicrobial lighting used for the sole purpose of disinfecting.
3. General service lamps complying with DOE 10 CFR, Part 430.32.
4. Luminaires with a rated electric input of not greater than 3.0 watts.

TABLE R404.1—LIGHTING POWER ALLOWANCES FOR BUILDING EXTERIORS

BASE SITE ALLOWANCE	280 WATTS
Uncovered parking areas and drives	0.026 W/ft^2
Building grounds	
Walkways and ramps	0.50 W/linear foot
Plaza areas	0.049 W/ft^2
Dining areas	0.273 W/ft^2
Stairways	Exempt
Pedestrian tunnels	0.110 W/ft^2
Landscaping	0.025 W/ft^2
Building entrances and exits	
Pedestrian and vehicular entrances and exits	9.8 W/linear foot of opening
Entry canopies	0.126 W/ft^2

For SI: 1 watt per square foot = 10.76 w/m^2, 1 foot = 304.8 mm.

R404.1.1 Exterior lighting. Connected exterior lighting for Group R-2, R-3 and R-4 *residential buildings* shall comply with Sections R404.1.2 through R404.1.5.

Exceptions:

1. Detached one- and two- family dwellings.
2. Townhouses.
3. Group R-3 *buildings* that do not contain more than two *dwelling units.*
4. Solar-powered lamps not connected to any electrical service.
5. Luminaires controlled by a motion sensor.
6. Lamps and luminaires that comply with Section R404.1.

R404.1.2 Exterior lighting power requirements. The total exterior connected lighting power shall be not greater than the exterior lighting power allowance calculated in accordance with Section R404.1.3. The total exterior connected lighting power shall be the total maximum rated wattage of all lighting that is powered through the energy service for the *building*.

Exceptions: Lighting used for the following applications shall not be included.

1. Lighting *approved* for safety reasons.
2. Emergency lighting that is automatically off during normal operations.
3. Exit signs.
4. Specialized signal, directional and marker lighting associated with transportation.
5. Lighting for athletic playing areas.
6. Temporary lighting.
7. Lighting used to highlight features of art, public monuments and the national flag.
8. Lighting for water features and swimming pools.
9. Lighting controlled from within *sleeping units* and *dwelling units.*
10. Lighting of the exterior means of egress as required by the *International Building Code*.

R404.1.3 Exterior lighting power allowance. The total area or length of each area type multiplied by the value for the area type in Table R404.1 shall be the lighting power (watts) allowed for each area type. For area types not listed, the area type that most closely represents the proposed use of the area shall be selected. The total exterior lighting power allowance (watts) shall be the sum of the base site allowance plus the watts from each area type.

R404.1.4 Additional exterior lighting power. Additional exterior lighting power allowances shall be available for the building facades at 0.075 W/ft^2 (0.807 w/m^2) of gross *above-grade wall* area. These additional power allowances shall be used only for the luminaires serving the facade and shall not be used to increase any other lighting power allowance.

R404.1.5 Gas lighting. Gas-fired lighting appliances shall not be equipped with a *continuous pilot* and shall be equipped with an *on-demand pilot*, *intermittent ignition* or *interrupted ignition* as defined by ANSI Z21.20.

R404.2 Interior lighting controls. All permanently installed luminaires shall be controlled as required in Sections R404.2.1 and R404.2.2.

Exception: Lighting controls shall not be required for safety or security lighting.

R404.2.1 Habitable spaces. All permanently installed luminaires in habitable spaces shall be controlled with a *manual dimmer* or with an *automatic* shutoff control that automatically turns off lights within 20 minutes after all occupants have left the space and shall incorporate a *manual* control to allow occupants to turn the lights on or off.

R404.2.2 Specific locations. All permanently installed luminaires in garages, unfinished basements, laundry rooms and utility rooms shall be controlled by an *automatic* shutoff control that automatically turns off lights within 20 minutes after all occupants have left the space and shall incorporate a *manual* control to allow occupants to turn the lights on or off.

R404.3 Exterior lighting controls. Exterior lighting controls shall comply with Section R404.3.1.

R404.3.1 Controls for individual dwelling units. Where the total permanently installed exterior lighting power is greater than 30 watts, the permanently installed exterior lighting shall comply with the following:

1. Lighting shall be controlled by a *manual* on and off switch which permits *automatic* shutoff actions.
2. Lighting shall be automatically shut off when daylight is present and satisfies the lighting needs.
3. Controls that override *automatic* shutoff actions shall not be allowed unless the override automatically returns *automatic* control to its normal operation within 24 hours.

R404.4 Renewable energy certificate (REC) documentation. Where renewable energy generation is used to comply with this code, documentation shall be provided to the *code official* by the property owner or owner's authorized agent demonstrating that where renewable energy certificates (RECs) or energy attributable certificates (EACs) are associated with that portion of renewable energy used to comply with this code, the RECs or EACs shall be retained, or retired, on behalf of the property owner.

SECTION R405—SIMULATED BUILDING PERFORMANCE

R405.1 Scope. This section establishes criteria for compliance using *simulated building performance* analysis. Such analysis shall include heating, cooling, mechanical *ventilation* and service water-heating energy only. Such analysis shall be limited to *dwelling units*. Spaces other than *dwelling units* in Group R-2, R-3 or R-4 buildings shall comply with Sections R402 through R404.

R405.2 Simulated building performance compliance. Compliance based on *simulated building performance* requires that a *building* comply with the following:

1. The requirements of the sections indicated within Table R405.2.
2. The proposed total *building thermal envelope* thermal conductance (TC) shall be less than or equal to the required total *building thermal envelope* TC using the prescriptive *U*-factors and *F*-factors from Table R402.1.2 multiplied by 1.08 in Climate Zones 0, 1 and 2, and 1.15 in Climate Zones 3 through 8, in accordance with Equation 4-2 and Section R402.1.5. The area-weighted maximum fenestration SHGC permitted in Climate Zones 0 through 3 shall be 0.30.

Equation 4-2 For Climate Zones 0–2: $TC_{Proposed\ design} \leq 1.08 \times TC_{Prescriptive\ reference\ design}$
For Climate Zones 3–8: $TC_{Proposed\ design} \leq 1.15 \times TC_{Prescriptive\ reference\ design}$

3. For each *dwelling unit* with one or more fuel-burning appliances for space heating, water heating, or both, the annual *energy cost* of the *dwelling unit* shall be less than or equal to 80 percent of the annual *energy cost* of the *standard reference design*. For all other *dwelling units*, the annual *energy cost* of the proposed design shall be less than or equal to 85 percent of the annual *energy cost* of the *standard reference design*. For each dwelling unit with greater than 5,000 square feet (465 m^2) of *living space* located above grade plane, the annual *energy cost* of the *dwelling unit* shall be reduced by an additional 5 percent of annual *energy cost* of the *standard reference design*. Energy prices shall be taken from an *approved* source, such as the US Energy Information Administration's State Energy Data System prices and expenditures reports. Code officials shall be permitted to require time-of-use pricing in *energy cost* calculations.

Exceptions:

1. The energy use based on source energy expressed in Btu or Btu per square foot of *conditioned floor area* shall be permitted to be substituted for the *energy cost*. The source energy multiplier for electricity shall be 2.51. The source energy multipliers shall be 1.09 for natural gas, 1.15 for propane, 1.19 for fuel oil, and 1.30 for imported liquified natural gas.
2. The energy use based on site energy expressed in Btu or Btu per square foot of conditioned floor area shall be permitted to be substituted for the energy cost.

TABLE R405.2—REQUIREMENTS FOR SIMULATED BUILDING PERFORMANCE	
SECTION[a]	TITLE
General	
R401.3	Certificate
Building Thermal Envelope	
R402.1.1	Vapor retarder
R402.1.6	Rooms containing fuel-burning appliances
R402.2.3	Attic knee wall
R402.2.4	Eave baffle
R402.2.5.1	Access hatches and door insulation installation and retention
R402.2.10	Slab-on-grade floors
R402.2.11	Crawl space walls
R402.5.1.1	Installation
R402.5.1.2	Air leakage testing
R402.5.1.3	Maximum air leakage rate
R402.5.2	Fireplaces
R402.5.3	Fenestration air leakage
R402.5.4	Recessed lighting
R402.5.5	Air-sealed electrical and communication outlet boxes
R402.6	Maximum fenestration *U*-factor and SHGC
Mechanical	
R403.1	Controls
R403.2	Hot water boiler temperature reset
R403.3	Duct systems
R403.4	Mechanical system piping insulation
R403.5	Service hot water systems
R403.6	Mechanical ventilation
R403.7, except Section R403.7.1	Equipment sizing and efficiency rating
R403.8	Systems serving multiple dwelling units
R403.9.2	Snow melt and ice system controls
R403.10	Energy consumption of pools and spas
R403.11	Portable spas
R403.12	Residential pools and permanent residential spas
R403.13	Gas fireplaces
Electrical Power and Lighting Systems	
R404.1	Lighting equipment
R404.2	Interior lighting controls

a. Reference to a code section includes all the relative subsections except as indicated in the table.

R405.3 Compliance documentation. The following compliance reports, which document that the performance of the *proposed design* and the performance of the as-built *dwelling unit* comply with the requirements of Section R405, shall be submitted to the *code official*.

1. A compliance report in accordance with Section R405.5.4.1 shall be submitted with the application for the building permit.
2. A compliance report in accordance with Section R405.5.4.2 shall be submitted before a certificate of occupancy is issued.

R405.4 Calculation procedure. Performance calculations shall be in accordance with Sections R405.4.1 through R405.4.3. Except as specified by this section, the *standard reference design* and *proposed design* shall be configured and analyzed using identical methods and techniques.

R405.4.1 General. Calculation procedures used to comply with Section R405 shall use a software tool, *approved* in accordance with Section R405.5, capable of calculating the annual energy consumption of all building elements that differ between the *standard reference design* and the *proposed design*.

R405.4.2 Residence specifications. The *standard reference design, proposed design* and as-built *dwelling unit* shall be configured and analyzed as specified by Table R405.4.2(1). Table R405.4.2(1) shall include, by reference, all notes contained in Table R402.1.3. Proposed *U*-factors and slab-on-grade *F*-factors shall be taken from Appendix RF, ANSI/ASHRAE/IES Standard 90.1 Appendix A, or determined using a method consistent with the ASHRAE *Handbook of Fundamentals* and shall include the thermal bridging effects of framing materials.

TABLE R405.4.2(1)—SPECIFICATIONS FOR THE STANDARD REFERENCE AND PROPOSED DESIGNS

BUILDING COMPONENT	STANDARD REFERENCE DESIGN	PROPOSED DESIGN
Above-grade walls	Type: mass where the proposed wall is a mass wall; otherwise wood frame.	As proposed.
	Gross area: same as proposed.	As proposed
	U-factor: as specified in Table R402.1.2.	As proposed.
	Solar reflectance = 0.25.	As proposed.
	Emittance = 0.90.	As proposed.
Basement and crawl space walls	Type: same as proposed.	As proposed.
	Gross area: same as proposed.	As proposed.
	U-factor: as specified in Table R402.1.2, with the insulation layer on the interior side of the walls.	As proposed.
Above-grade floors	Type: wood frame.	As proposed.
	Gross area: same as proposed.	As proposed.
	U-factor: as specified in Table R402.1.2.	As proposed.
Ceilings	Type: wood frame.	As proposed.
	Gross area: same as proposed.	As proposed
	U-factor: as specified in Table R402.1.2.	As proposed.
Roofs	Type: composition shingle on wood sheathing.	As proposed.
	Gross area: same as proposed.	As proposed.
	Solar reflectance = 0.25.	As proposed.
	Emittance = 0.90.	As proposed.
Attics	Type: vented with an aperture of 1 ft^2 per 300 ft^2 of ceiling area.	As proposed.
Foundations	Type: same as proposed.	As proposed.
	Foundation wall extension above and below grade: same as proposed. Foundation wall or slab perimeter length: same as proposed. Soil characteristics: same as proposed.	As proposed.
	Foundation wall *U*-factor and slab *F*-factor: as specified in Table R402.1.2.	
Opaque doors	Area: 40 ft^2.	As proposed.
	Orientation: North.	As proposed.
	U-factor: same as fenestration as specified in Table R402.1.2.	As proposed.
Vertical fenestration other than opaque doors	Total area[h] = (a) The proposed glazing area, where the proposed glazing area is less than 15 percent of the conditioned floor area. (b) 15 percent of the conditioned floor area, where the proposed glazing area is 15 percent or more of the conditioned floor area.	As proposed.
	Orientation: equally distributed to four cardinal compass orientations (N, E, S & W).	As proposed.
	U-factor: as specified in Table R402.1.2.	As proposed.
	SHGC: as specified in Table R402.1.2 except for climate zones without an SHGC requirement, the SHGC shall be equal to 0.40.	As proposed.
	Interior shade fraction: 0.92 – (0.21 × SHGC for the standard reference design).	Interior shade fraction: 0.92 – (0.21 × SHGC as proposed).
	External shading: none	As proposed.

TABLE R405.4.2(1)—SPECIFICATIONS FOR THE STANDARD REFERENCE AND PROPOSED DESIGNS—continued		
BUILDING COMPONENT	**STANDARD REFERENCE DESIGN**	**PROPOSED DESIGN**
Skylights	None	As proposed.
Thermally isolated sunrooms	None	As proposed.
Air leakage rate	For detached one-family dwellings, the air leakage rate at a pressure of 0.2 inch water gauge (50 Pa) shall be as follows: Climate Zones 0 through 2: 4.0 air changes per hour. Climate Zones 3, 4 and 5: 3.0 air changes per hour. Climate Zones 6 through 8: 2.5 air changes per hour. For detached one-family dwellings that are 1,500 ft^2 or smaller and attached dwelling units or sleeping units, the air leakage rate at a pressure of 0.2 inch water gauge (50 Pa) shall be 0.27 cfm/ft^2 of the testing unit enclosure area.	The measured air leakage rate.[a]
Mechanical ventilation rate	The mechanical ventilation rate shall be in addition to the air leakage rate and shall be the same as in the proposed design, but not greater than $B \times M$ where: B = $0.01 \times CFA + 7.5 \times (N_{br} + 1)$, cfm. M = 1.0 where the measured air leakage rate is ≥ 3.0 air changes per hour at 50 Pascals, and otherwise, M = minimum (1.7, Q/B). Q = the proposed mechanical ventilation rate, cfm. CFA = conditioned floor area, ft^2. N_{br} = number of bedrooms.	The measured mechanical ventilation rate[b] (Q) shall be in addition to the measured air leakage rate.
Mechanical ventilation fan energy	The mechanical ventilation system type shall be the same as in the proposed design. Heat recovery or energy recovery shall be modeled for mechanical ventilation where required by Section R403.6.1. Heat recovery or energy recovery shall not be modeled for mechanical ventilation where not required by Section R403.6.1. Where mechanical ventilation is not specified in the proposed design: None Where mechanical ventilation is specified in the proposed design, the annual vent fan energy use, in units of kWh/yr, shall equal $(8.76 \times B \times M)/e_f$ where: B and M are determined in accordance with the air exchange mechanical ventilation rate row of this table. e_f = the minimum fan efficacy, as specified in Table R403.6.2, corresponding to the system type at a flow rate of $B \times M$.	As proposed.
Internal gains	IGain, in units of Btu/day per dwelling unit, shall equal $17{,}900 + 23.8 \times CFA + 4{,}104 \times N_{br}$ where: CFA = conditioned floor area, ft^2. N_{br} = number of bedrooms.	Same as standard reference design.
Internal mass	Internal mass for furniture and contents: 8 pounds per square foot of floor area.	Same as standard reference design, plus any additional mass specifically designed as a thermal storage element[c] but not integral to the building thermal envelope or structure.
Structural mass	For masonry floor slabs: 80 percent of floor area covered by R-2 carpet and pad, and 20 percent of floor directly exposed to room air.	As proposed.
	For masonry basement walls: as proposed, but with insulation as specified in Table R402.1.3, located on the interior side of the walls.	As proposed.
	For other walls, ceilings, floors, and interior walls: wood-framed construction.	As proposed.
Heating systems[d, e, j, k]	Fuel type/capacity: same as proposed design.	As proposed.
	Product class: same as proposed design.	As proposed.
	Efficiencies:	
	Heat pump: complying with 10 CFR §430.32.	As proposed.
	Fuel gas and liquid fuel furnaces: complying with 10 CFR §430.32.	As proposed.
	Fuel gas and liquid fuel boilers: complying with 10 CFR §430.32.	As proposed.
Cooling systems[d, f, k]	Fuel type: electric. Capacity: Same as proposed design.	As proposed.
	Efficiencies: complying with 10 CFR §430.32.	As proposed.

<table>
<tr><th colspan="8">TABLE R405.4.2(1)—SPECIFICATIONS FOR THE STANDARD REFERENCE AND PROPOSED DESIGNS—continued</th></tr>
<tr><th>BUILDING COMPONENT</th><th colspan="4">STANDARD REFERENCE DESIGN</th><th colspan="3">PROPOSED DESIGN</th></tr>
<tr><td rowspan="12">Service water heating[d, g, k]</td><td colspan="4" rowspan="7">Use, in units of gal/day = 25.5 + (8.5 × N_{br})
where:
N_{br} = number of bedrooms.</td><td colspan="3">Use, in units of gal/day = 25.5 + (8.5 × N_{br}) × (1 – $HWDS$)
where:
N_{br} = number of bedrooms.
$HWDS$ = factor for the compactness of the hot water distribution system.</td></tr>
<tr><td colspan="2">Compactness ratio[i] factor</td><td rowspan="2">HWDS</td></tr>
<tr><td>1 story</td><td>2 or more stories</td></tr>
<tr><td>> 60%</td><td>> 30%</td><td>0</td></tr>
<tr><td>> 30% to ≤ 60%</td><td>> 15% to ≤ 30%</td><td>0.05</td></tr>
<tr><td>> 15% to ≤ 30%</td><td>> 7.5% to ≤ 15%</td><td>0.10</td></tr>
<tr><td>< 15%</td><td>< 7.5%</td><td>0.15</td></tr>
<tr><td colspan="4">Fuel type: same as proposed design.</td><td colspan="3">As proposed.</td></tr>
<tr><td colspan="4">Rated storage volume: same as proposed design.</td><td colspan="3">As proposed.</td></tr>
<tr><td colspan="4">Draw pattern: same as proposed design.</td><td colspan="3">As proposed.</td></tr>
<tr><td colspan="4">Efficiencies: Uniform Energy Factor complying with 10 CFR §430.32.</td><td colspan="3">As proposed.</td></tr>
<tr><td colspan="4">Tank temperature: 120°F (48.9°C).</td><td colspan="3">Same as standard reference design.</td></tr>
<tr><td rowspan="6">Thermal distribution systems</td><td colspan="4">Duct insulation: in accordance with Section R403.3.3.</td><td colspan="3">Duct insulation: as proposed.[m]</td></tr>
<tr><td colspan="4">Duct location:</td><td colspan="3">Duct location: as proposed.[l]</td></tr>
<tr><td>Foundation type</td><td>Slab on grade</td><td>Unconditioned crawl space</td><td>Basement or conditioned crawl space</td><td colspan="3">—</td></tr>
<tr><td>Duct location (supply and return)</td><td>One-story building: 100% in unconditioned attic.
All other: 75% in unconditioned attic and 25% inside conditioned space.</td><td>One-story building: 100% in unconditioned crawl space.
All other: 75% in unconditioned crawl space and 25% inside conditioned space.</td><td>75% inside conditioned space.
25% unconditioned attic.</td><td colspan="3" rowspan="2">Duct system leakage to outside: The measured total duct system leakage rate shall be entered into the software as the duct system leakage to outside rate.
Exceptions:
1. Where duct system leakage to outside is tested in accordance ANSI/RESNET/ICC 380 or ASTM E1554, the measured value shall be permitted to be entered.
2. Where total duct system leakage is measured without space conditioning equipment installed, the simulation value shall be 4 cfm per 100 ft^2 of conditioned floor area.</td></tr>
<tr><td colspan="4">Duct system leakage to outside: for duct systems serving > 1,000 ft^2 of conditioned floor area, the duct leakage to outside rate shall be 4 cfm per 100 ft^2 of conditioned floor area.
For duct systems serving ≤ 1,000 ft^2 of conditioned floor area, the duct leakage to outside rate shall be 40 cfm.</td></tr>
<tr><td colspan="4">Distribution system efficiency (DSE): for hydronic systems and ductless systems, a thermal DSE of 0.88 shall be applied to both the heating and cooling system efficiencies.</td><td colspan="3">Distribution system efficiency (DSE): for hydronic systems and ductless systems, DSE shall be as specified in Table R405.4.2(2).</td></tr>
<tr><td>Thermostat</td><td colspan="4">Type: Manual, cooling temperature setpoint = 75°F;
Heating temperature setpoint = 72°F.</td><td colspan="3">Same as standard reference design.</td></tr>
</table>

TABLE R405.4.2(1)—SPECIFICATIONS FOR THE STANDARD REFERENCE AND PROPOSED DESIGNS—continued		
BUILDING COMPONENT	**STANDARD REFERENCE DESIGN**	**PROPOSED DESIGN**
Dehumidistat	Where a mechanical ventilation system with latent heat recovery is not specified in the proposed design: None. Where the proposed design utilizes a mechanical ventilation system with latent heat recovery: Dehumidistat type: manual, setpoint = 60% relative humidity. Dehumidifier: whole-dwelling with integrated energy factor = 1.77 liters/kWh.	Same as standard reference design.

For SI: 1 square foot = 0.93 m^2, 1 British thermal unit = 1055 J, 1 pound per square foot = 4.88 kg/m^2, 1 gallon (US) = 3.785 L, °C = (°F – 32)/1.8, 1 degree = 0.79 rad, 1 cubic foot per minute = 28.317 L/min.

a. Hourly calculations as specified in the ASHRAE *Handbook of Fundamentals,* or the equivalent, shall be used to determine the energy loads resulting from infiltration.

b. The combined air exchange rate for infiltration and mechanical ventilation shall be determined in accordance with Equation 43 of 2001 ASHRAE *Handbook of Fundamentals,* page 26.24 and the "Whole-house Ventilation" provisions of 2001 ASHRAE *Handbook of Fundamentals,* page 26.19 for intermittent mechanical ventilation.

c. Thermal storage element shall mean a component that is not part of the floors, walls or ceilings that is part of a passive solar system, and that provides thermal storage such as enclosed water columns, rock beds, or phase-change containers. A thermal storage element shall be in the same room as fenestration that faces within 15 degrees (0.26 rad) of true south, or shall be connected to such a room with pipes or ducts that allow the element to be actively charged.

d. For a proposed design with multiple heating, cooling or water heating systems using different fuel types, the applicable standard reference design system capacities and fuel types shall be weighted in accordance with their respective loads as calculated by accepted engineering practice for each equipment and fuel type present.

e. For a proposed design without a proposed heating system, a heating system having the prevailing federal minimum efficiency shall be assumed for both the standard reference design and proposed design.

f. For a proposed design without a proposed cooling system, an electric air conditioner having the prevailing federal minimum efficiency shall be assumed for both the standard reference design and the proposed design.

g. For a proposed design without a proposed water heater, the following assumptions shall be made for both the proposed design and standard reference design. For a proposed design with a heat pump water heater, the following assumptions shall be made for the standard reference design, except the fuel type shall be electric.
Fuel type: same as the predominant heating fuel type
Rated storage volume: 40 gallons
Draw pattern: medium
Efficiency: Uniform Energy Factor complying with 10 CFR § 430.32

h. For residences with conditioned basements, R-2 and R-4 residences, and for townhouses, the following formula shall be used to determine glazing area:

$AF = A_s \times FA \times F$

where:

AF = Total glazing area.

A_s = Standard reference design total glazing area.

FA = (Above-grade thermal boundary gross wall area)/(above-grade boundary wall area + 0.5 × below-grade boundary wall area).

F = (above-grade thermal boundary wall area)/(above-grade thermal boundary wall area + common wall area) or 0.56, whichever is greater.

and where:

- Thermal boundary wall is any wall that separates conditioned space from unconditioned space or ambient conditions.
- Above-grade thermal boundary wall is any thermal boundary wall component not in contact with soil.
- Below-grade boundary wall is any thermal boundary wall in soil contact.
- Common wall area is the area of walls shared with an adjoining dwelling unit.

i. The factor for the compactness of the hot water distribution system is the ratio of the area of the rectangle that bounds the source of hot water and the fixtures that it serves (the "hot water rectangle") divided by the floor area of the dwelling.

1. Sources of hot water include water heaters, or in multiple-family buildings with central water heating systems, circulation loops or electric heat traced pipes.
2. The hot water rectangle shall include the source of hot water and the points of termination of all hot water fixture supply piping.
3. The hot water rectangle shall be shown on the floor plans and the area shall be computed to the nearest square foot.
4. Where there is more than one water heater and each water heater serves different plumbing fixtures and appliances, it is permissible to establish a separate hot water rectangle for each hot water distribution system and add the area of these rectangles together to determine the compactness ratio.
5. The basement or attic shall be counted as a story when it contains the water heater.
6. Compliance shall be demonstrated by providing a drawing on the plans that shows the hot water distribution system rectangle(s), comparing the area of the rectangle(s) to the area of the dwelling and identifying the appropriate compactness ratio and *HWDS* factor.

j. For a proposed design with electric resistance heating, a split system heat pump complying with 10 CFR §430.32 (2021) shall be assumed modeled in the standard reference design.

k. For heating systems, cooling systems, or water heating systems not included in this table, the standard reference design shall be the same as proposed design.

l. Only sections of ductwork that are installed in accordance with Section R403.3.4, Items 1 and 2 are assumed to be located completely inside conditioned space. All other sections of ductwork are not assumed to be located completely inside conditioned space.

m. Sections of ductwork installed in accordance with Section R403.3.5.1 are assumed to have an effective duct insulation *R*-value of R-25.

TABLE R405.4.2(2)—DEFAULT DISTRIBUTION SYSTEM EFFICIENCIES FOR PROPOSED DESIGNS[a]		
DISTRIBUTION SYSTEM CONFIGURATION AND CONDITION	**FORCED AIR SYSTEMS**	**HYDRONIC SYSTEMS[b]**
Distribution system components located in unconditioned space	NA	0.95
Distribution system components entirely located in conditioned space[c]	NA	1
Ductless systems[d]	1	NA

NA = Not Applicable.

a. Default values in this table are for untested distribution systems, which must still comply with Section R403.

b. Hydronic systems mean those systems that distribute heating and cooling energy directly to individual spaces using liquids pumped through closed-loop piping and that do not depend on ducted, forced airflow to maintain space temperatures.

c. Entire system in conditioned space means that no component of the distribution system is located outside of the conditioned space.

d. Ductless systems are allowed to have forced airflow across a coil but must not have any ducted airflow external to the space conditioning equipment.

R405.4.3 Input values. When calculations require input values not specified by Sections R402, R403, R404 and R405, those input values shall be taken from an *approved* source.

R405.5 Calculation software tools. Performance analysis tools meeting the applicable provisions of Sections R405.5.1 through R405.5.4 shall be permitted to be *approved*. Tools are permitted to be *approved* based on meeting a specified threshold for a jurisdiction. The *code official* shall be permitted to approve such tools for a specified application or limited scope.

R405.5.1 Minimum capabilities. *Approved* software tools shall include the following capabilities:

1. Computer generation of the *standard reference design* using only the input for the *proposed design.* The calculation procedure shall not allow the user to directly modify the building component characteristics of the *standard reference design.*
2. Calculation of whole-dwelling unit (as a single *zone*) sizing for the heating and cooling equipment in the *standard reference design* residence in accordance with Section R403.7.
3. Hourly calculations of building operation for a full calendar year (8,760 hours).
4. Calculations that account for hourly variations of indoor and outdoor temperatures and part-load ratios on the performance of heating, ventilating and air-conditioning equipment based on climate and equipment sizing.
4. Printing of a *code official* inspection checklist listing each of the *proposed design* component characteristics from Table R405.4.2(1) determined by the analysis to provide compliance, along with their respective performance ratings such as *R-value*, *U-factor*, SHGC, HSPF2, AFUE, SEER2 and UEF.

R405.5.2 Testing required by software vendors. Prior to approval, software tools shall be tested by the software vendor in accordance with ANSI/ASHRAE 140 Class II, Tier 1 test procedures. During testing, hidden inputs that are not normally available to the user shall be permitted to avoid introducing source code changes strictly used for testing. Software vendors shall publish, on a publicly available website, the following ANSI/ASHRAE 140 test results, input files and modeler reports for each tested version of a software tool:

1. Test results demonstrating the software tool was tested in accordance with ANSI/ASHRAE 140.
2. The modeler report in ANSI/ASHRAE 140, Annex A2, Attachment A2.7.

R405.5.3 Algorithms not tested. Algorithms not tested in accordance with Section R405.5.2 shall be permitted in accordance with ANSI/RESNET/ICC 301. Numerical settings not tested, such as timestep duration and tolerances, shall be permitted where they represent a higher resolution than the numerical settings used for testing.

R405.5.4 Compliance reports. *Approved* software tools shall generate compliance reports in accordance with Sections R405.5.4.1 and R405.5.4.2.

R405.5.4.1 Compliance report for permit application. A compliance report generated for submission with the application for building permit shall include the following:

1. Building street address or other *building site* identification.
2. The name of the individual performing the analysis and generating the compliance report.
3. The name and version of the compliance software tool.
4. Documentation of all inputs to the software used to produce the results for the *standard reference design* and the *proposed design*.
5. A certificate indicating that the *proposed design* complies with Section R405.2. The certificate shall document the building components' energy specifications that are included in the calculation including: component-level insulation *R*-values or *U*-factors; *duct system* and *building thermal envelope* air leakage testing assumptions; and the type and rated efficiencies of proposed heating, cooling, mechanical *ventilation* and service water-heating equipment to be installed. Where *on-site renewable energy* systems will be installed, the certificate shall report the type and production size of the proposed system.
6. Where a site-specific report is not generated, the *proposed design* shall be based on the worst-case orientation and configuration of the rated *dwelling unit*.

R405.5.4.2 Compliance report for certificate of occupancy. A compliance report generated for submission prior to obtaining the certificate of occupancy shall include the following:

1. Building street address, or other *building site* identification.
2. Declaration of the *simulated building performance* path on the title page of the energy report and the title page of the building plans.
3. A statement, bearing the name of the individual performing the analysis and generating the report, indicating that the as-built *building* complies with Section R405.2.
4. The name and version of the compliance software tool.
5. A site-specific *energy analysis* report that is in compliance with the requirements of Section R405.4, where all inputs for the *proposed design* have been replaced in the simulation with confirmed energy features of the as-built *dwelling unit*.
6. A final confirmed certificate indicating compliance based on inspection, and a statement indicating that the as-built *building* complies with Section R405.2. The certificate shall report the energy features that were confirmed to be in the *building*, including component-level insulation *R-values* or *U-factors*; results from any required *duct system* and

building thermal envelope air leakage testing; and the type and rated efficiencies of the heating, cooling, mechanical *ventilation* and service water-heating equipment installed.

7. When *on-site renewable energy* systems have been installed, the certificate shall report the type and production size of the installed system.

SECTION R406—ENERGY RATING INDEX COMPLIANCE ALTERNATIVE

R406.1 Scope. This section establishes criteria for compliance using an *Energy Rating Index* (ERI) analysis. Such analysis shall be limited to *dwelling units*. Spaces other than *dwelling units* in Group R-2, R-3 or R-4 buildings shall comply with Sections R402 through R404.

R406.2 ERI compliance. Compliance based on the *ERI* requires that the *rated design* and as-built *dwelling unit* meet all of the following:

1. The requirements of the sections indicated within Table R406.2.
2. Maximum *ERI* values indicated in Table R406.5.

TABLE R406.2—REQUIREMENTS FOR ENERGY RATING INDEX

SECTION[a]	TITLE
General	
R401.3	Certificate
Building thermal envelope	
R402.1.1	Vapor retarder
R402.1.6	Rooms containing fuel-burning appliances
R402.2.4	Eave baffle
R402.2.5.1	Access hatches and door insulation installation and retention
R402.2.10	Slab-on-grade floors
R402.2.11	Crawl space walls
R402.5.1.1	Installation
R402.5.1.2	Air leakage testing
R402.5.1.3	Maximum air leakage rate
R402.5.2	Fireplaces
R402.5.3	Fenestration air leakage
R402.5.4	Recessed lighting
R402.5.5	Air-sealed electrical and communication outlet boxes
R406.3	Building thermal envelope
Mechanical	
R403.1	Controls
R403.2	Hot water boiler temperature reset
R403.3	Duct systems
R403.4	Mechanical system piping insulation
R403.5	Service hot water systems
R403.6	Mechanical ventilation
R403.7, except Section R403.7.1	Equipment sizing and efficiency rating
R403.8	Systems serving multiple dwelling units
R403.9.2	Snow melt and ice system controls
R403.10	Energy consumption of pools and spas
R403.11	Portable spas
R403.12	Residential pools and permanent residential spas
R403.13	Gas fireplaces

TABLE R406.2—REQUIREMENTS FOR ENERGY RATING INDEX—continued	
SECTION[a]	TITLE
Electrical power and lighting systems	
R404.1	Lighting equipment
R404.2	Interior lighting controls

a. Reference to a code section includes all of the relative subsections except as indicated in the table.

R406.3 Building thermal envelope. The proposed total *building thermal envelope* thermal conductance (TC) shall be less than or equal to the required total *building thermal envelope* TC using the prescriptive *U*-factors and *F*-factors from Table R402.1.2 multiplied by 1.08 in Climate Zones 0, 1 and 2, and by 1.15 in Climates Zones 3 through 8, in accordance with Equation 4-2 and Section R402.1.5. The area-weighted maximum fenestration SHGC permitted in Climate Zones 0 through 3 shall be 0.30.

R406.4 Energy Rating Index. The *Energy Rating Index* (ERI) shall be determined in accordance with ANSI/RESNET/ICC 301. The mechanical *ventilation* rates used for the purpose of determining the *ERI* shall not be construed to establish minimum *ventilation* requirements for compliance with this code.

Energy used to recharge or refuel a vehicle used for transportation on roads that are not on the *building* site shall not be included in the *ERI reference design* or the *rated design.*

R406.5 ERI-based compliance. Compliance based on an *ERI* analysis requires that the *rated design* and each confirmed as-built *dwelling unit* be shown to have an *ERI* less than or equal to the appropriate value indicated in Table R406.5 where compared to the *ERI reference design* as follows:

1. Where on-site renewables are not installed, the values under ENERGY RATING INDEX NOT INCLUDING OPP apply.
2. Where on-site renewables are installed, the values under ENERGY RATING INDEX WITH OPP apply.

Exceptions:

1. Where the *ERI* analysis excludes on-site power production (OPP), the values under ENERGY RATING INDEX NOT INCLUDING OPP shall be permitted to be applied.
2. For buildings with 20 or more *dwelling units*, where *approved* by the *code official*, compliance shall be permitted using the Average Dwelling Unit *Energy Rating Index*, as calculated in accordance with ANSI/RESNET/ICC 301.

TABLE R406.5—MAXIMUM ENERGY RATING INDEX

CLIMATE ZONE	ENERGY RATING INDEX NOT INCLUDING OPP	ENERGY RATING INDEX WITH OPP
0 and 1	51	35
2	51	34
3	50	33
4	53	40
5	54	43
6	53	43
7	52	46
8	52	46

R406.6 Verification by approved agency. Verification of compliance with Section R406 as outlined in Sections R406.5 and R406.7 shall be completed by an *approved* third party. Verification of compliance with Section R406.2 shall be completed by the authority having jurisdiction or an *approved* third-party inspection agency in accordance with Section R107.4.

R406.7 Documentation. Documentation of the software used to determine the *ERI* and the parameters for the *ERI reference design* shall be in accordance with Sections R406.7.1 through R406.7.4.

R406.7.1 Compliance software tools. Software tools used for determining *ERI* shall be *approved* software rating tools as defined by ANSI/RESNET/ICC 301. Software vendors shall publish, on a publicly available website, documentation that the software tool has been validated using the Class II, Tier 1 test procedure in ANSI/ASHRAE 140.

R406.7.2 Compliance report. Compliance software tools shall generate a report that documents that the *ERI* of the *rated design* and as-built *dwelling unit* complies with Sections R406.2 through R406.5. Compliance documentation shall be created for the proposed design and shall be submitted with the application for the building permit. Confirmed compliance documents of the as-built *dwelling unit* shall be created and submitted to the *code official* for review before a certificate of occupancy is issued. Compliance reports shall include information in accordance with Sections R406.7.2.1 and R406.7.2.2.

R406.7.2.1 Proposed compliance report for permit application. Compliance reports submitted with the application for a building permit shall include the following:

1. Building street address, or other *building site* identification.
2. Declare *ERI* on title page and building plans.
3. The name of the individual performing the analysis and generating the compliance report.
4. The name and version of the compliance software tool.
5. Documentation of all inputs entered into the software used to produce the results for the *ERI reference design* and the *rated design*.
6. A certificate indicating that the proposed design has an *ERI* less than or equal to the appropriate score indicated in Table R406.5 when compared to the *ERI reference design*. The certificate shall document the building component energy specifications that are included in the calculation, including: component level insulation *R-values* or *U-factors*; assumed *duct system* and *building thermal envelope* air leakage testing results; and the type and rated efficiencies of proposed heating, cooling, mechanical *ventilation* and service water-heating equipment to be installed. Where *on-site renewable energy* systems will be installed, the certificate shall report the type and production size of the proposed system.
7. When a site-specific report is not generated, the proposed design shall be based on the worst-case orientation and configuration of the rated *dwelling unit*.

R406.7.2.2 Confirmed compliance report for a certificate of occupancy. A confirmed compliance report submitted for obtaining the certificate of occupancy shall be made site and address specific and include the following:

1. Building street address or other *building site* identification.
2. Declaration of *ERI* on title page and on building plans.
3. The name of the individual performing the analysis and generating the report.
4. The name and version of the compliance software tool.
5. Documentation of all inputs entered into the software used to produce the results for the *ERI reference design* and the as-built *dwelling unit*.
6. A final confirmed certificate indicating that the as-built *building* complies with Sections R406.2, R406.4 and R406.5. The certificate shall report the energy features that were confirmed to be in the *building*, including: component-level insulation *R-values* or *U-factors*; results from any required *duct system* and *building thermal envelope* air leakage testing; and the type and rated efficiencies of the heating, cooling, mechanical *ventilation*, and service water-heating equipment installed. Where *on-site renewable energy* systems have been installed on or in the *building*, the certificate shall report the type and production size of the installed system.

R406.7.3 Renewable energy certificate (REC) documentation. Where renewable energy power production is included in the calculation of an *ERI*, documentation shall comply with Section R404.4.

R406.7.4 Additional documentation. The *code official* shall be permitted to require the following documents:

1. Documentation of the building component characteristics of the *ERI reference design*.
2. A certification signed by the builder providing the building component characteristics of the *rated design*.
3. Documentation of the actual values used in the software calculations for the *rated design*.

R406.7.5 Specific approval. Performance analysis tools meeting the applicable subsections of Section R406 shall be *approved*. Documentation demonstrating the approval of performance analysis tools in accordance with Section R406.7.1 shall be provided.

R406.7.6 Input values. Where calculations require input values not specified by Sections R402, R403, R404 and R405, those input values shall be taken from ANSI/RESNET/ICC 301.

SECTION R407—TROPICAL CLIMATE REGION COMPLIANCE PATH

R407.1 Scope. This section establishes alternative criteria for *residential buildings* in the tropical region at elevations less than 2,400 feet (731.5 m) above sea level.

R407.2 Tropical climate region. Compliance with this section requires the following:

1. Not more than one-half of the *occupied* space is air conditioned.
2. The *occupied* space is not heated.
3. Solar, wind or other renewable energy source supplies not less than 80 percent of the energy for *service water heating*.
4. Glazing in *conditioned spaces* has a *solar heat gain coefficient* (SHGC) of less than or equal to 0.40, or has an overhang with a projection factor equal to or greater than 0.30.
5. Permanently installed lighting is in accordance with Section R404.
6. The exterior *low slope* roof surface complies with one of the options in Table R407.2 or the roof or ceiling has insulation with an *R-value* of R-15 or greater. Where attics are present, attics above the insulation are vented and attics below the insulation are unvented.

7. Roof surfaces have a slope of not less than $^1/_4$ unit vertical in 12 units horizontal (2 percent slope). The finished roof does not have water accumulation areas.
8. Operable *fenestration* provides a *ventilation* area of not less than 14 percent of the floor area in each room. Alternatively, equivalent *ventilation* is provided by a *ventilation* fan.
9. Bedrooms with *exterior walls* facing two different directions have operable *fenestration* on *exterior walls* facing two directions.
10. Interior doors to bedrooms are capable of being secured in the open position.
11. A ceiling fan or ceiling fan rough-in is provided for bedrooms and the largest space that is not used as a bedroom.

TABLE R407.2—MINIMUM LOW SLOPE ROOF REFLECTANCE AND EMITTANCE OPTIONS[a]

OPTIONS
3-year-aged solar reflectance[b] of 0.55 and 3-year-aged thermal emittance[c] of 0.75
3-year-aged solar reflectance index[d] of 64

a. The use of area-weighted averages to comply with these requirements shall be permitted. Materials lacking 3-year-aged tested values for either solar reflectance or thermal emittance shall be assigned both a 3-year-aged solar reflectance in accordance with Section R408.2.1.3.1and a 3-year-aged thermal emittance of 0.90.
b. Aged solar reflectance tested in accordance with ASTM C1549, ASTM E903, ASTM E1918 or CRRC S100.
c. Aged thermal emittance tested in accordance with ASTM C1371, ASTM E408 or CRRC S100.
d. Solar reflectance index (SRI) shall be determined in accordance with ASTM E1980 using a convection coefficient of 2.1 Btu/h × ft² × °F (12 W/m² × K). Calculation of aged SRI shall be based on aged-tested values of solar reflectance and thermal emittance.

SECTION R408—ADDITIONAL EFFICIENCY REQUIREMENTS

R408.1 Scope. This section provides additional efficiency measures and credits required to comply with Section R401.2.1.

R408.2 Additional energy efficiency credit requirements. Residential buildings shall earn not less than 10 credits from not less than two measures specified in Table R408.2. Five additional credits shall be earned for *dwelling units* with more than 5,000 square feet (465 m^2) of *living space* located above *grade plane*. To earn credit as specified in Table R408.2 for the applicable *climate zone*, each measure selected for compliance shall comply with the applicable subsections of Section R408. Each *dwelling unit* or *sleeping unit* shall comply with the selected measure to earn credit. Interpolation of credits between measures shall not be permitted.

TABLE R408.2—CREDITS FOR ADDITIONAL ENERGY EFFICIENCY

MEASURE NUMBER	MEASURE DESCRIPTION	CREDIT VALUE								
		Climate Zones 0 & 1	Climate Zone 2	Climate Zone 3	Climate Zone 4 Except Marine	Climate Zone 4 Marine	Climate Zone 5	Climate Zone 6	Climate Zone 7	Climate Zone 8
R408.2.1.1(1)	≥ 2.5% Reduction in total TC	0	0	0	1	1	1	1	1	1
R408.2.1.1(2)	≥ 5% reduction in total TC	0	1	1	2	1	2	2	2	2
R408.2.1.1(3)	> 7.5% reduction in total TC	0	1	2	2	2	2	3	3	3
R408.2.1.1(4)	> 10% reduction in total TC	1	1	2	3	3	4	4	5	5
R408.2.1.1(5)	> 15% reduction in total TC	1	2	2	4	4	5	6	7	8
R408.2.1.1(6)	> 20% reduction in total TC	2	4	4	5	6	7	8	9	11
R408.2.1.1(7)	> 30% reduction in total TC	3	6	6	8	8	11	12	13	16
R408.2.1.2(1)	*U*-factor and SHGC for vertical fenestration per Table R408.2.1.2	1	1	1	2	1	1	1	1	1
R408.2.1.3(1)	Roof solar reflectance index (roof is part of the building thermal envelope and directly above cooled, conditioned space)	1	0	0	0	0	0	0	0	0
R408.2.1.3(2)	Roof solar reflectance index (roof is above an unconditioned space that contains a duct system)	1	1	0	0	0	0	0	0	0
R408.2.1.4	Reduced air leakage	1	1	1	2	1	3	NA	NA	NA
R408.2.2(1)[b]	Ground source heat pump	14	14	14	15	10	15	17	18	21
R408.2.2(2)[b]	High Performance Cooling (Option 1)	5	4	3	2	1	1	1	1	1
R408.2.2(3)[b]	High Performance Cooling (Option 2)	6	4	3	2	1	1	1	1	1
R408.2.2(4)[b]	High Performance Gas furnace (Option 1)	0	1	2	5	3	6	7	7	9
R408.2.2(5)[b]	High Performance Gas furnace (Option 2)	0	1	2	4	3	5	6	7	8
R408.2.2(6)[b]	High Performance Gas furnace (Option 3)	0	1	1	NA	NA	NA	NA	NA	NA
R408.2.2(7)[b]	High Performance Gas furnace and cooling (Option 1)	5	5	4	NA	NA	NA	NA	NA	NA
R408.2.2(8)[b]	High Performance Gas furnace and cooling (Option 2)	6	5	5	NA	NA	NA	NA	NA	NA
R408.2.2(9)[b]	High Performance Gas furnace and heat pump (Option 1)	15	13	11	NA[e]	NA	NA	NA	NA	NA
R408.2.2(10)[b]	High Performance Heat pump with electric resistance backup (Option 1)	13	12	11	12	NA	NA	NA	NA	NA
R408.2.2(11)[b]	High Performance Gas furnace and cooling (Option 3)	NA	NA	NA	5	4	6	7	7	9
R408.2.2(12)[b]	High Performance Gas furnace and cooling (Option 4)	NA	NA	NA	6	5	7	8	8	10
R408.2.2(13)[b]	High Performance Gas furnace and heat pump (Option 2)	NA	NA	NA	12	8	11	11	12	12
R408.2.2(14)[b]	High Performance Heat pump with electric resistance backup (Option 2)	NA	NA	NA	12	8	12	13	14	16
R408.2.3(1)(a)[d]	Gas-fired storage water heaters (Option 1)	8	7	7	5	6	4	4	3	2

TABLE R408.2—CREDITS FOR ADDITIONAL ENERGY EFFICIENCY—continued

MEASURE NUMBER	MEASURE DESCRIPTION	CREDIT VALUE								
		Climate Zones 0 & 1	Climate Zone 2	Climate Zone 3	Climate Zone 4 Except Marine	Climate Zone 4 Marine	Climate Zone 5	Climate Zone 6	Climate Zone 7	Climate Zone 8
R408.2.3(1)(b)[d]	Gas-fired storage water heaters (Option 2)	9	8	8	6	7	5	4	4	3
R408.2.3(2)(a)[d]	Gas-fired instantaneous water heaters (Option 1)	10	9	9	6	7	5	5	4	3
R408.2.3(2)(b)[d]	Gas-fired instantaneous water heaters (Option 2)	11	10	9	6	7	6	5	4	3
R408.2.3(3)[d]	Electric water heaters (Option 1)	10	9	9	7	6	4	3	3	2
R408.2.3(4)[d]	Electric water heaters (Option 2)	8	8	8	6	5	4	3	3	2
R408.2.3(5)(a)[d]	Electric water heaters (Option 3)	7	8	8	6	7	5	4	3	3
R408.2.3(5)(b)[d]	Electric water heaters (Option 4)	8	9	10	7	8	5	5	4	3
R408.2.3(6)[d]	Electric water heaters (Option 5)	10	9	9	7	6	4	3	3	2
R408.2.3(7)(a)[d]	Solar hot water heating system (Option 1)	13	13	13	9	8	5	4	4	3
R408.2.3(7)(b)[d]	Solar hot water heating system (Option 2)	10	9	9	6	7	6	5	4	3
R408.2.3(8)[c]	Compact hot water distribution	2	2	2	2	2	2	2	2	2
R408.2.4(1)[c]	Ductless or hydronic thermal distribution	3	4	5	7	8	10	10	10	14
R408.2.4(2)[c]	100% of duct systems in conditioned space	2	3	4	6	7	9	9	9	13
R408.2.4(3)[c]	≥ 80% of ductwork inside conditioned space	2	3	3	5	6	7	7	7	9
R408.2.4(4)[c]	Reduced total duct system leakage	1	1	1	1	1	1	2	2	2
R408.2.5(1)[c]	ERV or HRV installed	0	0	0	0	1	3	2	2	2
R408.2.5(2)[c]	≤ 2.0 ACH50 with ERV or HRV installed	0	0	0	4	4	8	5	5	5
R408.2.5(3)[c]	≤ 2.0 ACH50 with a balanced ventilation system	0	0	0	0	0	0	4	4	4
R408.2.5(4)[c]	≤ 1.5 ACH50 with ERV or HRV installed	0	0	0	6	5	10	9	9	9
R408.2.5(5)[c]	≤ 1.0 ACH50 with ERV or HRV installed	0	0	1	7	6	12	12	12	12
R408.2.6[a]	Energy efficient appliances	1	1	1	1	1	1	0	0	0
R408.2.7	On-site renewable energy measures	17	16	17	11	11	9	8	7	4
R408.2.8[c]	Demand responsive thermostat	1	1	1	1	1	1	1	1	1
R408.2.10	Whole-home lighting control	1	1	1	0	0	0	0	0	0
R408.2.11	Higher efficacy lighting	0	0	0	0	0	0	0	0	0

NA = Not Applicable.

a. Where the measure is selected, each dwelling unit, sleeping unit and common area where the measure is applicable must have the measure installed.

b. Where multiple heating or cooling systems are installed, credits shall be determined using a weighted average of the square footage served by each system.

c. Where the measure is selected, each dwelling unit and sleeping unit must comply with the measure.

d. Where the measure is selected, each dwelling unit shall be served by a water heater meeting the applicable requirements. Where multiple service water heating systems are installed, credits shall be determined using a weighted average of the square footage served by each system.

e. Eleven credits are available for Climate Zone 4 where the following measure is used: gas furnace and heat pump (Option 3): greater than or equal to 95% AFUE fuel gas furnace and 7.8 HSPF2, 15.2 SEER2 and 10.0 EER2 air source heat pump.

R408.2.1 Enhanced building thermal envelope options. The *building thermal envelope* shall comply with one or more of the following:

1. Section R408.2.1.1 or R408.2.1.2. Credit shall be permitted from only one measure.
2. Section R408.2.1.3.
3. Section R408.2.1.4.

R408.2.1.1 Enhanced building thermal envelope performance. The total *building thermal envelope* thermal conductance (TC) shall be calculated for the proposed building in accordance with Section R402.1.5 and shall be reduced by not less than the percentage indicated in Table R408.2 in comparison to the reference building.

R408.2.1.2 Improved fenestration. The area weighted average *U-factor* and SHGC of all vertical *fenestration* shall be equal to or less than the values specified in Table R408.2.1.2.

TABLE R408.2.1.2—IMPROVED FENESTRATION

CLIMATE ZONE	*U*-FACTOR	SHGC
0	0.32	0.23
1	0.32	0.23
2	0.30	0.23
3	0.28	0.23
4 except Marine 4	0.25	0.40
5 and Marine 4	0.25	NR
6	0.25	NR
7 and 8	0.25	NR
NR = No Requirement.		

R408.2.1.3 Roof solar reflectance index. *Low slope* roofs in Climate Zones 0 through 2 shall earn credit for Table R408.2 measure numbers R408.2.1.3(1) and R408.2.1.3(2) where the 3-year-aged solar reflectance index (SRI) is greater than or equal to 75. To earn credit, not less than 95 percent of the roof area shall comply. The combined area of the following portions of roof shall not be greater than 5 percent of the roof area:

1. Portions that include or are covered by the following:
   - 1.1. Photovoltaic systems or components.
   - 1.2. Solar air or water-heating systems or components.
   - 1.3. Vegetative roofs or landscaped roofs.
   - 1.4. Above-roof decks or walkways.
   - 1.5. Skylights.
   - 1.6. HVAC systems and components, and other opaque objects mounted above the roof.
2. Portions shaded during the peak sun angle on the summer solstice by permanent features of the *building*, permanent features of adjacent buildings or natural objects.
3. Portions that are ballasted with a minimum stone ballast of 17 pounds per square foot (psf) (74 kg/m^2) or 23 psf (117 kg/m^2) pavers.

The 3-year-aged SRI shall be determined in accordance with ASTM E1980 using a convection coefficient of 2.1 Btu/h × ft^2 × °F (12 W/m^2 × K). Calculation of aged SRI shall be based on 3-year-aged solar reflectance values tested in accordance with ASTM C1549, ASTM E903, ASTM E1918 or CRRC S100 and 3-year-aged thermal *emittance* values tested in accordance with ASTM C1371, ASTM E408 or CRRC S100.

R408.2.1.3.1 Aged solar reflectance. Where a tested 3-year-aged solar reflectance value is not available, an assigned value shall be determined in accordance with Equation 4-3.

Equation 4-3 $R_{aged} = [0.2 + 0.7(R_{initial} - 0.2)]$

where:

R_{aged} = The aged solar reflectance.

$R_{initial}$ = The initial solar reflectance determined in accordance with ASTM C1549, ASTM E903, ASTM E1918 or CRRC S100.

R408.2.1.4 Reduced air leakage. The *building* shall have a measured air leakage rate not less than 2.0 ACH50 and not greater than 2.5 ACH50 or the *dwelling units* in the *building* shall have an average measured air leakage rate not greater than 0.24 cubic feet per minute per square foot [1.2 L/(s × m^2)].

R408.2.2 More efficient HVAC equipment performance options. Heating and cooling *equipment* shall meet one of the following measures as applicable for the *climate zone* where heating and cooling efficiencies are represented by Annual Fuel Utilization Efficiency (AFUE), Coefficient of Performance (COP), Energy Efficiency Ratio (EER and EER2), Heating Season Performance Factor (HSPF2) and Seasonal Energy Efficiency Ratio (SEER2). Where multiple heating or cooling systems are installed serving different *zones*, credits shall be earned based on the weighted average of square footage of the *zone* served by the system.

HVAC options applicable to all *climate zones*:

1. Ground source heat pump: Greater than or equal to 16.1 EER and 3.1 COP ground source heat pump.
2. Cooling (Option 1): Greater than or equal to 15.2 SEER2 and 12.0 EER2 air conditioner.
3. Cooling (Option 2): Greater than or equal to 16.0 SEER2 and 12.0 EER2 air conditioner.
4. Gas furnace (Option 1): Greater than or equal to 97 percent AFUE *fuel gas* furnace.
5. Gas furnace (Option 2): Greater than or equal to 95 percent AFUE *fuel gas* furnace.

HVAC options applicable to Climate Zones 0, 1, 2 and 3:

6. Gas furnace (Option 3): Greater than or equal to 90 percent AFUE *fuel gas* furnace.
7. Gas furnace and cooling (Option 1): Greater than or equal to 90 percent AFUE *fuel gas* furnace and 15.2 SEER2 and 10.0 EER2 air conditioner.
8. Gas furnace and cooling (Option 2): Greater than or equal to 95 percent AFUE *fuel gas* furnace and 16.0 SEER2 and 10.0 EER2 air conditioner.
9. Gas furnace and heat pump (Option 1): Greater than or equal to 90 percent AFUE *fuel gas* furnace and 7.8 HSPF2, 15.2 SEER2 and 10.0 EER2 air source heat pump.
10. Heat pump (Option 1): Greater than or equal to 7.8 HSPF2, 15.2 SEER2, and 11.7 EER2 air source heat pump.

HVAC options applicable to Climate Zones 4, 5, 6, 7 and 8:

11. Gas furnace and cooling (Option 3): Greater than or equal to 95 percent AFUE *fuel gas* furnace and 15.2 SEER2 and 12.0 EER2 air conditioner.
12. Gas furnace and cooling (Option 4): Greater than or equal to 97 percent AFUE *fuel gas* furnace and 16.0 SEER2 and 12.0 EER2 air conditioner.
13. Gas furnace and heat pump (Option 2): Greater than or equal to 95 percent AFUE *fuel gas* furnace and 8.1 HSPF2 and 15.2 SEER2 air source heat pump capable of meeting a capacity ratio ≥ 70 percent of heating capacity at 5°F (-15°C) versus rated heating capacity at 47°F (8.3°C).
14. Heat pump (Option 2): Greater than or equal to 8.1 HSPF2 and 15.2 SEER2 air source heat pump capable of meeting a capacity ratio ≥ 70 percent of heating capacity at 5°F (-15°C) versus rated heating capacity at 47°F (8.3°C).

R408.2.2.1 More efficient HVAC equipment for Climate Zone 4. For Climate Zone 4, the following HVAC options shall also apply:

1. Gas furnace and heat pump (Option 3): Greater than or equal to 95 percent AFUE *fuel gas* furnace and 7.8 HSPF2, 15.2 SEER2 and 10.0 EER2 air source heat pump.
2. Heat pump (Option 1): Greater than or equal to 7.8 HSPF2, 15.2 SEER2 and 11.7 EER2 air source heat pump.

R408.2.3 Reduced energy use in service water-heating options. For measure numbers R408.2.3(1) through R408.2.3(7), the installed hot water system shall meet one of the Uniform Energy Factors (UEF) or Solar Uniform Energy Factors (SUEF) in Table R408.2.3. For measure number R408.2.3(8), the hot water distribution system shall comply with Section R408.2.3.1.

TABLE R408.2.3—SERVICE WATER HEATING EFFICIENCIES

MEASURE NUMBER	WATER HEATER	SIZE AND DRAW PATTERN	TYPE	EFFICIENCY
R408.2.3(1)(a)	Gas-fired storage water heaters (Option 1)	All storage volumes, all draw patterns	—	UEF ≥ 0.81
R408.2.3(1)(b)	Gas-fired storage water heaters (Option 2)	≤ 55 gallons, high	—	UEF ≥ 0.86
		> 55 gallons, medium or high	—	UEF ≥ 0.86
		Rated input capacity > 75,000 Btu/h	—	UEF ≥ 0.86 or E_t ≥ 94%
R408.2.3(2)(a)	Gas-fired instantaneous water heaters (Option 1)	All storage volumes, medium or high	—	UEF ≥ 0.92
R408.2.3(2)(b)	Gas-fired instantaneous water heaters (Option 2)	All storage volumes, medium or high	—	UEF ≥ 0.95
R408.2.3(3)	Electric water heaters (Option 1)	All storage volumes, low, medium, or high	Integrated HPWH	UEF ≥ 3.30
R408.2.3(4)	Electric water heaters (Option 2)	All storage volumes, low, medium, or high	Integrated HPWH, 120 volt/15 amp circuit	UEF ≥ 2.20

TABLE R408.2.3—SERVICE WATER HEATING EFFICIENCIES—continued				
MEASURE NUMBER	WATER HEATER	SIZE AND DRAW PATTERN	TYPE	EFFICIENCY
R408.2.3(5)(a)	Electric water heaters (Option 3)	All storage volumes, low, medium, or high	Split-system HPWH	UEF ≥ 2.20
R408.2.3(5)(b)	Electric water heaters (Option 4)	All storage volumes, low, medium, or high	Split-system HPWH	UEF ≥ 3.75
R408.2.3(6)	Electric water heaters (Option 5)	Rated input capacity > 12 kW	—	COP ≥ 3.00
R408.2.3(7)(a)	Solar water heaters (Option 1)	All storage volumes, all draw patterns	Electric backup	SUEF ≥ 3.00
R408.2.3(7)(b)	Solar water heaters (Option 2)	All storage volumes, all draw patterns	Gas backup	SUEF ≥ 1.80

For SI: 1 British thermal unit per hour = 0.2931 W.
UEF = Uniform Energy Factor, E_t = Thermal Efficiency, COP = Coefficient of Performance.

R408.2.3.1 Compact hot water distribution system option. The pipe shall store not more than 16 ounces (0.47 L) of water between the nearest source of heated water and the termination of the fixture supply pipe when calculated using Section R408.2.3.1.1. Where the source of heated water is a circulation loop, the loop shall be primed with a *demand recirculation water system* that complies with Section R403.5.1.1.1. There shall be a dedicated return line for the loop that begins after the branch to the last fixture on the supply portion of the loop and runs back to the water heater.

TABLE R408.2.3.1—INTERNAL VOLUME OF VARIOUS WATER DISTRIBUTION TUBING									
OUNCES OF WATER PER FOOT OF TUBE									
Nominal Size (inches)	Copper Type M	Copper Type L	Copper Type K	CPVC CTS SDR 11	CPVC SCH 40	CPVC SCH 80	PE-RT SDR 9	Composite ASTM F1281	PEX CTS SDR 9
$^3/_8$	1.06	0.97	0.84	N/A	1.17	—	0.64	0.63	0.64
$^1/_2$	1.69	1.55	1.45	1.25	1.89	1.46	1.18	1.31	1.18
$^3/_4$	3.43	3.22	2.90	2.67	3.38	2.74	2.35	3.39	2.35
1	5.81	5.49	5.17	4.43	5.53	4.57	3.91	5.56	3.91
$1^1/_4$	8.70	8.36	8.09	6.61	9.66	8.24	5.81	8.49	5.81
$1^1/_2$	12.18	11.83	11.45	9.22	13.20	11.38	8.09	13.88	8.09
2	21.08	20.58	20.04	15.79	21.88	19.11	13.86	21.48	13.86

For SI: 1 foot = 304.8 mm, 1 inch = 25.4 mm, 1 liquid ounce = 0.030 L, 1 ounce per square foot = 305.15 g/m^2.
N/A = Not Available.

R408.2.3.1.1 Water volume determination. The water volume in the piping between a source of heated water and the termination of a fixture supply shall be calculated in accordance with this section. Water heaters, circulating water systems and heat trace temperature maintenance systems shall be considered to be sources of heated water. The volume shall be the sum of the internal volumes of pipe, fittings, valves, meters and manifolds between the nearest source of heated water and the termination of the fixture supply pipe. The volume in the piping shall be determined from Table R408.2.3.1. The volume contained within fixture shutoff valves, within flexible water supply connectors to a fixture fitting and within a fixture fitting shall not be included in the water volume determination. Where heated water is supplied by a recirculating system or heat-traced piping, the volume shall include the portion of the fitting on the branch pipe that supplies water to the fixture.

R408.2.4 More efficient thermal distribution system options. The thermal distribution system shall comply with one of the following:

1. The ductless thermal distribution system or hydronic thermal distribution system is located completely on the conditioned side of the *building thermal envelope*.
2. The *space conditioning equipment* is located inside *conditioned space*. In addition, 100 percent of the *ductwork* is located completely inside *conditioned space* as defined by Section R403.3.4, Items 1 and 2.
3. The *space conditioning equipment* is located inside *conditioned space* and not less than 80 percent of *ductwork* is located completely inside *conditioned space* as defined by Section R403.3.4, Items 1 and 2. In addition, not more than 20 percent of *ductwork* is contained within building assemblies separating unconditioned from *conditioned space* as defined by Section R403.3.4, Item 3.
4. Where *ductwork* is located outside conditioned space, the total leakage of the *duct system* measured in accordance with Section R403.3.7 is one of the following:
    4.1. Where the *space conditioning equipment* is installed at the time of testing, total leakage is not greater than 2.0 cubic feet per minute (0.94 L/s) per 100 square feet (9.29 m^2) of *conditioned floor area*.

4.2. Where the *space conditioning equipment* is not installed at the time of testing, total leakage is not greater than 1.75 cubic feet per minute (0.83 L/s) per 100 square feet (9.29 m^2) of *conditioned floor area*.

R408.2.5 Improved air sealing and efficient ventilation system options. The measured air leakage rate and *ventilation* system shall meet one of the following:

1. Either an Energy Recovery Ventilator (ERV) or a Heat Recovery Ventilator (HRV) installed.
2. Less than or equal to 2.0 ACH50, with either an ERV or HRV installed.
3. Less than or equal to 2.0 ACH50, with a *balanced ventilation system*.
4. Less than or equal to 1.5 ACH50, with either an ERV or HRV installed.
5. Less than or equal to 1.0 ACH50, with either an ERV or HRV installed.

In addition, for measures requiring either an ERV or HRV, HRV and ERV Sensible Recovery Efficiency (SRE) shall be not less than 75 percent at 32°F (0°C) at the lowest *listed* net airflow. ERV Latent Recovery/Moisture Transfer (LRMT) shall be not less than 50 percent at the lowest *listed* net airflow. In Climate Zone 8, recirculation shall not be used as a defrost strategy.

R408.2.6 Energy efficient appliances. Each appliance of a type listed in Table R408.2.6 installed in a residential *building* shall comply with the efficiency requirements specified in that table. Each appliance specified in Table R408.2.6 shall be installed. A clothes washer shall be installed at each location plumbed for a clothes washer.

Exception: In *dwelling units* of Group R-2 occupancies where a dishwasher is not installed in each unit, not fewer than two appliance types complying with Table R408.2.6 shall be installed.

TABLE R408.2.6—MINIMUM EFFICIENCY REQUIREMENTS: APPLIANCES

APPLIANCE TYPES	EFFICIENCY IMPROVEMENT	TEST PROCEDURE
Refrigerator	Maximum Annual Energy Consumption (AEC), not greater than 620 kWh/yr	10 CFR 430, Subpart B, Appendix A
Dishwasher	Maximum Annual Energy Consumption (AEC), not greater than 240 kWh/yr	10 CFR 430, Subpart B, Appendix C1
Clothes washer and clothes dryer	Clothes washer located within dwelling units: Maximum Annual Energy Consumption (AEC), not greater than 130 kWh/yr, and Integrated Modified Energy Factor (IMEF) > 1.84 cu ft/kWh/cycle Clothes washer not located within dwelling units and where dwelling units are not provided with rough-in plumbing for washers: Modified Energy Factor (MEF) > 2.0 cu ft/kWh/cycle	10 CFR 430, Subpart B, Appendices D1, D2 and J2
For SI: 1 cubic foot per kilowatt hour per cycle = 0.028 m^3/kWh/cycle.		

R408.2.7 Renewable energy. *Renewable energy resources* shall be permanently installed and have the rated capacity to produce not less than 1.0 watt of *on-site renewable energy* per square foot of *conditioned floor area*. To qualify for this option, *renewable energy certificate* (REC) documentation shall meet the requirements of Section R404.4.

R408.2.8 Demand response. The *thermostat* controlling the primary heating or cooling system of each *dwelling unit* shall be provided with a *demand responsive control* capable of communicating with the Virtual End Node (VEN) using a wired or wireless bi-directional communication pathway that provides the occupant the ability to voluntarily participate in utility demand response programs, where available. The *thermostat* shall be capable of executing the following actions in response to a *demand response signal*:

1. Automatically increasing the zone operating cooling set point by the following values: 1°F (0.5°C), 2°F (1°C), 3°F (1.5°C) and 4°F (2°C).
2. Automatically decreasing the zone operating heating set point by the following values: 1°F (0.5°C), 2°F (1°C), 3°F (1.5°C) and 4°F (2°C).

Thermostats controlling single-stage HVAC systems shall comply with Section R408.2.8.1. Thermostats controlling variable capacity systems shall comply with Section R408.2.8.2. Thermostats controlling multistage HVAC systems shall comply with either Section R408.2.8.1 or R408.2.8.2. Where a *demand response signal* is not available, the *thermostat* shall be capable of performing all other functions.

R408.2.8.1 Single-stage HVAC system controls. Thermostats controlling single-stage HVAC systems shall be provided with a *demand responsive control* that complies with one of the following:

1. Certified OpenADR 2.0a VEN, as specified under Clause 11, Conformance.
2. Certified OpenADR 2.0b VEN, as specified under Clause 11, Conformance.
3. Certified by the manufacturer as being capable of responding to a *demand response signal* from a certified OpenADR 2.0b VEN by automatically implementing the control functions requested by the VEN for the equipment it controls.
4. IEC 62746-10-1.

5. The communication protocol required by a controlling entity, such as a utility or service provider, to participate in an automated demand response program.
6. The physical configuration and communication protocol of CTA 2045-A or CTA-2045-B.

R408.2.8.2 Variable-capacity and two-stage HVAC system controls. Thermostats controlling variable-capacity and two-stage HVAC systems shall be provided with a *demand responsive control* that complies with the communication and performance requirements of AHRI 1380.

R408.2.9 Opaque walls. For *buildings* in Climate Zones 4 and 5, the maximum *U-factor* of 0.060 shall be permitted to be used for wood-framed walls for compliance with Table R402.1.2 where complying with one or more of the following:

1. Primary space heating is provided by a heat pump that meets one of the efficiencies in Section R408.2.2.
2. All installed water heaters are heat pumps that meet one of the efficiencies in Section R408.2.3.
3. In addition to the number of credits required by Section R408.2, three additional credits are achieved.
4. *Renewable energy resources* are installed to meet the requirements of Section R408.2.7.

R408.2.10 Whole-home lighting control. The *dwelling unit* shall have a *manual* control by the main entrance that turns off all the permanently installed interior lighting or a lighting control system that has the capability to turn off all permanently installed interior lighting from remote locations.

Exceptions:

1. Up to 5 percent of the total lighting power may remain uncontrolled.
2. Spaces where lighting is controlled by a count-down timer or *occupant sensor control*.

R408.2.11 Higher efficacy lighting. All spaces shall be provided with hardwired lighting with a lamp efficacy of 90 lumens per watt (lm/W) or a luminaire efficacy of 55 lm/W.

Exceptions:

1. Closets.
2. Other storage spaces.

CHAPTER

5 [RE] EXISTING BUILDINGS

User notes:

About this chapter: *Many buildings are renovated or altered in numerous ways that could affect the energy use of the building as a whole. Chapter 5 requires the application of certain parts of Chapter 4 in order to maintain, if not improve, the conservation of energy by the renovated or altered building.*

SECTION R501—GENERAL

R501.1 Scope. The provisions of this chapter shall control the *alteration*, *repair*, *addition* and change of occupancy of existing *buildings* and structures.

R501.1.1 General. Except as specified in this chapter, this code shall not be used to require the removal, *alteration* or abandonment of, nor prevent the continued use and maintenance of, an existing *building* or *building* system lawfully in existence at the time of adoption of this code. Unaltered portions of the *existing building* or *building* supply system shall not be required to comply with this code.

R501.2 Compliance. *Additions, alterations, repairs* or changes of occupancy to, or relocation of, an existing *building*, building system or portion thereof shall comply with Section R502, R503, R504 or R505, respectively, in this code and the provisions for alterations, repairs, additions and changes of occupancy or relocation, respectively, in the *International Building Code*, *International Existing Building Code*, *International Fire Code*, *International Fuel Gas Code*, *International Mechanical Code*, *International Plumbing Code*, *International Property Maintenance Code*, *International Private Sewage Disposal Code*, *International Residential Code* and NFPA 70, as applicable. Changes where unconditioned space is changed to *conditioned space* shall comply with Section R501.6.

R501.3 Maintenance. *Buildings* and structures, and parts thereof, shall be maintained in a safe and sanitary condition. Devices and systems that are required by this code shall be maintained in conformance to the code edition under which installed. The owner or the owner's authorized agent shall be responsible for the maintenance of *buildings* and structures. The requirements of this chapter shall not provide the basis for removal or abrogation of energy conservation, fire protection and safety systems and devices in existing structures.

R501.4 New and replacement materials. Except as otherwise required or permitted by this code, materials permitted by the applicable code for new construction shall be used. Like materials shall be permitted for *repairs*, provided that hazards to life, health or property are not created. Hazardous materials shall not be used where the code for new construction would not allow their use in *buildings* of similar occupancy, purpose and location.

R501.5 Historic buildings. Provisions of this code relating to the construction, *repair, alteration*, restoration and movement of structures, and *change of occupancy* shall not be mandatory for *historic buildings* provided that a report has been submitted to the *code official* and signed by the owner, a *registered design professional*, or a representative of the State Historic Preservation Office or the historic preservation authority having jurisdiction, demonstrating that compliance with that provision would threaten, degrade or destroy the historic form, fabric or function of the *building*.

R501.6 Change in space conditioning. Any unconditioned or low-energy space that is altered to become *conditioned space* shall be required to be brought into full compliance with Section R502.

Exception: Where the simulated performance option in Section R405 is used to comply with this section, the annual *energy cost* of the *proposed design* is permitted to be 110 percent of the annual *energy cost* otherwise allowed by Section R405.2.

SECTION R502—ADDITIONS

R502.1 General. *Additions* to an *existing building*, building system or portion thereof shall conform to the provisions of this code as those provisions relate to new construction. *Additions* shall not create an unsafe or hazardous condition or overload *existing building* systems.

R502.2 Prescriptive compliance. *Additions* shall comply with Sections R502.2.1 through R502.2.5.

R502.2.1 Building thermal envelope. New *building thermal envelope* assemblies that are part of the *addition* shall comply with Sections R402.1, R402.2, R402.4.1 through R402.4.5, and R402.5.

Exception: New *building thermal envelope* assemblies are exempt from the requirements of Section R402.5.1.2.

R502.2.2 Heating and cooling systems. HVAC *ductwork* newly installed as part of an *addition* shall comply with Section R403.

Exception: Where *ductwork* from an existing heating and cooling system is extended to an *addition*, Sections R403.3.7 and R403.3.8 shall not be required.

R502.2.3 Service hot water systems. New service hot water systems that are part of the *addition* shall comply with Section R403.5.

R502.2.4 Lighting. New lighting systems that are part of the *addition* shall comply with Section R404.1.

R502.2.5 Additional energy efficiency credit requirements for additions. *Additions* shall comply with sufficient measures from Table R408.2 to achieve not less than five credits. *Alterations* to the *existing building* that are not part of the *addition* but are permitted with an *addition* shall be permitted to be used to achieve this requirement.

Exceptions:

1. *Additions* that increase the *building*'s total *conditioned floor area* by less than 25 percent.
2. *Additions* that do not include the addition or replacement of equipment covered in Section R403.5 or R403.7.
3. *Additions* that do not increase *conditioned space*.
4. Where the *addition* alone or the *existing building* and *addition* together comply with Section R405 or R406.

SECTION R503—ALTERATIONS

R503.1 General. *Alterations* to any *building* or structure shall comply with the requirements of the code for new construction, without requiring the unaltered portions of the *existing building* or building system to comply with this code. *Alterations* shall be such that the *existing building* or structure is not less conforming to the provisions of this code than the *existing building* or structure was prior to the *alteration*.

Alterations shall not create an unsafe or hazardous condition or overload *existing building* systems. *Alterations* shall be such that the *existing building* or structure does not use more energy than the *existing building* or structure prior to the *alteration*. *Alterations* to *existing buildings* shall comply with Sections R503.1.1 through R503.1.5.

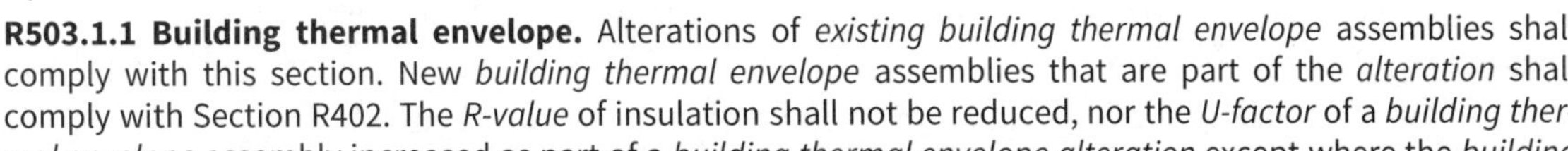
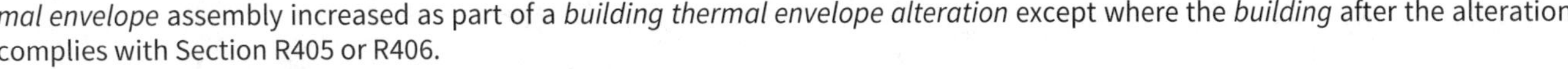

R503.1.1 Building thermal envelope. Alterations of *existing building thermal envelope* assemblies shall comply with this section. New *building thermal envelope* assemblies that are part of the *alteration* shall comply with Section R402. The *R-value* of insulation shall not be reduced, nor the *U-factor* of a *building thermal envelope* assembly increased as part of a *building thermal envelope alteration* except where the *building* after the alteration complies with Section R405 or R406.

Exception: The following alterations shall not be required to comply with the requirements for new construction provided that the energy use of the building is not increased:

1. Storm windows installed over existing *fenestration*.
2. *Roof recover*.
3. Surface-applied window film installed on existing single pane fenestration assemblies to reduce solar heat gain provided that the code does not require the glazing or fenestration assembly to be replaced.
4. *Roof replacement* where *roof assembly* insulation is integral to or located below the structural roof deck.

R503.1.1.1 Fenestration alterations. Where new *fenestration* area is added to an *existing building*, the new *fenestration* shall comply with Section R402.4. Where some or all of an existing *fenestration* unit is replaced with a new *fenestration* product, including sash and glazing, the replacement *fenestration* unit shall meet the applicable requirements for *U-factor* and *solar heat gain coefficient* (SHGC) as specified in Table R402.1.3. Where more than one replacement *fenestration* unit is to be installed, an area-weighted average of the *U-factor*, SHGC or both of all replacement *fenestration* units shall be an alternative that can be used to show compliance.

R503.1.1.2 Roof, ceiling and attic alterations. Roof, ceiling and attic insulation shall comply with Section R402.1. Alternatively, where limiting conditions prevent compliance with Section R402.1, an *approved* design that minimizes deviation from Section R402.1 shall be provided for the following *alterations*:

1. An *alteration* to roof/ceiling construction other than *reroofing* where existing insulation located below the roof deck or an attic floor above *conditioned space* does not comply with Table R402.1.3.
2. *Roof replacements* or a roof *alteration* that includes removing and replacing the roof covering where the *roof assembly* includes insulation entirely above the roof deck. Where limiting conditions require use of an *approved* design to minimize deviation from Section R402.1 for a Group R-2 *building*, a registered design professional or other *approved source* shall provide *construction documents* that identify the limiting conditions and the means to address them.
3. Conversion of an unconditioned attic space into *conditioned space*.
4. Replacement of ceiling finishes exposing cavities or surfaces of the roof/ceiling.

R503.1.1.3 Above-grade wall alterations. *Above-grade wall alterations* shall comply with the following as applicable:

1. Where wall cavities are exposed, the exposed cavities shall be filled with insulation complying with Section R303.1.4. New cavities created shall be insulated in accordance with Section R402.1 or an *approved* design that minimizes deviation from Section R402.1. An interior vapor retarder shall be provided where required in accordance with Section R702.7 of the *International Residential Code* or Section 1404.3 of the *International Building Code*, as applicable.
2. Where exterior wall coverings and *fenestration* are added or replaced for the full extent of any exterior facade of one or more elevations of the *building*, *continuous insulation* shall be provided where required in accordance with Section R402.1 or the wall insulation shall be in accordance with an *approved* design that minimizes deviation from Section R402.1. Where specified, the *continuous insulation* requirement also shall comply with Section R702.7 of the *International Residential Code*. Replacement exterior wall coverings shall comply with the water-resistance requirements of

Section R703.1.1 of the *International Residential Code* or Section 1402.2 of the *International Building Code*, as applicable, and manufacturers' instructions.

3. Where new interior finishes or exterior wall coverings are applied to the full extent of any exterior wall assembly of mass construction, insulation shall be provided in accordance with Section R402.1 or an *approved* design that minimizes deviation from Section R402.1.

R503.1.1.4 Floor alterations. Where cavities in a floor or floor overhang are exposed and the floor or floor overhang is part of the *building thermal envelope*, the floor or floor overhang shall comply with Section R402.1 or an *approved* design.

R503.1.1.5 Below-grade wall alterations. Where an unconditioned below-grade space is changed to *conditioned space*, the *building thermal envelope* walls enclosing such space shall be insulated in accordance with Section R402.1. Where the below-grade space is *conditioned space* and where *building thermal envelope* walls enclosing such space are altered, they shall be insulated in accordance with Section R402.1.

R503.1.1.6 Air barrier. Altered *building thermal envelope* assemblies shall be provided with an *air barrier* in accordance with Section R402.5. Such an *air barrier* need not be continuous with unaltered portions of the *building thermal envelope*. Testing requirements of Section R402.5.1.2 shall not be required.

R503.1.2 Heating and cooling systems. New heating and cooling systems and *ductwork* that are part of the *alteration* shall comply with Section R403 and this section. *Alterations* to existing heating and cooling systems and *ductwork* shall comply with this section.

Exception: Where *ductwork* from an existing heating and cooling system is extended.

R503.1.2.1 Ductwork. HVAC *ductwork* newly installed as part of an *alteration* shall comply with Section R403.

Exception: Where *ductwork* from an existing heating and cooling system is extended.

R503.1.2.2 System sizing. New heating and cooling equipment that is part of an *alteration* shall be sized in accordance with Section R403.7 based on the *existing building* features as modified by the *alteration*.

Exception: Where it has been demonstrated to the *code official* that compliance with this section would result in heating or cooling equipment that is incompatible with the remaining portions of the existing heating or cooling system.

R503.1.2.3 Duct system leakage. Where an *alteration* includes any of the following, *duct systems* shall be tested in accordance with Section R403.3.7 and shall have a total leakage less than or equal to 12.0 cubic feet per minute (339.9 L/min) per 100 square feet (9.29 m^2) of *conditioned floor area*:

1. Twenty-five percent or more of the registers that are part of the *duct system* are relocated.
2. Twenty-five percent or more of the total length of all *ductwork* in the *duct system* is relocated.
3. The total length of all *ductwork* in the *duct system* is increased by 25 percent or more.

Exception: *Duct systems* located entirely inside a *conditioned space* in accordance with Section R403.3.4.

R503.1.2.4 Controls. New heating and cooling equipment that is part of the *alteration* shall comply with Sections R403.1 and R403.2.

R503.1.3 Service hot water systems. New service hot water systems that are part of the *alteration* shall comply with Section R403.5.

R503.1.4 Lighting. New lighting systems that are part of the *alteration* shall comply with Section R404.1.

Exception: *Alterations* that replace less than 10 percent of the luminaires in a space, provided that such *alterations* do not increase the installed interior lighting power.

R503.1.5 Additional efficiency credit requirements for substantial improvements. *Substantial improvements* shall comply with sufficient measures from Table R408.2 to achieve not less than three credits.

Exceptions:

1. *Alterations* that are permitted with an *addition* complying with Section R502.2.5.
2. *Alterations* that comply with Section R405 or R406.
3. *Substantial improvements* that do not include the *addition* or replacement of equipment covered in either Section R403.5 or R403.7.

SECTION R504—REPAIRS

R504.1 General. *Buildings*, structures and parts thereof shall be repaired in compliance with Section R501.3 and this section. Work on nondamaged components necessary for the required *repair* of damaged components shall be considered to be part of the *repair* and shall not be subject to the requirements for *alterations* in this chapter. Routine maintenance required by Section R501.3, ordinary *repairs* exempt from *permit*, and abatement of wear due to normal service conditions shall not be subject to the requirements for *repairs* in this section.

R504.2 Application. For the purposes of this code, the following shall be considered to be *repairs*:

1. Glass-only replacements in an existing sash and frame.
2. Roof *repairs*.
3. *Repairs* where only the bulb, ballast or both within the existing luminaires in a space are replaced provided that the replacement does not increase the installed interior lighting power.

SECTION R505—CHANGE OF OCCUPANCY OR USE

R505.1 General. Any space that is converted to a *dwelling unit* or portion thereof from another use or occupancy shall comply with this chapter.

Exception: Where the *simulated building performance* option in Section R405 is used to comply with this section, the annual *energy cost* of the *proposed design* is permitted to be 110 percent of the annual *energy cost* allowed by Section R405.2.

R505.1.1 Unconditioned space. Any unconditioned or low-energy space that is altered to become a *conditioned space* shall comply with Section R501.6.

CHAPTER 6 [RE] REFERENCED STANDARDS

User notes:

About this chapter: *This code contains numerous references to standards promulgated by other organizations that are used to provide requirements for materials and methods of construction. Chapter 6 contains a comprehensive list of all standards that are referenced in this code. These standards, in essence, are part of this code to the extent of the reference to the standard.*

This chapter lists the standards that are referenced in various sections of this document. The standards are listed herein by the promulgating agency of the standard, the standard identification, the effective date and title, and the section or sections of this document that reference the standard. The application of the referenced standards shall be as specified in Section R108.

ACCA *Air Conditioning Contractors of America, 1330 Braddock Place, Suite 350, Alexandria, VA 22314*

ANSI/ACCA 1 Manual D—2023: Residential Duct Systems
R403.3.1

ANSI/ACCA 2 Manual J—2016: Residential Load Calculation
R403.7

ANSI/ACCA 3 Manual S—2023: Residential Equipment Selection
R403.7

ANSI/ACCA 5 QI—2010: HVAC Quality Installation Specification
R408.2.4

AHRI *Air-Conditioning, Heating, and Refrigeration Institute, 2311 Wilson Blvd, Suite 400, Arlington, VA 22201*

1380—2019: Demand Response through Variable Capacity HVAC Systems in Residential and Small Commercial Applications
R408.2.8.2

AISI *American Iron and Steel Institute, 25 Massachusetts Avenue, NW, Suite 800, Washington, DC 20001*

AISI S250—22: North American Standard for Thermal Transmittance of Building Envelopes with Cold-Formed Steel Framing, with Supplement 1, Dated 2022
R402.2.7

ANSI *American National Standards Institute, 25 West 43rd Street, 4th Floor, New York, NY 10036*

ANSI Z21.20—2005 (R2016): Automatic Gas Ignition Systems and Components
R403.13, R404.1.5

ANSI/AMCA 210-ANSI/ASHRAE 51—23: Laboratory Methods of Testing Fans for Aerodynamic Performance Rating
Table R403.6.2

ANSI/ASHRAE 140—2017 (2020): Standard Method of Test for the Evaluation of Building Energy Analysis Computer Programs
R405.5.2, R406.7.1

ANSI/CTA 2045-B—February 2021: Modular Communications Interface for Energy Management
R408.2.8.1

CSA/ANSI Z21.88—19/CSA 2.33—19: Vented Gas Fireplace Heaters
R403.13.1

Z21.50-19/CSA 2.22—2019: Vented Decorative Gas Appliances
R403.13.1

ASHRAE *ASHRAE, 180 Technology Parkway NW, Peachtree Corners, GA 30092*

ANSI/ASHRAE/IES 90.1—2022: Energy Standard for Sites and Buildings Except Low-Rise Residential Buildings
R402.1.5, R402.2.10.2, R402.2.11.2, R405.2

ASHRAE 193—2010(RA 2014): Method of Test for Determining the Airtightness of HVAC Equipment
R403.3.6.1

ASHRAE—2001: 2001 ASHRAE Handbook of Fundamentals
Table R405.4.2(1)

ASHRAE—2017: 2017 ASHRAE Handbook of Fundamentals
R402.1.5, R403.3.1, R405.4.2, Table R405.4.2(1)

ASTM *ASTM International, 100 Barr Harbor Drive, P.O. Box C700, West Conshohocken, PA 19428-2959*

C1313/C1313M—13(2019): Standard Specification for Sheet Radiant Barriers for Building Construction Applications
R303.2.2

C1363—19: Standard Test Method for Thermal Performance of Building Materials and Envelope Assemblies by Means of a Hot Box Apparatus
R303.1.4.1

C1371—15(2022): Standard Test Method for Determination of Emittance of Materials Near Room Temperature Using Portable Emissometers
Table R407.2, R408.2.1.3

C1549—16(2022): Standard Test Method for Determination of Solar Reflectance Near Ambient Temperature Using a Portable Solar Reflectometer
Table R407.2, R408.2.1.3, R408.2.1.3.1

C1743—19: Standard Practice for Installation and Use of Radiant Barrier Systems (RBS) in Residential Building Construction
R303.2.2, R402.3

E283/E283M—19: Standard Test Method for Determining the Rate of Air Leakage Through Exterior Windows, Skylights, Curtain Walls and Doors Under Specified Pressure Differences Across the Specimen
R402.5.4

E408—13(2019): Standard Test Methods for Total Normal Emittance of Surfaces Using Inspection-Meter Techniques
Table R407.2, R408.2.1.3

E779—19: Standard Test Method for Determining Air Leakage Rate by Fan Pressurization
R402.5.1.2, R402.5.1.3

E903—20: Standard Test Method for Solar Absorptance, Reflectance and Transmittance of Materials Using Integrating Spheres (Withdrawn 2005)
Table R407.2, R408.2.1.3, R408.2.1.3.1

E1554/E1554M—13(2018): Standard Test Methods for Determining Air Leakage of Air Distribution Systems by Fan Pressurization
R403.3.7, R403.3.8, Table R405.4.2(1)

E1827—22: Standard Test Methods for Determining Airtightness of Building Using an Orifice Blower Door
R402.5.1.2

E1918—21: Standard Test Method for Measuring Solar Reflectance of Horizontal and Low-Sloped Surfaces in the Field
Table R407.2, R408.2.1.3, R408.2.1.3.1

E1980—11(2019): Standard Practice for Calculating Solar Reflectance Index of Horizontal and Low-Sloped Opaque Surfaces
Table R407.2, R408.2.1.3

E2178—21a: Standard Test Method for Determining Air Leakage Rate and Calculation of Air Permanence of Building Materials
R303.1.5

E3158—18: Standard Test Method for Measuring the Air Leakage Rate of a Large or Multizone Building
R402.5.1.2

F1281—2017(2021)e1: Standard Specification for Electrofusion Type Polyethylene Fittings for Outside Diameter Controlled Polyethylene and Crosslinked Polyethylene Pipe and Tubing
Table R408.2.3.1

CRRC *Cool Roof Rating Council, 2435 North Lombard Street, Portland, OR 97217*

ANSI/CRRC S100—2021: Standard Test Methods for Determining Radiative Properties of Materials
Table R407.2, R408.2.1.3, R408.2.1.3.1

CSA *CSA Group, 8501 East Pleasant Valley Road, Cleveland, OH 44131-5516*

AAMA/WDMA/CSA 101/I.S.2/A440—22: North American Fenestration Standard/Specification for Windows, Doors, and Skylights
R402.5.3

CAN/CSA C439—18: Laboratory Methods of Test for Rating the Performance of Heat/Energy-Recovery Ventilators
Table R403.6.2

CSA B55.1—20: Test Method for Measuring Efficiency and Pressure Loss of Drain Water Heat Recovery Units
R403.5.3

CSA B55.2—20: Drain Water Heat Recovery Units
R403.5.3

CSA P.4.1—21: Testing Method for Measuring Fireplace Efficiency
R403.13.1

CTA
Consumer Technology Association, Technology & Standards Department; 1919 S Eads Street, Arlington, VA 22202

ANSI/CTA-2045-A—2018: Modular Communications Interface for Energy Management
R408.2.8.1

ANSI/CTA-2045-B—2018: Modular Communications Interface for Energy Management
R408.2.8.1

DASMA
Door & Access Systems Manufacturers Association, 1300 Sumner Avenue, Cleveland, OH 44115-2851

ANSI/DASMA 105—2020: Test Method for Thermal Transmittance and Air Infiltration of Garage Doors and Rolling Doors
R303.1.3

DOE
US Department of Energy, 1000 Independence Avenue SW, Washington, DC 20585

10 CFR, Part 430—2021: Energy Conservation Program for Consumer Products: Energy and Water Conservation Standards and Their Compliance Dates
Table R403.6.2, R404.1, Table R405.4.2(1), Table R408.2.6

FGIA
Fenestration & Glazing Industry Alliance (formerly American Architectural Manufacturers Association), 1900 E Golf Road, Suite 250, Schaumburg, IL 60173-4268

AAMA/WDMA/CSA 101/I.S.2/A440—22: North American Fenestration Standard/Specification for Windows, Doors, and Skylights
R402.5.3

ICC
International Code Council, Inc., 200 Massachusetts Avenue, NW, Suite 250, Washington, DC 20001

ANSI/APSP/ICC 14—2019: American National Standard for Portable Electric Spa Energy Efficiency
R403.11

ANSI/PHTA/ICC 15a—2021: American National Standard for Residential Swimming Pool and Spa Energy Efficiency
R403.12

ANSI/RESNET/ICC 301—2022: Standard for the Calculation and Labeling of the Energy Performance of Dwelling and Sleeping Units Using an Energy Rating Index—includes Addendum A, Approved July 28, 2022, and Addendum B, Approved October 12, 2022
R405.5.3, R406.4, R406.5, R406.7.1, R406.7.6,

ANSI/RESNET/ICC 380—2022: Standard for Testing Airtightness of Building, Dwelling Unit and Sleeping Unit Enclosures; Airtightness of Heating and Cooling Air Distribution Systems, and Airflow of Mechanical Ventilation Systems
R402.5.1.2, R403.3.7, R403.3.8, R403.6.3, Table R405.4.2(1)

IBC—24: International Building Code®
R201.3, R202, R303.1.1, R303.1.2, R303.2, R402.1.1, R402.2.11.1, R404.1.2, R501.2, R501.4, R503.1.1.3,

ICC 400—22: Standard on the Design and Construction of Log Structures
R402.1, Table R402.5.1.1

ICC 500—20: ICC/NSSA Standard for the Design and Construction of Storm Shelters
R402.6

IEBC—24: International Existing Building Code®
R501.2

IECC—06: 2006 International Energy Conservation Code®
R202

IFC—24: International Fire Code®
R201.3, R501.2

IFGC—24: International Fuel Gas Code®
R201.3, R501.2

IMC—24: International Mechanical Code®
R201.3, R403.3.5, R403.3.6, R403.6, R501.2

IPC—24: International Plumbing Code®
R201.3, R501.2

IPMC—24: International Property Maintenance Code®
R501.2

IPSDC—24: International Private Sewage Disposal Code®
R501.2

IRC—24: International Residential Code®
R201.3, R303.1.1, R303.1.2, R303.1.6, R303.2, R402.1.1, Table R402.1.2, Table R402.1.3, R402.1.6, R402.2.10.1, R402.2.11.1, R402.2.11.1, Table R402.5.1.1, R403.3.5, R403.3.6, R403.6, R501.2, R501.4, R503.1.1.3

IEC *IEC Regional Centre for North America, 446 Main Street, 16th Floor, Worcester, MA 01608*

IEC 62746-10-1—2018: Systems interface between customer energy management system and the power management system – Part 10-1: Open automated demand response
R408.2.8.1

IEEE *Institute of Electrical and Electronics Engineers, Inc., 3 Park Avenue, 17th Floor, New York, NY 10016-5997*

515.1—2012: IEEE Standard for the Testing, Design, Installation and Maintenance of Electrical Resistance Trace Heating for Commercial Applications
R403.5.1.2

NEMA *National Electrical Manufacturers Association, 1300 17th Street N #900, Arlington, VA 22209*

OS 4—2016: Requirements for Air-Sealed Boxes for Electrical and Communication Applications
R402.5.5

NFPA *National Fire Protection Association, 1 Batterymarch Park, Quincy, MA 02169-7471*

70—23: National Electrical Code
R501.2

NFRC *National Fenestration Rating Council, Inc., 6305 Ivy Lane, Suite 140, Greenbelt, MD 20770*

100—2023: Procedure for Determining Fenestration Products *U*-Factors
R303.1.3

200—2023: Procedure for Determining Fenestration Product Solar Heat Gain Coefficient and Visible Transmittance at Normal Incidence
R303.1.3

400—2023: Procedure for Determining Fenestration Product Air Leakage
R402.5.3

OpenADR *OpenADR Alliance, 111 Deerwood Road, Suite 200, San Ramon, CA 94583*

OpenADR 2.0a and 2.0b—2019: Profile Specification Distributed Energy Resources
R408.2.8.1

PHTA *Pool & Tub Alliance (formerly the APSP), 1650 King Street, Suite 602, Alexandria, VA 22314*

ANSI/APSP/ICC 14—2019: American National Standard for Portable Electric Spa Energy Efficiency
R403.11

ANSI/PHTA/ICC 15—2021: American National Standard for Residential Swimming Pool and Spa Energy Efficiency
R403.12

RESNET *Residential Energy Services Network, Inc., P.O. Box 4561, Oceanside, CA 92052-4561*

ANSI/RESNET/ICC 301—2022: Standard for the Calculation and Labeling of the Energy Performance of Dwelling and Sleeping Units using an Energy Rating Index—includes Addendum A, Approved July 28, 2022, and Addendum B, Approved October 12, 2022
R405.5.3, R406.4, R406.5, R406.7.1, R406.7.6

ANSI/RESNET/ICC 380—2022: Standard for Testing Airtightness of Building, Dwelling Unit and Sleeping Unit Enclosures; Airtightness of Heating and Cooling Air Distribution Systems, and Airflow of Mechanical Ventilation Systems
R402.5.1.2, R403.3.7, R403.3.8, R403.6.3, Table R405.4.2(1)

UL

UL LLC, 333 Pfingsten Road, Northbrook, IL 60062

127—2011: Standard for Factory-Built Fireplaces—with Revisions through February 2020
R402.5.2

515—2015: Electrical Resistance Trace Heating for Commercial Applications
R403.5.1.2

US-FTC

United States-Federal Trade Commission, 600 Pennsylvania Avenue NW, Washington, DC 20580

CFR Title 16 (2015): *R*-Value Rule
R303.1.4

WDMA

Window & Door Manufacturers Association, 2001 K Street NW, 3rd Floor North, Washington, DC 20006

AAMA/WDMA/CSA 101/I.S.2/A440—22: North American Fenestration Standard/Specification for windows, doors, and skylights
R402.5.3

APPENDIX RA BOARD OF APPEALS—RESIDENTIAL

The provisions contained in this appendix are not mandatory unless specifically referenced in the adopting ordinance.

User notes:

About this appendix: *Appendix RA provides criteria for board of appeals members. Also provided are procedures by which the board of appeals should conduct its business.*

SECTION RA101—GENERAL

RA101.1 Scope. A board of appeals shall be established within the jurisdiction for the purpose of hearing applications for modification of the requirements of this code pursuant to the provisions of Section R109. The board shall be established and operated in accordance with this section, and shall be authorized to hear evidence from appellants and the *code official* pertaining to the application and intent of this code for the purpose of issuing orders pursuant to these provisions.

RA101.2 Application for appeal. Any person shall have the right to appeal a decision of the *code official* to the board. An application for appeal shall be based on a claim that the intent of this code or the rules legally adopted hereunder have been incorrectly interpreted, the provisions of this code do not fully apply or an equally good or better form of construction is proposed. The application shall be filed on a form obtained from the *code official* within 20 days after the notice was served.

RA101.2.1 Limitation of authority. The board shall not have authority to waive requirements of this code or interpret the administration of this code.

RA101.2.2 Stays of enforcement. Appeals of notice and orders, other than Imminent Danger notices, shall stay the enforcement of the notice and order until the appeal is heard by the board.

RA101.3 Membership of board. The board shall consist of five voting members appointed by the chief appointing authority of the jurisdiction. Each member shall serve for **[INSERT NUMBER OF YEARS]** years or until a successor has been appointed. The board members' terms shall be staggered at intervals, so as to provide continuity. The *code official* shall be an ex officio member of said board but shall not vote on any matter before the board.

RA101.3.1 Qualifications. The board shall consist of five individuals, who are qualified by experience and training to pass on matters pertaining to building construction and are not employees of the jurisdiction.

RA101.3.2 Alternate members. The chief appointing authority is authorized to appoint two alternate members who shall be called by the board chairperson to hear appeals during the absence or disqualification of a member. Alternate members shall possess the qualifications required for board membership, and shall be appointed for the same term or until a successor has been appointed.

RA101.3.3 Vacancies. Vacancies shall be filled for an unexpired term in the same manner in which original appointments are required to be made.

RA101.3.4 Chairperson. The board shall annually select one of its members to serve as chairperson.

RA101.3.5 Secretary. The chief appointing authority shall designate a qualified clerk to serve as secretary to the board. The secretary shall file a detailed record of all proceedings, which shall set forth the reasons for the board's decision, the vote of each member, the absence of a member and any failure of a member to vote.

RA101.3.6 Conflict of interest. A member with any personal, professional or financial interest in a matter before the board shall declare such interest and refrain from participating in discussions, deliberations and voting on such matters.

RA101.3.7 Compensation of members. Compensation of members shall be determined by law.

RA101.3.8 Removal from the board. A member shall be removed from the board prior to the end of their term only for cause. Any member with continued absence from regular meeting of the board may be removed at the discretion of the chief appointing authority.

RA101.4 Rules and procedures. The board shall establish policies and procedures necessary to carry out its duties consistent with the provisions of this code and applicable state law. The procedures shall not require compliance with strict rules of evidence, but shall mandate that only relevant information be presented.

RA101.5 Notice of meeting. The board shall meet upon notice from the chairperson, within 10 days of the filing of an appeal or at stated periodic intervals.

RA101.5.1 Open hearing. All hearings before the board shall be open to the public. The appellant, the appellant's representative, the *code official* and any person whose interests are affected shall be given an opportunity to be heard.

RA101.5.2 Quorum. Three members of the board shall constitute a quorum.

RA101.5.3 Postponed hearing. When five members are not present to hear an appeal, either the appellant or the appellant's representative shall have the right to request a postponement of the hearing.

RA101.6 Legal counsel. The jurisdiction shall furnish legal counsel to the board to provide members with general legal advice concerning matters before them for consideration. Members shall be represented by legal counsel at the jurisdiction's expense in all matters arising from service within the scope of their duties.

RA101.7 Board decision. The board shall only modify or reverse the decision of the *code official* by a concurring vote of three or more members.

RA101.7.1 Resolution. The decision of the board shall be by resolution. Every decision shall be promptly filed in writing in the office of the *code official* within 3 days and shall be open to the public for inspection. A certified copy shall be furnished to the appellant or the appellant's representative and to the *code official*.

RA101.7.2 Administration. The *code official* shall take immediate action in accordance with the decision of the board.

RA101.8 Court review. Any person, whether or not a previous party of the appeal, shall have the right to apply to the appropriate court for a writ of certiorari to correct errors of law. Application for review shall be made in the manner and time required by law following the filing of the decision in the office of the chief administrative officer.

APPENDIX RB

SOLAR-READY PROVISIONS—DETACHED ONE- AND TWO-FAMILY DWELLINGS AND TOWNHOUSES

The provisions contained in this appendix are not mandatory unless specifically referenced in the adopting ordinance.

User notes:

About this appendix: *Harnessing the heat or radiation from the sun's rays is a method to reduce the energy consumption of a building. Although Appendix RB does not require solar systems to be installed for a building, it does require the space(s) for installing such systems, providing pathways for connections and requiring adequate structural capacity of roof systems to support the systems.*

SECTION RB101—SCOPE

RB101.1 General. These provisions shall be applicable for new construction where solar-ready provisions are required.

SECTION RB102—GENERAL DEFINITION

SOLAR-READY ZONE. A section or sections of the roof or *building* overhang designated and reserved for the future installation of a solar photovoltaic or solar thermal system.

SECTION RB103—SOLAR-READY ZONE

RB103.1 General. New detached one- and two-family dwellings, and townhouses with not less than 600 square feet (55.74 m^2) of roof area oriented between 110 degrees and 270 degrees of true north shall comply with Sections RB103.2 through RB103.8.

Exceptions:

1. New residential *buildings* with a permanently installed *on-site renewable energy* system.
2. A *building* where all areas of the roof that would otherwise meet the requirements of Section RB103 are in full or partial shade for more than 70 percent of daylight hours annually.

RB103.2 Construction document requirements for solar-ready zone. *Construction documents* shall indicate the *solar-ready zone*.

RB103.3 Solar-ready zone area. The total *solar-ready zone* area shall be not less than 300 square feet (27.87 m^2) exclusive of mandatory access or setback areas as required by the *International Fire Code*. New townhouses three stories or less in height above *grade plane* and with a total floor area less than or equal to 2,000 square feet (185.8 m^2) per dwelling shall have a *solar-ready zone* area of not less than 150 square feet (13.94 m^2). The *solar-ready zone* shall be composed of areas not less than 5 feet (1524 mm) in width and not less than 80 square feet (7.44 m^2) exclusive of access or setback areas as required by the *International Fire Code*.

RB103.4 Obstructions. *Solar-ready zones* shall be free from obstructions, including but not limited to vents, chimneys, and roof-mounted equipment.

RB103.5 Shading. The *solar-ready zone* shall be set back from any existing or new permanently affixed object on the *building* or site that is located south, east or west of the solar zone a distance not less than two times the object's height above the nearest point on the roof surface. Such objects include, but are not limited to, taller portions of the *building* itself, parapets, chimneys, antennas, signage, rooftop equipment, trees and roof plantings.

RB103.6 Capped roof penetration sleeve. A capped roof penetration sleeve shall be provided adjacent to a *solar-ready zone* located on a roof slope of not greater than 1 unit vertical in 12 units horizontal (8-percent slope). The capped roof penetration sleeve shall be sized to accommodate the future photovoltaic system conduit, but shall have an inside diameter of not less than 1$^1/_4$ inches (32 mm).

RB103.7 Roof load documentation. The structural design loads for roof dead load and roof live load shall be clearly indicated on the *construction documents*.

RB103.8 Interconnection pathway. *Construction documents* shall indicate pathways for routing of conduit or plumbing from the *solar-ready zone* to the electrical service panel or service hot water system.

RB103.9 Electrical service reserved space. The main electrical service panel shall have a reserved space to allow installation of a dual pole circuit breaker for future solar electric installation and shall be labeled "For Future Solar Electric." The reserved space shall be positioned at the opposite (load) end from the input feeder location or main circuit location.

RB103.10 Construction documentation certificate. A permanent certificate, indicating the *solar-ready zone* and other requirements of this section, shall be posted near the electrical distribution panel, water heater or other conspicuous location by the builder or registered design professional.

ZERO NET ENERGY RESIDENTIAL BUILDING PROVISIONS

The provisions contained in this appendix are not mandatory unless specifically referenced in the adopting ordinance.

User notes:

About this appendix: *This appendix provides requirements for residential buildings intended to result in zero net energy consumption over the course of a year. Where adopted by ordinance as a requirement, Sections RC101.1, RC101.2, RC103.2, RC103.4 and RC103.5 are intended to replace Sections R401.1, R401.2, R406.2, R406.4 and R406.5, respectively. Where adopted by ordinance as a requirement, Sections R401.3 (Certificate), R406.1 (Scope), R406.3 (Building thermal envelope), R406.6 (Verification by approved agency) and R406.7 (Documentation) are not replaced.*

SECTION RC101—GENERAL

RC101.1 Scope. This appendix applies to new *residential buildings*.

RC101.2 Application. *Residential buildings* shall comply with Section R406.

Exception: *Additions*, *alterations*, *repairs* and changes of occupancy to *existing buildings* complying with Chapter 5.

RC101.3 Certificate. (No change, same as Section R401.3.)

SECTION RC102—GENERAL DEFINITIONS

COMMUNITY RENEWABLE ENERGY FACILITY (CREF). A facility that produces energy from *renewable energy resources* and that is qualified as a community energy facility under applicable jurisdictional statutes and rules.

FINANCIAL RENEWABLE ENERGY POWER PURCHASE AGREEMENT (FPPA). A financial arrangement between a renewable electricity generator and a purchaser wherein the purchaser pays or guarantees a price to the generator for the project's renewable generation. Also known as a financial power purchase agreement and virtual power purchase agreement.

PHYSICAL RENEWABLE ENERGY POWER PURCHASE AGREEMENT (PPPA). A contract for the purchase of renewable electricity from a specific renewable electricity generator by a purchaser of renewable electricity.

SECTION RC103—ZERO NET ENERGY RESIDENTIAL BUILDINGS

RC103.1 Scope. (No change, same as Section R406.1.)

RC103.2 ERI compliance. Compliance based on the *ERI* requires that the *rated design* meets one of the following:

1. The requirements of the sections indicated within Table R406.2 and Sections R406.3 through R406.7, or
2. The requirements of ASHRAE/IES Standard 90.2, including:
   2.1. The *ERI* requirements of ASHRAE/IES Standard 90.2 Table 6-1 without the use of on-site power production (OPP).
   2.2. The requirements of Sections R402.5.1.1, R402.5.1.2, and R406.3.
   2.3. The maximum *ERI* including adjusted OPP of Table RC103.5 determined in accordance with Section RC103.4.

RC103.3 Building thermal envelope. (No change, same as Section R406.3.)

RC103.4 Energy Rating Index. The *Energy Rating Index* (ERI) not including *renewable energy resources* shall be determined in accordance with ANSI/RESNET/ICC 301. The *ERI* including *renewable energy resources* shall be determined in accordance with ANSI/RESNET/ICC 301, except where electrical energy is provided from a community renewable energy facility (CREF) or contracted from a physical or financial renewable energy power purchase agreement that meets requirements of Section RC103.4.1, on-site power production (OPP) shall be adjusted in accordance with Equation RC-1.

Equation RC-1 $\text{Adjusted OPP} = OPP_{kWh} + CREF_{kWh} + PPPA_{kWh} + FPPA_{kWh}$

where:

OPP_{kWh} = Annual electrical energy from *on-site renewable energy*, in units of kilowatt-hours (kWh).

$CREF_{kWh}$ = Annual electrical energy from a community renewable energy facility (CREF), in units of kilowatt-hours (kWh).

$PPPA_{kWh}$ = Where not included as OPP, the annual electrical energy contracted from a physical renewable energy power purchase agreement, in units of kilowatt-hours (kWh).

$FPPA_{kWh}$ = Where not included as OPP, the annual electrical energy contracted from a financial renewable energy power purchase agreement (FPPA), in units of kilowatt-hours (kWh).

RC103.4.1 Renewable energy contract. The renewable energy shall be delivered or credited to the *building site* under an energy contract with a duration of not less than 15 years. The contract shall be structured to survive a partial or full transfer of ownership of the building property.

RC103.5 ERI-based compliance. Compliance based on an *ERI* analysis requires that the *rated design* and confirmed built dwelling be shown to have an *ERI* less than or equal to both values indicated in Table RC103.5 when compared to the *ERI reference design*.

TABLE RC103.5—MAXIMUM ENERGY RATING INDEX

CLIMATE ZONE	ENERGY RATING INDEX NOT INCLUDING RENEWABLE ENERGY	ENERGY RATING INDEX INCLUDING ADJUSTED OPP
0	42	0
1	42	0
2	42	0
3	42	0
4	42	0
5	42	0
6	42	0
7	42	0
8	42	0

RC103.6 Verification by approved agency. (No change, same as Section R406.6.)

RC103.7 Documentation. (No change, same as Section R406.7.)

SECTION RC104—REFERENCED STANDARDS

RC104.1 General. See Table RC104.1 for standards that are referenced in various sections of this appendix. Standards are listed by the standard identification with the effective date, standard title, and the section or sections of this appendix that reference the standard.

TABLE RC104.1—REFERENCED STANDARDS

STANDARD ACRONYM	STANDARD NAME	SECTIONS HEREIN REFERENCED
ASHRAE/IES 90.2—2018	*Energy-Efficient Design of Low-Rise Residential Buildings, including approved addenda* [Addenda A (approved Jan 2021), B (June 2021) and D (February 2022)]	RC103.2

ELECTRIC ENERGY STORAGE PROVISIONS

The provisions contained in this appendix are not mandatory unless specifically referenced in the adopting ordinance.

User notes:

About this appendix: *This voluntary appendix provides requirements for electric energy storage readiness provisions.*

SECTION RD101—SCOPE

RD101.1 General. These provisions shall be applicable for new construction where solar-ready measures or an on-site solar PV system is required.

SECTION RD102—GENERAL DEFINITION

ENERGY STORAGE SYSTEM (ESS). One or more devices, assembled together, capable of storing energy in order to supply electrical energy at a future time.

SECTION RD103—ELECTRICAL ENERGY STORAGE

RD103.1 Electrical energy storage. One- and two-family *dwellings*, townhouse units and Group R-3 occupancies shall comply with either Section RD103.2 or RD103.3. *Buildings* with Group R-2 and R-4 occupancies shall comply with Section RD103.4.

RD103.2 Electrical energy storage energy capacity. Each *building* shall have an *ESS* with a rated energy capacity of not less than 5 kWh with not fewer than four *ESS*-supplied branch circuits.

RD103.3 Electrical energy storage system readiness. Each *building* shall be energy-storage ready in accordance with Sections RD103.3.1 through RD103.3.4.

RD103.3.1 Energy storage system space. Interior or exterior space with dimensions and locations in accordance with Section R330 of the *International Residential Code* and Section 110.26 of NFPA 70 shall be reserved to allow for the future installation of an *ESS*.

RD103.3.2 System isolation equipment space. Space shall be reserved to allow for the future installation of a transfer switch within 3 feet (305 mm) of the main panelboard. Raceways shall be installed between the panelboard and the transfer switch location to allow the connection of an *ESS*.

RD103.3.3 Panelboard with backed-up load circuits. A dedicated raceway shall be provided from the main service to a panelboard that supplies the branch circuits served by the *ESS*. All branch circuits are permitted to be supplied by the main service panel prior to the installation of an *ESS*. The trade size of the raceway shall be not less than 1 inch (25 mm). The panelboard that supplies the branch circuits shall be labeled, "Subpanel reserved for future battery energy storage system to supply essential loads."

RD103.3.4 Branch circuits served by the ESS. Not fewer than four branch circuits shall be identified and have their source of supply collocated at a single panelboard supplied by the *ESS*. The following end uses shall be served by the branch circuits:

1. A refrigerator.
2. One lighting circuit near the primary egress.
3. A sleeping room receptacle outlet.

RD103.4 Electrical energy storage system. *Buildings* with Group R-2 and R-4 occupancies shall comply with Appendix CJ.

APPENDIX RE

ELECTRIC VEHICLE CHARGING INFRASTRUCTURE

The provisions contained in this appendix are not mandatory unless specifically referenced in the adopting ordinance.

User notes:

About this appendix: *This appendix provides requirements for electric vehicle charging infrastructure for adopting jurisdictions.*

SECTION RE101—ELECTRIC VEHICLE POWER TRANSFER

RE101.1 Definitions.

AUTOMOBILE PARKING SPACE. A space within a *building* or private or public parking lot, exclusive of driveways, ramps, columns, office and work areas, for the parking of an automobile.

ELECTRIC VEHICLE (EV). An automotive-type vehicle for on-road use, such as passenger automobiles, buses, trucks, vans, neighborhood electric vehicles and electric motorcycles, primarily powered by an electric motor that draws current from a building electrical service, *electric vehicle supply equipment (EVSE)*, a rechargeable storage battery, a fuel cell, a photovoltaic array or another source of electric current.

ELECTRIC VEHICLE CAPABLE SPACE (EV CAPABLE SPACE). A designated *automobile parking space* that is provided with electrical infrastructure such as, but not limited to, raceways, cables, electrical capacity, a panelboard or other electrical distribution equipment space necessary for the future installation of an *EVSE*.

ELECTRIC VEHICLE READY SPACE (EV READY SPACE). An *automobile parking space* that is provided with a branch circuit and an outlet, junction box or receptacle that will support an installed *EVSE*.

ELECTRIC VEHICLE SUPPLY EQUIPMENT (EVSE). Equipment for plug-in power transfer, including ungrounded, grounded and equipment grounding conductors; electric vehicle connectors; attached plugs; any personal protection system; and all other fittings, devices, power outlets or apparatus installed specifically for the purpose of transferring energy between the premises wiring and the *electric vehicle*.

ELECTRIC VEHICLE SUPPLY EQUIPMENT INSTALLED SPACE (EVSE SPACE). An *automobile parking space* that is provided with a dedicated *EVSE* connection.

RE101.2 Electric vehicle power transfer infrastructure. New residential *automobile parking spaces* for residential *buildings* shall be provided with *electric vehicle power* transfer infrastructure in accordance with Sections RE101.2.1 through RE101.2.5.

RE101.2.1 Quantity. New one- and two-family dwellings and townhouses with a designated attached or detached garage or other on-site private parking provided adjacent to the *dwelling unit* shall be provided with one *EV capable*, *EV ready* or *EVSE* space per *dwelling unit*. R-2 occupancies or allocated parking for R-2 occupancies in mixed-use *buildings* shall be provided with an *EV capable space*, *EV ready space* or *EVSE* space for 40 percent of the *dwelling units* or *automobile parking spaces*, whichever is less.

Exceptions:

1. Where the local electric distribution entity certifies in writing that it is not able to provide 100 percent of the necessary distribution capacity within 2 years after the estimated certificate of occupancy date, the required *EV* charging infrastructure shall be reduced based on the available existing electric distribution capacity.
2. Where substantiation is *approved* that meeting the requirements of Section RE101.2.5 will alter the local utility infrastructure design requirements on the utility side of the meter so as to increase the utility side cost to the builder or developer by more than $450 per *dwelling unit*.

RE101.2.2 EV capable spaces. Each *EV capable space* used to meet the requirements of Section RE101.2.1 shall comply with all of the following:

1. A continuous raceway or cable assembly shall be installed between a suitable panelboard or other on-site electrical distribution equipment and an enclosure or outlet located within 6 feet (1828 mm) of the *EV capable space*.
2. The installed raceway or cable assembly shall be sized and rated to supply a minimum circuit capacity in accordance with Section RE101.2.5.
3. The electrical distribution equipment to which the raceway or cable assembly connects shall have sufficient dedicated space and spare electrical capacity for a two-pole circuit breaker or set of fuses.
4. The electrical enclosure or outlet and the electrical distribution equipment directory shall be marked: "For future electric vehicle supply equipment (EVSE)."

RE101.2.3 EV ready spaces. Each branch circuit serving *EV ready spaces* shall comply with all of the following:

1. Termination at an outlet or enclosure, located within 6 feet (1828 mm) of each *EV ready space* it serves and marked "For electric vehicle supply equipment (EVSE)."
2. Service by an electrical distribution system and circuit capacity in accordance with Section RE101.2.5.

3. Designation on the panelboard or other electrical distribution equipment directory as "For electric vehicle supply equipment (EVSE)."

RE101.2.4 EVSE spaces. An installed *EVSE* with multiple output connections shall be permitted to serve multiple *EVSE spaces*. Each *EVSE* serving either a single *EVSE space* or multiple *EVSE spaces* shall comply with the following:

1. Be served by an electrical distribution system in accordance with Section RE101.2.5.
2. Have a nameplate charging capacity of not less than 6.2 kVA (or 30A at 208/240V) per *EVSE space* served. Where an *EVSE* serves three or more *EVSE spaces* and is controlled by an energy management system in accordance with Section RE101.2.5, the nameplate charging capacity shall be not less than 2.1 kVA per *EVSE space* served.
3. Be located within 6 feet (1828 mm) of each *EVSE space* it serves.
4. Be installed in accordance with NFPA 70 and be *listed* and *labeled* in accordance with UL 2202 or UL 2594.

RE101.2.5 Electrical distribution system capacity. The branch circuits and electrical distribution system serving each *EV capable space*, *EV ready space* and *EVSE space* used to comply with Section RE101.2.1 shall comply with one of the following:

1. Sized for a calculated *EV* charging load of not less than 6.2 kVA per *EVSE*, *EV ready* or *EV capable space*. Where a circuit is shared or managed, it shall be in accordance with NFPA 70.
2. The capacity of the electrical distribution system and each branch circuit serving multiple *EVSE spaces*, *EV ready spaces* or *EV capable spaces* designed to be controlled by an energy management system in accordance with NFPA 70 shall be sized for a calculated *EV* charging load of not less than 2.1 kVA per space. Where an energy management system is used to control *EV* charging loads for the purposes of this section, it shall not be configured to turn off electrical power to *EVSE* or *EV ready spaces* used to comply with Section RE101.2.1.

SECTION RE102—REFERENCED STANDARDS

RE102.1 General. See Table RE102.1 for standards that are referenced in various sections of this appendix. Standards are listed by the standard identification with the effective date, standard title, and the section or sections of this appendix that reference the standard.

TABLE RE102.1—REFERENCED STANDARDS

STANDARD ACRONYM	STANDARD NAME	SECTIONS HEREIN REFERENCED
UL 2202—2009	*Electric Vehicle (EV) Charging System Equipment*—with revisions through February 2018	RE101.2.4
UL 2594—2016	*Standard for Electric Vehicle Supply Equipment*	RE101.2.4

APPENDIX RF ALTERNATIVE BUILDING THERMAL ENVELOPE INSULATION R-VALUE OPTIONS

The provisions contained in this appendix are not mandatory unless specifically referenced in the adopting ordinance.

User notes:

About this appendix: *The purpose of this appendix is to provide expanded R-value options for determining compliance with the U-factor criteria prescribed in Section R402.1.2. It also supplements the limited selection of common insulation conditions addressed in the R-value approach of Table R402.1.3.*

SECTION RF101—GENERAL

RF101.1 General. This appendix shall be used as a basis to determine alternative building assembly and insulation component *R-value* solutions that comply with the maximum *U*-factors and *F*-factors in Table R402.1.2. Alternative building assembly insulation solutions determined in accordance with this appendix also shall comply with the requirements of Section R702.7 of the *International Residential Code*.

SECTION RF102—ABOVE-GRADE WALL ASSEMBLIES

RF102.1 Wood-framed walls. Wood-framed *above-grade wall* assemblies shall comply with both the *cavity insulation* and *continuous insulation R-values* and framing conditions specified by Table RF102.1 where the tabulated *U*-factors are less than or equal to those needed for compliance with Section R402.1.2. For assemblies not addressed by the conditions of Table RF102.1, *U*-factors shall be determined by using accepted engineering practice or by testing in accordance with ASTM C1363 and shall be subject to approval by the *code official* in accordance with Section R104.1. Use of a lesser framing fraction than the indicated maximums in Table RF102.1 shall require wall framing layout details on *approved construction documents* for each *above-grade wall* elevation and shall be inspected for compliance.

TABLE RF102.1—ASSEMBLY *U*-FACTORS FOR WOOD-FRAMED WALLS[a, b, c, d, e, f]

WOOD STUD SIZE AND SPACING	CAVITY INSULATION INSTALLED *R*-VALUE	CONTINUOUS INSULATION *R*-VALUE																		
		0	1	2	3	4	5	6	7	8	9	10	11	12	13	14	15	20	25	30
2 × 4 (12 inches o.c)	0	0.324	0.239	0.190	0.158	0.136	0.119	0.106	0.096	0.087	0.080	0.074	0.069	0.064	0.060	0.057	0.054	0.042	0.035	0.030
	11	0.094	0.085	0.078	0.072	0.067	0.062	0.059	0.055	0.052	0.050	0.047	0.045	0.043	0.041	0.040	0.038	0.032	0.027	0.024
	12	0.090	0.082	0.075	0.069	0.064	0.060	0.057	0.054	0.051	0.048	0.046	0.044	0.042	0.040	0.039	0.037	0.031	0.027	0.024
	13	0.087	0.079	0.072	0.067	0.063	0.059	0.055	0.052	0.049	0.047	0.045	0.043	0.041	0.039	0.038	0.036	0.031	0.027	0.023
	14	0.084	0.076	0.070	0.065	0.061	0.057	0.054	0.051	0.048	0.046	0.044	0.042	0.040	0.038	0.037	0.036	0.030	0.026	0.023
	15	0.082	0.074	0.068	0.063	0.059	0.055	0.052	0.049	0.047	0.045	0.043	0.041	0.039	0.038	0.036	0.035	0.030	0.026	0.023
	16	0.079	0.072	0.066	0.062	0.058	0.054	0.051	0.048	0.046	0.044	0.042	0.040	0.038	0.037	0.036	0.034	0.029	0.025	0.022
	17	0.077	0.070	0.065	0.060	0.056	0.053	0.050	0.047	0.045	0.043	0.041	0.039	0.038	0.036	0.035	0.034	0.029	0.025	0.022
	18	0.076	0.069	0.063	0.059	0.055	0.052	0.049	0.046	0.044	0.042	0.040	0.038	0.037	0.036	0.034	0.033	0.028	0.025	0.022
	19	0.074	0.067	0.062	0.058	0.054	0.051	0.048	0.045	0.043	0.041	0.039	0.038	0.036	0.035	0.034	0.032	0.028	0.024	0.022
	20	0.072	0.066	0.061	0.056	0.053	0.050	0.047	0.044	0.042	0.040	0.039	0.037	0.036	0.034	0.033	0.032	0.027	0.024	0.021
2 × 6 (12 inches o.c.)	0	0.0313	0.230	0.183	0.153	0.131	0.115	0.102	0.093	0.084	0.078	0.072	0.067	0.063	0.059	0.055	0.053	0.041	0.034	0.029
	18	0.065	0.060	0.056	0.053	0.050	0.048	0.045	0.043	0.041	0.040	0.038	0.037	0.035	0.034	0.033	0.032	0.027	0.024	0.021
	19	0.063	0.059	0.055	0.052	0.049	0.047	0.044	0.042	0.040	0.039	0.037	0.036	0.035	0.033	0.032	0.031	0.027	0.024	0.021
	20	0.062	0.057	0.054	0.051	0.048	0.046	0.043	0.041	0.040	0.038	0.037	0.035	0.034	0.033	0.032	0.031	0.026	0.023	0.021
	21	0.060	0.056	0.053	0.050	0.047	0.045	0.043	0.041	0.039	0.037	0.036	0.035	0.033	0.032	0.031	0.030	0.026	0.023	0.021
	22	0.059	0.055	0.052	0.049	0.046	0.044	0.042	0.040	0.038	0.037	0.035	0.034	0.033	0.032	0.031	0.030	0.026	0.023	0.020
	23	0.058	0.054	0.051	0.048	0.045	0.043	0.041	0.039	0.038	0.036	0.035	0.033	0.032	0.031	0.030	0.029	0.025	0.022	0.020
	24	0.057	0.053	0.050	0.047	0.044	0.042	0.040	0.039	0.037	0.035	0.034	0.033	0.032	0.031	0.030	0.029	0.025	0.022	0.020
	25	0.056	0.052	0.049	0.046	0.044	0.042	0.040	0.038	0.036	0.035	0.034	0.032	0.031	0.030	0.029	0.028	0.025	0.022	0.020
	30	0.052	0.048	0.045	0.043	0.041	0.039	0.037	0.035	0.034	0.033	0.031	0.030	0.029	0.028	0.027	0.027	0.023	0.021	0.019
	35	0.049	0.046	0.043	0.040	0.038	0.036	0.035	0.033	0.032	0.031	0.030	0.029	0.028	0.027	0.026	0.025	0.022	0.020	0.018
2 × 8 (12 inches o.c.)	0	0.308	0.226	0.179	0.149	0.128	0.112	0.100	0.091	0.083	0.076	0.070	0.066	0.061	0.058	0.054	0.052	0.041	0.034	0.029
	20	0.056	0.053	0.050	0.047	0.045	0.043	0.041	0.039	0.038	0.036	0.035	0.034	0.033	0.032	0.031	0.030	0.026	0.023	0.020
	21	0.055	0.052	0.049	0.046	0.044	0.042	0.040	0.039	0.037	0.036	0.034	0.033	0.032	0.031	0.030	0.029	0.025	0.022	0.020
	22	0.053	0.050	0.048	0.045	0.043	0.041	0.039	0.038	0.036	0.035	0.034	0.033	0.032	0.031	0.030	0.029	0.025	0.022	0.020
	23	0.052	0.049	0.047	0.044	0.042	0.040	0.039	0.037	0.036	0.034	0.033	0.032	0.031	0.030	0.029	0.028	0.025	0.022	0.020
	24	0.051	0.048	0.046	0.044	0.042	0.040	0.038	0.037	0.035	0.034	0.033	0.032	0.031	0.030	0.029	0.028	0.024	0.022	0.019
	25	0.050	0.047	0.045	0.043	0.041	0.039	0.037	0.036	0.035	0.033	0.032	0.031	0.030	0.029	0.028	0.027	0.024	0.021	0.019
	30	0.046	0.044	0.041	0.039	0.038	0.036	0.035	0.033	0.032	0.031	0.030	0.029	0.028	0.027	0.026	0.026	0.023	0.020	0.018
	35	0.043	0.041	0.039	0.037	0.035	0.034	0.032	0.031	0.030	0.029	0.028	0.027	0.026	0.026	0.025	0.024	0.021	0.019	0.017
	40	0.041	0.039	0.037	0.035	0.033	0.032	0.031	0.030	0.029	0.028	0.027	0.026	0.025	0.024	0.024	0.023	0.020	0.018	0.017

TABLE RF102.1—ASSEMBLY *U*-FACTORS FOR WOOD-FRAMED WALLS[a, b, c, d, e, f]—continued

WOOD STUD SIZE AND SPACING	CAVITY INSULATION INSTALLED *R*-VALUE	CONTINUOUS INSULATION *R*-VALUE																		
		0	1	2	3	4	5	6	7	8	9	10	11	12	13	14	15	20	25	30
2 × 4 (16 inches o.c.)	0	0.331	0.243	0.193	0.161	0.138	0.120	0.107	0.097	0.088	0.081	0.075	0.069	0.065	0.061	0.057	0.054	0.043	0.035	0.030
	11	0.092	0.083	0.076	0.071	0.066	0.061	0.058	0.054	0.052	0.049	0.047	0.045	0.043	0.041	0.039	0.038	0.032	0.027	0.024
	12	0.088	0.080	0.073	0.068	0.063	0.059	0.056	0.053	0.050	0.048	0.045	0.043	0.041	0.040	0.038	0.037	0.031	0.027	0.024
	13	0.084	0.077	0.071	0.066	0.061	0.057	0.054	0.051	0.049	0.046	0.044	0.042	0.040	0.039	0.037	0.036	0.030	0.026	0.023
	14	0.081	0.074	0.068	0.064	0.059	0.056	0.053	0.050	0.047	0.045	0.043	0.041	0.039	0.038	0.037	0.035	0.030	0.026	0.023
	15	0.079	0.072	0.066	0.062	0.058	0.054	0.051	0.049	0.046	0.044	0.042	0.040	0.039	0.037	0.036	0.034	0.029	0.025	0.023
	16	0.077	0.070	0.065	0.060	0.056	0.053	0.050	0.047	0.045	0.043	0.041	0.039	0.038	0.036	0.035	0.034	0.029	0.025	0.022
	17	0.075	0.068	0.063	0.058	0.055	0.052	0.049	0.046	0.044	0.042	0.040	0.039	0.037	0.036	0.034	0.033	0.028	0.025	0.022
	18	0.073	0.066	0.061	0.057	0.053	0.050	0.048	0.045	0.043	0.041	0.039	0.038	0.036	0.035	0.034	0.033	0.028	0.024	0.022
	19	0.071	0.065	0.060	0.056	0.052	0.049	0.047	0.044	0.042	0.040	0.039	0.037	0.036	0.034	0.033	0.032	0.027	0.024	0.021
	20	0.069	0.063	0.059	0.055	0.051	0.048	0.046	0.043	0.041	0.039	0.038	0.036	0.035	0.034	0.032	0.031	0.027	0.024	0.021
2 × 6 (16 inches o.c.)	0	0.322	0.236	0.187	0.156	0.133	0.117	0.104	0.094	0.086	0.079	0.073	0.068	0.063	0.059	0.056	0.053	0.042	0.034	0.029
	18	0.063	0.059	0.055	0.052	0.049	0.047	0.044	0.042	0.041	0.039	0.037	0.036	0.035	0.034	0.032	0.031	0.027	0.024	0.021
	19	0.061	0.057	0.054	0.051	0.048	0.046	0.043	0.042	0.040	0.038	0.037	0.035	0.034	0.033	0.032	0.031	0.027	0.023	0.021
	20	0.060	0.056	0.052	0.050	0.047	0.045	0.042	0.041	0.039	0.037	0.036	0.035	0.033	0.032	0.031	0.030	0.026	0.023	0.021
	21	0.058	0.055	0.051	0.048	0.046	0.044	0.042	0.040	0.038	0.037	0.035	0.034	0.033	0.032	0.031	0.030	0.026	0.023	0.020
	22	0.057	0.053	0.050	0.047	0.045	0.043	0.041	0.039	0.037	0.036	0.035	0.033	0.032	0.031	0.030	0.029	0.025	0.022	0.020
	23	0.056	0.052	0.049	0.046	0.044	0.042	0.040	0.038	0.037	0.035	0.034	0.033	0.032	0.031	0.030	0.029	0.025	0.022	0.020
	24	0.055	0.051	0.048	0.046	0.043	0.041	0.039	0.038	0.036	0.035	0.033	0.032	0.031	0.030	0.029	0.028	0.025	0.022	0.020
	25	0.054	0.050	0.047	0.045	0.042	0.040	0.039	0.037	0.035	0.034	0.033	0.032	0.031	0.030	0.029	0.028	0.024	0.022	0.019
	30	0.050	0.046	0.044	0.046	0.039	0.037	0.036	0.034	0.033	0.032	0.031	0.029	0.029	0.028	0.027	0.026	0.023	0.020	0.018
	35	0.047	0.043	0.041	0.039	0.037	0.035	0.033	0.032	0.031	0.030	0.029	0.028	0.027	0.026	0.025	0.025	0.022	0.019	0.017
2 × 8 (16 inches o.c.)	0	0.317	0.232	0.184	0.152	0.131	0.115	0.102	0.092	0.084	0.077	0.071	0.066	0.062	0.058	0.055	0.052	0.041	0.034	0.029
	20	0.055	0.052	0.049	0.046	0.044	0.042	0.040	0.039	0.037	0.036	0.035	0.033	0.032	0.031	0.030	0.029	0.026	0.023	0.020
	21	0.053	0.050	0.048	0.045	0.043	0.041	0.040	0.038	0.037	0.035	0.034	0.033	0.032	0.031	0.030	0.029	0.025	0.022	0.020
	22	0.052	0.049	0.047	0.044	0.042	0.040	0.039	0.037	0.036	0.034	0.033	0.032	0.031	0.030	0.029	0.028	0.025	0.022	0.020
	23	0.051	0.048	0.046	0.043	0.041	0.040	0.038	0.036	0.035	0.034	0.033	0.032	0.031	0.030	0.029	0.028	0.024	0.022	0.020
	24	0.050	0.047	0.045	0.043	0.041	0.039	0.037	0.036	0.034	0.033	0.032	0.031	0.030	0.029	0.028	0.027	0.024	0.021	0.019
	25	0.049	0.046	0.044	0.042	0.040	0.038	0.037	0.035	0.034	0.033	0.032	0.031	0.030	0.029	0.028	0.027	0.024	0.021	0.019
	30	0.045	0.042	0.040	0.038	0.037	0.035	0.034	0.032	0.031	0.030	0.029	0.028	0.027	0.027	0.026	0.025	0.022	0.020	0.018
	35	0.042	0.039	0.037	0.036	0.034	0.033	0.031	0.030	0.029	0.028	0.027	0.027	0.026	0.025	0.024	0.024	0.021	0.019	0.017
	40	0.039	0.037	0.035	0.034	0.032	0.031	0.030	0.029	0.028	0.027	0.026	0.025	0.024	0.024	0.023	0.022	0.020	0.018	0.016

TABLE RF102.1—ASSEMBLY *U*-FACTORS FOR WOOD-FRAMED WALLS[a, b, c, d, e, f]—continued

WOOD STUD SIZE AND SPACING	CAVITY INSULATION INSTALLED *R*-VALUE	CONTINUOUS INSULATION *R*-VALUE																		
		0	1	2	3	4	5	6	7	8	9	10	11	12	13	14	15	20	25	30
2 × 4 (24 inches o.c.)	0	0.339	0.248	0.196	0.163	0.139	0.122	0.108	0.098	0.089	0.081	0.075	0.070	0.065	0.061	0.058	0.055	0.043	0.035	0.030
	11	0.089	0.081	0.075	0.069	0.065	0.061	0.057	0.054	0.051	0.048	0.046	0.044	0.042	0.040	0.039	0.037	0.031	0.027	0.024
	12	0.085	0.078	0.072	0.067	0.062	0.058	0.055	0.052	0.049	0.047	0.045	0.043	0.041	0.039	0.038	0.036	0.031	0.027	0.023
	13	0.082	0.075	0.069	0.064	0.060	0.056	0.053	0.050	0.048	0.046	0.044	0.042	0.040	0.038	0.037	0.036	0.030	0.026	0.023
	14	0.079	0.072	0.067	0.062	0.058	0.055	0.052	0.049	0.047	0.044	0.042	0.041	0.039	0.037	0.036	0.035	0.030	0.026	0.023
	15	0.076	0.070	0.065	0.060	0.056	0.053	0.050	0.048	0.045	0.043	0.041	0.040	0.038	0.037	0.035	0.034	0.029	0.025	0.022
	16	0.074	0.068	0.063	0.058	0.055	0.052	0.049	0.046	0.044	0.042	0.040	0.039	0.037	0.036	0.034	0.033	0.028	0.025	0.022
	17	0.072	0.066	0.061	0.057	0.053	0.050	0.048	0.045	0.043	0.041	0.039	0.038	0.036	0.035	0.034	0.033	0.028	0.024	0.022
	18	0.070	0.064	0.059	0.055	0.052	0.049	0.046	0.044	0.042	0.040	0.039	0.037	0.036	0.034	0.033	0.032	0.027	0.024	0.021
	19	0.068	0.062	0.058	0.054	0.051	0.048	0.045	0.043	0.041	0.039	0.038	0.036	0.035	0.034	0.032	0.031	0.027	0.024	0.021
	20	0.066	0.061	0.056	0.053	0.050	0.047	0.044	0.042	0.040	0.039	0.037	0.036	0.034	0.033	0.032	0.031	0.027	0.023	0.021
2 × 6 (24 inches o.c.)	0	0.330	0.241	0.191	0.159	0.136	0.119	0.106	0.095	0.087	0.080	0.074	0.068	0.064	0.060	0.057	0.053	0.042	0.035	0.030
	18	0.061	0.057	0.054	0.051	0.048	0.046	0.044	0.042	0.040	0.038	0.037	0.036	0.034	0.033	0.032	0.031	0.027	0.024	0.021
	19	0.060	0.056	0.052	0.050	0.047	0.045	0.043	0.041	0.039	0.037	0.036	0.035	0.034	0.032	0.031	0.030	0.026	0.023	0.021
	20	0.058	0.054	0.051	0.048	0.046	0.044	0.042	0.040	0.038	0.037	0.035	0.034	0.033	0.032	0.031	0.030	0.026	0.023	0.020
	21	0.057	0.053	0.050	0.047	0.045	0.043	0.041	0.039	0.037	0.036	0.035	0.033	0.032	0.031	0.030	0.029	0.025	0.022	0.020
	22	0.055	0.052	0.049	0.046	0.044	0.042	0.040	0.038	0.037	0.035	0.034	0.033	0.032	0.031	0.030	0.029	0.025	0.022	0.020
	23	0.054	0.051	0.048	0.045	0.043	0.041	0.039	0.037	0.036	0.035	0.033	0.032	0.031	0.030	0.029	0.028	0.025	0.022	0.020
	24	0.053	0.049	0.047	0.044	0.042	0.040	0.038	0.037	0.035	0.034	0.033	0.032	0.031	0.030	0.029	0.028	0.024	0.022	0.019
	25	0.052	0.048	0.046	0.043	0.041	0.039	0.038	0.036	0.035	0.033	0.032	0.031	0.030	0.029	0.028	0.027	0.024	0.021	0.019
	30	0.047	0.044	0.042	0.040	0.038	0.036	0.035	0.033	0.032	0.031	0.030	0.029	0.028	0.027	0.026	0.025	0.022	0.020	0.018
	35	0.044	0.041	0.039	0.037	0.035	0.034	0.032	0.031	0.030	0.029	0.028	0.027	0.026	0.025	0.025	0.024	0.021	0.019	0.017
2 × 8 (24 inches o.c.)	0	0.326	0.238	0.188	0.156	0.133	0.117	0.104	0.094	0.085	0.078	0.072	0.067	0.063	0.059	0.056	0.053	0.042	0.034	0.029
	20	0.054	0.051	0.048	0.046	0.043	0.042	0.040	0.038	0.037	0.035	0.034	0.033	0.032	0.031	0.030	0.029	0.025	0.022	0.020
	21	0.052	0.049	0.047	0.044	0.042	0.041	0.039	0.037	0.036	0.035	0.033	0.032	0.031	0.030	0.029	0.029	0.025	0.022	0.020
	22	0.051	0.048	0.046	0.043	0.041	0.040	0.038	0.037	0.035	0.034	0.033	0.032	0.031	0.030	0.029	0.028	0.024	0.022	0.020
	23	0.050	0.047	0.044	0.042	0.041	0.039	0.037	0.036	0.034	0.033	0.032	0.031	0.030	0.029	0.028	0.028	0.024	0.021	0.019
	24	0.048	0.046	0.044	0.041	0.040	0.038	0.036	0.035	0.034	0.033	0.032	0.031	0.030	0.029	0.028	0.027	0.024	0.021	0.019
	25	0.047	0.045	0.043	0.041	0.039	0.037	0.036	0.034	0.033	0.032	0.031	0.030	0.029	0.028	0.027	0.027	0.023	0.021	0.019
	30	0.043	0.041	0.039	0.037	0.035	0.034	0.033	0.032	0.030	0.029	0.029	0.028	0.027	0.026	0.025	0.025	0.022	0.020	0.018
	35	0.040	0.038	0.036	0.034	0.033	0.032	0.030	0.029	0.028	0.027	0.027	0.026	0.025	0.024	0.024	0.023	0.021	0.018	0.017
	40	0.037	0.035	0.034	0.032	0.031	0.030	0.029	0.028	0.027	0.026	0.025	0.024	0.024	0.023	0.022	0.022	0.019	0.018	0.016

TABLE RF102.1—ASSEMBLY *U*-FACTORS FOR WOOD-FRAMED WALLS[a, b, c, d, e, f]—continued

For SI: 1 British thermal unit per hour per square foot per °Fahrenheit = 5.6783 W/m^2 × K.

a. Linear interpolation of *U*-factors shall be permitted between continuous insulation and cavity insulation *R*-values. For nonstandard stud spacing, use the next-lesser stud spacing shown in the table.

b. Table values are based on the parallel path calculation procedure as applicable to wood-framed assemblies and require compliance with the following assembly conditions:

1. Framing fractions of not greater than 28 percent (assumed for 12-inch o.c. studs), 25 percent (assumed for 16-inch o.c. studs), and 22 percent (assumed for 24-inch o.c. studs) with 4 percent attributed to headers in all cases. The framing fraction is the percentage of overall opaque wall area occupied by framing members.
2. Wood framing materials or species with a thermal resistivity of not less than R-1.25 per inch.
3. Exterior sheathing with an *R*-value of not less than R-0.62 as based on wood structural panel. For walls having no exterior sheathing or sheathing of lesser *R*-value, Note d shall be used to adjust the tabulated *U*-factor.
4. Siding of not less than R-0.62 as based on the assumption of vinyl siding. For walls with siding having a lower *R*-value, Note d shall be used to adjust the tabulated *U*-factor.
5. Interior finish of not less than R-0.45 based on $^1/_2$-inch gypsum. For walls having no interior finish or a finish of lesser *R*-value, Note d shall be used to adjust the tabulated *U*-factor.
6. Cavity insulation with a rated *R*-value installed as required by the manufacturer's installation instructions to satisfy the indicated installed *R*-value, considering a reduced *R*-value for compression in an enclosed cavity where applicable.
7. Continuous insulation specified in accordance with the indicated rated *R*-value and installed continuously over all exterior wood framing, including studs, plates, headers and rim joists.
8. Indoor air film *R*-value of 0.68 and outdoor air-film *R*-value of 0.17.

c. Where any of the building materials that are continuous over the interior or exterior wall surface vary from those stated in Note b, it is permissible to adjust the *U*-factor as follows: $Uadj = 1/\ [1/U + Rd]$ where *U* is the *U*-factor from the table and *Rd* is the increase (positive) or decrease (negative) in the cumulative *R*-value of building material layers on the outside and inside faces of the wall, excluding the continuous insulation *R*-value if present.

d. For a specific continuous insulation *R*-value not addressed in this table, the *U*-factor of the assembly shall be permitted to be determined as follows: $Uadj = 1/[1/Unci + Rci]$ where *Unci* is the *U*-factor from the table for no continuous insulation (0 *R*-value column) and *Rci* is the specific rated *R*-value of continuous insulation added to the assembly.

e. For double wall framing, the *U*-factor shall be permitted to be determined by combining the *U*-factors for single-wall framing from the table as follows: $Ucombined = 1/[1/U1 + 1/U2]$ where *U*1 and *U*2 are the *U*-factors from the table for each of the adjacent parallel walls in the double-wall assembly.

f. The use of insulation in accordance with this table does not supersede requirements in Section R702.7 for use of insulation and water vapor retarders to control water vapor.

RF102.2 Mass walls. Reserved.

RF102.3 Cold-formed steel frame walls. Reserved.

SECTION RF103—ROOF AND CEILING ASSEMBLIES—RESERVED

SECTION RF104—FLOOR ASSEMBLIES—RESERVED

SECTION RF105—BASEMENT AND CRAWL SPACE WALLS

RF105.1 Basement and crawl space walls. *U*-factors for basement and *crawl space walls* shall be as specified in accordance with Table RF105.1. Effective *U*-factors for the proposed and reference foundation wall design must be used to demonstrate compliance with Section R402.1.5. Effective *U*-factors shall not be used for other compliance methods referenced in Section R401.2.1.

TABLE RF105.1—*U*-FACTORS FOR BASEMENT AND CRAWL SPACE WALLS[a]

INSULATION CONFIGURATIONS[b]	WALL *U*-FACTOR[c] (Btu/h × ft² × °F)	WALL EFFECTIVE *U*-FACTOR[d] BY PERCENTAGE OF WALL HEIGHT PROJECTING ABOVE GRADE (Btu/h × ft² × °F) FOR USE ONLY WITH SECTION R402.1.5			
—	—	50%	35%	20%	5%
Basement walls					
Uninsulated and unfinished basement wall	0.360	0.324	0.288	0.252	0.216
Continuous insulation	—	—	—	—	—
R-5ci	0.122	0.109	0.097	0.085	0.073
R-7.5ci	0.093	0.084	0.075	0.065	0.056
R-10ci	0.076	0.068	0.060	0.053	0.045
R-15ci	0.055	0.049	0.044	0.038	0.033
R-20ci	0.043	0.039	0.034	0.030	0.026
R-25ci	0.035	0.032	0.028	0.025	0.021
Cavity insulation	—	—	—	—	—
R-11	0.076	0.068	0.060	0.053	0.045
R-13	0.067	0.060	0.054	0.047	0.040
R-15	0.060	0.054	0.048	0.042	0.036
R-19	0.050	0.045	0.040	0.035	0.030
R-21	0.045	0.041	0.036	0.032	0.027
Cavity + continuous insulation	—	—	—	—	—
R-13 + R-5ci	0.050	0.045	0.040	0.035	0.030
R-13 + R-7.5ci	0.045	0.040	0.036	0.031	0.027
R-13 + R-10ci	0.040	0.036	0.032	0.028	0.024
R-19 + R-5ci	0.040	0.036	0.032	0.028	0.024
R-19 + R-7.5ci	0.036	0.033	0.029	0.025	0.022
R-19 + R-10ci	0.033	0.030	0.027	0.023	0.020
Crawl space walls					
Uninsulated crawl space wall	0.477	0.429	0.382	0.334	N/A
Continuous insulation	—	—	—	—	—
R-5ci	0.141	0.127	0.113	0.099	N/A
R-7.5ci	0.104	0.094	0.083	0.073	—
R-10ci	0.083	0.074	0.066	0.058	—
R-15ci	0.058	0.053	0.047	0.041	—
R-20ci	0.045	0.041	0.036	0.032	—
R-25ci	0.037	0.033	0.030	0.026	—

TABLE RF105.1—*U*-FACTORS FOR BASEMENT AND CRAWL SPACE WALLS[a]—continued

INSULATION CONFIGURATIONS[b]	WALL *U*-FACTOR[c] (Btu/h × ft² × °F)	WALL EFFECTIVE *U*-FACTOR[d] BY PERCENTAGE OF WALL HEIGHT PROJECTING ABOVE GRADE (Btu/h × ft² × °F) FOR USE ONLY WITH SECTION R402.1.5			
Crawl space walls					
Cavity insulation	—	—	—	—	—
R-11	0.083	0.074	0.066	0.058	N/A
R-13	0.072	0.065	0.058	0.051	—
R-15	0.065	0.058	0.052	0.045	—
R-19	0.054	0.049	0.043	0.038	—
R-21	0.048	0.043	0.038	0.033	—
Cavity + continuous insulation	—	—	—	—	—
R-13 + R-5ci	0.053	0.048	0.043	0.037	N/A
R-13 + R-7.5ci	0.047	0.042	0.038	0.033	—
R-13 + R-10ci	0.042	0.038	0.034	0.029	—
R-19 + R-5ci	0.043	0.038	0.034	0.030	—
R-19 + R-7.5ci	0.039	0.035	0.031	0.027	—
R-19 + R-10ci	0.035	0.032	0.028	0.025	—

N/A = Not Applicable.

For SI: 1 British thermal unit per hour per square foot per °Fahrenheit = 5.6783 W/m2 × K.

a. The wall *U*-factor excludes exterior the air-film *R*-value and, for insulated assemblies, includes the following: R-0.68 for interior air film, R-0.45 for $^1/_2$-inch gypsum panel finish (insulated basement walls only), and R-2.1 for 12-inch block basement wall or R-1.4 for 8-inch block crawl space wall, both with empty cells. Where cavity insulation is included between 2 × 4 or 2 × 6 framing on the interior side of a foundation wall, wood stud material with thermal resistivity of R-1.25/in is assumed to be spaced at not less than 16 inches on center with an assumed framing factor not greater than 0.15.

b. All insulation configurations extend from the top of the foundation wall to the floor of the basement or crawl space. Extrapolation to partial height insulation shall not be permitted; *U*-factors for such insulation configurations shall be determined by accepted engineering practice for modeling of thermal bridging and ground-coupled assemblies.

c. Applicable to Sections R402.1.2, R405 and R406.

d. Effective *U*-factors are adjusted to account for ground-coupling effects to provide equivalency to *U*-factors used for above-grade building thermal envelope assemblies. The effective *U*-factors are provided for use with Section R402.1.5 for evaluation of trade-offs with above-grade assemblies and other components of the building thermal envelope. The effective *U*-factor shall apply to the foundation wall area from the interior floor or ground surface to the top of the wall. Interpolation between *R*-values and percentage of wall height projecting above grade within a given insulation configuration type is permitted.

SECTION RF106—SLABS-ON-GRADE

RF106.1 Slabs-on-grade. *F*-factors for unheated and heated slabs-on-grade shall be as specified in Table RF106.1. All applicable adjustment factors in the table notes shall apply. *F*-factors for basement floor slabs and crawl space ground surfaces located below exterior grade shall be adjusted in accordance Note f as applicable.

TABLE RF106.1—*F*-FACTORS FOR SLABS-ON-GRADE[a, b, c, d, e, f]

UNHEATED SLABS-ON-GRADE: INSULATION CONFIGURATIONS	*F*-FACTOR (Btu/h × ft × °F)
Uninsulated slab	—
Horizontal insulation under slab at slab perimeter—slab edge not insulated	—
≥ R-5 for 2 ft	0.70
R-5 for 4 ft	0.67
≥ R-10 for 4 ft	0.64
Vertical insulation on exterior face[g]—slab edge insulated[h]	—
R-2.5 for 2 ft	0.66
R-5 for 2 ft	0.58
R-7.5 for 2 ft	0.56
R-10 for 2 ft	0.54
R-15 for 3 ft	0.52
R-5 for 3 ft	0.56
R-7.5 for 3 ft	0.54
R-10 for 3 ft	0.51
R-15 for 3 ft	0.49
R-5 for 4 ft	0.54

TABLE RF106.1—*F*-FACTORS FOR SLABS-ON-GRADE[a, b, c, d, e, f]—continued

UNHEATED SLABS-ON-GRADE: INSULATION CONFIGURATIONS	*F*-FACTOR (Btu/h × ft × °F)
R-7.5 for 4 ft	0.51
R-10 for 4 ft	0.48
R-15 for 4 ft	0.45
Fully insulated slab—full slab area and slab edge continuously insulated	—
R-5 entire slab area and R-3.5 edge	0.48
R-5 entire slab area and edge	0.46
R-7.5 entire slab area and R-3.5 edge	0.45
R-7.5 entire slab area and edge	0.41
R-10 entire slab area and R-5 edge	0.40
R-10 entire slab area and edge	0.36
R-15 entire slab area and R-5 edge	0.35
R-15 entire slab area and edge	0.30
R-10 slab edge and under slab perimeter inward 4 ft; R-5 remaining slab area	0.42
R-15 slab edge and under slab perimeter inward 4 ft; R-5 remaining slab area	0.40
R-15 slab edge and under slab perimeter inward 4 ft; R-10 remaining slab area	0.34
HEATED SLABS-ON-GRADE: INSULATION CONFIGURATIONS	***F*-FACTOR (Btu/h × ft × °F)**
Uninsulated	1.35
Fully insulated slab—full slab area and slab edge continuously insulated	—
R-5 entire slab area and R-3.5 edge	0.77
R-5 entire slab area and edge	0.74
R-7.5 entire slab area and R-3.5 edge	0.71
R-7.5 entire slab area and edge	0.64
R-10 entire slab area and R-5 edge	0.62
R-10 entire slab area and edge	0.55
R-15 entire slab area and R-5 edge	0.54
R-15 entire slab area and edge	0.44
R-20 entire slab area and R-7.5 edge	0.44
R-20 entire slab area and edge	0.37
R-5 entire slab area and R-10 slab edge extending downward for minimum 3 ft	0.66
R-10 slab edge and under slab perimeter inward 4 ft; R-5 remaining slab area	0.66
R-15 slab edge and under slab perimeter inward 4 ft; R-5 remaining slab area	0.62
R-15 slab edge and under slab perimeter inward 4 ft; R-10 remaining slab area	0.51

For SI: 1 British thermal unit per hour per square foot per °Fahrenheit = 5.6783 W/m^2 × K.

a. For alternative slab-on-grade insulation configurations, *F*-factors shall be determined in accordance with accepted engineering practice for modeling three-dimensional ground-coupled building assemblies using project-specific building and site conditions to estimate annual energy use attributed to foundation heat transfer and converting the result to an equivalent air-to-air *F*-factor basis.

b. Interpolation between *R*-values for a given insulation configuration type is permitted.

c. Tabulated *F*-factors are based on a typical soil thermal conductivity of 0.75 Btu/h × ft × °F and shall be multiplied by one of the following adjustment factors as applicable to site soil conditions: (1) rock or any soil on sites with poor drainage or high water table, 1.2; (2) sandy soils, 1.1; (3) loam or clay soils on well-drained sites in dry climate zones, 0.85; and (4) for all other soil or site conditions, 1.00. Where soil conditions are unknown, use of 1.00 is permitted.

d. Tabulated *F*-factors are based on a slab area to perimeter length ratio of 9:1 and shall be multiplied by one of the following adjustment factors as applicable to a slab's area to perimeter length ratio: 5:1, 0.7; 6:1, 0.8; 7:1, 0.9; 8:1, 0.95; 9:1, 1.0; 10:1, 1.05; 15:1, 1.2; 20:1, 1.35; 30:1, 1.5; and for ≥ 40:1, 1.7.

e. Tabulated *F*-factors are based on a slab perimeter edge projection above exterior finish grade of 6 inches. For portions of slab perimeter projecting 12 inches or more above grade, multiply the tabulated *F*-factors by one of the following adjustment factors as applicable: less than 12 inches, 1.0; 12 inches, 1.05; 18 inches, 1.1; 24 inches, 1.15; and 30 inches, 1.2.

f. For basement floor slabs, crawl space slabs or gravel floors, the tabulated *F*-factors shall be multiplied by one of the following adjustment factors based on the depth of the floor surface below exterior finish grade: less than 1 foot, 1.0; 1 foot, 0.95; 3 feet, 0.9; and 6 feet or more, 0.8.

g. Vertical insulation on the exterior shall extend for the indicated depth below finish grade and above grade to the top of the slab or stem wall. Where insulation is placed on the interior side of a foundation stem wall, it shall extend from the top of the slab to the indicated depth below the exterior finish grade and the applicable tabulated *F*-factor shall be multiplied by 1.05.

h. The *R*-value of the vertical insulation located on the interior side of a stem wall shall be permitted to be reduced to R-2.5 at the slab edge, not exceeding 6 inches thick, provided that the applicable *F*-factor is multiplied by 1.15 where R-5 vertical insulation is specified, 1.2 where R-10 vertical insulation is specified, or 1.25 where R-15 vertical insulation is specified.

APPENDIX

RG 2024 IECC STRETCH CODE

The provisions contained in this appendix are not mandatory unless specifically referenced in the adopting ordinance.

User notes:

About this appendix: *This appendix provides requirements for residential buildings intended to result in lower energy consumption compared to adoption of the residential provisions of this code. Where adopted by ordinance as a requirement, Section RG101.1 language is intended to replace Section R405.2, Section RG101.2 language is intended to replace Section R406.5, and Section RG101.3 language is intended to replace Section R408.2. Where those sections of the code have been amended for other purposes, this appendix is only intended to increase the number of credits required in the Prescriptive path, to increase the energy cost savings in the Simulated Performance path, and to lower the maximum ERI in the ERI path.*

ICC Council Policy-49 Note: *This voluntary appendix is intended for adopting authorities that wish to extend beyond the mandatory provisions of this code toward Zero Net Energy goals. For jurisdictions in the United States, compliance options appear to be available but may be limited in Climate Zones 0-3 if using only minimum efficiency mechanical and service water heating equipment. Adopting authorities may need to consider alternative means to expand methods for compliance under these conditions (see Section R104.1).*

SECTION RG101—COMPLIANCE

RG101.1 (R405.2) Simulated building performance compliance. Compliance based on *simulated building performance* requires that a *building* comply with the following:

Scan for Changes

60c8b96

1. The requirements of the sections indicated within Table R405.2.
2. The proposed total *building thermal envelope* thermal conductance, TC, shall be less than or equal to the *building thermal envelope* TC using the prescriptive *U*-factors and *F*-factors from Table R402.1.2 multiplied by 1.08 in Climate Zones 0, 1 and 2, and 1.15 in Climate Zones 3 through 8 in accordance with Equation 4-2 and Section R402.1.5. The area-weighted maximum fenestration SHGC permitted in Climate Zones 0 through 3 shall be 0.30.

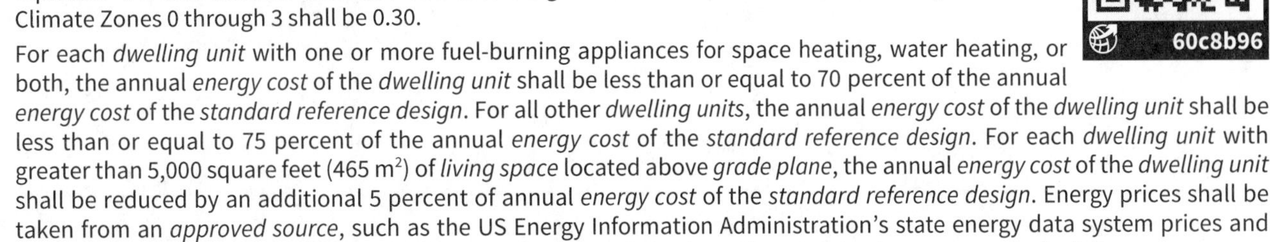

3. For each *dwelling unit* with one or more fuel-burning appliances for space heating, water heating, or both, the annual *energy cost* of the *dwelling unit* shall be less than or equal to 70 percent of the annual *energy cost* of the *standard reference design*. For all other *dwelling units*, the annual *energy cost* of the *dwelling unit* shall be less than or equal to 75 percent of the annual *energy cost* of the *standard reference design*. For each *dwelling unit* with greater than 5,000 square feet (465 m^2) of *living space* located above *grade plane*, the annual *energy cost* of the *dwelling unit* shall be reduced by an additional 5 percent of annual *energy cost* of the *standard reference design*. Energy prices shall be taken from an *approved source*, such as the US Energy Information Administration's state energy data system prices and expenditures reports. *Code officials* shall be permitted to require time-of-use pricing in *energy cost* calculations.

Exceptions:

1. The energy use based on source energy expressed in Btu or Btu per square foot of *conditioned floor area* shall be permitted to be substituted for the *energy cost*. The source energy multiplier for electricity shall be 2.51. The source energy multiplier for fuels other than electricity shall be 1.09.
2. The energy use based on site energy expressed in Btu or Btu per square foot of *conditioned floor area* shall be permitted to be substituted for the *energy cost*.

RG101.2 (R406.5) ERI-based compliance. Compliance based on an *Energy Rating Index* (ERI) analysis requires that the *rated design* and each confirmed as-built *dwelling unit* be shown to have an *ERI* less than or equal to the applicable value indicated in Table RG101.2 where compared to the *ERI reference design* as follows:

1. Where on-site renewables are not installed, the maximum ENERGY RATING INDEX NOT INCLUDING OPP applies.
2. Where on-site renewables are installed, the maximum ENERGY RATING INDEX WITH OPP applies.

Exceptions:

1. Where the ERI analysis excludes on-site power production (OPP), the maximum ENERGY RATING INDEX NOT INCLUDING OPP shall be permitted.
2. For *buildings* with 20 or more *dwelling units*, where *approved* by the *code official*, compliance shall be permitted using the Average Dwelling Unit *Energy Rating Index*, as calculated in accordance with ANSI/RESNET/ICC 301.

TABLERG101.2 (R406.5)—MAXIMUM ENERGY RATING INDEX

CLIMATE ZONE	ENERGY RATING INDEX NOT INCLUDING OPP	ENERGY RATING INDEX WITH OPP
0 and 1	46	27
2	46	26

TABLERG101.2 (R406.5)—MAXIMUM ENERGY RATING INDEX—continued		
CLIMATE ZONE	ENERGY RATING INDEX NOT INCLUDING OPP	ENERGY RATING INDEX WITH OPP
3	45	24
4	48	32
5	49	37
6	48	39
7	47	43
8	47	43

RG101.3 (R408.2) Additional energy efficiency credit requirements. *Residential buildings* shall earn not less than 20 credits from not less than two measures specified in Table R408.2. Five additional credits shall be earned for *dwelling units* with more than 5,000 square feet (465 m^2) of *living space* located above *grade plane*. To earn credit as specified in Table R408.2 for the applicable *climate zone*, each measure selected for compliance shall comply with the applicable subsections of Section R408. Each *dwelling unit* or *sleeping unit* shall comply with the selected measure to earn credit. Interpolation of credits between measures shall not be permitted.

APPENDIX RH OPERATIONAL CARBON RATING AND ENERGY REPORTING

The provisions contained in this appendix are not mandatory unless specifically referenced in the adopting ordinance.

User notes:

About this appendix: *This appendix provides a means to evaluate a building's greenhouse gas performance determined in accordance with ANSI/RESNET/ICC 301.*

SECTION RH101—GENERAL DEFINITIONS

CO_2e INDEX. A numerical integer value, calculated in accordance with ANSI/RESNET/ICC 301, that represents the relative Carbon Dioxide equivalence (CO_2e) emissions of a *rated design* as compared with the CO_2e emissions of the CO_2e reference design, where an Index value of 100 represents the CO_2e performance of the CO_2e reference design and an Index value of 0 (zero) represents a home that emits zero net CO_2e annually.

SECTION RH102—COMPLIANCE

RH102.1 Application (replaces Section R401.2). *Residential buildings* shall comply with Section R406.

Exception: *Additions*, *alterations*, *repairs* and changes of occupancy to *existing buildings* complying with Chapter 5.

RH102.2 Certificate (replaces Section R401.3). A permanent certificate shall be completed by the builder or other *approved* party and posted on a wall in the space where the furnace is located, a utility room or an *approved* location inside the *building*. Where located on an electrical panel, the certificate shall not cover or obstruct the visibility of the circuit directory label, service disconnect label or other required labels. The certificate shall indicate the following:

1. The predominant *R-values* of insulation installed in or on ceilings, roofs, walls, foundation components such as slabs, *basement walls*, *crawl space walls* and floors and *ducts* outside *conditioned spaces*.
2. *U*-factors of *fenestration* and the *solar heat gain coefficient* (SHGC) of *fenestration*. Where there is more than one value for any component of the *building thermal envelope*, the certificate shall indicate both the value covering the largest area and the area weighted average value if available.
3. The results from any required *duct system* and *building thermal envelope* air leakage testing performed on the *building*.
4. The types, sizes and efficiencies of heating, cooling and service water-heating equipment. Where a gas-fired unvented room heater, electric furnace or baseboard electric heater is installed in the residence, the certificate shall indicate "gas-fired unvented room heater," "electric furnace" or "baseboard electric heater," as appropriate. An efficiency is not required to be indicated for gas-fired unvented room heaters, electric furnaces or electric baseboard heaters.
5. Where on-site photovoltaic panel systems have been installed, the array capacity, inverter efficiency, panel tilt and orientation shall be noted on the certificate.
6. For *buildings* where an *Energy Rating Index* score is determined in accordance with Section R406, the *Energy Rating Index* score and *CO_2e Index*, both with and without any on-site generation, shall be listed on the certificate.
7. The code edition under which the structure was permitted.
8. Where a *solar-ready zone* is provided, the certificate shall indicate the location and dimensions.

RH102.3 ERI and CO_2e Index compliance (replaces Section R406.2). Compliance based on the *ERI* and *CO_2e Index* requires that the *rated design* and as-built *dwelling unit* meet all of the following:

1. The requirements of the sections indicated in Table R406.2.
2. Maximum *ERI* values indicated in Table R406.5.
3. For all-electric *dwelling units*, maximum *CO_2e Index* of 65, not including OPP, determined in accordance with ANSI/RESNET/ICC 301. For mixed-fuel *dwelling units*, a maximum *CO_2e Index* established at the time of adoption of this appendix by the authority having jurisdiction based on the CO_2e emissions data specific to the jurisdiction.

RH102.4 Confirmed compliance report for a certificate of occupancy (replaces Section R406.7.2.2). A confirmed compliance report submitted for obtaining the certificate of occupancy shall be made site and address specific and include the following:

1. Building street address or other *building site* identification.
2. Declaration of *ERI* and *CO_2e Index* on title page and on *building* plans.
3. The name of the individual performing the analysis and generating the report.
4. The name and version of the compliance software tool.
5. Documentation of all inputs entered into the software used to produce the results for the *ERI reference design* and the as-built *dwelling unit*.
6. A final confirmed certificate indicating that the as-built *building* has been verified to comply with Sections R406.2, R406.4 and R406.5. The certificate shall report the energy features that were confirmed to be in the *building*, including component-

level insulation *R*-values or *U*-factors; results from any required *duct system* and *building thermal envelope* air leakage testing; and the type and rated efficiencies of the heating, cooling, mechanical *ventilation* and service water-heating equipment installed. The certificate shall report the estimated *dwelling unit* energy use by fuel type, inclusive of all end uses. Where *on-site renewable energy* systems have been installed on or in the *building*, the certificate shall report the type and production size of the installed system.

APPENDIX RI ON-SITE RENEWABLE ENERGY

The provisions contained in this appendix are not mandatory unless specifically referenced in the adopting ordinance.

User notes:

About this appendix: *This proposal describes requirements for prescriptive solar PV that must be installed at the time of construction.*

SECTION RI101—GENERAL

RI101.1 Scope. These provisions shall apply where *on-site renewable energy* is required.

SECTION RI102—GENERAL DEFINITIONS

ANNUAL SOLAR ACCESS. The ratio of annual solar insolation with shade to the annual solar insolation without shade. Shading from obstructions located on the roof or any other part of the *building* are not included in the determination of annual solar access. Shading from existing permanent natural or person-made obstructions that are external to the *building*, including but not limited to trees, hills and adjacent structures, are included in annual solar access calculations.

PHYSICAL RENEWABLE ENERGY POWER PURCHASE AGREEMENT. A contract for the purchase of renewable electricity from a specific renewable electricity generator to a purchaser of renewable electricity.

POTENTIAL SOLAR AREA ZONE. The combined area of any *steep slope* roofs oriented between 90 degrees and 300 degrees of true north and any *low slope* roofs where the *annual solar access* is 70 percent or greater.

SECTION RI103—ON-SITE RENEWABLE ENERGY

RI103.1 General. *Buildings* shall comply with Section R401.2 and the requirements of this section.

RI103.1.1 Installed capacity. An *on-site renewable energy* system shall be installed on, or at the site of, the *building* with a peak rated capacity, measured under standard test conditions, in accordance with one of the following:

1. For one- and two- family dwellings, townhouses and other Group R-3 occupancies, the peak rated capacity shall be not less than 2 kW.
2. For Group R-2 or R-4 residential buildings, the peak rated capacity shall be not less than 0.75 watts per square foot (8.07 W/m^2) multiplied by the gross *conditioned floor area*.
3. Where a *building* includes both commercial occupancies and R-2 or R-4 occupancies required to comply with this code, the peak capacity shall be not less than 0.75 watts per square foot (8.07 W/m^2) multiplied by the gross *conditioned floor area* of the Group R-2 and R-4 occupancies.

The capacity of installed *on-site renewable energy* systems used to comply with this appendix shall be in addition to the total capacity of installed *on-site renewable energy* systems used to comply with all other requirements of this code.

Exceptions:

1. A *building* with a permanently installed domestic solar water heating system sized with a solar savings fraction of not less than 0.5 based on the total *service water heating* load of all residential occupancies.
2. One- and two-family dwellings, townhouses and other Group R-3 Occupancies in Climate Zone 4C, 5C or 8.
3. Group R-2 or R-4 occupancies in Climate Zone 8.
4. *Buildings* where the potential solar zone area is less than 300 square feet (28 m^2).
5. *Buildings* with a *physical renewable energy power purchase agreement* with a duration of not less than 15 years from a utility or a community renewable energy facility and for not less than 80 percent of the estimated whole-building electric use on an annual basis. This exception shall not apply where off-site renewable energy credits are used to comply with the requirements of Section R408.
6. *Buildings* that demonstrate compliance in accordance with Section RI103.1.1.1.

RI103.1.1.1 Alternate capacity determination. Where compliance is demonstrated in accordance with Section R405 and the *proposed design* and *standard reference design* are adjusted in accordance with Items 1 and 2, the required capacity of the installed renewable energy systems shall be permitted to differ.

1. *Proposed Design*. Where applicable, the *proposed design* shall comply with one of the following:
   1.1. Where one or more systems providing *on-site renewable energy* are included in the *construction documents*, the systems shall be modeled in the *proposed design* with a design capacity not greater than the required capacity in accordance with Section RI103.1.1. A combination of *on-site renewable energy* systems shall be permitted to be included in the *proposed design*.

1.2. Where no *on-site renewable energy* systems are specified in the *construction documents*, no *on-site renewable energy* systems shall be modeled in the *proposed design*.

2. *Standard Reference Design*. Where applicable, the *standard reference design* shall comply with one of the following:

2.1. Where a *proposed design* includes one or more *on-site renewable energy* systems, the same systems shall be modeled identically in the *standard reference design* except the total rated capacity of all systems shall be equal to the required capacity in accordance with Section RI103.1.1. Where more than one type of *on-site renewable energy* system is modeled, the total capacity of each system shall be allocated in the same proportion as in the *proposed design*.

2.2. Where the *proposed design* does not include any *on-site renewable energy* systems, an unshaded photovoltaic system shall be modeled in the *standard reference design* in accordance with the performance criteria in Table RI103.1.1.1.

TABLE RI103.1.1.1—PERFORMANCE CRITERIA FOR STANDARD REFERENCE DESIGN PHOTOVOLTAIC SYSTEMS

CRITERIA	DESIGN MODEL
Size	Rated capacity not less than required in accordance with Section RI103.1.1.
Module type	Crystalline silicon panel with a glass cover, 19.1% nominal efficiency and temperature coefficient (Tc Power) of -0.37%/°C.
Array type	Rack-mounted array with installed nominal operating cell temperature (INOCT) of 103°F (45°C).
Total system losses (DC output)	11.3%
Tilt	0 degrees (mounted horizontally)
Azimuth	180 degrees

For SI: °C = [(°F) – 32]/1.8.

RI103.1.2 ERI with OPP requirements. Where compliance is demonstrated in accordance with Section R406.5 using the *Energy Rating Index* with OPP, a project shall comply with the requirements of this appendix if the rated *proposed design* and confirmed built dwelling are shown to have an *ERI* less than or equal to the values in Table RI103.1.2.

TABLE RI103.1.2—MAXIMUM ENERGY RATING INDEX INCLUDING OPP

CLIMATE ZONE	ENERGY RATING INDEX WITH OPP
0 and 1	35
2	34
3	33
4	40
5	43
6	43
7 and 8	46

RI103.2 Renewable energy certificate (REC) documentation. Where *renewable energy certificates* (RECs) are associated with renewable energy power production required documentation shall comply with Section R404.4.

APPENDIX RJ

DEMAND RESPONSIVE CONTROLS

The provisions contained in this appendix are not mandatory unless specifically referenced in the adopting ordinance.

User notes:

About this appendix: *This appendix can by adopted by authorities having jurisdiction seeking demand responsive controls to be integrated into water heating systems.*

SECTION RJ101—DEMAND RESPONSIVE WATER HEATING

RJ101.1 Demand responsive water heating. Electric storage water heaters with a rated water storage volume of 40 gallons (150 L) to 120 gallons (450 L) and a nameplate input rating equal to or less than 12 kW shall be provided with *demand responsive controls* in accordance with Table RJ101.1.

Exceptions:

1. Water heaters that are capable of delivering water at a temperature of 180°F (82°C) or greater.
2. Water heaters that comply with Section IV, Part HLW or Section X of the *ASME Boiler and Pressure Vessel Code.*
3. Water heaters that use three-phase electric power.

TABLE RJ101.1—DEMAND RESPONSIVE CONTROLS FOR WATER HEATING

EQUIPMENT TYPE	CONTROLS	
	Manufactured before 7/1/2025	Manufactured on or after 7/1/2025
Electric storage water heaters	AHRI 1430 (I-P) or ANSI/CTA-2045-B Level 1 and also capable of initiating water heating to meet the temperature set point in response to a demand response signal.	AHRI 1430 (I-P).

SECTION RJ102—REFERENCED STANDARDS

RJ102.1 General. See Table RJ102.1 for standards that are referenced in various sections of this appendix. Standards are listed by the standard identification with the effective date, the standard title, and the section or sections of this appendix that reference this standard.

TABLE RJ102.1—REFERENCED STANDARDS

STANDARD ACRONYM	STANDARD NAME	SECTIONS HEREIN REFERENCED
AHRI 1430—2022 (I-P)	*Demand Flexible Electric Storage Water Heaters*	Table RJ101.1
ASME BPVC	*ASME Boiler and Pressure Vessel Code*	RJ101.1

APPENDIX RK ELECTRIC-READY RESIDENTIAL BUILDING PROVISIONS

The provisions contained in this appendix are not mandatory unless specifically referenced in the adopting ordinance.

User notes:

About this appendix: *This appendix can by adopted by authorities having jurisdiction seeking electrification readiness.*

SECTION RK101—ELECTRIC READINESS

RK101.1 Electric readiness. Water heaters, household clothes dryers and cooking appliances that use *fuel gas* or *liquid fuel* shall comply with Sections RK101.1.1 through RK101.1.4.

RK101.1.1 Cooking appliances. A dedicated branch circuit outlet with a rating not less than 240 volts and not less than 40 amperes shall be installed and terminate within 3 feet (914 mm) of conventional cooking tops, conventional ovens or cooking appliances combining both.

Exception: Cooking appliances not installed in an individual *dwelling unit*.

RK101.1.2 Household clothes dryers. A dedicated branch circuit with a rating not less than 240 volts and not less than 30 amperes shall be installed and terminate within 3 feet (914 mm) of each household clothes dryer.

Exception: Clothes dryers not installed in an individual *dwelling unit*.

RK101.1.3 Water heaters. A dedicated branch circuit with a rating either not less than 240 volts and not less than 30 amperes, or not less than 120 volts and not less than 20 amperes, shall be installed and terminate within 3 feet (914 mm) of each water heater.

Exception: Water heaters serving multiple *dwelling units* in a R-2 occupancy.

RK101.1.4 Electrification-ready circuits. The unused conductors required by Sections RK101.1.1 through RK101.1.3 shall be labeled with the word "spare." Space shall be reserved in the electrical panel in which the branch circuit originates for the installation of an overcurrent device. Capacity for the circuits required by Sections RK101.1.1 through RK101.1.3 shall be included in the load calculations of the original installation.

APPENDIX

RL RENEWABLE ENERGY INFRASTRUCTURE

The provisions contained in this appendix are not mandatory unless specifically referenced in the adopting ordinance.

User notes:

About this appendix: *This appendix provides readiness requirements for renewable energy infrastructure.*

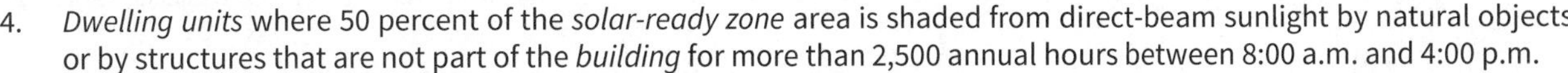

SECTION RL101—RENEWABLE ENERGY INFRASTRUCTURE

RL101.1 Renewable energy infrastructure. The *building* shall comply with the requirements of Section RL101.1.1 or RL101.1.2.

RL101.1.1 One- and two-family dwellings and townhouses. One- and two-family dwellings and townhouses shall comply with Sections RL101.1.1.1 through RL101.1.1.4.

Exceptions:

1. A *dwelling unit* with a permanently installed on-site renewable energy system.
2. A *dwelling unit* with a *solar-ready zone* area that is less than 500 square feet (46 m^2) of roof area oriented between 110 degrees (1.92 rad) and 270 (4.71 rad) degrees of true north.
3. A *dwelling unit* with less than 500 square feet (46 m^2) of roof area oriented between 110 degrees and 270 degrees of true north.
4. *Dwelling units* where 50 percent of the *solar-ready zone* area is shaded from direct-beam sunlight by natural objects or by structures that are not part of the *building* for more than 2,500 annual hours between 8:00 a.m. and 4:00 p.m.
5. A *dwelling unit* that complies with Appendix RC.
6. A *dwelling unit* with a renewable energy power purchase agreement with a duration of not less than 15 years from a utility or a community renewable energy facility and for not less than 80 percent of the estimated *dwelling unit* whole-building electric use on an annual basis.
7. A *dwelling unit* with less than or equal to 1,500 square feet (139 m^2) of *living space* located above *grade plane*.

RL101.1.1.1 Solar-ready zone area. The total area of the *solar-ready zone* shall not be less than 250 square feet (23.2 m^2) and shall be composed of areas not less than 5.5 feet (1676 mm) in one direction and not less than 80 square feet (7.4 m^2) exclusive of access or set back areas as required by the *International Residential Code*.

Exception: *Dwelling units* in townhouses three stories or less in height above grade plane and with a total floor area less than or equal to 2,000 square feet (186 m^2) per dwelling shall be permitted to have a solar-ready zone area of not less than 150 square feet (14 m^2).

RL101.1.1.2 Obstructions. *Solar-ready zones* shall be free from obstructions, including but not limited to vents, chimneys and roof-mounted equipment.

RL101.1.1.3 Electrical service reserved space. The main electrical service panel shall have a reserved space for a dual-pole circuit breaker and shall be labeled "For Future Renewable Electric." The reserved space shall be at the opposite (load) end of the busbar from the primary energy source.

RL101.1.1.4 Electrical interconnection. An electrical junction box shall be installed within 24 inches (610 mm) of the main electrical service panel and shall be connected to a capped roof penetration sleeve or a location in the attic that is within 3 feet (914 mm) of the *solar-ready zone* by a nonflexible metallic conduit not less than 1 inch (25 mm) in diameter or by permanently installed wire as *approved*. Where the interconnection terminates in the attic, the location shall be not less than 12 inches (35 mm) above ceiling insulation. Both ends of the interconnection shall be labeled "For Future Renewable Electric."

RL101.1.2 Group R occupancies. Residential *buildings* other than one- and two-family dwellings and townhouses shall comply with Sections RL101.1.2.1 through RL101.1.2.8.

RL101.1.2.1 General. A *solar-ready zone* shall be located on the roof of residential *buildings* that are oriented between 110 degrees and 270 degrees of true north or have *low slope* roofs. *Solar-ready zones* shall comply with Sections RL101.1.2.2 through RL101.1.2.8.

Exceptions:

1. A *building* with a permanently installed on-site renewable energy system.
2. A *building* with a *solar-ready zone* area that is shaded for more than 70 percent of daylight hours annually.
3. A *building* where an *approved* party certifies that the incident solar radiation available to the *building* is not suitable for a *solar-ready zone*.
4. A *building* where an *approved* party certifies that the *solar-ready zone* area required by Section RL101.1.2.3 cannot be met because of rooftop equipment, skylights, vegetative roof areas or other obstructions.
5. A *building* that complies with Appendix RC.

6. A *building* with a renewable energy power purchase agreement with a duration of not less than 15 years from a utility or a community renewable energy facility and for not less than 80 percent of the estimated electric use of the residential occupancy portion of the building on an annual basis.

RL101.1.2.2 Construction document requirements for a *solar-ready zone*. *Construction documents* shall indicate the *solar-ready zone*.

RL101.1.2.3 Solar-ready zone area. The total *solar-ready zone* area shall be not less than 40 percent of the roof area calculated as the horizontally projected gross roof area less the area covered by penthouses, mechanical equipment, rooftop structures, skylights, occupied roof decks, vegetative roof areas and mandatory access or set back areas as required by the *International Fire Code*. The *solar-ready zone* shall be a single area or smaller, separated sub-zone areas. Each sub-zone shall be not less than 5 feet (1524 mm) in width in the narrowest dimension.

RL101.1.2.4 Obstructions. *Solar-ready zones* shall be free from obstructions, including pipes, vents, ducts, HVAC equipment, skylights and roof-mounted equipment.

RL101.1.2.5 Roof loads and documentation. A collateral dead load of not less than 5 pounds per square foot (24.41 kg/m^2) shall be included in the gravity and lateral design calculations for the *solar-ready zone*. The structural design loads for roof dead load and roof live load shall be indicated on the *construction documents*.

RL101.1.2.6 Interconnection pathway. *Construction documents* shall indicate pathways for routing of conduit or plumbing from the *solar-ready zone* to the electrical service panel or service hot water system.

RL101.1.2.7 Electrical service reserved space. The main electrical service panel shall have a reserved space to allow installation of a dual-pole circuit breaker for future solar electric and shall be labeled "For Future Renewable Electric." The reserved spaces shall be positioned at the end of the panel that is opposite from the panel supply conductor connection.

RL101.1.2.8 Construction documentation certificate. A permanent certificate, indicating the *solar-ready zone* and other requirements of this section, shall be posted near the electrical distribution panel, water heater or other conspicuous location.

INDEX

ALL-ELECTRIC RESIDENTIAL BUILDINGS

Resources are related information that are not part of the code.

User notes:

About this resource: *This resource provides code compliance pathways for residential buildings intended to result in all-electric buildings for adopting jurisdictions or individual projects.*

ICC Council Policy-49 Note: *In considering whether to adopt the content in this resource, jurisdictions in the United States should note that federal law might be found to preempt the provisions it prescribes. See the Public Health and Welfare Act, 42 U.S.C. § 6297: Effect on other law. Whether the content of this resource or a modification thereof is subject to preemption may depend on court decisions or whether a waiver has been issued by the US Department of Energy pursuant to 42 U.S.C. § 6297(d).*

SECTION RRA101—GENERAL

RRA101.1 Intent. The intent of this resource is to amend the *International Energy Conservation Code* to reduce greenhouse gas emissions and improve the safety and health of buildings by not permitting combustion equipment in buildings.

RRA101.2 Scope. This resource applies to new *residential buildings*.

SECTION RRA102—GENERAL DEFINITIONS

ALL-ELECTRIC BUILDING. A building that contains no *combustion equipment*, or plumbing for combustion *equipment*, installed within the building or building site.

APPLIANCE. A device or apparatus that is manufactured and designed to utilize energy and for which this code provides specific requirements.

COMBUSTION EQUIPMENT. Any *equipment* or *appliance* used for space heating, service water heating, cooking, clothes drying or lighting that uses fuel gas or liquid fuel.

EQUIPMENT. Piping, ducts, vents, control devices and other components of systems other than *appliances* that are permanently installed and integrated to provide control of environmental conditions for buildings. This definition shall also include other systems specifically regulated in this code.

SECTION RRA103—ALL-ELECTRIC RESIDENTIAL BUILDINGS

RRA103.1 Application. Residential buildings shall be *all-electric buildings* and comply with Section R401.2.1, R401.2.2, R401.2.3 or R401.2.4.